CARDIOVASCULAR PHYSIOLOGY IN THE GENETICALLY ENGINEERED MOUSE SECOND EDITION

Developments in Cardiovascular Medicine

213. Juan Carlos Kaski (ed.): *Chest Pain with Normal Coronary Angiograms: Pathogenesis, Diagnosis and Management.* 1999. ISBN: 0-7923-8421-0
214. P.A. Doevendans, R.S. Reneman and M. Van Bilsen (eds): *Cardiovascular Specific Gene Expression.* 1999 ISBN:0-7923-5633-0
215. G. Pons-Lladó, F. Carreras, X. Borrás, Subirana and L.J. Jiménez-Borreguero (eds.): *Atlas of Practical Cardiac Applications of MRI.* 1999 ISBN: 0-7923-5636-5
216. L.W. Klein, J.E. Calvin, *Resource Utilization in Cardiac Disease.* 1999. ISBN:0-7923-8509-8
217. R. Gorlin, G. Dangas, P. K. Toutouzas, M.M Konstadoulakis, *Contemporary Concepts in Cardiology, Pathophysiology and Clinical Management.* 1999 ISBN:0-7923-8514-4
218. S. Gupta, J. Camm (eds.): *Chronic Infection, Chlamydia and Coronary Heart Disease.* 1999. ISBN:0-7923-5797-3
219. M. Rajskina: *Ventricular Fibrillation in Sudden Coronary Death.* 1999. ISBN:0-7923-8570-5
220. Z. Abedin, R. Conner: *Interpretation of Cardiac Arrhythmias: Self Assessment Approach.* 1999. ISBN:0-7923-8576-4
221. J. E. Lock, J.F. Keane, S. B. Perry: *Diagnostic and Interventional Catheterization In Congenital Heart Disease.* 2000. ISBN: 0-7923-8597-7
222. J.S. Steinberg: *Atrial Fibrillation after Cardiac Surgery.* 2000. ISBN: 0-7923-8655-8
223. E.E. van der Wall, A. van der Laarse, B.M. Pluim, A.V.G. Bruschke: Left Ventricular Hypertrophy: Physiology versus Pathology. 2000 ISBN: 0-7923-6038-9
224. J.F. Keaney, Jr. (ed.): *Oxidative Stress and Vascular Disease.* 2000. ISBN: 0-7923-8678-7
228. B.E. Jaski: *Basics of Heart Failure.* 2000 ISBN: 0-7923-7786-9
229. H.H. Osterhues, V. Hombach, A.J. Moss (eds.): *Advances in Non-Invasive Electrocardiographic Monitoring Techniques.* 2000. ISBN: 0-7923-6214-4
230. K. Robinson (ed.): *Homocysteine and Vascular Disease.* 2000 ISBN: 0-7923-6248-9
231. C.I. Berul, J.A. Towbin (eds.): *Molecular Genetics of Cardiac Electrophysiology.* 2000. ISBN: 0-7923-7829-6
232. A. Bayés de Luna, F. Furlanello, B.J. Maron and D.P. Zipes (eds.): *Arrhythmias and Sudden Death in Athletes.* 2000 ISBN: 0-7923-6337-X
233. J-C. Tardif and M.G. Bourassa (eds): *Antioxidants and Cardiovascular Disease.* 2000. ISBN: 0-7923-7829-6
234. J. Candell-Riera, J. Castell-Conesa, S. Aguadé Bruiz (eds): *Myocardium at Risk and Viable Myocardium Evaluation by SPET.* 2000.ISBN: 0-7923-6724-3
235. M.H. Ellestad and E. Amsterdam (eds): *Exercise Testing: New Concepts for the* New Century. 2001. ISBN: 0-7923-7378-2
236. Douglas L. Mann (ed.): *The Role of Inflammatory Mediators in the Failing Heart.* 2001 ISBN: 0-7923-7381-2
237. Donald M. Bers (ed.): *Excitation-Contraction Coupling and Cardiac Contractile Force, Second Edition.* 2001 ISBN: 0-7923-7157-7
238. Brian D. Hoit, Richard A. Walsh (eds.): *Cardiovascular Physiology in the Genetically Engineered Mouse, Second Edition.* 2001 ISBN 0-7923-7536-X

Previous volumes are still available

CARDIOVASCULAR PHYSIOLOGY IN THE GENETICALLY ENGINEERED MOUSE SECOND EDITION

edited by

Brian D. Hoit
Case Western Reserve University and University Hospitals of Cleveland

And

Richard A. Walsh
Case Western Reserve University and University Hospitals of Cleveland

KLUWER ACADEMIC PUBLISHERS
Boston / Dordrecht / London

Distributors for North, Central and South America:
Kluwer Academic Publishers
101 Philip Drive
Assinippi Park
Norwell, Massachusetts 02061 USA
Telephone (781) 871-6600
Fax (781) 681-9045
E-Mail <kluwer@wkap.com>

Distributors for all other countries:
Kluwer Academic Publishers Group
Distribution Centre
Post Office Box 322
3300 AH Dordrecht, THE NETHERLANDS
Telephone 31 78 6392 392
Fax 31 78 6546 474
E-Mail <services@wkap.nl>

Electronic Services <http://www.wkap.nl>

Library of Congress Cataloging-in-Publication Data

Cardiovascular physiology in the genetically engineered mouse / edited by Brian D. Hoit and Richard A. Walsh.—2nd ed.
p.; cm. — (Developments in cardiovascular medicine; 238)
Includes bibliographical references and index.
ISBN 0-7923-7536-X (alk. paper)
1. Heart—Physiology. 2. Transgenic mice. 3. Heart—Diseases—Animal models. I. Hoit, Brian D. II. Walsh, Richard A., 1946- III. Developments in cardiovascular medicine; v. 238.
[DNLM: 1. Cardiovascular Physiology. 2. Genetic Engineering. 3. Mice, Transgenic—physiology. WG 102 C267152 2001]
QP112.5.A55 C37 2001
573.1'9353—dc21

2001038763

Printed on acid-free paper.
Printed in the United States of America

DEDICATION

To our wives, Valerie Hoit and Donna Walsh.

TABLE OF CONTENTS

CONTRIBUTING AUTHORS

Susan Acton, Ph.D.
Senior Scientist
Millennium Pharmaceuticals, Inc.
75 Sydney St.
Cambridge, Mass 02139

Kuniya Asai, M.D.
Assistant Professor of Medicine
Cardiovascular Research Institute
University of Medicine & Dentistry
of New Jersey
185 South Orange Street
Newark, New Jersey 07103

William Barry, M.D.
Professor of Medicine
University of Utah Medical Center
50 North Medical Drive
Salt Lake City, UT 84132-0001

Daniel Bernstein, M.D.
Chief, Division of Pediatric Cardiology
Stanford University Medical Center
750 Welch Road, Suite 305
Palo Alto, CA 94304

C.I. Berul, M.D.
Associate Professor of Pediatrics
Department of Cardiology
Children's Hospital
300 Longwood Avenue
Boston, MA 02115

Roger E. Breitbart, M.D.
Research Scientist
Millennium Pharmaceuticals, Inc.
75 Sydney St.
Cambridge, Ma. 02139

Nipavan Chiamvimonvat, M.D.
Assistant Professor of Medicine
University of California, Davis
808 Kestrel Place
Davis, CA 94616

Kavin H. Desai, M.D.
Assistant Professor of Medicine
Department of Pediatric Cardiology
Stanford University Medical Center
750 Welch Road, Suite 305
Palo Alto, CA 94304

Benjamin C. Eloff, M.S.
The Heart & Vascular Research Center
MetroHealth Campus and
Case Western Reserve University
2500 MetroHealth Drive
Cleveland, OH 44109

Steven Fisher, M.D.
Assistant Professor of Medicine
Case Western University and
University Hospitals of Cleveland
11100 Euclid Avenue
Cleveland, OH 44109

Dimitrios Georgakopoulos, Ph.D.
Johns Hopkins Medical Institution
Johns Hopkins University
725 N Wolfe St
Baltimore, MD 21205-2100

Ingrid L. Grupp, M.D.
Professor of Pharmacol and Cell Biophysics
University of Cincinnati
P. O. Box 670575
Cincinnati, OH 45267-0575

Gunter L. Grupp, M.D. Ph.D.
Professor Emeritus
Molecular and Cell Physiology
University of Cincinnati
P .O. Box 670576
Cincinnati, OH 45267 0576

Elizabeth Hapke, B.S.
Medical University of South Carolina
Gazes Cardiac Research Center
Two Richland Medical Park
Columbia, SC 29203

Anne Hauet, M.D.
Fellow in Cardiology
Medical University of South Carolina
Gazes Cardiac Research Center
Two Richland Medical Park
Columbia, SC 29203

Brian D. Hoit, M.D.
Professor of Medicine
Case Western Reserve University and
University Hospitals of Cleveland
11100 Euclid Ave
Cleveland, OH 44106-5038

Yasuhiro Ikeda, M.D. Ph.D.
Department of Medicine
University of California San Diego
200 W. Arbor Street
San Diego, CA 92103-9000

Joanne S. Ingwall, Ph.D.
Director, NMR Lab Physiology Chemistry
Brigham and Women's Hospital
221 Longwood Avenue, Room 209
Boston, MA 02115-5817

Maryam M. Javadpour, Ph.D.
Brigham and Women's Hospital
221 Longwood Avenue, Room 209
Boston, MA 02115-5817

Raju Jeyaseelan, Ph.D.
Senior Scientist
Millennium Pharmaceuticals, Inc.
75 Sydney St.
Cambridge, MA. 02139

Vivek J. Kadambi, Ph.D.
Scientist
Millennium Pharmaceuticals, Inc.
75 Sydney Street
Cambridge, MA 02139

David Kass, M.D.
Professor of Medicine
& Biomedical Engineering
Johns Hopkins University
725 N Wolfe St
Baltimore, MD 21205-2100

Bradley B. Keller, M.D.
Jennifer Gill Associate Professor of Pediatrics
Chief, Pediatric Cardiology
Director, Ped. CV Research Program
University of Kentucky, Chandler Medical Center
800 Rose Street, Room MN472
Lexington, KY 40536-0298

Evangelia G. Kranias, Ph.D.
Professor of Pharmacology and
Cell Biophysics
University of Cincinnati
P. O. Box 670575
Cincinnati, OH 45267 0575

M. Jane Lalli, Ph.D.
Research Associate
Hackensack University Medical Center
30 Prospect Street
Hackensack, NJ 07601

Roee S. Lazebnik
Department of Biomedical Eng.
Case Western Reserve University
11100 Euclid Avenue
Cleveland, Ohio 44109

Martin LeWinter, M.D.
Professor of Medicine and Director,
Cardiology Unit
University of Vermont
Fletcher Allen Health Care,
MCHV Campus
111 Colchester Ave.
Burlington, Vt. 05401-1473

William Lewis, M.D.
Professor of Pathology
Emory University
1364 Clifton Road NE
Atlanta, GA 30303-3033

Wenfeng Miao, M.D. Ph.D.
Brigham and Women's Hospital
221 Longwood Avenue, Room 209
Boston, MA 02115-5817

Tetsuo Minamino, M.D. PhD.
Associate Professor
Center for Cardiovascular Development
Baylor College of Medicine
6550 Fannin, Suite 6629
Houston, TX 77030

Richard J. Paul, Ph.D.
Professor of Molecular and
Cellular Physiology
University of Cincinnati
P. O. Box 670576
Cincinnati, OH 45267 0576

David S. Rosenbaum, M.D.
Director, Heart & Vascular Research Center
Associate Professor of Medicine, Biomedical Engineering, Physiology
MetroHealth Campus
2500 MetroHealth Drive
Cleveland, OH 44106

John Ross, Jr., M.D.
Professor of Medicine
Assoc. Director for Scientific Affairs
University of California, San Diego
9500 Gilman Drive 0613B
La Jolla, CA 92093-5003

Albrecht G. Schmidt, M.D.
Department of Pharmacology
and Cell Biophysics
University of Cincinnati
College of Medicine
P. O. Box 67075
Cincinnati, OH 45267-0575

Jo El Schultz, Ph.D.
Assistant Professor of Pharmacology
University of Cincinnati
P. O. Box 670575
Cincinnati, OH 45267 0575

Michael Schneider, M.D.
Professor of Medicine
Center for Cardiovascular Development
1206-A Farrot Hall
Bellows Lane
Houston, TX 77030

George Sfyris
Research Associate
Department of Molecular and
Cellular Physiology
University of Cincinnati
P. O. Box 670576
Cincinnati, OH 45267 0576

Richard P. Shannon, M.D.
Chairman, Department of Medicine
Allegheny University Hospital
320 East North Avenue
Pittsburgh, PA 15212 4772

Janet Simsic, M.D.
Assistant Professor, Pediatric Cardiology
Medical University of South Carolina
Gazes Cardiac Research Institute
Ralph H. Johnson Department of Veteran
Affairs Medical Center
Two Richland Medical Park
Columbia, SC 29203

Zhi Su, M.D., Ph.D.
Research Associate,
Cardiology Division
University of Utah Health Sciences Center
50 North Medical Drive
Salt Lake City, UT 84132

Roy L. Sutliff, Ph.D.
Department of Pathology
Emory University
1639 Pierce Drive, Rm 7120 WMB
Atlanta, GA 30322

Gen Takagi, M.D.
Post Doctorate Fellow
University of Medic ine & Dentistry
185 South Orange St
Newark, NJ 07103-2714

Kimimasa Tobita, Ph.D.
Cardiovascular Development Research
Program
University of Kentucky
740 South Limestone Street
Lexington, KY 40536-0284

Joseph P. Tinney
Cardiovascular Research Program
University of Kentucky
740 South Limestone Street
Lexington, KY 40536-0284

Stephen Vatner, M.D.
Professor, Department of Medicine
Cardiovascular Research Institute
University of Medicine &
Dentistry of New Jersey
New Jersey Medical School
185 S. Orange Street
Newark, New Jersey 07103-2757

Richard A. Walsh, M.D.
Professor of Medicine
Case Western Reserve University &
University Hospitals of Cleveland
11100 Euclid Avenue
Cleveland, OH 44106

Paul J. Wang, M.D.
Associate Director of Cardiac
Electrophysiology and Pacing Lab
New England Medical Center
Tufts University School of Medicine
750 Washington Street
Boston, MA 02111-1526

David Wilson, Ph.D.
Associate Professor of Biomedical Engineering
Department of Biomedical Engineering and
Radiology
Case Western Reserve University
11100 Euclid Avenue
Cleveland, OH 44106

Atsuko Yatani, Ph.D.
Professor, Department of Medicine
New Jersey Medical School
Hackensack Medical Center
30 Prospect Street
Hackensack, NJ 07601

Karen B. Young, M.S.
University of Cincinnati
P .O. Box 670575
Cincinnati, OH 45267-0575

Michael R. Zile, M.D.
Professor of Medicine
Medical University of South Carolina
Gazes Cardiac Research Institute
Ralph H. Johnson Department of Veterans
Affairs Medical Center
2 Richland Medical Park
Charleston, SC 23909

PREFACE TO THE SECOND EDITION

The enormous advances in molecular biology and genetics coupled with the progress in instrumentation and surgical techniques have produced a voluminous and often bewildering quantity of data. The need for a second edition *of Cardiovascular Physiology in the Genetically Engineered Mouse* is underscored not only by these rapid advances, but by the increasing numbers of scientists who have focussed their research on genetically engineered mice. It is the primary objective of this second edition to interpret critically the literature and to provide a framework for the enormous amount of information in this burgeoning field. As in the first edition, the monograph serves as a practical guide for the investigator interested in the functional methods used to characterize the murine cardiovascular phenotype. However, this guidebook is a more comprehensive text than its predecessor; although the major objectives enumerated in the first edition have not substantially changed, they have been refined in keeping with the increased sophistication of the molecular biologist, geneticist, and physiologist in each other's discipline.

Each chapter has been expanded and updated, richly enhanced with original tables and figures, and in many cases, extensively rewritten. Eight chapters written by internationally recognized experts have been added; this represents a 43 % increase from the first edition. New topics include conditional transgenesis (Chapter 3), array transcription profiling (Chapter 4), cardiovascular development (Chapter 5), optical mapping (Chapter 11), pressure-volume relations (Chapter 14), assessment of diastolic function (Chapter 15), blood chemistry analysis (Chapter 19), and the principles and applications of microtechnologies (Chapter 23).

The monograph is organized into three parts. The first deals with principles of transgenesis and homologous recombination. The second part, which again is the largest, discusses the various techniques used to assess the cardiovascular mechanical, metabolic, and electrophysiologic phenotype. This section is organized in a hierarchical manner- i.e. from isolated myocyte to isolated heart to the intact, anesthetized and conscious mouse. The third part examines techniques used to evaluate murine smooth muscle function, genetic mouse models of hypertrophy and heart failure, and the methods to assess the cardiovascular phenotype in the developing mouse embryo. In addition, newer methods that "push the envelope", such as magnetic resonance imaging (MRI), positron emission tomography (PET), ultrasound biomicroscopy (UBM), and computed tomography (microCT) are discussed.

An essential and enduring goal of this second edition is to continue to facilitate interactions between the basic science disciplines and help bridge the gap between molecular biology and physiology.

Brian D Hoit
Richard A Walsh

ACKNOWLEDGMENTS

We wish to thank the contributors to this text for their exceptional efforts and the assistance of Joanne Tracy, Dianne Wuori, Mary Panarelli, and Melissa A. Ramondetta of Kluwer.

1

THE HIERARCHICAL APPROACH TO ANALYSIS OF CARDIOVASCULAR FUNCTION IN MICE: AN OVERVIEW

Richard A. Walsh, M.D.
Case Western Reserve University and University Hospitals of Cleveland
11100 Euclid Avenue, Cleveland OH, 44106

"I cannot know the whole without understanding the parts; neither can I understand the parts without knowing the whole."

Blase Pascal from the Pensées

INTRODUCTION

Over the past decade remarkable progress has been made in our knowledge of the pathobiology of human disease through the use of genetically engineered mice. The ability to manipulate the murine genome by use of gain of function or loss of function strategies is providing fundamental insights into the mechanisms which underlie atherosclerosis (1), hypertension (2), cardiac hypertrophy and failure (3). The power of this approach derives from the ability to precisely overexpress or ablate a gene of interest and examine the phenotypic consequences during embryogenesis, development, maturation and senescence. In contrast to conventional experimental animal models of human disease which employ some form of environmental stress and elicit a complex array of molecular and biochemical events, genetic engineering involves a single known molecular perturbation which produces a phenotype that may or may not recapitulate human disease. Gene targeting and transgenic analysis involves a candidate gene approach.

The gene of interest is believed to potentially be involved in the pathogenesis of a disease state. These approaches are now being complemented by viral or chemical mutagenesis studies which produce random disruptions in the mouse genome. This strategy holds the promise of discovering new genes responsible for a pathologic state or a previously unrecognized role in the pathogenesis of disease for known genes.

Advantages of the Mouse

The mouse is and will be for some time to come the preferred mammalian system for genetic manipulation for a variety of reasons. Its small size, relatively low maintenance costs, rapid gestation period, and large litter sizes are obvious practical advantages. There is extensive knowledge of the mouse genome with over 10,000 genetic markers including 6000 microsatellite sequences that are critical landmarks for genetic studies (4). The entire mouse genome will likely be sequenced by 2004 and there are many hundreds of inbred strains to facilitate study of the influence of genetic background upon gene expression. Finally germ line transmission has only been achieved with mouse embryonic stem cells (5). This approach permits more precise and subtle alterations of the mammalian genome than are possible using microinjection techniques. The major problem with pronuclear microinjections as utilized for the creation of transgenic as opposed to gene targeted animals is the random integration of exogenous DNA into chromosomal DNA. Positional effects, whereby the transgene is influenced by its site of integration into the host chromosome can have major independent consequences on gene expression of both the exogenously introduced material and endogenous genes affected by the insertion site. This potential problem is minimized by the creation of multiple lines of mice with variable copy numbers of the transgene so that a gene dosage effect may be observed and quantitated.

Disadvantages of the Mouse

Despite the multiple advantages of the mouse as a reagent for studying the *in vivo* effects of genetic manipulation there are certain limitations that must be acknowledged. Physical size relative to other larger mammals challenges biochemical sampling and physiological analysis making regional study of cardiac function difficult. Species dependent differences in isoform expression of contractile and regulatory proteins may confound the analysis of the functional consequences cardiac specific transgenesis. For example, part of the fetal gene program which is elicited by the hypertrophy of mechanical overload or transgenic pertubation in the mouse involves a switch from the fast to the slow myosin heavy chain isoform and upregulation of alpha skeletal actin. These compensatory changes have intrinsic functional consequences independent of the expression of the transgene that must be accounted for. Notably the isoform shifts do not occur with pathologic hypertrophy in higher mammals and man since the slow myosin heavy chain is the predominant isoform (6) and there is an equal abundance of cardiac and skeletal actins. Finally transgene expression may have differing phenotypic consequences which are species dependent. Examples include the modulation of systemic arterial pressure by the mouse Ren-2 gene and the modeling of inflammatory disease. In both instances no phenotype was observed in respective transgenic mice but useful information was derived when the genes were introduced into the rat genome (5).

Table 1-1. Advantages and limitations of genetically engineered mice in the study of cardiovascular disease

Advantages
- Small size (25 gms)
- Relatively low maintenance costs
- Rapid gestation period (21 days)
- Large litter size
- Genome extensively characterized
- Multiple inbred lines
- Germ line transmission only possible in this species

Disadvantages
- Small heart size (160 mg) limits regional biochemical and physiologic analysis (currently)
- Species dependent cardiac protein isoform expression
- Potential species dependent phenotypic consequences of transgene expression (e.g. Ren-2)

These limitations notwithstanding valuable information is, and will continue to be, derived from analysis of genetically engineered mice. This requires the interdisciplinary interaction among molecular biologists, pathologists and physiologists. Conventional wisdom holds that molecular biologists use a reductionist approach to problem solving in scientific discovery while physiologists employ an integrated perspective to study cell, organ and organismal function. The *in vivo* analysis of gene targeting and transgenic perturbations has proven these notions incorrect. Molecular biologists have discovered the complex and pleiotropic effects of a single manipulation of the mammalian genome that involves frequently unanticipated and unknown primary and compensatory molecular and biochemical effects. Similarly physiologists have been challenged to miniaturize and to deconstruct the analysis of cardiovascular function from the whole animal to the isolated organ, cellular and molecular levels. This transference and reciprocity of traditional perspectives between disciplines has been the catalyst for most of the novel insights derived from mouse genetic engineering and the resultant phenotypic consequences (Figure 1).

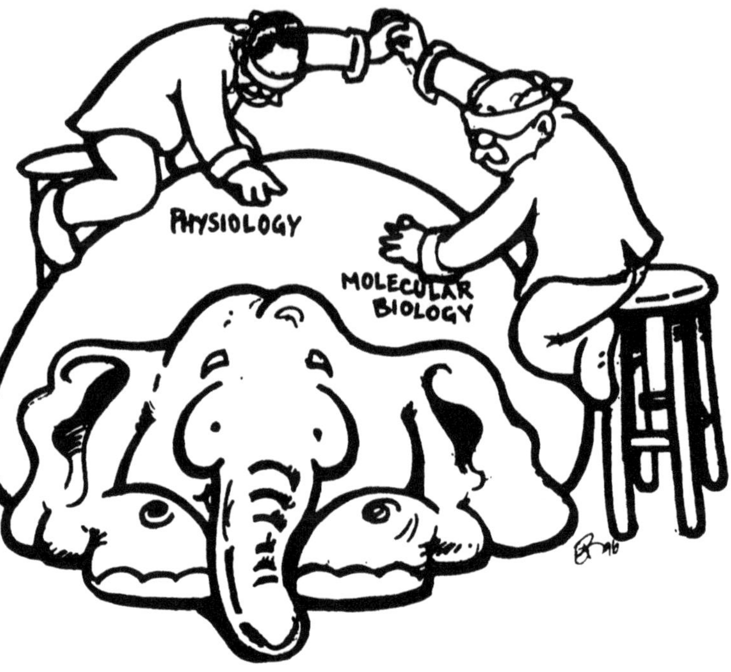

Figure 1-1. The interdisciplinary approach to characterization of genetically engineered mice.

Deconstructing Cardiovascular Physiology

The principal function of the cardiovascular system is to deliver oxygen and nutrients and to remove carbon dioxide and waste from metabolizing tissues. Integrated cardiovascular performance results from a complex interplay among the intrinsic properties of the cardiomyocytes, isolated muscle and chamber; the loading conditions of the heart and modulation of cardiac performance by neural, endocrine and renal compensatory mechanisms (Figure 2). Abnormalities of any, a combination or all of these determinants may adversely affect cardiac performance. *In vivo* analysis of each of these components of cardiovascular function in the awake unsedated animal represents arguably the most physiologic and most challenging approach to understanding the phenotypic consequences of the cardiac specific overexpression or ablation of a candidate gene of interest. Unfortunately current technical limitations preclude accurate analysis of cardiac chamber pressure, volume and output in the preinstrumented awake mouse as has been employed in larger animals. However, even in the latter circumstance dissection of the relative roles of each of the determinants of cardiovascular function involves many simplifying assumptions and is imprecise. Furthermore measurable alterations in cardiovascular performance in response to various conventional pathologic stimuli are dependent upon the intensity and duration of the stress, the development stage at which it is introduced and the success of compensatory mechanisms which may offset abnormal cell, muscle, isolated heart or intact cardiovascular performance.

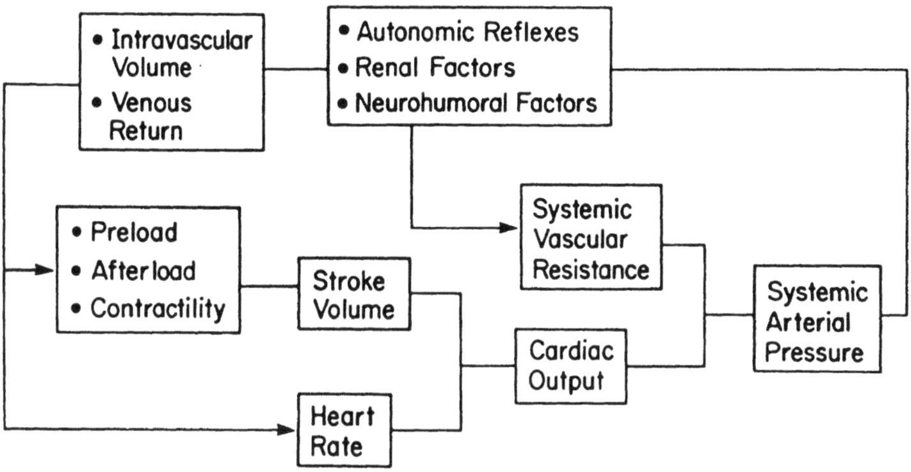

Figure 1-2. Schematic diagram of the cardiovascular system demonstrating the closed loop interaction among the major determinants of *in vivo* circulatory dynamics.

Fortunately more sophisticated, sensitive and definitive approaches to dissecting the relative importance of a transgenic or gene targeting pertubation have been developed and modified for mouse physiology studies. Clear insight into the impact of genetic modifications involves complementary studies at the isolated myocyte, isolated heart and intact organism levels. Each of these approaches has relative advantages and disadvantages and the order of choice among them depends upon the question asked. For example, consider the cardiac specific effects of the overexpression of a growth factor which may alter cardiomyocyte size and function as well as fibroblast number and abundance of the extracellular matrix. Directionally opposite effects may occur in cardiomyocyte and left ventricular chamber behavior. Under these circumstances a complementary approach involving analysis of both isolated myocyte function and *in vivo* cardiac performance will yield a clearer picture of the impact of the transgenic pertubation than either approach taken in isolation. The following section will catalogue the available strategies for analysis of cardiac function in the mouse from the simplest (most reductionist) to the most complex (most integrated) with emphasis on the relative advantages and disadvantages of each approach.

Ex Vivo Techniques

In vitro motility assays were initially developed for structure function analysis of contractile protein mutagenesis in slime mold, fly and worm. The approach involves the measurement of the velocity of filamentous actin as it travels over a bed of myosin (7). Recently this technique has been applied to quantitate the effects of alterations in mouse myosin heavy chains. The advantage of the approach is that it permits analysis of the effects of site directed mutagenesis on actin myosin interactions independent of potential compensatory alterations in the myofilament regulatory proteins or the rest of the subcellular environment of the myocyte. Alternatively, effects of the myosin light isoform alterations produced by transgenesis have been observed (7). Such changes may or may not be manifest by functional alterations at the cellular or organ levels.

Isolated Cardiomyocyte Mechanics

This approach involves the enzymatic disaggregation of mouse cardiomyocytes by retrograde perfusion with collagenalytic proteases (8-14). Analysis of sarcomere length and shortening can then be performed using laser diffraction microscopy. Alternatively whole cell length and shortening can be quantitated using microscopic video-edge detection. In principle both approaches should provide similar information. However, it is possible that alterations in cytoskeletal proteins with resultant changes in internal resistive and viscous load may result in quantitative differences between the two techniques. Isolated myocyte mechanics under either condition directly quantitates extent and rate of unloaded shortening and relengthening. The resultant measurements are exquisitely sensitive to alterations in contractility or relaxation which are intrinsic to the cardiomyocyte and devoid of the effects of potential alterations in the extracellular matrix, chamber mass or geometry, loading conditions, ventricular vascular coupling and neurohormonal changes which may complicate *in vivo* analysis. An additional advantage is the capacity to load the cells with fluorescent dyes to monitor intracellular calcium transients and pH. This permits the assessment of alterations in calcium kinetics and proton density upon myocyte function. A disadvantage of the approach is the limited number of cells which can be studied per heart and the need to apply strict morphologic criteria to selection of cells which are studied to avoid the potentially confounding effects of damage during the isolation procedure.

Limited experience has also permitted adaptation of techniques to measure isolated cardiomyocyte force development by impaling the cells on microtransducers but the technical difficulty of the approach coupled with concerns regarding the effects of impalement upon mechanical performance has limited widespread utilization of the approach (16, 17).

Whole cell patch clamp techniques have been adopted to mouse cardiomyocytes to permit accurate detection and quantification of altered sarcolemmal ion channel activity which may occur consequent to transgenic or gene targeting pertubations (18).

Isolated Muscle Mechanics

Most of the fundamental principles of cardiac muscle behavior have been derived from studies of papillary muscles extracted from large animals. The papillary muscle is ideal for this purpose since uniform longitudinal fiber orientation permits assessment of isometric or isotonic muscle performance independent of site of selection of the preparation. This approach which has been adopted to mouse physiology has the advantage of the ability to quantitate the integrated effects of alterations in cardiomyocyte function and the extracellular matrix upon muscle behavior independent of chamber mass, geometry, fiber orientation or loading condition (19). The principal disadvantage is the unpredictable effect of damaged ends of the muscle and differential core muscle hypoxia on the observed result which largely led to the abandonment of this approach in conventional experimental large animal models of cardiovascular disease.

Atrial strip preparations have been employed widely in pharmacologic studies. Their use in genetically engineered mice and larger animal models of disease has been limited by the complex muscle fiber orientation of the atria which makes muscle strip behavior highly dependent on the selection site and hence limits the interpretability of intergroup comparisons.

Detergent treatment of isolated muscle with resultant generation of a "skinned fiber" which is devoid of sarcolemmal or SR membrane permits the direct analysis of calcium sensitivity of the myofilament independent of the effects of electromechanical coupling (EC). This approach has largely been superceded by the use of isolated myocyte mechanics and measurement of calcium transients to assess independently the effects of myofilament calcium sensitivity and EC coupling on myocyte function devoid of the effects of the saponification process (13).

Isolated Heart Preparations

Isolated heart preparations, particularly from rodents where oxygenation with buffer or blood perfusate is less problematic than in larger animals have been employed widely to study normal and pathologic cardiac function. These approaches provide the opportunity to quantitate isolated left ventricular *chamber* performance without the effects of neural control, ventricular vascular coupling and with defined loading conditions. Cardiac performance results from the interplay among cardiomyocyte function, extracellular matrix composition, fiber orientation and chamber mass.

The *working heart preparation* was among the first approaches utilized to quantitate chamber performance in genetically engineered mice (20). The approach involves retrograde coronary perfusion with oxygenated buffer, left ventricular filling to variable levels of minimum diastolic pressure and ejection against an artificial resistance. The advantage of the preparation is the capacity to measure ejection performance of the heart against a precisely controlled afterload resistance. A disadvantage is the inability to measure ventricular volume and the resultant necessity of using ventricular pressure as a surrogate for LV chamber size. The curvilinear relation between diastolic pressure and volume makes it difficult to distinguish between abnormal systolic and/or diastolic function.

The *isovolumically contracting Langendorff preparation* historically antedated the working heart (21) but modification of the approach for isolated LV mechanics in the mouse heart has been slower (22, 23). The approach involves retrograde coronary perfusion with oxygenated buffer and placement of a latex or cellophane balloon across the mitral orifice into the left ventricular chamber. Developed pressure (the difference between minimum diastolic and peak LV pressure) can be measured with high fidelity miniaturized micromanometer catheters at variable and precisely known ventricular volumes. The slower acceptance of this simpler approach compared with the working heart preparation is the technical challenge of creating a balloon catheter assembly dead volume which is considerably smaller than the end diastolic volume of the adult mouse left ventricle. Some investigators have avoided the use of the intraventricular balloon. The use of pressure as a surrogate for volume in this circumstance poses the same problems as does the working heart preparation.

Magnetic resonance spectroscopy has recently been adapted to the study of high energy phosphate metabolism in genetically engineered mice (24). This development permits the examination of myocardial energetics, mechanics and other aspects of intermediary metabolism at the isolated heart and intact animal levels.

In Vivo Techniques

Whole animal analysis of integrated cardiovascular function in genetically engineered mice has been accomplished by miniaturization and refinement of methodologies used in larger animals and clinical investigation. The evolution of *in vivo* analysis has proceeded temporally from anesthetized open chest, anesthetized closed chest and, more recently, limited conscious preparations. The progression is a consequence of the increasing technical challenges and refinements in microsurgery and physiologic instrumentation required for each stage.

Comprehensive and accurate analysis of cardiovascular functions in the adult mouse required the development of miniaturized techniques to measure cardiac chamber pressure, volume and flow as a function of time; the capacity to vary or fix heart rate by electrical pacing and the ability to simultaneously measure chamber pressure and volume and to relate them to each other.

Pressure Measurement

1.4 and 1.8F micromanometer tipped catheters are now widely available for the measurement of intracavitary pressures (25, 26). Although expensive, they are essential to acquire accurate measurements of pressures and their first derivatives because of the rapid heart rates which exist in the anesthetized (200-450 bpm) or conscious (500-800 bpm) mouse. Fluid filled catheters have much lower damping frequencies and will give misleading estimates of peak positive and negative dP/dt under these conditions. The latter isovolumic indices of contraction and relaxation respectively are exquisitely sensitive to changes in cardiac inotropic or lusitropy. Peak positive dP/dt is sensitive to preload and comparisons between groups should be made at common ventricular diastolic pressures (or ideally volumes). Negative dP/dt is very sensitive to afterload and inferences regarding differential rates of isovolumic relaxation can only be made at equivalent systemic arterial pressures. A more accurate and load insensitive estimate of dissipation of LV pressure can be made using a time constant (27).

Volume Measurement

Chamber volumes in the mouse have been measured accurately using echo Doppler, magnetic resonance imaging, conductance catheter and epicardial piezoelectric crystal techniques. Initial attempts to utilize quantitative cineangiography have been abandoned due to the effects of contrast on hemodynamics and the slow frame rates relative to the spontaneous heart rates in the mouse (28). Echo Doppler analysis of mouse LV performance is the most widespread approach because of its widespread availability and noninvasive nature (29, 30). M Mode echo has sufficient temporal resolution to provide a highly accurate estimated "ice pick" view of left ventricular volume while Doppler signals are used to quantitate aortic outflow and mitral inflow. By contrast 2 dimensional echo measurements of chamber volumes are imprecise because of the insufficiently high video frame rate of commercial machines. The necessary reliance on M mode echocardiography for mouse studies precludes the ability to quantitate accurately regional ventricular function. Other techniques for volume measurement have met with less acceptance because of relative expense, limited availability and technical limitations.

In vivo assessment of ventricular volume permits quantitation of ejection phase indices of cardiac performance such as the ejection fraction and velocity of circumferential fiber shortening. These parameters are affected by preload (estimated by ventricular end diastolic volume), afterload (estimated by systemic arterial pressure), contractility (ejection performance independent of loading conditions) and heart rate. Repetitive, noninvasive screening of LV volume, mass and performance by transthoracic M mode echo Doppler interrogation represents arguably the most efficient tool for initial screening of genetically engineered mice for a cardiovascular phenotype. It should be emphasized that normal resting LV performance can occur despite depressed cardiomyocyte function by the operation of acute (e.g. heart rate and sympathetic stimulation) and chronic (e.g. LV hypertrophy) compensatory mechanisms. Conversely ejection performance may be enhanced (e.g. decreased systemic arterial pressure) or depressed (e.g. elevated systemic arterial pressure) independent of alterations in the contractile state of the cardiomyocyte.

Assessment of In Vivo Myocardial Contractility

The capacity to simultaneously measure pressure and volume in the mouse heart permits an estimate of intrinsic contractile function by the use of so called "load independent" measures of contractility. These include end systolic elastance, peak dP/dt-end diastolic volume relations and stress shortening or stress velocity relations. These analytic approaches to ventricular function have in common the incorporation of alterations in loading conditions to dissect potential changes in contractility. They are beginning to be utilized for this purpose in *in vivo* mouse studies (25, 31).

In Vivo Electrophysiology Studies

In addition to the repertoire of techniques available for *in vivo* characterization of ventricular performance in genetically engineered mice, it is possible to perform whole animal electrophysiologic studies in this species (32). Detection and quantitation of abnormalities of SA nodal, intraatrial, AV nodal and ventricular conduction can be studied by intravascular placement of multiple and multipolar electrode catheters. Monophasic action potentials can be analyzed along with the propensity to provoke atrial and ventricular arrhythmias in a way analogous to human diagnostic electrophysiology studies. This *in vivo* approach complements and extends information derived from single cell patch clamp analysis of ionic currents performed on mice with potential arrhythmogenic phenotypes.

Table 1-2. Available approaches to hierarchical analysis of cardiovascular function in genetically modified mice

Simplest ↑

EX VIVO

- *In vitro* motility assay
- Isolated myocyte
- Skinned fiber
- Intact muscle
- Isolated heart
  - Langendorff
  - Working

IN VIVO

Open and closed chest anesthetized

- Pressure measurement (fluid filled and micromanometer catheters)
- Volume measurement
  Echocardiography
  Cineangiography
  Magnetic Resonance Imaging
  Ultrasonic crystal imaging
  Conductance catheter
- Flow measurement
  Transthoracic
  Doppler
  Radiolabeled microsphere
- Electrophysiology

↓ **Most Complex**

Awake unsedated

- Heart rate level and variability
- Blood pressure
- Exercise and oxygen consumption

Experimental Pitfalls

False positive or false negative information regarding the existence and nature of a cardiovascular phenotype in genetically engineered mice may be derived from studies with inadequate attention to several variables. Initial *in vivo* hemodynamic studies were performed in *open chest anesthetized mice* because of the ease of catheter insertion, manipulation and electrical stimulation. Subsequently it has become possible to passage high fidelity catheters and pacing/sensing electrodes in the closed chest mouse through peripheral arteries and veins. This approach permits acquisition of hemodynamic data with lighter levels of anesthesia and without the confounding effects of mechanical ventilation. The result is higher spontaneous heart rates and substantially less preparation dependent depression of cardiac contraction and relaxation. For these reasons the *closed chest anesthetized* mouse preparation is preferred (26).

The cardiovascular effects of *anesthesia* have been extensively studied in large animals and provided the impetus to develop awake preinstrumented preparations for physiologic investigation. Available information suggests that anesthetic effects are at least as important in the mouse. In general, all anesthetics have variable negative inotropic, chronotropic and peripheral vasodilatory effects. They profoundly alter autonomic function and redistribute regional blood flow. Additional effects upon respiratory drive and thermal regulation make careful monitoring of acid base balance and body temperature essential. Administration of equivalent levels of anesthesia and attention to the possibility of differential sensitivity to the same anesthetic plane are critical for interpretation of hemodynamic data between groups of normal and genetically altered mice.

Strain differences in normal cardiovascular performance are an important and frequently overlooked confounding variable in mouse physiologic investigations. For example, we observed substantial differences in heart rates, peak positive and negative dP/dt between FVB/N and 129SvJ mice. For this reason, it is essential for cardiovascular studies at any level to be performed using age matched littermates as controls for transgenic mice.

Conscious Hemodynamic Measurements

Highly sophisticated hemodynamic measurements of cardiac pressures, dimensions and flows in chronic preinstrumented awake unanesthetized large animals and in man have been possible for the past two decades (e.g. 33, 34). The impetus for this technology was the desire to measure integrated circulatory dynamics in the absence of the confounding effects of anesthesia and surgical instrumentation. In particular, an immense amount was learned regarding the reflex control of the circulation and the effects of physiologic and pathologic interventions upon left ventricular performance in the conscious state. A similar rationale for awake preinstrumentation exists in the mouse but success in chronic measurement of ventricular pressures, dimensions and flows have been thwarted by the limitations of miniaturization of transducers and the ability to condition behavior of small rodents in a manner analogous to large animals. Despite these limitations considerable progress has been made in the development of conscious mouse cardiovascular physiology. It is now possible to accurately measure heart rate, QT interval (35), blood pressure (36, 37) and respiratory rate using indwelling catheters and a tether system or a free ranging telemetry approach. These variables permit an assessment of autonomic function using pharmacologic blocking agents. Methods also exist for the measurement of cardiac output and oxygen consumption in the conscious mouse (38).

Physiologic Stress and Conventional Experimental Models of Cardiovascular Disease in the Mouse

The ability to create genetically engineered mice with cardiovascular phenotypes stimulated the adaptation of methods to physiologically and pathologically stress this species in a manner analogous to that utilized in higher mammals. This permits observation of the interplay between genetic manipulation and environmental factors. Physiologic stress in the form of swimming and treadmill running have been used for exercise conditioning, measurement of exercise duration and quantitation of maximal oxygen consumption in mice (38). Latent or subtle defects in mice with cardiac specific overexpression of a gene of interest may be uncovered or abnormalities of exercise performance in mice with manifestly altered cardiac performance may be quantitated and followed serially.

Pressure overload hypertrophy of the left and right ventricles has been produced by acute transverse aortic or pulmonary artery banding respectively in the mouse (8, 28, 39). Both maneuvers result in a form of compensated concentric hypertrophy characterized by largely normal chamber function or subtle abnormalities of myocyte function. These models have been exploited to examine the mechanism for transition to cardiac chamber depression and congestive heart failure. Infrarenal aortocaval fistulae have been created in mice to produce left ventricular combined pressure and volume overload (40). There have been no reports of reproducible surgical production of aortic and mitral regurgitation in the adult mouse. Coronary artery ligation with resultant myocardial ischemia and infarction in the mouse are being exploited by a number of groups to study molecular mechanisms of myocardial stunning, preconditioning and the role of adhesion molecules in genetically altered mice (41). One remarkable general feature of the imposition of this variety of pathologic stimuli in the mouse is the relative resistance of their species to the development of heart failure in contrast to what is observed in larger rodents and higher mammals.

Table 1-3. Surgical modeling of cardiovascular disease in the mouse

- Pressure overload
    - Pulmonary artery banding
    - Aortic banding
- Volume overload
    - Infrarenal aortocaval fistula
- Myocardial ischemia and infarction

Interpreting Cardiovascular Phenotypes in Genetically Engineered Mice

The stimuli for creation of the large number of cardiac specific genetically altered mice have been 1) to understand the functional role of various proteins in the *in vivo* context, 2) to define functionally specific roles of various isoforms of cardiac proteins, 3) to examine whether, and to what extent, altered protein abundance creates a phenotype resembling human disease (13, 42, 43, 44, 45). In the latter context, it is critically important to answer several questions: 1) Is the phenotype due to the altered expression of the transgene? 2) Are similar alterations in protein abundance observed in conventional experimental animal models or human disease? 3) Is the phenotype due to an insertional effect of the transgene? 4) Is the effect due to ectopic expression of the transgene protein in a subcellular compartment where it does not usually reside? 5) Is the effect due to nonspecific overabundance of protein in the cell? 6) Is the effect a consequence of compensatory changes evoked by gene ablation or overexpression? If the answers to questions one and two are yes, to questions three, four and five no and to question six maybe - the result is relevant and important.

Future Directions

Genetic engineering of mammals will continue to be a powerful tool to elucidate mechanisms responsible for cardiovascular development, normal cardiovascular function and disease. The continued development of organ specific, temporally programmed transgenic and knockout mouse technology will refine our knowledge of cardiovascular mechanisms in health and disease (46, 47). For example, most mouse models of human disease use promoters that are active throughout development or shortly after birth. By contrast most cardiovascular disease occurs during adult life. Continued efforts will be made to apply transgenesis and germ line transmission of knockout levels to larger mammals closer to man (48, 49). Increased attention will be focused upon the critical role of genetic background on the variable phenotypic expression of single gene engineered events and random mutagenesis in mice will be applied to new gene discovery (50, 51). Mass screening by highly automated analysis of differential display of mRNA and protein analysis will become widespread in the analysis of changes consequent to overexpression or knockout mice and will be applied to conventional experimental models of disease.

Major innovations will occur in the ability to further miniaturize physiologic transducers so that study of the cardiovascular performance in the conscious mouse will be possible. Similar developments will occur in the ability to study cardiovascular function during embryogenesis and fetal life. Advances will also occur at the level of molecular physiology using confocal microscopy and new fluorescent dyes to examine biochemistry and function in the intact cell and subcellular compartments.

Genetically engineered mice have united molecular biologists, pathologists and physiologists in a synergistic research effort which is intrinsic and somewhat unique to this technology. As molecular biologists learn and do physiology, and as physiologists learn and do molecular biology, there is a danger of a dilutional effect and a loss of rigor. Standardized protocols, meaningful continuous interactions, and vigorous in depth cross training can temper this effect and keep us all on the right track.

REFERENCES

1. Breslow JL. Mouse models of atherosclerosis. Science 1996;272:685-688
2. Thompson MW, Merrill DC, Yang G, Robillard JE, Sigmund CD. Invited Review. Transgenic animals in the study of blood pressure regulation and hypertension. Am J Physiol 1995;269:E793-E803
3. Chien KR. Cardiac muscle diseases in genetically engineered mice: evolution of molecular physiology. Am J Physiol 1995;269:H755-H766

4. Paigen K. A miracle enough: the power of mice. Nature Medicine 1995;1:215-220
5. Mullins LJ, Mullins JJ. Perspectives series: molecular medicine in genetically engineered animals. J Clin Invest 1996;97:1557-1560
6. Henkel RD, VandeBerg JL, Shade RE, Leger JJ, Walsh RA. Cardiac beta myosin heavy chain diversity in normal and chronically hypertensive baboons. J Clin Invest 1989;83:1487-1493
7. Fewell JG, Hewett TE, Sanbe A, Klevitsky R, Hayes E, Warshaw D, Maughan D, Robbins J. Functional significance of cardiac myosin essential light chain PU300,4 isoform switching in transgenic mice. J Clin Invest 1998;101:2630-2639
8. Dorn GW II, Robbins J, Ball N, Walsh RA. Myosin heavy chain regulation and myocyte contractile depression after left ventricular hypertrophy in aortic banded mice. Am J Physiol 1994;267:H400-H405
9. Kadambi VJ, Ponniah S, Koss KL, Harrer JM, Dorn GW II, Walsh RA, Kranias EG. Cardiac-specific overexpression of phospholamban alters calcium kinetics and resultant cardiomyocyte mechanics in transgenic mice. J Clin Invest 1996;97:533-539
10. Gunteski-Hamblin AM, Song G, Walsh RA, Frenzke M, Boivin GP, Dorn GW II, Kaetzel MA, Horseman ND, Dedman JR. Annexin VI overexpression targeted to heart alters cardiomyocyte function in transgenic mice. Am J Physiol 1996;270:H1091-1100
11. Chu G, Dorn GW II, Luo W, Harrrer JM, Kadambi VJ, Walsh RA, Kranias EG. Monomeric phospholamban overexpression in transgenic mouse hearts. Circ Res 1997;81:485-492
12. Pawloski-Dahm C, Song G, Kirkpatrick DL, Palermo J, Gulick J, Dorn GW II, Robbins J, Walsh RA. Effects of total replacement of atrial myosin light chain-2 with the ventricular isoform in atrial myocytes of transgenic mice. Circ 1998;97:1508-1513
13. Takeishi Y, Chu G, Kirkpatrick DL, Li Z, Wakasaki H, Kranias EG, King GL, Walsh RA. *In vivo* phosphorylation of cardiac troponin I by protein kinase C ∃2 decreases cardiomyocyte calcium responsiveness and contractility in transgenic mouse hearts. J Clin Invest 1998;102:72-78
14. Loukianov E, Ji Y, Grupp IL, Kirkpatrick DL, Baker DL, Loukianov T, Grupp G, Lytton J, Walsh RA, Periasamy M. Enhanced myocardial contractility and increased Ca^{2+} transport function in transgenic hearts expressing the fast-twitch skeletal muscle sarcoplasmic reticulum Ca^{2+} ATPase. Circ Res – In Press
15. Sato Y, Ferguson DG, Sako H, Dorn GW II, Kadambi VJ, Yatani A, Hoit BD, Walsh RA, Kranias EG. Cardiac specific overexpression of mouse cardiac calsequestrin is associated with hypertrophy and altered cardiovascular function in transgenic mice. J Biol Chem – In Press
16. McDonald KS, Field LJ, Parmacek MS, Soonpaa M, Leiden JM, Moss RL. Length dependence of Ca^{2+} sensitivity of tension in mouse cardiac myocytes expressing skeletal troponin C. J Physiol 1995;483:131-139
17. Lin G, Pister KSJ, Roos KP. Micro-scale force-transducer system to quantify isolated heart cell contractile characteristics. Sensors and Actuators 1995;A46-47:233-236
18. Albitz R, Kammermeier H, Nilius B. Free energy of ATP-hydrolysis fails to affect ATP-dependent potassium channels in isolated mouse ventricular cells. J Mol Cell Cardiol 1990;22:183-190
19. Capasso J, Robinson T, Anversa P. Alterations in collagen cross-linking impair myocardial contractility in the mouse heart. Circ Res 1989;65:1657-1664
20. Grupp IL, Subramaniam A, Hewett TE, Robbins J, Grupp G. Comparison of normal, hypodynamic, and hyperdynamic mouse hearts using isolated work-performing heart preparations. Am J Physiol 1993;265:H1401-1410
21. Langendorff O. Untersuchungen am ueberlebenden Saeugetierherzen. Pfluegers Arch 1985;61:291-332
22. Brooks WW, Apstein CS. Effect of treppe on isovolumic function in the isolated blood-perfused mouse heart. J Mol Cell Cardiol 1996;28:1817-1822
23. Kameyama T, Chen Z, Bell SP, Fabian J, LeWinter MM. Mechanoenergetic studies in isolated mouse heart. Am J Physiol 1998;274:H366-H374
24. Chu G, Luo W, Slack JP, Tilgmann C, Sweet WE, Spindler M, Saupe KW, Boivin GP, Moravec CS, Matlib MA, Grupp IL, Ingwall JS, Kranias EG. Compensatory mechanisms associated with the hyperdynamic function of phospholamban-deficient mouse hearts. Circ Res 1996;78:1064-1076
25. Hoit BD, Khan ZU, Pawloski-Dahm CM, Walsh RA. Special Communication. *In vivo* determination of the left ventricular wall stress-shortening relationship in normal mice. Am J Physiol 1997;272:H1047-H1052
26. Hoit BD, Ball N, Walsh RA. Invasive hemodynamics and force-frequency relations in open- versus closed-chest mice. Am J Physiol 1997;273:H2528-2533

27. Starling MR, Montgomery DG, Mancini GBJ, Walsh RA. Load independence of the rate of isovolumic relaxation in man. Circ 1987;76:1274-1281
28. Rockman HA, Ono S, Ross RS, Jones LR, Karimi M, Bhargava V, Ross J Jr., Chien, KR. Molecular and physiological alterations in murine ventricular dysfunction. Proc Natl Acad Sci 1994;91:2694-2698
29. Hoit BD, Khoury SF, Kranias EG, Ball N, Walsh RA. *In vivo* echocardiographic detection of enhanced left ventricular function in gene targeted mice with phospholamban deficiency. Circ Res 1995;77:632-637
30. Hoit BD, Walsh RA. *In vivo* echocardiographic assessment of left ventricular function in transgenic and gene-targeted mice. Trends Cardiovasc Med 1997;7:129-134
31. Georgakopoulos D, Mitzner WA, Chen CH, Byrne BJ, Millar HD, Hare JM, Kass DA. Special Communication. *In vivo* murine left ventricular pressure-volume relations by miniaturized conductance micromanometry. Am J Physiol 1998;274:H1416-H1422
32. Berul CI, Christe ME, Aronovitz MJ, Seidman CE, Seidman JG, Mendelsohn ME. Electrophysiological abnormalities and arrhythmias in ∀MHC mutant familial hypertrophic cardiomyopathy mice. J Clin Invest 1997;99:570-576
33. Walsh RA, O'Rourke RA. Direct and indirect effects of calcium entry blocking agents on isovolumic left ventricular relaxation in conscious dogs. J Clin Invest 1985;75:1426-1435
34. Walsh RA, Cleary FX, O'Rourke RA. The effects of calcium entry blockade on left ventricular dynamics in exercising dogs. Am J Physiol 1986;20:H656-H663
35. Mitchell GF, Jeron A, Koren G. Measurement of heart rate and Q-T interval in the conscious mouse. Am J Physiol 1998;274:H747-H751
36. Krege JH, Hodgin JB, Hagaman JR, Smithies O. A noninvasive computerized tail-cuff system for measuring blood pressure in mice. Hypertension 1995;25:1111-1115
37. Ito M, Oliviero MI, Mannon PJ, Best CF, Maeda N, Smithies O, Coffman TM. Regulation of blood pressure by the type 1A angiotensin II receptor gene. Proc Natl Acad Sci 1995;92:3521-3525
38. Desai KH, Sato R, Schauble E, Barsh GS, Kobilka BK, Bernstein D. Special Communication. Cardiovascular indexes in the mouse at rest and with exercise: new tools to study models of cardiac disease. Am J Physiol 1997;272:H1053-H1061
39. Rockman HA, Ross RS, Harris AN, Knowlton KU, Steinhelper ME, Field LJ, Ross J Jr., Chien KR. Segregation of atrial-specific and inducible expression of an atrial natriuretic factor transgene in an *in vivo* murine model of cardiac hypertrophy. Proc Natl Acad Sci 1991;88:8277-8281
40. Tanaka N, Dalton N, Mao L, Rockman HA, Peterson KL, Gottshall KR, Hunter JJ, Chien KR, Ross J Jr. Transthoracic echocardiography in models of cardiac disease in the mouse. Circ 1996;94:1109-1117
41. Michael LH, Entman ML, Hartley CJ, Youker KA, Zhu J, Hall SR, Hawkins HK, Berens K, Ballantyne CM. Myocardial ischemia and reperfusion: a murine model. Am J Physiol 1995;269:H2147-H2154
42. D'Angelo DD, Sakata Y, Lorenz JN, Boivin GP, Walsh RA, Liggett SB, Dorn GW II. Transgenic G∀q overexpression induces cardiac contractile failure in mice. Proc Natl Acad Sci 1997;94:8121-8126
43. Wakasaki H, Koya D, Schoen FJ, Hoit BD, Jirousek MR, Ways DK, Walsh RA, King GL. Targeted overexpression of protein kinase C ∃2 isoform in myocardium causes cardiomyopathy. Proc Natl Acad Sci 1997;94:9320-9325
44. Sakata Y, Hoit BD, Liggett SB, Walsh RA, Dorn GW II. Decompensation of pressure-overload hypertrophy in G∀q-overexpressing mice. Circ 1998;97:1488-1495
45. Bowling N, Walsh RA, Song G, Estridge T, Sandusky GE, Fouts RL, Mintze K, Pickard T, Roden R, Bristow MR, Sabbah HN, Mizrahi JL, Gromo G, King GL, Vlahos CJ. Increased protein kinase C activity and expression of Ca^{2+}-sensitive isoforms in the failing human heart. Circ – In Press
46. Furth PA, St. Onge L, Böger H, Gruss P, Gossen M, Kistner A, Bujard H, Hennighausen L. Temporal control of gene expression in transgenic mice by a tetracycline-responsive promotor. Proc Natl Acad Sci 1994;91:9302-9306
47. Yu Z, Redfern CS, Fishman GI. Conditional transgene expression in the heart. Circ Res 1996;79:691-697
48. Hammer RE, Pursel VG, Rexroad CE Jr., Wall RJ, Bolt DJ, Ebert KM, Palmiter RD, Brinster RL. Production of transgenic rabbits, sheep and pigs by microinjection. Nature 1985;315:680-683
49. Kass DA, Hare JM, Georgakopoulos D. Murine cardiac function. A cautionary tail. Circ Res 1998;82:519-522

50. Rubin EM, Barsh GS. Perspectives series: molecular medicine in genetically engineered animals. J Clin Invest 1996;97:275-280
51. Rijkers T, Peetz A, Rüther U. Insertional mutagenesis in transgenic mice. Transgenic Research 1994;3:203-215

2

GENETIC ALTERATIONS AND MODELING OF CARDIOVASCULAR PHYSIOLOGY

Albrecht G. Schmidt, MD, Vivek J. Kadambi, PhD, Karen B. Young and Evangelia G. Kranias, PhD
University of Cincinnati, Cincinnati, OH

INTRODUCTION

Animal models have provided invaluable tools in improving our understanding of cardiac physiology and the mechanisms underlying heart disease. Until recently, available models were generated by surgical or pharmacological interventions or were the result of random genetic events, which made it difficult to establish a direct relationship between structural changes and functional alterations (1). However, the advancements in our ability to genetically manipulate the mammalian genome through transgenesis and gene targeting have provided important insights into the *in vivo* structure-function relation and interaction of specific gene products. Furthermore, genetically altered animals offer unique opportunities to identify genes that are causative for certain diseases, to evaluate molecular mechanisms responsible for the development and progression of diseases, and finally, to test new therapeutic strategies (2). In this regard, the mouse has emerged as the premier mammalian model system since it allows precise manipulation of its genome, has a short gestation period and its genetics and development have been well characterized (3). Initially, the small size of the murine heart was a challenging and limiting factor in the physiological and pathophysiological evaluation of phenotypic cardiac alterations. However, advances in "microphysiology" provided us with various techniques to assess cardiac function in an integrative approach from the cellular to the intact animal levels (4).

The contraction-relaxation cycle of the heart is controlled by the sequential rise and fall of cytosolic Ca^{2+} concentration (5). Ca^{2+} homeostasis is maintained by proteins of the sarcolemma, the myofilaments, and the sarcoplasmic reticulum (SR), which is the major intracellular Ca^{2+} storage organelle. Perturbations of this finely tuned Ca^{2+} cycling are associated with alterations in calcium transients and contractile dysfunction, which are characteristics of heart failure. The development of genetic techniques to specifically modify the function of key Ca^{2+} handling proteins and the assessment of the physiological consequences of such manipulations greatly advanced our understanding of the physiological role of these gene products and the pathophysiological mechanisms in heart failure.

This chapter will address three broad topics covering the range from genotype to phenotype. The first section will focus on how transgenic animals are produced and will outline the basic steps in gene targeting used to disrupt, or knock-out, the expression of endogenous genes. The second section will briefly summarize standard techniques employed in the physiological characterization of genetically modified mice. And the third section will highlight a selection of genetically engineered mouse models with alterations in Ca^{2+} handling proteins.

GENERATION OF GENETICALLY ALTERED MICE

Two procedures have been optimized during the last 20 years to stably manipulate the mouse genome and to generate genetically engineered animals: a) Transgenesis; and b) Gene targeting.

A. Transgenesis

The classical transgenic procedure is based on the ability to microinject naked DNA (transgenic construct) into the pronucleus of one-celled embryos (6,7). This process of creating transgenic animals can be divided into five stages (schematically depicted in Fig. 1): 1) Preparation of the DNA to be transferred; 2) Pronuclear injection; 3) Implantation of injected eggs; 4) Identification of transgenic pups; and 5) Propagation of the transgenic colony.

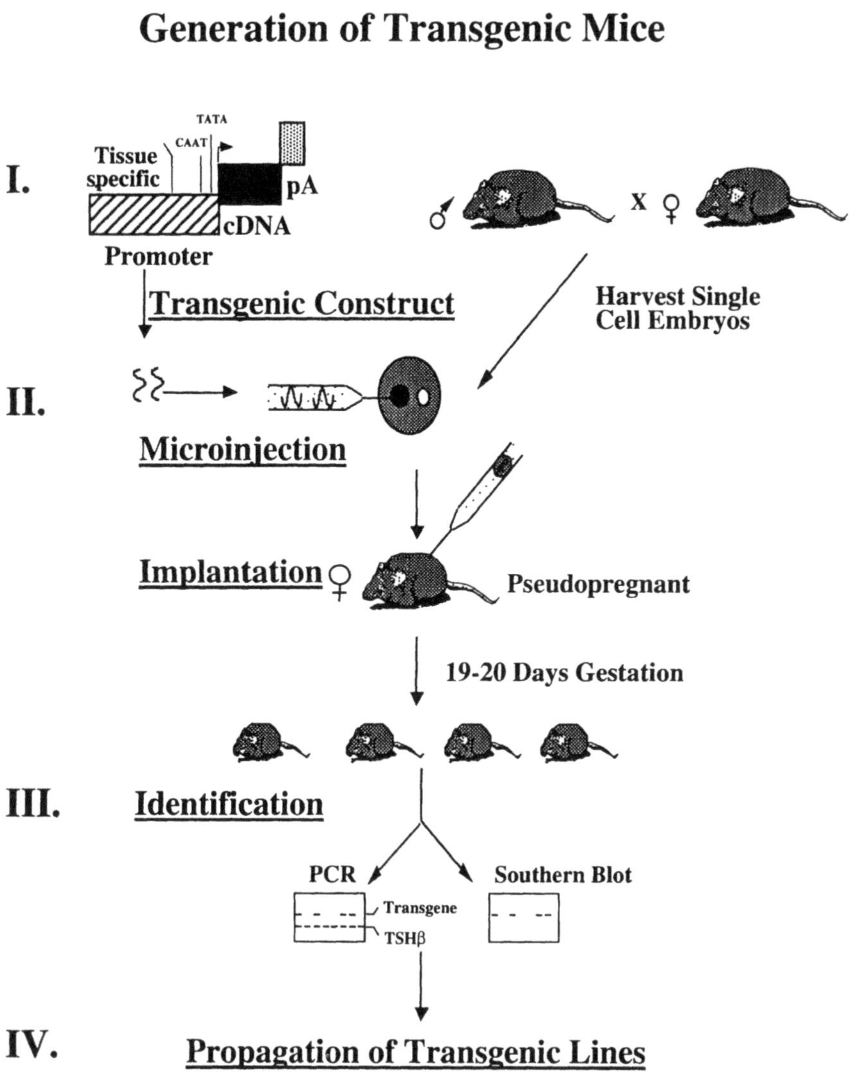

Figure 1. Generation of transgenic mice by injection of the transgenic construct into the male pronucleus of fertilized mouse eggs. "TATA", "CAAT" and "Tissue specific" boxes represent cis/acting regulatory elements located in the promoter region capable of conferring spatio-temporal specificity. The cDNA corresponds to the coding region of the gene under investigation and pA represents the polyadenylation signal sequence.

1) Generation of the transgenic construct: The first step in producing transgenic animals is to construct the DNA to be transferred – the transgene. Typical transgenes consist of a functional promoter, capable of conferring spatio-temporal resolution, the nucleotide sequence of the gene of interest (including the 5' untranslated region, a translation-initiation codon, the coding region, a stop codon, and a 3'untranslated region), and a polyadenylation signal sequence, fused to the genomic DNA/cDNA to enhance mRNA stability. Conventionally, three polyadenylation signal sequences have been used to generate transgenic mice: a) simian virus (SV 40) late signal; b) human growth hormone (HGH) signal; and c) bovine growth hormone (BGH). A critical component of the transgene is the promoter, or regulatory region, that drives transcription. The transgene can be expressed in many tissues of the transgenic animal by using a promoter from a ubiquitously expressed gene, such as that of β-actin or SV 40 large T antigen. Alternatively, tissue specific promoters can be used to express the transgene in a given location. A selection of cardiac specific promoters is summarized in Table I (8). The transgenic construct containing the promoter – coding region – polyadenylation signal sequence is digested away from vector sequences and purified using CsCl gradient centrifugation or exchange resin chromatography.

2) Pronuclear injection: The efficiency of producing transgenic mice is critically dependent on the availability of fertilized one-celled eggs. Female donor mice have been induced to superovulate by pregnant mare serum gonadotropin (PMSG) and human chorionic gonadotropin (hCG) hormones. The day after the mating, mice are sacrificied and fertilized eggs are retrieved from the oviducts. The single celled embryos are enzymatically denuded of the associated cumulus cells and cultured for microinjection. The purified DNA in Tris/HCL buffer (pH 7.5) is diluted to about 2000 copies/pl and microinjected into the male pronucleus using a flame polished glass needle under a microscope.

3) Implantation of injected eggs: DNA that is microinjected into eggs is incorporated into the genome in a random location within the chromosomal DNA (usually on autosomes - only occasionally on X- or Y-chromosomes). Typically, more than one copy (1 to 200) is incorporated at a single random site (usually as a tandem, head-to-tail-orientation) in the mouse genome. The eggs are transferred either immediately after microinjection to the oviduct of day 1 pseudopregnant female mice (recipients), or cultured overnight to the two-cell stage and then transferred. Female mice are made pseudopregnant by mating them with vasectomized males.

4) Screening for founder animals: Typically, 15 to 30% of the injected embryos will proceed to term. The resulting pups (15 days old) are screened for the presence of the transgene using polymerase chain reaction (PCR) and Southern blot methods. Genomic DNA for genotyping is isolated from tail biopsies. For PCR analysis, an aliquot of the isolated DNA is amplified using a set of primers specific for the transgene as well as a set of primers

Table 1. Cardiac specific promoters in transgenic mice (adopted from Franz et al. (8))

Cardiac-specific promoters	Further activity in	Reference
Atrium		
Atrial natriuretic factor	Hypothalamus	Seidman et al. 1991
Ventricle		
Myosin light chain-2	None	Lee et al. 1992
		Franz et al. 1993
β-Myosin heavy chain	Skeletal muscle	Rindt et al. 1993
Atrium and ventricle		
α-Myosin heavy chain	Lung	Subramaniam et al. 1991
α-Skeletal actin	Skeletal muscle (high), stomach (low)	Petropoulos et al. 1989
		Brennan et al. 1993
M-creatine kinase	Skeletal muscle (high), lung, brain	Johnson et al. 1989
Myoglobin	Skeletal muscle	Parsons et al. 1993
Myosin light chain-3F	Skeletal muscle	Kelly et al. 1995
α-B-crystallin	Skeletal muscle, lens	Gopal-Srivastava et al. 1995
Cardiac actin	Skeletal muscle, blood vessels	Biben et al. 1996
Conductive tissue (AV node)		
Troponin I	Skeletal muscle, cartilage	Zhu et al. 1995

for the single copy thyrotropin stimulating hormone β-subunit (TSHβ). Following PCR, the samples are electrophoresed onto agarose gels containing ethidium bromide and the PCR amplified bands are visualized using ultraviolet light. The TSHβ serves as a positive control for each PCR reaction. Thus, transgenic pups exhibit two bands, one corresponding to the transgene and the other corresponding to TSHβ. On the other hand, non-transgenic pups exhibit only a single band corresponding to TSHβ. This method is extremely rapid and can be used to genotype multiple transgenes in a single analysis. The presence of the transgene is also confirmed by Southern blot analysis, in which genomic tail DNA is digested with restriction endonucleases. The restricted DNA is electrophoresed onto agarose gels, transferred to a nylon membrane and hybridized with a [^{32}P]-labeled transgene specific probe. A major advantage of this technique over PCR analysis is that Southern blotting provides information about the pattern of integration of the transgene into the genome, the intactness of the transgene, and the number of copies of the transgene integrated into the genome. The yield of pups, harboring the transgene, is highly dependent on the size and purity of the microinjected DNA. On an average, 10-40% of the resulting pups exhibit the presence of the transgene.

5) Propagation of the transgenic colony: The pups, which score positive for the presence of the transgene by both PCR and Southern blot analysis, referred to as founders, are mated with nontransgenic cohorts to determine germline propagation of the transgene. Approximately 80% of the founder mice transmit the transgene to their offspring in a Mendelian manner. Founder mice, which have successfully transmitted the transgene to their progeny in a Mendelian fashion, are selected for propagation of the transgenic colony. Conventionally, breeder mice of both sexes are mated with non-transgenic cohorts and individual lines are established. It is critical to determine whether the transgene is being expressed in the appropriate location. A quantitative assay of the tissue of interest for messenger RNA and/or protein allows to identify founder animals with suitable expression levels. The level of expression depends on - among many factors – the intrinsic activity of the transgene promoter, the copy-number of the transgene, and the site of integration within the mouse genome. Furthermore, it is important to establish transgenic lines from founder mice as soon as possible since transgene expression/integration may compromise fertility. In approximately 30%, founder mice may be mosaic and harbor the transgene in only a small subset of their somatic and germ cells. In this case, several litters may be required to segregate the transgene and to establish a transgenic line.

Transgenic technology has moved from being an experimental approach to a routine methodology. A big advantage of this technique is its speed and simplicity. DNA sequences corresponding to any gene (<50 kb) can be introduced into the mouse genome by microinjection into the fertilized egg. In addition, the progeny transgenic mice carry the microinjected DNA in every cell, making it very easy to propagate transgenic lines. However, this technique is not without drawbacks. First and foremost, an appropriate promoter must be chosen such that expression of the transgene mimics the temporal and spatial expression pattern of the gene of interest. However, several eukaryotic promoters have been shown to be dependent on chromatin structure for their activity, such that random integration of the transgene in a transcriptionally silent area of the mouse genome may result in lack of expression. Transgene expression may be also dependent on the copy number of the microinjected DNA, which usually integrates into the mouse genome as tandem arrays. To overcome these limitations, multiple transgenic lines are utilized for establishing a causal relationship between an observed phenotype and transgene expression. In addition, a limitation of this technique is that transgene expression is entirely dependent on the expression pattern of the promoter used and initiation of transgene expression in the intact animal is essentially an irreversible process. Thus, this technique offers very little or no temporal control on transgene expression. This situation is particularly important when transgene expression results in either embryonic or neonatal lethality and precludes the propagation and functional investigation of the transgenic mice.

Conditional Transgenics

Therefore, it would be highly desirable to be able to control transgene expression in the intact animal. The emerging new technology on conditional gene expression offers several advantages over traditional transgenesis, described above: a) the ability to control transgene expression in situations of embryonic/neonatal lethality; b) the ability to developmentally determine the physiological effects of transgene expression; and c) the ability to generate models recapitulating the onset and progression of disease. The identification of several eukaryotic and viral promoters and prokaryotic operon systems, which are capable of transcriptional inducibility, have led to the development of "conditional transgenic mice." The transcriptionally silent promoter/operons can be induced in response to treatment with specific factors such as heavy metals, hormones, steroids, and endogenous metabolities. One promising system for achieving temporal regulation ("on/off" nature) of transgene expression in the cardiovascular system is the tetracycline-inducible system, which uses a tetracycline-controlled transactivator protein that can bind with high specificity to a promoter containing the *tet* operator, thereby activating gene expression. In the presence of tetracycline, the binding of the transactivator to *tet* operator sequences is blocked, preventing the transcription of the gene of interest (9).

B. Gene targeting

Gene targeting is most commonly used to generate null-mutant animals in which there is loss of function of a specific gene product. These so called knockout mice are created by means of homologous recombination, entailing replacement of part or the whole of the gene of interest by a foreign gene, such as the neomycin resistance gene. Generating gene knock-out mice has proved to be a powerful tool for many areas of cardiovascular research (10). However, the process is more labor intensive and time consuming than most conventional transgenic experiments. The complete procedure consists of four major steps and is illustrated in Fig. 2: 1) Preparation of the targeting construct; 2) Electroporation of the targeting construct into embryonic stem cells, which are derived from the inner cell mass of a mouse blastocyst; and selection of embryonic stem cells, which have undergone homologous recombination; 3) Microinjection of targeted embryonic stem cells into blastocysts, followed by implantation into pseudopregnant females; and 4) Identification of germ-line chimeras and generation of animals that are heterozygous and homozygous for the targeted allele.

1) Preparation of the targeting construct: The goal of the gene targeting method is to replace a specific gene of interest with one that is inactive or irrelevant (Figure 3). Both ends of the replacement gene (e.g. the neomycin resistance gene) are flanked by long DNA sequences, homologous to the sequences flanking the endogenous gene. When the DNA breaks and rejoins, gene constructs of this type permit corresponding stretches of DNA to be exchanged, which is termed "homologous recombination". This rare event results in a replacement of an important exon (or several exons) with a selectable marker within the targeting vector. The elimination of such an exon leads to an inactive gne product or a null mutation. It is desirable to include a minimum of 6 kb of genomic DNA sequences from the gene of interest into the flanking long and short arms of the construct in order to improve targeting frequencies. Furthermore, utilizing isogenic DNA to that from which the ES cells are derived, increases the frequency of homologous recombination.

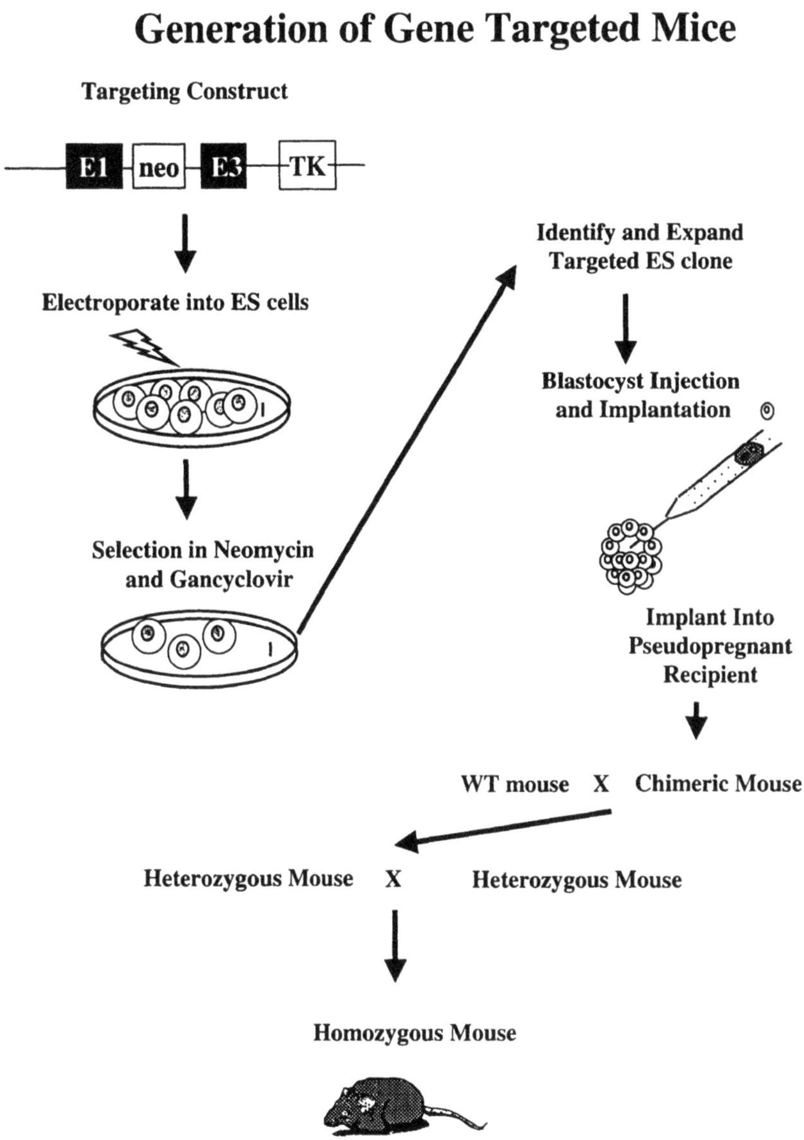

Figure 2. Generation of gene targeted mice. Shown is the experimental outline for generating heterozygous and homozygous gene knockout mice by homologous recombination in mouse embryonic stem (ES) cells.

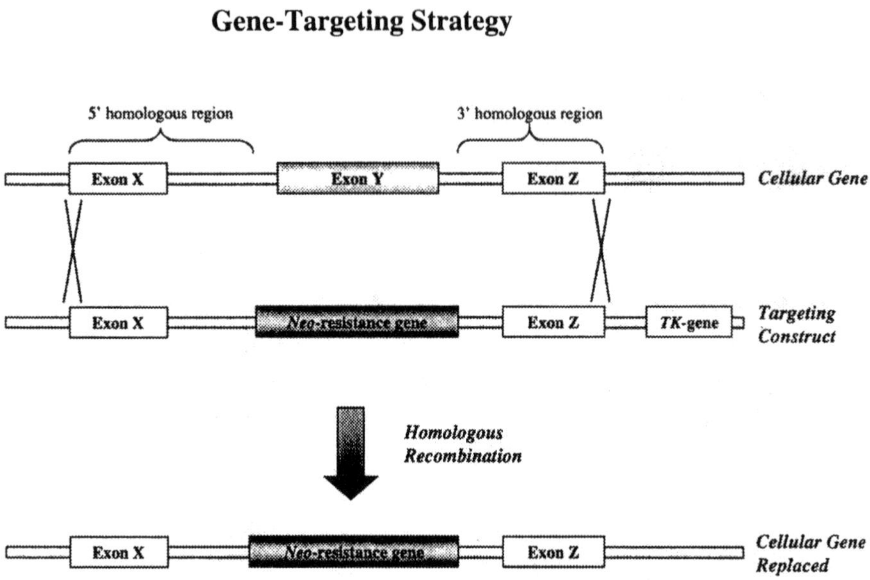

Figure 3. Gene targeting strategy using the bacterial neomycin resistance gene (neo) for positive selection and the viral thymidine kinase gene (TK) for negative selection. The desired homologous recombination event will result in exchange at the targeting site such that *neo* is inserted where exon Y was present.

2) Electroporation of the targeting construct into ES cells: ES cell lines are derived by culturing blastocysts in media of mitotically inactive mouse embryonic feeder layers (MEFs) or on specialized mouse fibroblast cells. Linearized gene-targeting vectors are introduced into trypsinized ES cells by electroporation. Since the frequency of homologous recombination is low, there must be a way of selecting the rare cells in which the target gene has been replaced by the constructed gene. Two selectable markers meet this purpose. The neo replacement-gene between the flanking homologous sequences confers resistance to the antibiotic neomycin (positive selection for ES cell clones in which the targeting vector has entered the cell and become integrated into the genome). The targeting vector also contains a gene that encodes viral thymidine kinase (tk gene), which confers sensitivity to ganciclovir. This second selectable marker lies outside the two homologous flanking regions of the replacement gene. It allows negative selection against ES cell clones in which the vector has integrated nonhomologously (random integration of the targeting vector). Appropriately targeted cells are not killed by the drug, because the tk gene is lost when the targeting vector recombines with the cognate gene. To identify targeted ES cell clones, Southern blot screening or PCR is used.

3) Generation of chimeric mice: Embryonic stem cells, which have undergone recombination, are microinjected into the cavity of 3.5 days old blastocysts (usually from strain C57BL/6), which are then impanted into 2.5-day pseudopregnant females. Since the ES cells used for gene-targeting are generally taken from strain 129 mice with distinct coat color alleles, chimeric pups can be identified by examining the coat color (patches of yellowish brown fur).

4) Breeding of chimeras to obtain heterozyogous and homozygous gene knockout mice: Chimeric male mice are typically bred with female C57BL/6 mice. Chimeras with germ cells derived from the altered embryonic stem cells transmit the targeted embryonic stem cell genome to their offsprings. Of these pups, 50% should be heterozygous for the targeted allele, as assessed by Southern blot analysis of tail DNA. Heterozygous mice are mated with one another to generate homozygous or knockout mice.

PHYSIOLOGICAL CHARACTERIZATION OF THE CARDIAC PHENOTYPE

The development of molecular techniques to modify the murine genome has provided a powerful tool to test the physiological consequences of single gene perturbations. Initially, the small heart size of the mouse was a challenging and limiting factor in the physiological assessment of phenotypic cardiac alterations. However, the availability of genetically altered mouse models prompted the development of methodologies capable of assessing cardiac function from the cellular to the intact animal level. (11,12). The following section attempts to survey standard techniques, used for the phenotypic characterization of mouse models (a full and detailed description of these techniques can be found in specific chapters of this book).

Isolated cardiomyocytes allow reliable measurements of mechanical function (e.g. resting cell length, shortening fraction, rate of shortening and relengthening using video edge detection systems), intracellular signaling (e.g. Ca^{2+} transients using fluorescence microscopy), and membrane physiology (electrophysiological characterization via patch clamp techniques) (13). Several of these measurements can be performed simultaneously, providing important insights into the subcellular mechanisms underlying alterations in excitation-contraction coupling of genetically engineered mice. The obtained results are devoid of the effects of potential alterations in extracellular matrix or chamber geometry. However, a major disadvantage is the heterogeneity of the myocytes and the selection of possibly unrepresentative cells during the isolation procedure.

At the whole organ level, the ***isolated perfused heart*** (14) allows assessment of contractile parameters without the influence of neurohumoral factors in a multicellular *in vitro* system, using either a Langendorff or a work performing mode. The isovolumic Langendorff preparation involves retrograde coronary perfusion with oxygenated buffer and a latex or cellophane balloon in the left ventricular chamber is used to measure developed pressure as a surrogate for volume. The work-performing heart is a technically more sophisticated system in which the cardiac muscle performs work at defined loading conditions. This system has been successfully adapted for measuring intraventricular pressure, rates of pressure development, and time indices for contraction and relaxation at the organ level of genetically engineered mice.

Isolated papillary muscles are another multicellular in vitro system used to measure cardiac muscle function (isometric and isotonic). This technique offers the possibility to control preload and afterload conditions. Principle disadvantages are potential damage to the ends of the muscle and the influence of core muscle hypoxia.

In vivo analysis of integrated cardiovascular function in mice has been accomplished by miniaturization and refinement of methodologies used in larger animals and clinical investigation. High fidelity ***measurements of left ventricular and arterial pressure*** in anesthetized and conscious mice can be obtained with Millar catheter-tip microtransducers (15-17). However, one has to interpret the measurements carefully, as these parameters are load dependent. At the non-invasive level, ***transthoracic 2-D-guided M-Mode and Doppler-echocardiography*** have been shown to be reliable techniques in assessing wall thickness, chamber size and cardiac function in the mouse (18). These techniques are perfectly suited for serial/longitudinal studies, and they are considered the most efficient tests for initial screening of the cardiac phenotype in genetically engineered mouse models.

In summary, this form of inter-disciplinary analysis has led to a better understanding of cardiac physiology and the pathways involved in cardiac hypertrophy and failure. The following section will summarize examples of such approaches, which provided unique insights into the genetic basis of cardiovascular physiology and heart disease.

GENETICALLY ENGINEERED MODELS WITH ALTERATIONS IN CARDIAC MEMBRANE CALCIUM HANDLING PROTEINS

Although a large number of murine models with altered expression of key regulatory proteins have been generated, this section will primarily focus on models with alterations in the levels of key Ca^{2+} handling proteins, which demonstrate the power of using loss-of-function and gain-of-function approaches to define the physiological role of single gene product *in vivo*. The contraction-relaxation cycle in the heart is controlled by the sequential rise and fall of cytosolic Ca^{2+} concentration. Ca^{2+} enters the cell through the L-type Ca^{2+} channels and causes the release of Ca^{2+} from the SR Ca^{2+} store via the SR release channels or ryanodine receptors. The released Ca^{2+} binds to troponin C and this results in a conformational change and the initiation of contraction of the myofilaments. Subsequently, relaxation is mediated by the rapid removal of cytosolic Ca^{2+}, which is predominantly re-sequestered into the SR by a Ca^{2+} ATPase (SERCA) and – to a lesser extent - extruded through the sarcolemma via the Na^+/ Ca^{2+} exchanger and a sarcolemmal Ca^{2+} ATPase. Table 2 briefly summarizes findings in murine genetic models with alterations in the expression levels of the sarcolemmal Ca^{2+} ATPase, the Na^+/ Ca^{2+} exchanger, and the SR proteins, which are involved in Ca^{2+} sequestration (SERCA and its regulator, phospholamban), Ca^{2+} storage (calsequestrin), and Ca^{2+} release (ryanodine receptor, FK506-binding protein and junctin) during excitation-contraction coupling.

The SR Ca^{2+} uptake complex

The rate of Ca^{2+} sequestration by the SR Ca^{2+} ATPase is a major determinant of the rate of muscle relaxation. Two independent studies overexpressed the cardiac isoform SERCA2a and observed relatively small increases in protein expression, but significant elevation of cardiac contractile parameters (19,20) Conversely, reduction in SR Ca^{2+} transport in SERCA2 heterozygous mice (~65% of wild-type SERCA2 protein) resulted in decreases in systolic ventricular pressure, rates of contraction and relaxation, and mean arterial blood pressure. Homozygous SERCA2 null mutants died in utero, indicating that two functional copies of the SERCA2 gene are required to maintain SERCA2 levels and Ca^{2+} homeostasis in the heart (21).

Table 2: **Examples of genetically engineered animal models with alterations in key Ca^{2+} handling proteins**

Genotype	Phenotype	Reference
SR Ca^{2+} uptake complex		
SERCA 2-overexpression	Contractility increased	He et al. 1997
SERCA 2 overexpression	Contractility increased	Baker et al. 1998
SERCA 1 overexpression	Contractility increased, can substitute for SERCA 2	Loukianov et al. 1998
SERCA 2 knock-out	Heterozygous: contractility decreased homozygous: embryonic lethality	Periasamy et al. 1999
Phospholamban knock-out	Viable, hyperdynamic function	Luo et al. 1996
Phospholamban overexpression	Contractility decreased	Kadambi et al. 1996
SR Ca^{2+} storage		
Calsequestrin overexpression	Increased SR Ca^{2+} storage, heart failure	Jones et al. 1998
Calsequestrin overexpression	Increased SR Ca^{2+} storage, depressed contractility, compensated hypertrophy	Sato et al. 1998
	Depressed force-frequency; pulsus alternans	Schmidt et al. 2000
SR Ca^{2+} release complex		
Ryanodine receptor knock-out	Homozygous: intra uterine death heterozygous: normal morphology and physiology	Takeshima et al. 1998
FKBP 12 knock-out	Severe juvenile dilated cardiomyopathy, VSD, normal skeletal muscle morphology	Shou et al. 1999
Junctin overexpression	May facilitate association between SR and T-tubule	Zhang et al. 2001
The sarcolemmal Ca^{2+} ATPase and Na^{+}/Ca^{2+} exchanger		
SL Ca^{2+} ATPase overexpression	Normal cardiac function and morphology	Hammes et al. 1998
Na^{+}/Ca^{2+} exchanger overexpression	Post-ischemic cardiac function depressed in male mice	Cross et al. 1998

Phospholamban (PLB) is a small (52 amino acids) phosphoprotein, which has been shown to regulate the activity of the SR Ca^{2+} ATPase. Dephosphorylated PLB is an inhibitor of SERCA, whereas phosphorylation during β-adrenergic stimulation relieves its inhibitory effects. Although considerable information on the mechanism by which PLB regulates SR Ca^{2+} ATPase activity was generated at the biochemical level (22), the physiological and pathophysiological role of PLB in modulating cardiac contractility *in vivo* was only recently elucidated. Genetically engineered animal models with diagonally opposed levels of PLB were generated, and the resulting cardiac phenotype was assessed using and integrated hierarchical approach (13,23)(Fig. 4). In the first model, the murine PLB gene was ablated by homologous recombination technology. Ablation of PLB was compatible with life in homozygous mice. At the morphological and histological levels, these mice were indistinguishable from their wild-type littermates. Biochemical studies indicated that the affinity of the SR Ca^{2+} ATPase for Ca^{2+} was significantly increased in the phospholamban knock-out (PLB-KO) hearts compared to wild-type (WT) controls. Physiological measurements in isolated cardiac myocytes revealed significant increases in the rates of myocyte shortening and relengthening. This hyperdynamic contractile function was also observed at the intact organ level. Hearts from PLB-KO and WT littermates were perfused in parallel in a Langendorff or work-performing mode. The rates of contraction and relaxation were significantly increased in the PLB-KO hearts compared to WTs. These findings could also be confirmed in the intact animal, using left ventricular catheter techniques and echocardiography. Thus, ablation of PLB was associated with significant increases in contractile parameters at the cellular, organ, and intact animal levels. The enhanced contractile parameters could be only minimally stimulated by β-agonists. These findings indicate that PLB is a major regulator of contractility and the β-agonist response in the mammalian heart. On the contrary, in order to determine whether all the SR Ca^{2+} ATPases in the native cardiac sarcoplasmic reticulum are subject to regulation by PLB, transgenic mice with overexpression of PLB were generated, using the cardiac α-myosin heavy chain promoter (13). Transgenic mice with a two-fold overexpression of PLB in their hearts, compared to wild-type littermates, showed no morphological alterations. However, isolated cardiac myocytes from transgenic animals exhibited diminished shortening fraction and decreased rates of shortening and relengthening compared with wild types. This impaired contractile function was reflected in diminished and prolonged Ca^{2+} transients, which were associated with a significant decrease in the apparent affinity of the SR Ca^{2+} ATPase for Ca^{2+}. Again, these findings were consistent with depressed contractile parameters at the intact organ and whole animal levels. However, the observed differences in contractile parameters and Ca^{2+} transients were abolished upon isoproterenol stimulation. These results suggest that there exists a fraction of spare SERCA molecules, which are not subject to regulation by PLB in the native SR. Furthermore, previous studies have demonstrated that: a) the variation of heart rate is a powerful modulator of cardiac contractility and myocardial function in the anesthetized mouse (force frequency relation); and b) the biphasic force-frequency relation can be characterized by the heart rate at which the peak rate of left ventricular isovolumic contraction is maximal (critical heart rate) (15). In order to determine whether and to what extend PLB levels regulate the force-frequency relation *in vivo*, transgenic mice overexpressing PLB, gene-targeted PLB-KO mice, and their isogenic wild-type controls were studied in parallel (24). The results indicate that the level of PLB is an important negative determinant of the force-frequency relation, such that ablation of PLB enhances, whereas overexpression of PLB depresses the critical heart rate in anesthetized closed –chest mice (Fig. 5). Based on these studies in genetically engineered PLB mouse models, it is interesting to propose that increases in the PLB /SR Ca^{2+} ATPase, such as those observed in human heart failure, may be at least partially responsible for slowed kinetics of Ca^{2+} transients, depressed contractile parameters and functional alterations associated with the disease.

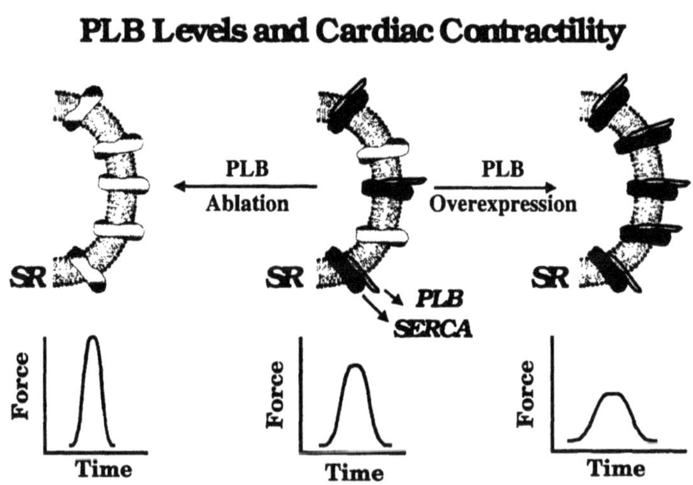

Figure 4. Schematic representation of the influence of phospholamban gene expression on cardiac contractility in genetically engineered animals. Reduction in PLB levels is associated with an increase in contractility, whereas an increase in PLB levels depresses rates of contraction and relengthening.

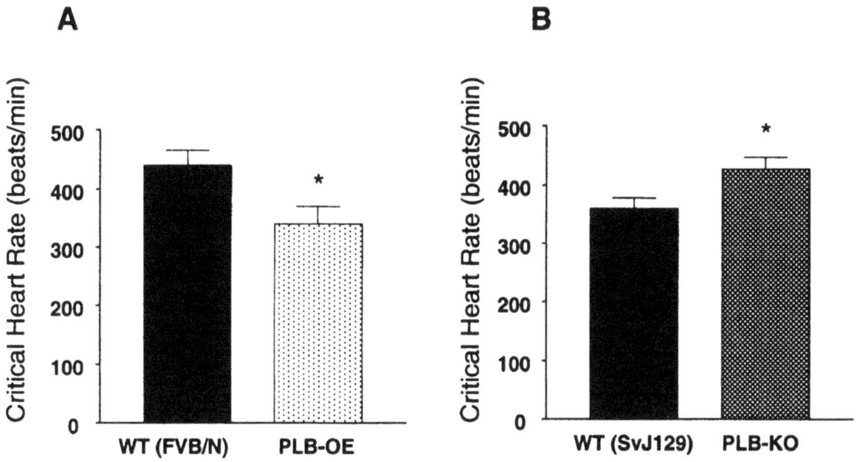

Figure 5. Effects of phospholamban levels on the force-frequency relation. (A) Overexpression of phospholamban (PLB-OE) is associated with a decrease in the critical heart rate (HRcrit), and (B) Ablation of phospholamban is associated with an increase in the critical heart rate, compared with their respective controls.

SR Ca^{2+} storage

Calsequestrin is the major Ca^{2+} storage protein in the junctional portion of the SR and has been postulated to form a functional complex with the Ca^{2+} release complex. Valuable insights on the physiological role of calsequestrin have been gained from the generation of two transgenic mouse models with the specific overexpression of calsequestrin in the heart (25,26). In both models, overexpression of calsequestrin resulted in cardiac hypertrophy. Although SR Ca^{2+} storage capacity was increased, the SR Ca^{2+} was not available for release during excitation contraction coupling, leading to depressed amplitude of the Ca^{2+} transient and depressed contractile parameters in isolated cardiomyocytes (Fig. 6). Furthermore, this was also associated with a depressed force-frequency relation (Fig. 7) and mechanical pulsus alternans in vivo (Fig. 6), supporting the hypothesis that tight regulation of the SR Ca^{2+} storage capacity is a central element of cardiac calcium-homeostasis.

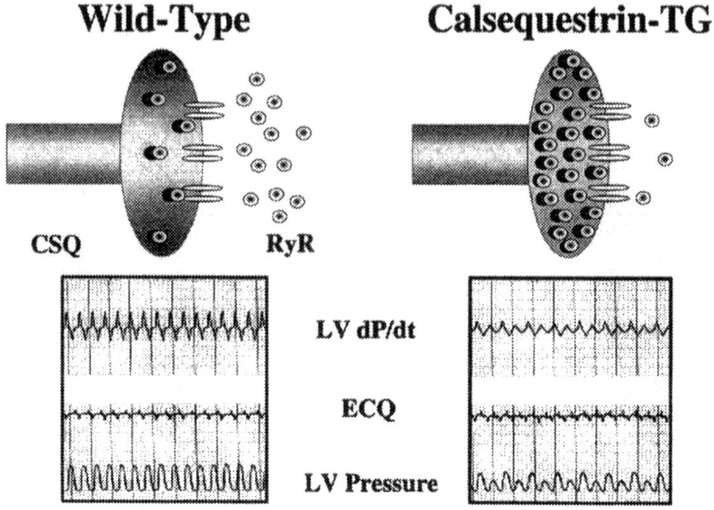

Figure 6. Schematic illustration of Ca2+ release from the sarcoplasmic reticulum in wild-type and calsequestrin overexpression (CSQ-OE) hearts. Field-stimulation resulted in lower levels of Ca^{2+} released from the SR in isolated cardiomyocytes from CSQ-OE hearts versus cells from WT hearts. In contrast, caffeine induced Ca^{2+} release was much higher in CSQ-OE, indicating an increase in SR Ca^{2+} load. This was associated with depressed contractility *in vivo* and the induction of mechanical pulsus alternans at pacing rates > 410 bpm in mice overexpressing CSQ.

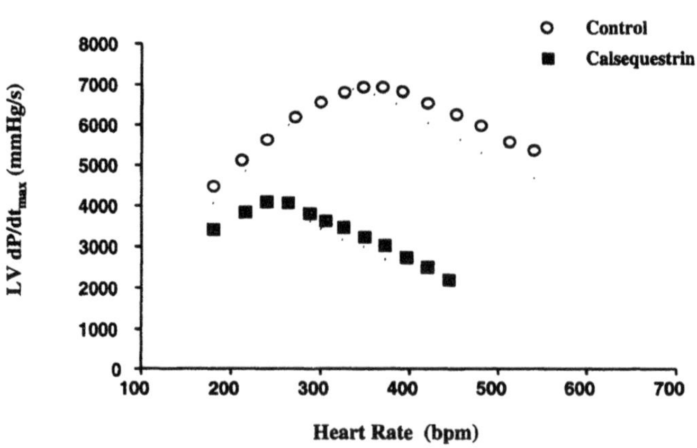

Figure 7. Representative force-frequency relationship from a transgenic mouse overexpressing calsequestrin and isogenic wild/type control. The relationship in calsequestrin overexpressing mice is shifted downward and leftward, and the critical heart rate is significantly lower.

The SR Ca^{2+} release complex
The ryanodine receptor (RyR) forms the key element of the SR Ca^{2+} release complex and is associated with a number of proteins, namely FKBP, junctin, triadin, and calsequestrin. *RyR2* was recently targeted by Takeshima et al (27). Mice completely lacking the cardiac ryanodine receptor died in utero, prior to the receptor undertaking its principal role during excitation-contraction coupling. It is interesting to note that heterozygous mice for the targeted allele were asymptomatic and had normal morphological and physiological characteristics. However, mice generated using a recombinant mutant gene (crr^{m1}), may be a valuable model to study SR Ca^{2+} release during E-C coupling in the adult.

FKBP12 is known to bind to the skeletal ryanodine receptor and to modulate coordinate gating of the receptor, while FKBP12.6 was believed to be important in cardiac muscle. Surprisingly, ablation of FKBP12 was not associated with any alterations in skeletal muscle morphology (28), but the majority of mice died between embryonic day 14.5 and birth, due to severe dilated cardiomyopathy and ventricular septal defects. Thus, the absence of detrimental consequences in skeletal muscle, although cardiac muscle was severely affected, may be due to differences in excitation-contraction coupling between these muscles.

Data from a recently generated transgenic model overexpressing junctin, another member of the release complex, suggests that junctin may facilitate the association between the SR and transverse tubules (29).

The sarcolemmal Ca^{2+} ATPase and Na^+/Ca^{2+} exchanger
Recent overexpression of the sarcolemmal Ca^{2+} ATPase in the rat did not appear to have any effects on cardiac function except when the SR function was abolished by thapsigargin or ryanodine, which led to a significantly smaller time constant of the Ca^{2+} decline in transgenic than in WT hearts. Nevertheless, while the physiological relevance of this change in E-C coupling is doubtful, the sarcolemmal Ca^{2+} ATPase may modulate myocardial growth through adjusting caveolar transduction (30).

The major physiological role of the Na^+/ Ca^{2+} exchanger is to extrude Ca^{2+} from the cell, by operating in the efflux or forward mode. However, the exchanger can operate in either direction. Transgenic mice with cardiac-specific overexpression of the Na^+/ Ca^{2+} exchanger exhibited similar basal contraction of isolated atrial preparations as wild-type controls (31). However, application of the Na^+ channel agonist BDF 9148 resulted in a concentration dependent increase in the force of contraction in transgenic animals, which was attributed to the presence of elevated levels of the Na^+/Ca^{2+} exchanger (32). Interestingly, post-ischemic cardiac function was markedly depressed in male transgenic hearts but not female transgenics compared to wild-type controls (33).

FUTURE DIRECTIONS

The advances in transgenic and gene-targeting methodology, coupled with the developments in miniaturized technology to assess cardiac function in the mouse, have added a wealth of information to our understanding of the physiological and pathophysiological role of Ca^{2+} homeostasis. The ability to manipulate the particular time-point at which a gene is switched on or off in a tissue-specific manner will further help to expand this knowledge. However, it is important to keep in mind that the mouse myocardium is considerably different from the human myocardium regarding excitation-contraction coupling, and that findings in the mouse cannot be readily extrapolated to other species and humans. Nevertheless, genetically modified animals are an invaluable tool, which allows the identification of genes that are causative for a certain disease (e.g. heart failure) and evaluation of molecular mechanisms responsible for the development and progression of the disease. Furthermore, it is hoped that this will lead to new therapeutic strategies, which can be tested on models resembling more closely the human disease.

REFERENCES

1. Hasenfuss G. Animal models of human cardiovascular disease, heart failure and hypertrophy. *Cardiovasc Res.* 1998;39:60-76.
2. Field LJ. Transgenic mice in cardiovascular research. *Annu Rev Physiol.* 1993;55:97-114.
3. Paigen K. A miracle enough: the power of mice. *Nat Med.* 1995;1:215-20.
4. James JF, Hewett TE, Robbins J. Cardiac physiology in transgenic mice. *Circ Res.* 1998;82:407-15.
5. Bers DM. Ca transport during contraction and relaxation in mammalian ventricular muscle. *Basic Res Cardiol.* 1997;92:1-10.
6. Gordon JW, Scangos GA, Plotkin DJ, Barbosa JA, Ruddle FH. Genetic transformation of mouse embryos by microinjection of purified DNA. *Proc Natl Acad Sci U S A.* 1980;77:7380-4.
7. Gordon JW, Ruddle FH. Gene transfer into mouse embryos: production of transgenic mice by pronuclear injection. *Methods Enzymol.* 1983;101:411-33.

8. Franz WM, Mueller OJ, Hartong R, Frey N, Katus HA. Transgenic animal models: new avenues in cardiovascular physiology. *J Mol Med.* 1997;75:115-29.
9. Yu Z, Redfern CS, Fishman GI. Conditional transgene expression in the heart. *Circ Res.* 1996;79:691-7.
10. Robbins J. Gene targeting. The precise manipulation of the mammalian genome. *Circ Res.* 1993;73:3-9.
11. Chien KR. Genes and physiology: molecular physiology in genetically engineered animals. *J Clin Invest.* 1996;97:901-9.
12. Christensen G, Wang Y, Chien KR. Physiological assessment of complex cardiac phenotypes in genetically engineered mice. *Am J Physiol.* 1997;272:H2513-24.
13. Kadambi VJ, Ponniah S, Harrer JM, Hoit BD, Dorn GW, 2nd, Walsh RA, Kranias EG. Cardiac-specific overexpression of phospholamban alters calcium kinetics and resultant cardiomyocyte mechanics in transgenic mice. *J Clin Invest.* 1996;97:533-9.
14. Grupp IL, Subramaniam A, Hewett TE, Robbins J, Grupp G. Comparison of normal, hypodynamic, and hyperdynamic mouse hearts using isolated work-performing heart preparations. *Am J Physiol.* 1993;265:H1401-10.
15. Hoit BD, Ball N, Walsh RA. Invasive hemodynamics and force-frequency relationships in open- versus closed-chest mice. *Am J Physiol.* 1997;273:H2528-33.
16. Lorenz JN, Kranias EG. Regulatory effects of phospholamban on cardiac function in intact mice. *Am J Physiol.* 1997;273:H2826-31.
17. Schmidt AG, Kadambi VJ, Ball N, Sato Y, Walsh RA, Kranias EG, Hoit BD. Cardiac-specific overexpression of calsequestrin results in left ventricular hypertrophy, depressed force-frequency relation and pulsus alternans In vivo. *J Mol Cell Cardiol.* 2000;32:1735-44.
18. Hoit BD, Khoury SF, Kranias EG, Ball N, Walsh RA. In vivo echocardiographic detection of enhanced left ventricular function in gene-targeted mice with phospholamban deficiency. *Circ Res.* 1995;77:632-7.
19. He H, Giordano FJ, Hilal-Dandan R, Choi DJ, Rockman HA, McDonough PM, Bluhm WF, Meyer M, Sayen MR, Swanson E, Dillmann WH. Overexpression of the rat sarcoplasmic reticulum Ca2+ ATPase gene in the heart of transgenic mice accelerates calcium transients and cardiac relaxation. *J Clin Invest.* 1997;100:380-9.
20. Baker DL, Hashimoto K, Grupp IL, Ji Y, Reed T, Loukianov E, Grupp G, Bhagwhat A, Hoit B, Walsh R, Marban E, Periasamy M. Targeted overexpression of the sarcoplasmic reticulum Ca2+-ATPase increases cardiac contractility in transgenic mouse hearts. *Circ Res.* 1998;83:1205-14.
21. Periasamy M, Reed TD, Liu LH, Ji Y, Loukianov E, Paul RJ, Nieman ML, Riddle T, Duffy JJ, Doetschman T, Lorenz JN, Shull GE. Impaired cardiac performance in heterozygous mice with a null mutation in the sarco(endo)plasmic reticulum Ca2+-ATPase isoform 2 (SERCA2) gene. *J Biol Chem.* 1999;274:2556-62.
22. Kranias EG. Regulation of Ca2+ transport by cyclic 3',5'-AMP-dependent and calcium-calmodulin-dependent phosphorylation of cardiac sarcoplasmic reticulum. *Biochim Biophys Acta.* 1985;844:193-9.
23. Luo W, Grupp IL, Harrer J, Ponniah S, Grupp G, Duffy JJ, Doetschman T, Kranias EG. Targeted ablation of the phospholamban gene is associated with markedly enhanced myocardial contractility and loss of beta-agonist stimulation. *Circ Res.* 1994;75:401-9.
24. Kadambi VJ, Ball N, Kranias EG, Walsh RA, Hoit BD. Modulation of force-frequency relation by phospholamban in genetically engineered mice. *Am J Physiol.* 1999;276:H2245-50.

25. Jones LR, Suzuki YJ, Wang W, Kobayashi YM, Ramesh V, Franzini-Armstrong C, Cleemann L, Morad M. Regulation of Ca2+ signaling in transgenic mouse cardiac myocytes overexpressing calsequestrin. *J Clin Invest.* 1998;101:1385-93.
26. Sato Y, Ferguson DG, Sako H, Dorn GW, Kadambi VJ, Yatani A, Hoit BD, Walsh RA, Kranias EG. Cardiac-specific overexpression of mouse cardiac calsequestrin is associated with depressed cardiovascular function and hypertrophy in transgenic mice. *J Biol Chem.* 1998;273:28470-7.
27. Takeshima H, Komazaki S, Hirose K, Nishi M, Noda T, Iino M. Embryonic lethality and abnormal cardiac myocytes in mice lacking ryanodine receptor type 2. *Embo J.* 1998;17:3309-16.
28. Shou W, Aghdasi B, Armstrong DL, Guo Q, Bao S, Charng MJ, Mathews LM, Schneider MD, Hamilton SL, Matzuk MM. Cardiac defects and altered ryanodine receptor function in mice lacking FKBP12. *Nature.* 1998;391:489-92.
29. Zhang L, Franzini-Armstrong C, Ramesh, Jones LR. Structural Alterations in Cardiac Calcium Release Units Resulting from Overexpression of Junctin. *J Mol Cell Cardiol.* 2001;33:233-247.
30. Hammes A, Oberdorf-Maass S, Rother T, Nething K, Gollnick F, Linz KW, Meyer R, Hu K, Han H, Gaudron P, Ertl G, Hoffmann S, Ganten U, Vetter R, Schuh K, Benkwitz C, Zimmer HG, Neyses L. Overexpression of the sarcolemmal calcium pump in the myocardium of transgenic rats. *Circ Res.* 1998;83:877-88.
31. Adachi-Akahane S, Lu L, Li Z, Frank JS, Philipson KD, Morad M. Calcium signaling in transgenic mice overexpressing cardiac Na(+)-Ca2+ exchanger. *J Gen Physiol.* 1997;109:717-29.
32. Baumer AT, Flesch M, Kilter H, Philipson KD, Bohm M. Overexpression of the Na(+)-Ca2+ exchanger leads to enhanced inotropic responsiveness to Na(+)-channel agonist without sarcoplasmic reticulum protein changes in transgenic mice. *Biochem Biophys Res Commun.* 1998;249:786-90.
33. Cross HR, Lu L, Steenbergen C, Philipson KD, Murphy E. Overexpression of the cardiac Na+/Ca2+ exchanger increases susceptibility to ischemia/reperfusion injury in male, but not female, transgenic mice. *Circ Res.* 1998;83:1215-23.

3
CONDITIONAL TRANSGENESIS

**Tetsuo Minamino, MD, PhD
and Michael D. Schneider, MD**
*Center for Cardiovascular Development
Baylor College of Medicine
Houston, TX*

INTRODUCTION

Gain- and loss-of-function mutations allow a direct assessment of gene function in vivo and the creation of useful animal models of human disease. In recent years, both methods have been fruitfully applied to many of the fundamental and translational questions central to cardiovascular biology and pathobiology, the two themes of this volume (1-4). Incorporation of an exogenous gene into mouse genome can be accomplished directly, e.g., by microinjection of fertilized oocytes, to produce a transgenic animal that is, most typically, a gain-of-function mutant. Alteration of endogenous genes (gene targeting) can be accomplished by homologous recombination in embryonic stem (ES) cells, most typically to generate a loss-of-function mutation in the germline, also known as a "knockout." Whereas exogenous transgenes already are applicable to larger mammals as well (5), gene targeting by homologous recombination is feasible only in mice, or cultured cells. Descriptions of these two contrasting methods, in their conventional form, are detailed elsewhere (6). More baroque variations can be useful, such as transgenes that encode a dominant-interfering protein to ablate protein function rather than expression (7), or "knockin" mutations, using homologous recombination to systematically compare protein structure and function in vivo (8).

The fundamental difference between these two approaches thus is not the direction toward which gene function is altered, but rather the underlying technology. As a cautionary note to guard against experimental hubris (excessive claims that protein function is best discerned from knockout mutations alone), it is useful to remember that more two-thirds of the genes even in *C. elegans*, *Drosophila*, and yeast show no loss-of-function phenotype (9), and all three genomes have lesser risk of redundancy than holds true for mammals. Saturation mutagenesis in model organisms has been useful not merely for the large number of genes affected, or the lack of a priori genetic bias, but also because the informative alleles that result range from null mutations to hypomorphic and activated, neomorphic ones.

Classically, the second key difference between exogenous transgenes versus gene targeting by homologous recombination has been whether the resulting mutation modifies all cells that express the gene, equivalently (as with a null allele in the germline), versus a subset (as can be done with an exogenous gene, driven by an exogenous control region). Although promoterless transgenes have been inserted into mice as a means to pinpoint tissue-specific enhancers ("gene trap" experiments), including cardiac and vascular ones (10,11), defined control regions are used far more commonly. Among those in widest use for cardiac applications is the mouse α–myosin heavy chain (MHC) promoter, which is more potent than that of myosin light chain-2v, and more stringently specific than that of beta-MHC (2).

Together, these inherent technical differences contribute to the fact that germline deletions have had profound impact on developmental biology, answering one question exceedingly well by their nature — what a protein's earliest essential function might be — but have achieved more sporadic success in explaining the physiology and pathophysiology of adult animals. The reasons for this "utility gap" are twofold:

First, it was impossible, until lately, to control the timing of gene deletion. A null mutation in the germline affects all cells that would ordinarily express the gene, *ab initio*, from the fertilized oocyte onward. To illustrate, endothelin-1 is implicated in blood pressure regulation, but homozygous endothelin-deficient mice die at birth with failure to form the myenteric plexus (12). Early lethality thus renders it difficult for any biological function of endothelin to be studied in adult mice, by conventional gene deletion. Difficult, but not impossible: chimeric mice containing a mix of mutant and wild-type cells provide a means to this end, albeit laborious (13).

Second, the targeted gene is altered in all cells and cell types concurrently. For genes whose expression is more widespread than one cell type alone, this means that one cannot examine the biological impact of gene targeting on any given lineage, without taking into account potential indirect effectors including paracrine, endocrine, inflammatory, and metabolic circuits. In the jargon of developmental biology, intrinsic defects are referred to as "cell-autonomous," and indirect ones as "non-cell-autonomous." For example, mice lacking the insulin-sensitive glucose transporter GLUT4 exhibit cardiac hypertrophy, but it is ambiguous whether this change relates to GLUT4 deficiency in cardiac muscle or instead to the many metabolic changes that ensue from the absence of GLUT4 in other tissues, including post-prandial hyperinsulinemia and impaired fat storage (14). Furthermore, some lineage-specific promoters are active enough early in embryogenesis, that even effects arising in adult animals could be contingent on antecedent effects, occurring before birth.

Even stringent tissue-specificity might not suffice to overcome this set of obstacles. The dual need arose to regulate expression of exogenous genes or disruption of endogenous ones conditionally — in a specific cell type, at a given time, or, ideally, both.

CONDITIONAL TRANSGENE EXPRESSION AND FUNCTION

Conceptually, the simplest way to regulate expression of a transgene is to control its transcription, using tetracycline (15,16), synthetic steroid hormones (17), or ecdysone (18). The feature common to all these drug-regulated systems is their use of heterologous, chimeric transcription factors whose action on the target gene is controlled by administration of a drug. Notable examples of their use for cardiovascular applications include inducible signaling by an engineered G-protein (16), and inducible myocyte death by conditional expression of a toxin (15).

A very different strategy to activate a transgene would control protein function, rather than expression. One of the generic approaches to accomplish this exploits protein-protein interactions via drug-induced protein dimerization. The actual molecular steps involved here include protein auto-activation by proximity, heterodimerization, targeting to the membrane or other sites, or recruitment of separable protein domains (19,20).

An especially refined application of this principle to cardiovascular studies was to control the function of epsilon protein kinase C in myocardium by controlling its translocation to epsilon RACK, the receptor for its activated form (21,22). By this means, epsilon protein kinase C was shown to confer protection from ischemic injury and to promote concentric hypertrophy, whereas inhibition of endogenous epsilon protein kinase C resulted in a lethal cardiomyopathy (21,22) .

A dimerization-dependent transcriptional switch bridges the two complementary strategies discussed here, regulating transgene expression by drug-dependent reunification of a transcription factor that has been split into its DNA-binding and activation domains (20).

SITE-SPECIFIC MUTAGENESIS USING DNA RECOMBINASES

Confining gene deletion with analogous precision has been made feasible by the use of site-specific DNA recombination, the two components for which are an exogenous DNA recombinase, whose expression can be regulated by a variety of means, and a gene that has been engineered to harbor the enzyme's target site (23,24). Both yeast and phage recombinases, Flp and Cre respectively, have been used for regulated recombination of mouse genes in vivo (23,24), although more extensive experience is reported thus far for Cre. This 38 kDa protein splices DNA between specific 34 bp sequences known as loxP sites. Each loxP site contains two palindromic copies of a 13 bp repeat, separated by an 8 bp spacer that gives 5' to 3' orientation to the motif.

Whether Cre triggers an excision (the most widely used application), inversion, integration, or translocation is determined by the position and orientation of loxP

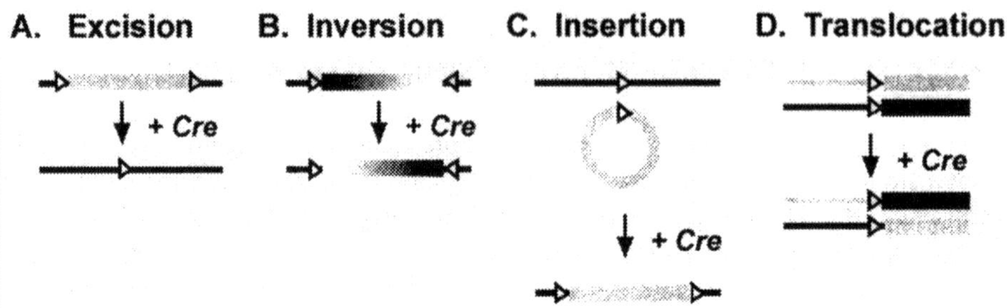

Figure 1. Site-specific DNA recombination using the Cre/loxP system. (A) Excision, using two loxP sites (shown as triangles) in the same orientation. (B) Inversion, using paired loxP sites in the same DNA molecule, but in opposite orientation. (C) Insertion and (D) translocation, using loxP sites on separate DNA molecules.

sites, relative to each other (Figure 1). Excision and inversion occur when the loxP sites exist on the same molecule of DNA in *cis*, as direct or inverted repeats, respectively. When loxP sites are located on separate molecules, in *trans*, integration or translocation can be achieved. Notably. this system does not require additional host factors or specific DNA topology for efficient recombination.

Apart from conditional deletion of endogenous genes, which will be detailed subsequently, site-specific excision also is useful for other purposes, discussed immediately below: to delete drug-selectable markers from conventional targeting vectors, to activate a latent transgene that is 3' to a stop signal, or to consolidate integration sites (which will contain variable copy number, from founder to founder) into a more uniform single-copy event.

Deleting the selection cassette is desirable, after successful homologous recombination into ES cells, to minimize any fortuitous consequences of the markers themselves or the control regions used to drive their expression (25,26), most especially when these are positioned between the exons of an endogenous gene. This gives much greater flexibility to the design of targeting vectors (27), and helps preclude a large category of potential confounding effects. Even brief expression of Cre suffices for marker excision (28). In addition, deleting the markers also facilitates "marker recycling," allowing homozygous deletions to be made in ES cells, using the identical targeting vector for a second round of homologous recombination (29). Because extended manipulations in cell culture can reduce the totipotency of ES cells, alternative routes to marker removal are attractive, including the use of CMV-Cre or EIIa-Cre mice (30,31), or direct injection of zygotes with a Cre expression vector (32).

Activating a latent transgene, using Cre to excise a loxP-flanked stuffer lying between the promoter and transgene of interest (33), is one of several binary systems that might be used to sidestep early lethality. Death in utero, in the perinatal period, or at any later time before reaching reproductive age can complicate propagation of transgenic lines, necessitating a reliance on F1 or even F0 ("transient transgenic") mice (34,35). Even where lines are propagated successfully, biological selection against high-expressers must be considered, which would also be obviated by this approach. By comparison to the drug-inducible systems cited earlier, a theoretical advantage of Cre/lox methods is the complete absence of "leak" in the absence of Cre. Hence, lines with silent transgenes can be readily established and expanded, and the potential phenotypes can be readily provoked in the progeny, by mating to appropriate Cre mice (33). In addition, after removing the stop signal, the 3' transgene would be expressed under the direct control of the promoter used. Cre-dependent activation of a LacZ reporter gene, as a readily assayable marker of cell lineage, is another application of this approach (36,37).

Consolidating the integration site into a single copy, in lieu of the usual concatamer, has merit for reasons both intuitive and counterintutive. Some targets (notably, including sarcomeric proteins) accommodate a wide array of copy number, at least within limits (38), perhaps by auto-regulation of proteins' abundance at a translational or post-translational level. Although dose-response relations can be informative (39) or give rise to models that reflect differing aspects of a disease, comparing structure and function systematically in vivo, by transgenesis, can be complex, as a consequence of variable dosage. Side-by-side comparison of single-copy transgenes would be advantageous, towards this goal. Moreover, although expression levels and the severity of phenotypes often track with copy number (38,39), this is not always the case. A less obvious purpose for engineering single-copy transgenes is, therefore, the chance to abrogate repeat-induced gene silencing, an important contributor to variegation and mosaic expression (40).

TISSUE-SPECIFIC GENE DELETION

To use the Cre/loxP system for gene targeting in vivo (or, to use analogous methods with the Flp recombinase), two lines of mice are generated. To create the loxP-tagged ("floxed") allele, the endogenous target gene is modified by homologous recombination in ES cells, flanking a genomic segment that is critical for the protein's expression or function with two loxP sites in the same orientation. Often, the loxP sites are placed within the introns, distant from splice donor or acceptor sites, to minimize their risk of interfering by themselves with the gene's normal expression. To create the Cre "deleter" mice, the recombinase is expressed either as a transgene under the control of a suitable promoter (41-43), or is knocked into a suitable endogenous locus (37). In the latter case, it is desirable that haploinsufficiency does not occur. Mating mice that contain a floxed gene with mice that express Cre will result in Cre-mediated gene disruption in the progeny, exclusively in those that co-inherit both genes and restricted to cells in which Cre is transcribed.

A major advantage of conditional strategies for gene inactivation using the Cre/loxP system is that the same loxP-tagged mouse can be used for gene deletion in a variety of biological contexts — different tissues or developmental stages — simply mating to Cre mice that display the recombinase with the desired tissue or temporal program of expression.

Reported Cre genes with the most direct applicability to cardiac myocytes include those driven by the control regions for α–MHC (41,44-47), ventricular myosin light chain-2 (42,48), cardiac or smooth muscle α–actin (49), and muscle creatine kinase (50,51). For vascular biology, in addition to the actin promoters cited above, reported systems use SM22 (52), smooth muscle MHC (43), platelet endothelial cell adhesion molecule-1 (53), Tie1 (54) and Tie2 (37,55). With some systems for Cre expression, such as α–MHC (41), high fidelity to the native pattern and a high prevalence for recombination are achievable using a transgene. With others, lesser fidelity or less complete recombination result, which can potentially be improved by knocking Cre into the corresponding endogenous locus (52,54,56). Although more demanding initially, the knockin also may obviate the need to compare multiple Cre lines, for any given promoter. However, few side-by-side comparisons exist (46).

Several complementary criteria are useful to evaluate conditional deletion. Because excision is produced only in the particular cell type(s) where Cre is expressed, less than complete recombination should be anticipated for any heterogeneous tissue. Hence, monitoring the prevalence for recombination needs to take the biologically relevant denominator into account. For cardiac muscle cells, this analysis can be improved using myocytes enriched by centrifugation through a Percoll gradient. Mosaic or ectopic expression of Cre, in a pattern different from the expected, also can complicate the interpretation of experiments. Thus, it is crucial first to know the expression pattern for Cre, in cellular detail.

Methods to monitor Cre protein expression cell-by-cell include immunostaining for Cre itself (57), staining for an epitope tag (58), or a chimeric fusion of Cre incorporating green fluorescent protein (28). A convenient assay for Cre protein function is obtained by mating Cre mice with mice harboring a ubiquitously transcribed floxed cassette which, when deleted, enables the expression of a 3' reporter gene such as *E. coli* β-galactosidase (36,59). In that circumstance, blue staining signifies the presence of Cre in the labeled cells, or can serve as a marker for cell fate, based on its earlier expression in their precursors.

CRE/LOX SYSTEMS FOR CHROMOSOME ENGINEERING

In addition to biological limitations like early lethality and the inability to resolve cell-autonomous versus non-cell-autonomous defects, as discussed above, conventional methods for gene targeting by homologous recombination have other, purely technical limitations. Homologous recombination is poorly suited or unable to engineer complex chromosomal alterations such as large deletions, inversions, and translocations, as occur in a number of human disorders. By contrast, because Cre will catalyze recombination between large distances within a chromosome, and even in *trans* between separate chromosomes, this system can be used to produce precise chromosome translocations, deletions, and inversions in ES cells (60,61).

Cre-mediated engineering of a megabase deletion syntenic with the DiGeorge syndrome locus in humans was used to model this disorder in mice (62). Using retrovirus-mediation insertion of loxP sites for nested deletions within the region (63), and a series of duplication alleles, this ultimately led to the identification of a T-box transcription factor, Tbx1, as the haploinsufficient gene responsible at least for the aortic arch defects in this syndrome (64).

Cre-mediated inversion also has been used successfully in chromosome engineering for mice (60). Akin to the use of balancer chromosomes to expedite gene mapping after mutagenesis in flies, genome-wide saturation mutagenesis in mice will likely be facilitated using analogous large-scale inversions to suppress recombination between the mutated and wild-type chromosomes (60). These recent advances using site-specific recombination for large-scale manipulation of chromosomal DNA offer especially exciting possibilities for studying the mouse genome systematically and generating novel animal models of human disease.

SPECIFYING DNA RECOMBINATION IN BOTH TIME AND SPACE

Notwithstanding constraints that are specifically germane to large-scale chromosome engineering — such as a much lesser efficiency for recombination in trans — Cre-mediated recombination is dependent only on the availability of Cre recombinase in the desired cells. Thus, gene ablation by Cre can be regulated directly, by controlling the timing and spatial distribution of Cre expression. As is true for transgenes in general, three general approaches can be taken.

Tissue-restricted promoters can potentially provide specificity not merely for a lineage defined in its broadest terms (cardiac myocytes, vascular smooth muscle cells), but also for more selective compartments, specializations, and origins (atria versus ventricles, trabecular versus compact myocardium, smooth muscle that is neural crest-derived or not). Even differences in cell position can be the basis for selective expression of Cre, as genes have emerged that define the left and right heart or even distinguish sidedness at the symmetric linear heart tube stage (65). Analogously, the sequential program of transcription during cardiac and vascular development can be exploited to defer Cre expression to particular stages, to the extent that endogenous genes or individual enhancers exhibit the requisite timing.

Ligand-inducible promoters come in the flavors described for inducible transgenes in general, and in principle can provide more selective timing to the induced recombination event, e.g., at defined intervals before or after infarction, aortic banding, or other surgical interventions in mice. One early example that was applied successfully to Cre induction in vivo is the interferon-responsive Mx1 promoter (66), which has been used for post-natal deletion of the cytokine receptor subunit gp130 (67), among other applications. However, use of inteferon itself, double-stranded RNA to up-regulate endogenous interferon, or other biologically active inducers could be complicated by pleiotropic effects unrelated to Cre or the target floxed allele. Alternatives whose application seems more straightforward to regulate Cre expression include the transcriptional switches cited earlier, involving tetracycline (68,69) or RU486 (70). Two pivotal differences apply to conditional recombination, however, by contrast to other conditional systems. First, leak is cumulative and irreversible, unlike any ordinary drug-dependent gene. Even a low frequency for spontaneous recombination can result in substantial recombination over time, due to the baseline levels of Cre in the absence of ligand. Second, it could be cumbersome if two independent transgenes were needed to trigger recombination of a floxed allele (one to express Cre, and a second to control the first's expression), and either gene's expression might be limited as the result of mosaicism.

Ligand-dependent Cre fusion proteins, whose function can be activated selectively by drugs, are therefore attractive as an alternative route to achieve both spatial and temporal control over gene recombination, concurrently, given the limitation of other methods just mentioned. Reported ways to control Cre function in cells fuse the Cre recombinase to the ligand-binding domain (LBD) of a steroid hormone receptor. Fusion with Cre has been done for estrogen, glucocorticoid, and progesterone receptor LBDs (71-75). Because mutated derivates exist for many of these domains, chimeric Cre recombinases have been engineered that are activated by synthetic ligands, but not by the natural ligands of the corresponding LBD. Thus, chimeric Cre recombinases that are selectively activated by synthetic drugs can provide useful tools to control the timing of recombination, using a single transgene to achieve tissue-specificity as well.

Illustrated in Figure 2, the authors have developed transgenic mice for heart-specific, drug-inducible recombination (76). These express a Cre fusion protein along the lines described above, with a mutated progesterone receptor LBD (PR1), driven by the α–MHC promoter. The Cre fusion protein translocates from the cytoplasm to the nucleus upon administration of the synthetic antiprogestin RU486, and becomes functionally active. Unlike the glucocorticoid receptor (77), wild-type progesterone receptor is constitutively nuclear, and therefore additional mechanisms apart from translocation alone likely contribute to the activation of chimeric proteins through this heterologous domain (78).

This conditional system bypasses all embryonic and perinatal effects of αMHC-Cre on the heart, and permits inducible recombination in cardiac muscle, e.g., at weaning (76). Especially at later ages, leaky recombination was detected (independent of the exogenous ligand), which roughly paralleled the levels of the fusion protein in the myocardium, across three independent transgenic lines. This could reflect partial sensitivity of the fusion protein to endogenous steroids, or hormone-independent Cre function, overriding the constraints imposed by the progesterone receptor LBD.

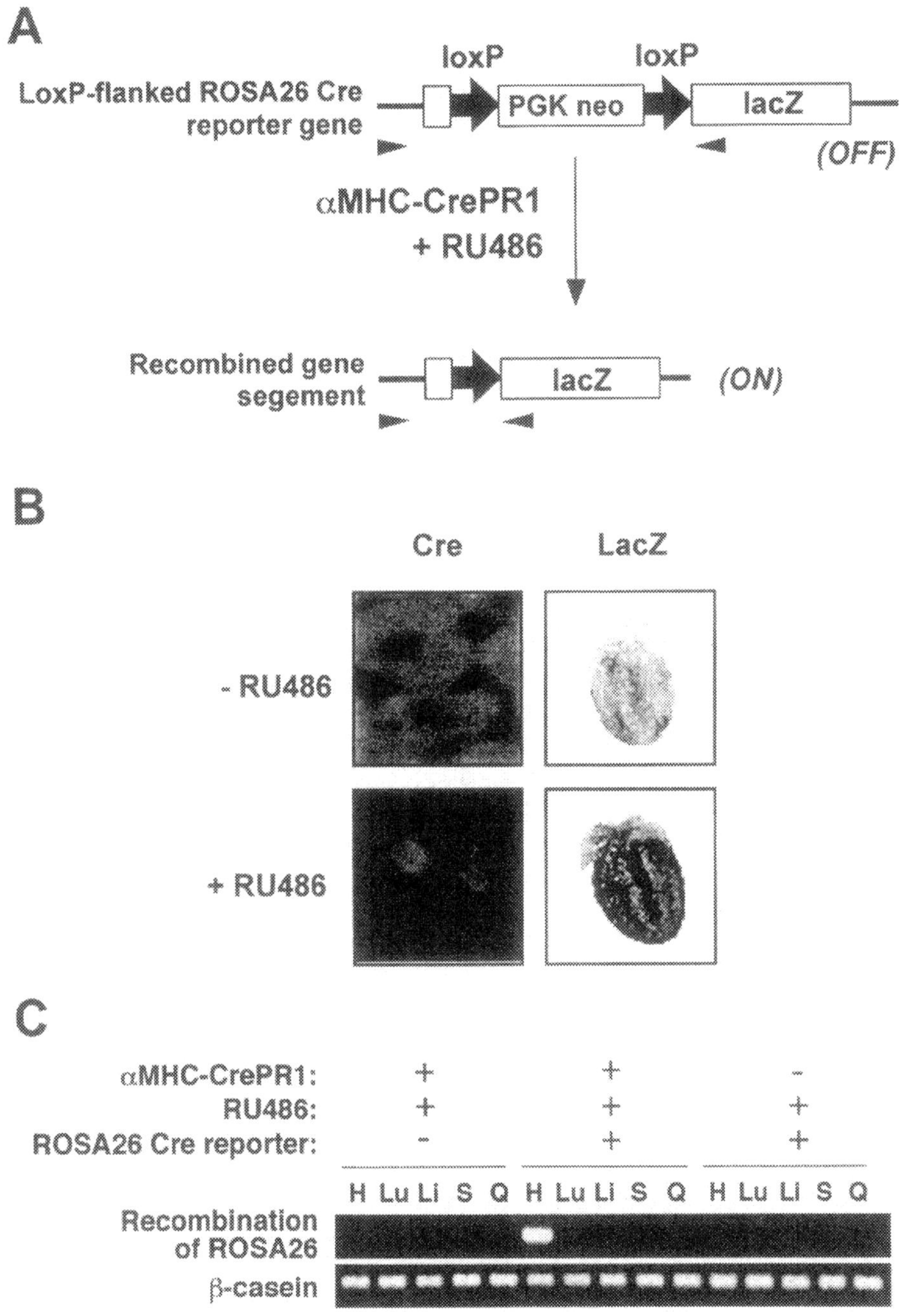

Figure 2. RU486-inducible recombination in post-natal myocardium. (A) Diagram of the Cre-dependent ROSA26 lacZ reporter gene, before and after recombination. The PCR primers used in panel C are shown as triangles. (B) RU486-induced translocation of CrePR1 to cardiac myocyte nuclei (left) and Cre-dependent lacZ activation in cardiac muscle, as a readout of gene recombination (right). (C) Tissue-specificity for recombination using αMHC-CrePR1. Adapted from Minamino et al (76).

Potential refinements of this system therefore include further modification of the motifs controlling protein translocation, or conditional expression of the existing fusion protein. Even in its current form, this system overcomes the embryonic activity of existing cardiac-specific Cre lines (42,47). An analogous system has been reported for conditional deletion in smooth muscle (52). Viral delivery of Cre for de novo gene deletion

Although viral delivery of exogenous genes is not strictly germane to this chapter's theme of conditional transgenesis, viral delivery of Cre recombinase can complement the systems described above for temporal and spatial control of gene deletion (41). Adenoviral gene transfer with Cre is a straightforward and potentially powerful means to examine gene function in culture, following the infection of cardiac myocytes that contain a floxed allele (lox/- or lox/lox). This strategy may be applicable in vivo, too, for de novo gene deletion after birth. Cre-expressing adenoviral vectors can be administered by direct injection and cause localized recombination at the site of application, even in the post-mitotic adult myocardium (41). This finding highlights the fact that on-going DNA synthesis is not needed for efficient recombination to take place.

The continued development of improved systems for more homogeneous delivery to myocardium bears mentioning, as implemented at least for rats, if not yet mice (79). Other relevant methods that may prove useful in connection with viral delivery of recombinase to the cardiovascular system include less immunogenic adenoviral vectors, as well as alternatives like adeno-associated virus and lentivirus, that also offer diminished immunogenicity (79).

CONCLUSIONS

Site-specific DNA recombination has become a powerful addition to the experimentalist's armamentarium, to link genes to function in the mammalian genome. Using this system, one can ablate endogenous genes in a tissue-specific and temporally defined manner, overcoming key limitations of conventional germline deletion; activate or excise an exogenous transgene with especially tight control; and even engineer chromosomes to model human disorders, or expedite gene mapping in mice. Further refinements are expected for optimizing the Cre and Flp recombinase (80), including the reliability and precision for their conditional function in mice. In addition, more widespread use can be foreseen for site-specific recombination, applied to in vivo promoter mapping (81). The combinatorial use of Cre and Flp recombinases together has simplified the mechanics of gene targeting, allowing independent recombinases to be used to delete the selectable marker cassette versus the conditional allele, and has expanded the efficiency of gene targeting, allowing a allelic series of mutations to be made at a given locus, from a single targeted mouse line (82).

REFERENCES

1. MacLellan WR, Schneider MD. Genetic dissection of cardiac growth control pathways. Annu. Rev. Physiol. 2000;62:289-320.
2. Robbins J. Remodeling the cardiac sarcomere using transgenesis. Annu. Rev. Physiol. 2000;62:261-87.
3. Chien KR. Stress pathways and heart failure. Cell. 1999;98:555-558.
4. Frey N, McKinsey TA, Olson EN. Decoding calcium signals involved in cardiac growth and function. Nat Med. 2000;6:1221-7.
5. Marian AJ, Wu Y, Lim DS, McCluggage M, Youker K, Yu QT, Brugada R, DeMayo F, Quinones M, Roberts R. A transgenic rabbit model for human hypertrophic cardiomyopathy. J. Clin. Invest. 1999;104:1683-92.
6. Hogan B, Beddington R, Costantini F, Lacy E, eds. Manipulating the Mouse Embryo: A Laboratory Manual. second ed. Cold Spring Harbor, N. Y.: Cold Spring Harbor Laboratory; 1994.
7. Shioi T, Kang PM, Douglas PS, Hampe J, Yballe CM, Lawitts J, Cantley LC, Izumo S. The conserved phosphoinositide 3-kinase pathway determines heart size in mice. EMBO J. 2000;19:2537-48.
8. Tallquist MD, Klinghoffer RA, Heuchel R, Mueting-Nelsen PF, Corrin PD, Heldin CH, Johnson RJ, Soriano P. Retention of PDGFR-beta function in mice in the absence of phosphatidylinositol 3'-kinase and phospholipase Cgamma signaling pathways. Genes Dev. 2000;14:3179-90.
9. Rorth P, Szabo K, Bailey A, Laverty T, Rehm J, Rubin GM, Weigmann K, Milan M, Benes V, Ansorge W, Cohen SM. Systematic gain-of-function genetics in Drosophila. Development. 1998;125:1049-57.
10. Chen Z, Friedrich GA, Soriano P. Transcriptional enhancer factor 1 disruption by a retroviral gene trap leads to heart defects and embryonic lethality in mice. Genes Dev. 1994;8:2293-2301.
11. Hidai C, Zupancic T, Penta K, Mikhail A, Kawana M, Quertermous EE, Aoka Y, Fukagawa M, Matsui Y, Platika D, Auerbach R, Hogan BL, Snodgrass R, Quertermous T. Cloning and characterization of developmental endothelial locus-1: an embryonic endothelial cell protein that binds the alphavbeta3 integrin receptor. Genes Dev. 1998;12:21-33.
12. Kurihara Y, Kurihara H, Oda H, Maemura K, Nagai R, Ishikawa T, Yazaki Y. Aortic arch malformations and ventricular septal defect in mice deficient in endothelin-1. J. Clin. Invest. 1995;96:293-300.
13. Tran CM, Sucov HM. The RXR alpha gene functions in a non-cell-autonomous manner during mouse cardiac morphogenesis. Development. 1998;125:1951-1956.
14. Katz EB, Stenbit AE, Hatton K, DePinho R, Charron MJ. Cardiac and adipose tissue abnormalities but not diabetes in mice deficient in GLUT4. Nature. 1995;377:151-5.
15. Lee P, Morley G, Huang Q, Fischer A, Seiler S, Horner JW, Factor S, Vaidya D, Jalife J, Fishman GI. Conditional lineage ablation to model human diseases. Proc. Natl. Acad. Sci. U. S. A. 1998;95:11371-11376.
16. Redfern CH, Coward P, Degtyarev MY, Lee EK, Kwa AT, Hennighausen L, Bujard H, Fishman GI, Conklin BR. Conditional expression and signaling of a specifically designed G(I)-coupled receptor in transgenic mice. Nat. Biotechnol. 1999;17:165-169.
17. Wang XJ, Liefer KM, Tsai S, O'Malley BW, Roop DR. Development of gene-switch transgenic mice that inducibly express transforming growth factor beta1 in the epidermis. Proc. Natl. Acad. Sci. U. S. A. 1999;96:8483-8.
18. No D, Yao TP, Evans RM. Ecdysone-inducible gene expression in mammalian cells and transgenic mice. Proc. Natl. Acad. Sci. U. S. A. 1996;93:3346-51.
19. Spencer DM. Creating conditional mutations in mammals. Trends Genet. 1996;12:181-7.
20. Amara JF, Clackson T, Rivera VM, Guo T, Keenan T, Natesan S, Pollock R, Yang W, Courage NL, Holt DA, Gilman M. A versatile synthetic dimerizer for the regulation of protein-protein interactions. Proc. Natl. Acad. Sci. U. S. A. 1997;94:10618-23.

21. Dorn GW, Souroujon MC, Liron T, Chen CH, Gray MO, Zhou HZ, Csukai M, Wu G, Lorenz JN, Mochly-Rosen D. Sustained in vivo cardiac protection by a rationally designed peptide that causes epsilon protein kinase C translocation. Proc. Nat. Acad. Sci. U. S. A. 1999;96:12798-803.
22. Mochly-Rosen D, Wu G, Hahn H, Osinska H, Liron T, Lorenz JN, Yatani A, Robbins J, Dorn GW. Cardiotrophic effects of protein kinase C epsilon: analysis by in vivo modulation of PKCepsilon translocation. Circ. Res. 2000;86:1173-9.
23. Rossant J, McMahon A. "Cre"-ating mouse mutants-a meeting review on conditional mouse genetics. Genes Dev. 1999;13:142-5.
24. Rodriguez CI, Buchholz F, Galloway J, Sequerra R, Kasper J, Ayala R, Stewart AF, Dymecki SM. High-efficiency deleter mice show that FLPe is an alternative to Cre-loxP. Nat. Genet. 2000;25:139-40.
25. Olson EN, Arnold HH, Rigby PWJ, Wold BJ. Know your neighbors: three phenotypes in null mutants of the myogenic bHLH gene MRF4. Cell. 1996;85:1-4.
26. Artelt P, Grannemann R, Stocking C, Friel J, Bartsch J, Hauser H. The prokaryotic neomycin-resistance-encoding gene acts as a transcriptional silencer in eukaryotic cells. Gene. 1991;99:249-54.
27. Baudoin C, Goumans MJ, Mummery C, Sonnenberg A. Knockout and knockin of the beta(1) exon D define distinct roles for integrin splice variants in heart function and embryonic development (Vol 12, pg 1209, 1998). Genes Dev. 1998;12:1749-1749.
28. Gagneten S, Le YZ, Miller J, Sauer B. Brief expression of a GFPcre fusion gene in embryonic stem cells allows rapid retrieval of site-specific genomic deletions. Nucleic Acids Res. 1997;25:3326-3331.
29. Abuin A, Bradley A. Recycling selectable markers in mouse embryonic stem cells. Mol. Cell. Biol. 1996;16:1851-1856.
30. Schwenk F, Kuhn R, Angrand PO, Rajewsky K, Stewart AF. Temporally and spatially regulated somatic mutagenesis in mice. Nucleic Acids Res. 1998;26:1427-1432.
31. Lakso M, Pichel JG, Gorman JR, Sauer B, Okamoto Y, Lee E, Alt FW, Westphal H. Efficient in vivo manipulation of mouse genomic sequences at the zygote stage. Proc Natl Acad Sci USA. 1996;93:5860-5865.
32. Araki K, Araki M, Miyazaki J, Vassalli P. Site-specific recombination of a transgene in fertilized eggs by transient expression of Cre recombinase. Proc. Natl. Acad. Sci. U. S. A. 1995;92:160-164.
33. Lakso M, Sauer B, Mosinger B, Jr., Lee EJ, Manning RW, Yu SH, Mulder KL, Westphal H. Targeted oncogene activation by site-specific recombination in transgenic mice. Proc. Natl. Acad. Sci. U. S. A. 1992;89:6232-6236.
34. Zhang D, Gaussin V, Taffet GE, Belaguli NS, Yamada M, Schwartz RJ, Michael LH, Overbeek PA, Schneider MD. TAK1 is activated in the myocardium after pressure overload and is sufficient to provoke heart failure in transgenic mice. Nat. Med. 2000;6:556-563.
35. Charng MJ, Frenkel PA, Lin Q, Yumada M, Schwartz RJ, Olson EN, Overbeek P, Schneider MD. A constitutive mutation of ALK5 disrupts cardiac looping and morphogenesis in mice. Dev. Biol. 1998;199:72-79.
36. Soriano P. Generalized lacZ expression with the ROSA26 Cre reporter strain. Nat. Genet. 1999;21:70-71.
37. Kisanuki YY, Hammer RE, Miyazaki Ji J, Williams SC, Richardson JA, Yanagisawa M. Tie2-Cre transgenic mice: A new model for endothelial cell-lineage analysis in vivo. Dev. Biol. 2001;230:230-242.
38. James J, Osinska H, Hewett TE, Kimball T, Klevitsky R, Witt S, Hall DG, Gulick J, Robbins J. Transgenic over-expression of a motor protein at high levels results in severe cardiac pathology. Transgenic Res. 1999;8:9-22.
39. Dorn GW, Tepe NM, Lorenz JN, Koch WJ, Liggett SB. Low- and high-level transgenic expression of beta(2)-adrenergic receptors differentially affect cardiac hypertrophy and function in G alpha q-overexpressing mice. Proc. Natl. Acad. Sci. U. S. A. 1999;96:6400-6405.
40. Garrick D, Fiering S, Martin DI, Whitelaw E. Repeat-induced gene silencing in mammals. Nat. Genet. 1998;18:56-9.
41. Agah R, Frenkel PA, French BA, Michael LH, Overbeek PA, Schneider MD. Gene recombination in postmitotic cells: Targeted expression of Cre recombinase provokes cardiac-restricted, site-specific rearrangement in adult ventricular muscle in vivo. J. Clin. Invest. 1997;100:169-179.

42. Chen J, Kubalak SW, Chien KR. Ventricular muscle-restricted targeting of the RXR alpha gene reveals a non-cell-autonomous requirement in cardiac chamber morphogenesis. Development. 1998;125:1943-1949.
43. Regan CP, Manabe I, Owens GK. Development of a smooth muscle-targeted cre recombinase mouse reveals novel insights regarding smooth muscle myosin heavy chain promoter regulation. Circ. Res. 2000;87:363-369.
44. MacLellan WR, Frenkel PA, Vooijs M, Berns A, Schneider MD. Cardiac-restricted gene targeting of the retinoblastoma gene using the Cre/LoxP system. Circulation. 1998;98:I-608 (Abstr.).
45. Abel ED, Kaulbach HC, Tian R, Hopkins JC, Duffy J, Doetschman T, Minnemann T, Boers ME, Hadro E, Oberste-Berghaus C, Quist W, Lowell BB, Ingwall JS, Kahn BB. Cardiac hypertrophy with preserved contractile function after selective deletion of GLUT4 from the heart. J Clin Invest. 1999;104:1703-1714.
46. Gutstein DE, Morley GE, Tamaddon H, Vaidya D, Schneider MD, Chen J, Chien KR, Stuhlmann H, Fishman GI. Conduction slowing and sudden arrhythmic death in mice with cardiac-restricted inactivation of connexin43. Circ. Res. 2001;88:333-339.
47. Li H, Wang J, Wilhelmsson H, Hansson A, Thoren P, Duffy J, Rustin P, Larsson NG. Genetic modification of survival in tissue-specific knockout mice with mitochondrial cardiomyopathy. Proc. Natl. Acad. Sci. U. S. A. 2000;97:3467-3472.
48. Hirota H, Chen J, Betz UA, Rajewsky K, Gu Y, Ross J, Jr., Muller W, Chien KR. Loss of a gp130 cardiac muscle cell survival pathway is a critical event in the onset of heart failure during biomechanical stress. Cell. 1999;97:189-198.
49. Miwa T, Koyama T, Shirai M. Muscle specific expression of Cre recombinase under two actin promoters in transgenic mice. Genesis. 2000;26:136-138.
50. Bruning JC, Michael MD, Winnay JN, Hayashi T, Horsch D, Accili D, Goodyear LJ, Kahn CR. A muscle-specific insulin receptor knockout exhibits features of the metabolic syndrome of NIDDM without altering glucose tolerance. Mol. Cell. 1998;2:559-569.
51. Wang J, Wilhelmsson H, Graff C, Li H, Oldfors A, Rustin P, Bruning JC, Kahn CR, Clayton DA, Barsh GS, Thoren P, Larsson NG. Dilated cardiomyopathy and atrioventricular conduction blocks induced by heart-specific inactivation of mitochondrial DNA gene expression. Nat Genet. 1999;21:133-137.
52. Kühbandner S, Brummer S, Metzger D, Chambon P, Hofmann F, Feil R. Temporally controlled somatic mutagenesis in smooth muscle. Genesis. 2000;28:15-22.
53. Terry RW, Kwee L, Baldwin HS, Labow MA. Cre-mediated generation of a VCAM-1 null allele in transgenic mice. Transgenic Res. 1997;6:349-356.
54. Gustafsson E, Brakebusch C, Hietanen K, Fassler R. Tie-1-directed expression of Cre recombinase in endothelial cells of embryoid bodies and transgenic mice. J Cell Sci. 2001;114:671-676.
55. Theis M, de Wit C, Schlaeger TM, Eckardt D, Kruger O, Doring B, Risau W, Deutsch U, Pohl U, Willecke K. Endothelium-specific replacement of the connexin43 coding region by a lacZ reporter gene. Genesis. 2001;29:1-13.
56. Minamisawa S, Gu YS, Ross J, Chien KR, Chen J. A post-transcriptional compensatory pathway in heterozygous ventricular myosin light chain 2-deficient mice results in lack of gene dosage effect during normal cardiac growth or hypertrophy. J. Biol. Chem. 1999;274:10066-10070.
57. Tsien JZ, Chen DF, Gerber D, Tom C, Mercer EH, Anderson DJ, Mayford M, Kandel ER, Tonegawa S. Subregion- and cell type-restricted gene knockout in mouse brain. Cell. 1996;87:1317-1326.
58. Stricklett PK, Nelson RD, Kohan DE. Site-specific recombination using an epitope tagged bacteriophage P1 Cre recombinase. Gene. 1998;215:415-423.
59. Mao XH, Fujiwara Y, Orkin SH. Improved reporter strain for monitoring Cre recombinase-mediated DNA excisions in mice. Proc. Nat. Acad. Sci. U. S. A. 1999;96:5037-5042.
60. Zheng BH, Sage M, Cai WW, Thompson DM, Tavsanli BC, Cheah YC, Bradley A. Engineering a mouse balancer chromosome. Nat. Genet. 1999;22:375-378.
61. Zheng B, Sage M, Sheppeard EA, Jurecic V, Bradley A. Engineering mouse chromosomes with Cre-loxP: range, efficiency, and somatic applications. Mol. Cell. Biol. 2000;20:648-655.

62. Lindsay EA, Botta A, Jurecic V, CarattiniRivera S, Cheah YC, Rosenblatt HM, Bradley A, Baldini A. Congenital heart disease in mice deficient for the DiGeorge syndrome region. Nature. 1999;401:379-383.
63. Su H, Wang X, Bradley A. Nested chromosomal deletions induced with retroviral vectors in mice. Nat. Genet. 2000;24:92-95.
64. Lindsay E, Vitelli F, Su H, Morishima M, Huynh T, Prampano T, Jurecic V, Ogunrinu G, Sutherland HF, Scambler PJ, Bradley A, Baldini A. Tbx1 haploinsufficiency in the DiGeorge syndrome region causes aortic arch defects in mice. Nature. 2001;410:97-101.
65. Harvey RP, Rosenthal N, eds. Heart Development. San Diego, CA: Academic Press; 1998.
66. Kuhn R, Schwenk F, Aguet M, Rajewsky K. Inducible gene targeting in mice. Science. 1995;269:1427-1429.
67. Betz UAK, Bloch W, vandenBroek M, Yoshida K, Taga T, Kishimoto T, Addicks K, Rajewsky K, Muller W. Postnatally induced inactivation of gp130 in mice results in neurological, cardiac, hematopoietic, immunological, hepatic, and pulmonary defects. J. Exp. Med. 1998;188:1955-1965.
68. Utomo ARH, Nikitin AY, Lee WH. Temporal, spatial, and cell type-specific control of Cre-mediated DNA recombination in transgenic mice. Nat. Biotechnol. 1999;17:1091-1096.
69. Holzenberger M, Zaoui R, Leneuve P, Hamard G, Le Bouc Y. Ubiquitous postnatal LoxP recombination using a doxycycline auto-inducible Cre transgene (DAI-Cre). Genesis. 2000;26:157-159.
70. Wang Y, DeMayo FJ, Tsai SY, O'Malley BW. Ligand-inducible and liver-specific target gene expression in transgenic mice. Nat. Biotechnol. 1997;15:239-243.
71. Kellendonk C, Tronche F, Monaghan AP, Angrand PO, Stewart F, Schutz G. Regulation of cre recombinase activity by the synthetic steroid RU 486. Nucleic Acids Res. 1996;24:1404-1411.
72. Kellendonk C, Troche F, Casanova E, Anlag K, Opherk C, Schutz G. Inducible site-specific recombination in the brain. J. Mol. Biol. 1999;285:175-182.
73. Tsujita M, Mori H, Watanabe M, Suzuki M, Miyazaki J, Mishina M. Cerebellar granule cell-specific and inducible expression of Cre recombinase in the mouse. J. Neurosci. 1999;19:10318-10323.
74. Brocard J, Warot X, Wendling O, Messaddeq N, Vonesch JL, Chambon P, Metzger D. Spatio-temporally controlled site-specific somatic mutagenesis in the mouse. Proc. Natl. Acad. Sci. U. S. A. 1997;94:14559-14563.
75. Indra AK, Warot X, Brocard J, Bornert JM, Xiao JH, Chambon P, Metzger D. Temporally-controlled site-specific mutagenesis in the basal layer of the epidermis: comparison of the recombinase activity of the tamoxifen-inducible Cre-ER(T) and Cre-ER(T2) recombinases. Nucleic Acids Res. 1999;27:4324-4327.
76. Minamino T, Gaussin V, DeMayo F, Schneider MD. Inducible gene targeting in post-natal myocardium by cardiac-specific expression of a hormone-activated Cre fusion protein. Circ. Res. 2001:(in press).
77. Wan Y, Coxe KK, Thackray VG, Housley PR, Nordeen SK. Separable features of the ligand-binding domain determine the differential subcellular localization and ligand-binding specificity of glucocorticoid receptor and progesterone receptor. Mol. Endocrinol. 2001;15:17-31.
78. Rinaudo D, Lamartina S, Roscilli G, Ciliberto G, Toniatti C. Conditional site-specific integration into human chromosome 19 by using a ligand-dependent chimeric adeno-associated virus/Rep protein. J. Virol. 2000;74:281-294.
79. Hajjar RJ, del Monte F, Matsui T, Rosenzweig A. Prospects for gene therapy for heart failure. Circ Res. 2000;86:616-621.
80. Buchholz F, Angrand PO, Stewart AF. Improved properties of FLP recombinase evolved by cycling mutagenesis. Nat. Biotechnol. 1998;16:657-662.
81. Fiering S, Bender MA, Groudine M. Analysis of mammalian cis-regulatory DNA elements by homologous recombination. Methods Enzymol. 1999;306:42-66.
82. Meyers EN, Lewandoski M, Martin GR. An Fgf8 mutant allelic series generated by Cre- and Flp-mediated recombination. Nat. Genet. 1998;18:136-141.

4

ARRAY TRANSCRIPTION PROFILING: MOLECULAR PHENOTYPING OF RODENT CARDIOVASCULAR MODELS

Susan Acton, Ph.D., Raju Jeyaseelan, Ph.D., Vivek J. Kadambi, Ph.D., and Roger E. Breitbart, M.D.
Millennium Pharmaceuticals, Inc., Cambridge, MA 02139 USA

1. INTRODUCTION

Since its introduction just a few years ago, array transcription profiling (TP) has emerged as a powerful tool for the study of gene expression in cells and tissues, normal and diseased, across biological systems from bacteria to humans[1-5]. In principle, TP measures expression levels for thousands of genes simultaneously and allows comparisons of these levels across different experimental conditions. Among the important applications of this technology is the molecular phenotyping of rodent genetic models. Gene expression profiling in engineered animals can be used in conjunction with cardiovascular anatomy, histology, and physiology to develop hypotheses regarding the roles of key genes in normal and disease biology. It has the potential to uncover not only individual genes but entire molecular pathways that operate in the cells of the cardiovascular system. In this chapter we will review the fundamentals of array TP, outline basic methodology, and discuss applications with a focus on rodent models of cardiovascular disease.

2. METHODS

Array TP is fundamentally similar to northern blot analysis to the extent that it measures relative abundance, although not the size, of particular RNA transcripts; however, the techniques differ in key respects. Northern analysis quantifies immobilized cellular RNA by hybridization to a single soluble cDNA probe, whereas TP hybridizes complex RNA-derived probes to immobilized unlabeled cDNAs. Northern analysis permits evaluation of one or at most a few mRNAs at a time, while TP measures thousands in a single experiment.

Array TP, as a hybridization technology, is also distinct from a variety of strategies that rely on quantitative or semi-quantitative PCR-based cDNA amplification and/or sequencing for determination of differential gene expression, e.g. differential display[6], representation difference analysis (RDA)[7], and serial analysis of gene expression (SAGE)[8]. While each of these approaches has merit, array TP is now seeing the widest application because it is at once rapid, quantitive, and comprehensive in its capacity to interrogate a very large number of genes simultanously.

An overview of methodological considerations follows. The intent is to provide an introduction to techniques employed for array TP; however, detailed experimental methods are beyond the scope of this brief review.

2.1 Array Platforms

DNA arrays comprise hundreds to thousands of printed "spots" of amplified DNA, each with a defined address in a high-density grid (see Figure 1). Three principal array platforms are currently in wide use: silicon chips, glass slides, and nylon membranes. In chip technology, oligonucleotides representing specific genes are synthesized in situ on the chip surface (Affymetrix, Inc., Santa Clara, CA, USA). In glass and nylon technologies, PCR-amplified cDNA fragments can instead be laid down on the respective surface using automated robotics. In contrast to oligonucleotide arrays where only known sequences can be arrayed, cDNA arrays can accommodate both known and anonymous clones. For example, entire cDNA libraries can be spotted on these surfaces without need of prior sequencing.

Probes for these arrays are prepared from cell or tissue RNA by reverse transcriptase incorporation of labeled deoxynucleotides. Fluorescent labels are used for chip and glass arrays[9]. Typically, two RNA populations are labeled with two different fluorescent tags and hybridized competetively to the same array to compare relative transcript abundance. Nylon technology typically uses ^{33}P radiolabeled probes and measures signal intensity at each spot in separate hybridizations on identical arrays. This approach is not limited to two-point comparisons, i.e. three or more RNA populations can be profiled, for example, in a time course experiment.

Hybridization of probe to array is accomplished under standardized conditions. Automated high-throughput technologies for each of the three platforms have been developed. Similarly, sophisticated hardware and software for each have been developed for signal acquisition, image processing, and signal quantitation. More recently, novel platforms employing coated bead arrays on fiberoptic bundles have been developed (Illumina, Inc., San Diego, CA, USA).

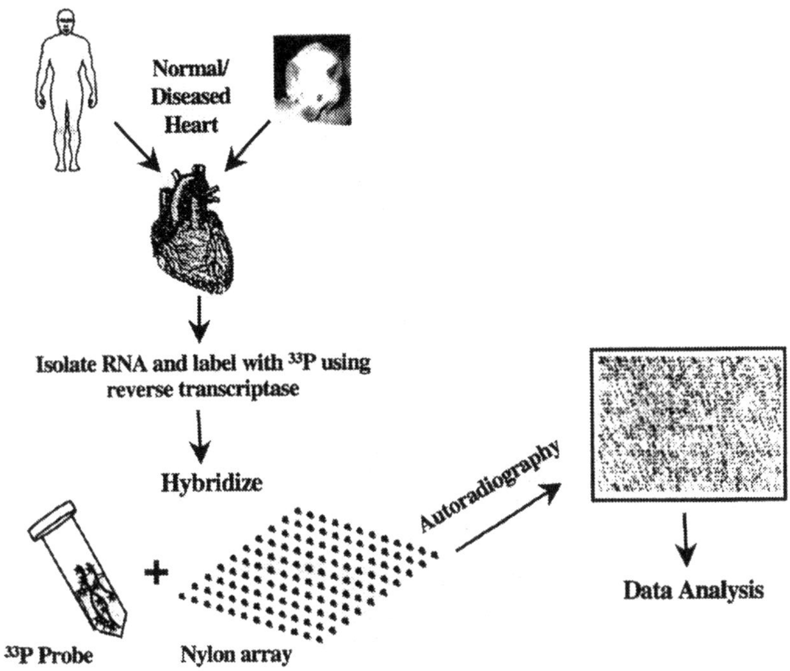

Figure 1: TP flow diagram. Radiolabeled probes prepared from cardiac RNA from normal and cardiomyopathic human or genetically engineered mouse hearts are hybridized to arrays comprising thousands of spotted cDNAs. When two or more such probes are compared on identical arrays, differential intensity at a given spot represents differential expression of the corresponding gene.

2.2 Experimental Design

RNA for profiling can be derived from any cell or tissue. Labeled probe is prepared directly from total RNA or, if quantities are limiting, from linearly amplified RNA. For applications in engineered mice, TP experiments would typically involve comparisons between respective tissues from transgenic or knockout mice and age-matched littermate controls to identify up- and down-regulated genes. They can also involve comparisons of primary cell cultures from these animals that may be subjected to in vitro stimuli. In the case of whole tissues, multiple cell types contribute to the total RNA pool; thus, in-migrating or locally proliferating cell populations may confound the analysis. In any TP analysis the most robust results, in terms of statistical significance, are obtained using multiple individually hybridized probes for each experimental point, e.g. from multiple animals

The ideal TP experiment would employ an array comprising every gene in the genome, whether rodent or human. While the field is on the threshold of this ideal, it remains necessary at present to choose an array suitable to a given experiment. In some cases this may be the subtotal set of identified genes and ESTs; in others it may be a direct or subtracted cDNA library pertinent to the paradigm to be profiled. In any case, relevant control cDNAs are included as markers for expected transcriptional regulation, e.g. atrial natriuretic factor for analysis of cardiac hypertrophy. Such markers—key signaling molecules, enzymes, transcription factors, and others—can also be critical guideposts for interpreting the regulation of other genes on the array by providing a conceptual framework of cellular pathways into which novel co-regulated factors may be fit.

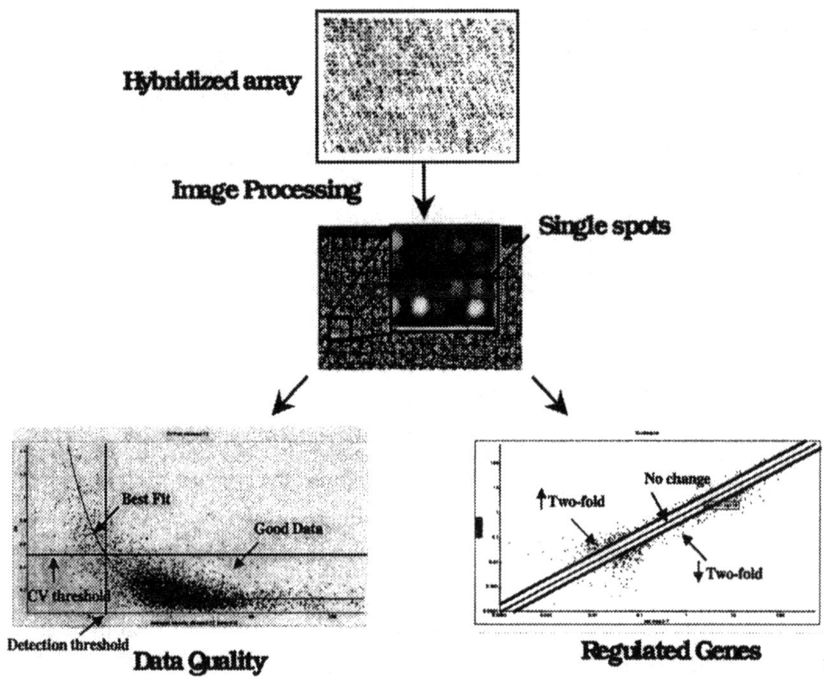

Figure 2: TP data analysis. Advanced image-processing is used to acquire and quantify the signal from each spot on the hybridized array. Data quality is assessed by a variety of means including the coefficient of variation at each spot in experimental replicates (left plot). Fold regulation is determined by comparing normalized spot intensities between two probes, e.g. transgenic and wild-type ventricle. A scatterplot (right) of spot intensities from control (x axis) and experimental (y axis) probes shows unregulated genes along a line of slope=1 and up- and down-regulated genes significantly above and below that line, respectively.

2.3 Data Analysis

Beyond the technical requirements for TP execution, thoughtful and critical analysis of the generated data is the most important challenge in these experiments (see Figure 2). It is essential that rigorous statistical methods be brought to bear to distinguish true differential regulation from false positive and false negative results. To this end, the greatest validity is obtained using multiple samples, e.g. multiple animals for any given experimental point, and two or more replicate filters for each sample[10]. A variety of parameters impact the outcome of these analyses. Again, space does not permit detailed consideration of these in this review, but they include: choice of putative unregulated control genes; methods for normalization of signals across multiple filters; statistical tests for pairwise and groupwise analysis; self-organizing map algorithms; and decisions regarding the statistical and, ultimately, biological significance of small (e.g. less than two-fold) changes in expression level.

3. TP IN RODENT CARDIOVASCULAR MODELS

Burgeoning transgenic and knockout mouse models of cardiovascular disease, reviewed in this volume, and the acceleration of the murine genome project create critical new opportunities to use array TP in the elucidation of cardiovascular molecular pathways. RNA prepared from whole tissues or even isolated primary cells from these animals and their wild-type littermates can be used to probe arrays of mouse ESTs, cDNAs, and/or gene sequences. The first reports of such experiments have been published only recently while other potential applications have yet to appear in the literature. We consider some of these here.

3.1 Transgenic mice

The availability of cardiomyocyte-specific promoters, including α-myosin heavy chain and myosin light chain 2v, has permitted transgenic overexpression of a number of critical genes in the heart. Many such engineered mice manifest cardiac phenotypes of hypertrophy and failure but their similarities and differences at the molecular level have been difficult to resolve. Without TP, molecular investigation of these phenotypes has been limited to analysis of small numbers of candidate genes known or suspected to be involved in relevant pathways. The advent of array TP, however, provides the opportunity in these animals to interrogate thousands of genes—and eventually the entire genome—at once. Transcriptional up- or down-regulation of numbers of cardiac genes may result directly or indirectly from over-expression of a particular transgene in the heart. In this manner, genes not thought *a priori* to be involved may emerge, and with them novel mechanisms. In addition, the molecular signatures indicating activation of known cellular pathways can be recognized.

Redfern et al.[11] have used a chip array of over 6,000 mouse gene sequences to profile cardiac ventricle RNAs from cardiomyopathic mice with conditional transgenic expression and activation of a modified G_i-coupled receptor, compared to littermate controls. Focusing first on known G protein signalling cascades, they found expected regulation of genes in the G_i pathway and little regulation of those downstream of G_q. Consistent with the marked myocardial fibrosis in these animals, genes encoding extracellular matrix proteins such as collagens, laminins, and fibronectin were up-regulated. In addition, activation of the fetal gene expression program typical of many cardiomyopathies was evident in the upregulation of genes such as atrial natriuretic factor and β-myosin heavy chain. However, many of the most highly regulated genes in these experiments have not been previously connected to cardiomyopathy and, as such, represent key insights into novel molecular pathways in the heart. Therein lies the power of array transcription profiling.

3.2 Knockout mice

Targeted mutations of critical genes in the mouse have produced diverse murine models of cardiovascular disease ranging from atherosclerosis to heart failure to congenital structural heart defects. In some of these the phenotype could be anticipated based on prior understanding, e.g. atherosclerosis in the apolipoprotein E (ApoE) knockout[12] or dilated cardiomyopathy in the myosin-binding protein-C mutant knock-in that recapitulates a heritable human mutation[13]. In many others, however, cardiac phenotypes have appeared quite unexpectedly, e.g. structural defects in the neurotrophin 3 knockout[14]. As in transgenics, array TP has the potential in knockout and knock-in mice to elucidate unanticipated cardiac molecular mechanisms by identifying changes in other gene expression resulting directly or indirectly from loss of the targeted gene. Furthermore, TP provides a new opportunity to glean critical information from certain knockout mice that have been otherwise uninformative, e.g. those that are embryonic lethal or, alternatively, those that produce no overt cardiac phenotype where one had been hypothesized. In the latter case, TP of knockout and wild-type cardiac RNAs could reveal intersecting molecular pathways that compensate for loss of the targeted gene product.

3.3 Strain-specific phenotypes

Certain transgenic and knockout phenotypes are modulated depending on the inbred mouse strain in which they are propagated, the result of strain-specific differences in putative modifying genetic loci. For example, the atherosclerotic phenotype in the ApoE knockout mouse is more severe in the C57Bl/6J background while the FVB/NJ strain is relatively resistant[12]. Analogously, the MLP knockout has more severe heart failure in the 129/Sv background than in the C57Bl/6J background[15]. Array TP can be used to advantage in these and other models to identify candidate modifying genes whose expression in relevant tissues differs according to strain. These candidate genes can then be scanned for strain-specific polymorphisms which, if confirmed, may point to novel molecular pathways and mechanisms.

3.4 Non-genetic rodent models of cardiovascular disease

In addition to engineered animals, induced models in rodents have been important in understanding cardiovascular pathophysiology. Examples include coronary artery ligation for the study of ischemia/reperfusion, myocardial infarction, remodeling, and heart failure, and aortic banding for the study of pressure overload hypertrophy, remodeling, and the transition to failure. TP provides critical opportunities for molecular analysis in these paradigms just as it does in the genetic models. Several groups have identified changes in gene expression in the ventricle following coronary occlusion and infarction, as compared with sham-operated controls[16-18]. Others have profiled changes in gene expression in the mouse ventricle during induction and regression of cardiac hypertrophy by administration and withdrawal of agents including isoproterenol and angiotensin II[19]. Indeed, TP also can also be applied in the development of new cardiovascular drugs tested in rodent models. Both drug efficacy and toxicity can be better understood at the molecular level through TP analysis, leading to more informed and effective lead optimization.

4. LIMITATIONS

In practice, array TP analysis has certain limitations, both conceptual and technical. First, not all gene regulation is transcriptional, and important regulation clearly occurs post-transcriptionally. Thus, TP is not a comprehensive analysis of cellular regulation, and not all genes involved in a particular response or phenotype will be revealed by TP. Nor for that matter will all transcriptionally regulated genes participate mechanistically in that response or phenotype, i.e. some will be drivers but others passengers in the process. Still, TP provides a major window into certain critical biology.

Technical limitations derive in part from issues inherent in nucleic acid hybridization. For example, certain spots on an array may cross-hybridize with homologous but non-identical transcripts from related but distinct genes. This is particularly true in cross-species TP experiments, e.g. murine probes on human arrays. In addition, different alternatively spliced transcripts may not be resolved in TP. While these limitations need to be kept in mind, the power of the technology remains undiminished.

5. FUTURE DIRECTIONS

The completion of the human and, soon, mouse genome projects will make whole genome TP possible in the near future. Even now, using subtotal arrays, vast amounts of profiling data are being generated in many laboratories. Certainly, the challenge going forward will be to extract maximum information from this effort, not simply one experiment at a time, but through communal databasing and iterative re-analysis of TP results. A critical advance will be algorithms that capture and display whole sets of genes related by function and pathway. Pioneering work in this area is being pursued by Conklin and colleagues[11] (see *www.GenMAPP.org*). Improved technology and more powerful analysis strategies will greatly enhance molecular phenotyping by TP.

ACKNOWLEDGEMENTS

The authors recognize key contributions by colleagues at Millennium Pharmaceuticals, Inc., including Amélie Rodrigue-Way, Nga Tang, and others too numerous to name. The authors also thank Greg Kirk, Shane Weber, and Elena Grigorenko for critical reading of the manuscript.

REFERENCES

1. Schena M, Shalon D, Davis RW, Brown PO. Quantitative monitoring of gene expression patterns with a complementary DNA microarray. *Science*. 1995;270:467-470.
2. Schena M, Shalon D, Heller R, Chai A, Brown PO, Davis RW. Parallel human genome analysis: microarray-based expression monitoring of 1000 genes. *Proc Natl Acad Sci USA*. 1996;93:10614-10619.
3. Ramsay G. DNA chips: state-of-the art. *Nat Biotechnol*. 1998;16:40-44.

4. Watson A, Mazumder A, Stewart M, Balasubramanian S. Technology for microarray analysis of gene expression. *Curr Opin Biotechnol.* 1998;9:609-614.
5. Zweiger G. Knowledge discovery in gene-expression-microarray data: mining the information output of the genome. *Trends Biotechnol.* 1999;17:429-436.
6. Liang P, Pardee AB. Differential display of eukaryotic messenger RNA by means of the polymerase chain reaction. *Science.* 1992;257:967-971.
7. Hubank M, Schatz DG. Identifying differences in mRNA expression by representational difference analysis of cDNA. *Nucleic Acids Res.* 1994;22:5640-5648.
8. Velculescu VE, Zhang L, Vogelstein B, Kinzler KW. Serial analysis of gene expression. *Science.* 1995;270:484-487.
9. Winzeler EA, Schena M, Davis RW. Fluorescence-based expression monitoring using microarrays. *Methods Enzymol.* 1999;306:3-18.
10. Lee ML, Kuo FC, Whitmore GA, Sklar J. Importance of replication in microarray gene expression studies: statistical methods and evidence from repetitive cDNA hybridizations. *Proc Natl Acad Sci USA.* 2000;97:9834-9839.
11. Redfern CH, Degtyarev MY, Kwa AT, Salomonis N, Cotte N, Nanevicz T, Fidelman N, Desai K, Vranizan K, Lee EK, Coward P, Shah N, Warrington JA, Fishman GI, Bernstein D, Baker AJ, Conklin BR. Conditional expression of a Gi-coupled receptor causes ventricular conduction delay and a lethal cardiomyopathy. *Proc Natl Acad Sci USA.* 2000;97:4826-4831.
12. Dansky HM, Charlton SA, Sikes JL, Heath SC, Simantov R, Levin LF, Shu P, Moore KJ, Breslow JL, Smith JD. Genetic background determines the extent of atherosclerosis in ApoE-deficient mice. *Arterioscler Thromb Vasc Biol.* 1999;19:1960-1968.
13. McConnell BK, Jones KA, Fatkin D, Arroyo LH, Lee RT, Aristizabal O, Turnbull DH, Georgakopoulos D, Kass D, Bond M, Niimura H, Schoen FJ, Conner D, Fischman DA, Seidman CE, Seidman JG, Fischman DH. Dilated cardiomyopathy in homozygous myosin-binding protein-C mutant mice. *J Clin Invest.* 1999;104:1235-1244.
14. Donovan MJ, Hahn R, Tessarollo L, Hempstead BL. Identification of an essential nonneuronal function of neurotrophin 3 in mammalian cardiac development. *Nat Genet.* 1996;14:210-213.
15. Arber S, Hunter JJ, Ross J, Jr., Hongo M, Sansig G, Borg J, Perriard JC, Chien KR, Caroni P. MLP-deficient mice exhibit a disruption of cardiac cytoarchitectural organization, dilated cardiomyopathy, and heart failure. *Cell.* 1997;88:393-403.
16. Lyn D, Liu X, Bennett NA, Emmett NL. Gene expression profile in mouse myocardium after ischemia. *Physiol Genomics.* 2000;2:93-100.
17. Sehl PD, Tai JT, Hillan KJ, Brown LA, Goddard A, Yang R, Jin H, Lowe DG. Application of cDNA microarrays in determining molecular phenotype in cardiac growth, development, and response to injury. *Circulation.* 2000;101:1990-1999.
18. Stanton LW, Garrard LJ, Damm D, Garrick BL, Lam A, Kapoun AM, Zheng Q, Protter AA, Schreiner GF, White RT. Altered patterns of gene expression in response to myocardial infarction. *Circ Res.* 2000;86:939-945.
19. Friddle CJ, Koga T, Rubin EM, Bristow J. Expression profiling reveals distinct sets of genes altered during induction and regression of cardiac hypertrophy. *Proc Natl Acad Sci USA.* 2000;97:6745-6750.

5
MURINE CARDIOVASCULAR DEVELOPMENT

Steven A. Fisher, M.D.
Case Western Reserve University, Cleveland OH, 44106

Introduction

The fascination of developmental processes is how the complex structures that characterize advanced organisms can arise from a single cell. As with any system, the key to its understanding is to first define its component parts and then devise experimental strategies to test their function. Experiments in the early days of the discipline manipulated cells and tissues in frog eggs and chick embryos to demonstrate the role of cell inductive or determined behaviors in the formation of embryonic structures. The advent of the molecular era of science witnessed a renaissance in experimentation in developmental processes. Various cloning and gene expression-profiling techniques have allowed for the identification of many gene products that are expressed in the heart early in development. The completion of the sequencing of a number of genomes means that most or all of the genes will have been identified. The challenge now is to understand how a gene-based regulatory program controls the proliferation, differentiation, death and migration of cells in a regional and temporal-specific manner so as to form the mature four-chambered heart.

Method

A key to studying the roles of regulatory genes in developmental processes was the discovery of techniques to manipulate gene expression. Two strategies emerged in the 1980s to delete genes or introduce genes into the mouse germ line[1]. Homologous recombination of a synthetic gene construct with the target gene in mouse embryonic stem cells *in vitro* inactivates the target gene in these cells. The cells with the inactivated gene are selected *in vitro* by their acquired antibiotic resistance, and injected into an early stage mouse embryo that is placed into a pregnant mouse. The first generation of offspring are chimeric with respect to the gene inactivation, but subsequent inbreeding leads to mice that are heterozygous or homozygous null for the gene of interest. To introduce genes into the mouse germ line, eggs are directly microinjected with numerous copies of linear recombinant DNA. The foreign DNA contains a promoter of transcription upstream of the inserted cDNA that targets its expression to the tissue of interest. The recombinant

DNA will integrate into the chromosomes, usually in multiple copies, and when it does so in germ line cells, is passed on through generations of breeding. Here I will review experiments that have used these techniques to provide insights into heart development.

Role of a homeobox transcription factor in heart development

It is the hope of many investigators that a gene transcriptional regulatory program of heart development will be identified. This is no small task, given the dramatic structural, functional and molecular changes that occur (Fig. 1). A number of transcription factors have now been identified in the developing heart and their patterns of expression determined. Their role in heart development has been interrogated by loss-of-function experiments.

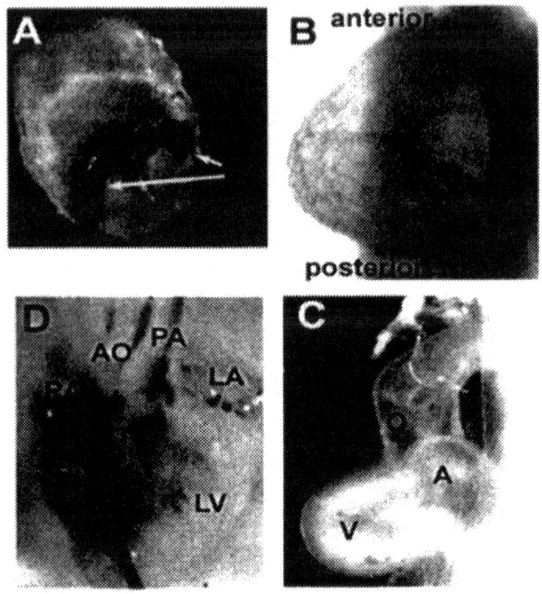

Fig. 1 Key developmental stages in the formation of the 4-chambered heart.
A) cardiac precursors in the anterolateral mesoderm **B)** the looped heart tube **C)** distinct chambers; a=atria, v=ventricle, o=OFT **D)** the mature configuration of the heart.

Loss-of-function experiments The knock-out strategy was used to examine the role of the homeobox protein Nkx2.5 [2] (also designated Csx [3]) in heart development. The homeobox class of transcription factors contain a 60 amino acid conserved helix-turn-helix DNA binding motif. They were given their names based on the naturally occurring homeotic transformations in body parts that resulted from their gene duplications [4]. These factors were subsequently shown to determine the segmental organization of the body along the anterior-posterior axis, as well as the segmental organization of certain organs, e.g. the intestines [5]. While exactly how the homeodomain family of proteins regulate cell behavior is not clear, a number of target genes that control diverse cellular functions have been identified [6;7]. It was surmised that the homeodomain proteins might play a similar role in the patterning of the heart during development [8]. The knock-out of the fruitfly Nkx2.5 homologue caused an absence in the formation of heart muscle cells (as well as other mesoderm-derived muscle lineages), lending itself to the name *tinman* [9;10]. The homozygous Nkx2.5 null mouse had a different phenotype [11]. The embryonic mice became nonviable coincident with an arrest of heart development at ED 9.5-10.5. The primitive linear heart tube formed with the presence of differentiated myocardial cells, as evidenced by the expression of muscle-specific genes and heart function (beating), but the heart failed to undergo subsequent looping morphogenesis.

This result demonstrates the importance of Nkx2.5 in mouse heart development as well as illustrates limitations of the technique. The knockout experiment does not precisely define the molecular role of Nkx in mouse heart development. It would be incorrect to think of Nkx2.5 as a looping gene, and in fact it is common for perturbations of the cardiac transcriptional program to result in the arrest of development at this complex juncture (e.g. MEF-2C [12]). This leads to the principle of "first phenotype", i.e. that using the standard gene knock-out approach enables the investigator to see the first point at which absence of the gene has an effect, without providing evidence that the gene product is directly involved in the process, nor indicating the role it may have in later developmental processes.

However, an important clue to the Nkx2.5 homozygous null phenotype is the misregulated expression of genes in the developing hearts of these embryos. The approach that is commonly used is to assess the expression of other genes that regulate heart development. In this way, a hierarchical regulatory pathway is established in which genes are assigned as lying upstream, if their expression is unaffected, or downstream, if their expression is affected, of the gene of interest. The expression of a number of transcripts are misregulated in the Nkx2.5 null heart, including: Irx4 [13], a homeobox gene that appears to specify the ventricular muscle phenotype [14], the HAND genes [15;16], that appear to be involved in left/right specification of the ventricles [17], and a number of muscle-specific structural proteins [15;16]. These results support the hypothesis that the Nkx gene family may have evolved from its ancestral founder to direct patterning of the mouse heart through the regional regulation of muscle cell differentiation. Experimental evidence that Nkx2.5 directly regulates the transcription of these genes will require promoter-reporter assays, as has been shown for α-cardiac actin and atrial natriuretic factor [18;19].

It is also possible that the failure of regional specification of the heart and looping is not a specific effect of Nkx deletion, but a general response to an earlier and perhaps subtle perturbation of development. This leads to the alternative explanation for the Nkx2 null phenotype and points to a second limitation of the experiment. Nkx2.5, as is true of many other genes, is a member of a multi-gene family that has evolved from a primitive ancestor [20]. In many instances the genes in a family may partially or completely compensate for an absence or defect in another family member. Thus, the mouse Nkx2.5-null phenotype could be a reflection of the redundancy in this gene family and indicate the point at which another family member was no longer able to compensate for the loss of Nkx2.5.

Gain-of-function experiments A complementary approach is Nkx2.5 gain-of-function experiments. This approach has been extensively used to study the postnatal mouse heart by targeting transgenes to the ventricles with an α-MHC promoter construct (Chapter 2). However, gain-of-function transgenic mouse experiments to study heart development have been performed on a limited basis, perhaps due to a historical lack of promoters that would provide spatial and temporal targeting of a gene product in the developing heart. This is likely to change in the near future as a number of promoters have recently been identified for which reporter protein expression is temporally and spatially restricted. Gain of function experiments with Nkx2.5 have been performed in fish and frog embryos by microinjection of the mRNA, and led to the conclusion that Nkx2.5 somehow causes cells to commit to the cardiac muscle lineage without being able by itself to induce overt differentiation [21-23]. This result would be consistent with in vitro experiments that indicate that Nkx2.5 must synergize with other regulators of muscle-specific gene expression.

Complementary approaches Research conducted in the Seidmans' human genetics laboratory found that mutations that affected the DNA-binding homeodomain of Nkx2.5 occurred in families that suffered from an autosomal dominant form of secundum ASD and conduction block [24]. In one of the four families studied there were associated defects in cardiac structures, including Tetralogy of Fallot, VSD and subvalvular aortic stenosis. The human genetic linkage studies cannot determine how mutations in Nkx cause the disease, for which investigations must return to the mouse models for genetic manipulations. The human Nkx studies, as well as the study of other genes in human populations, also point out the importance of the total genotype and genetic variability in determining the effect of a mutation on the phenotype. This effect of the genetic background on genotype/phenotype relationships has become well appreciated in the mouse knockout studies, many of which were conducted on the highly inbred 129/Sv strain. The phenotype may be modified or lost when the genetic mutation is outbred or crossed into a different strain of mouse.

Re-examination of the mouse Nkx2.5 phenotype The observations of Nkx2.5 mutations in human populations led to a re-examination of the mouse models. More detailed analyses of the Nkx2.5 heterozygotes by echocardiography and dissection revealed defects in the atrial septum that were highly strain-dependent [25]. These defects included patent foramen ovale and septal aneurysm, but ASDs and significant conduction block were not present. This result is not directly comparable to the human mutations, since the mice experiments did not recreate the human mutations Thr178Met,Gln170ter and Gln198ter. However, two important general principles are illustrated. One, that there are differences between mouse and human development that could prevent modeling human genetic defects in the mouse. Second, the findings in the Nkx2-5 heterozygotes demonstrate the need to develop and utilize sophisticated methods for the study of cardiovascular structure and function in the developing mouse (see Chapter 22). In fact, it is the more subtle phenotypes that are likely to contribute most to our understanding of congenital human heart disease, since the more severe cardiac phenotypes cannot support the fetal circulation and are spontaneously aborted early in pregnancy.

Other considerations: the external milieu. The heart from very early in development must couple with the maternal circulation through the placenta to support development of the embryo. Many of the genes that have been mutated in mice also affect placental development (reviewed in [26]), or may affect other vascular structures, leading to an indirect effect on the heart. This may be anticipated by an initial analysis of the pattern of expression of the gene of interest. While, unlike the case in skeletal muscle, no master transcription factors have been identified that are cardiac-specific, some such as Nkx2-5, are highly restricted in their expression. Others, such as the more ubiquitous retinoic acid receptors (RAR), regulate transcription in a number of developing cell types and organs. In this case, two approaches may by used to determine if an observed cardiac phenotype results from a direct effect on the myocardial cells. In a chimera experiment, the behavior of labeled mutant myocardial cells is observed in a wild-type background. If their behavior (e.g. proliferation, differentiation) is normal, the mutant phenotype is not cell autonomous, i.e. it is due to an influence external to the myocardial cell in the mutant embryo. A second method to address this is the use of tissue-specific deletion of genes via the Cre-lox system (Chapter 3). These techniques have established that the developmental effects of RXRα knockout on the ventricular myocytes are non-cell autonomous [27], while the effects on the development of other cardiovascular structures are most likely direct [28]. In contrast the Nkx2.5 null phenotype is cardiomyocyte autonomous, though defects in the mutant yolk sac vasculature and blood components apparently also contribute to the phenotype [16].

Future directions

The ability to manipulate the mouse genome has proved to be a powerful tool to study the role of regulatory genes in the formation of the specialized structures of the heart. Refinements of the techniques will allow an even more powerful and precise investigation of questions that have interested scientists through the ages. Key methodological issues that only more experimentation will be able to answer are 1) will manipulations of the mouse genome be possible to model human congenital heart syndromes that are multigenic? 2) Will we be able to use the mouse models to investigate genetic susceptibility to environmentally induced congenital heart disease? 3) Will the methods for diagnosing congenital heart defects in the mouse advance to the level of that in humans? If the answers to these questions are yes, then we can be confident that in the near future we will have a more complete understanding of the molecular basis of congenital human heart disease.

REFERENCES

1. Alberts,B. Recombinant DNA technology. In: Molecular Biology of the Cell. Alberts,B, Bray,D, Lewis,J, Raff,M, Roberts,K, Watson,JD, eds. 1994. Garland Press, New York.
2. Lints,TJ, Parsons,LM, Hartley,L, Lyons,I, Harvey,RP. *NKX-2.5:* a novel murine homeobox gene expressed in early heart progenitor cells and their myogenic descendants. Development 119:419-431.
3. Komuro,I, Izumo,S. *Csx*: A murine homeobox-containing gene specifically expressed in the developing heart. Proc Natl Acad Sci USA 90:8145-8149.
4. Krumlauf,R. *Hox* genes in vertebrate development. Cell 78:191-201.
5. Beck,F, Tata,F, Chawengsaksophak,K. Homeobox genes and gut development. Bioessays 22:431-441.
6. Burke,AC. Hox genes and the global patterning of the somitic mesoderm. Curr.Top.Dev.Biol. 47:155-181.
7. Veraksa,A, Del Campo,M, McGinnis,W. Developmental patterning genes and their conserved functions: from model organisms to humans. Mol.Genet.Metab 69:85-100.
8. Gorski,DH, Patel,CV, Walsh,K. Homeobox transcription factor regulation in the Cardiovascular System. TCM 3:184-190.
9. Azpiazu,N, Frasch,M. *tinman* and *bagpipe*: Two homeobox genes that determine cell fates in the dorsal mesoderm of *Drosophila*. Genes & Development 7:1325-1340.
10. Bodmer,R. The gene tinman is required for specification of the heart and visceral muscles in Drosophila. Development 118:719-729.
11. Lyons,I, Parsons,LM, Hartley,L, Li,R, Andrews,JE, Robb,L, Harvey,RP. Myogenic and morphogenetic defects in the heart tubes of murine embryos lacking the homeo box gene *Nkx2-5*. Genes & Development 9:1654-1666.
12. Lin,Q, Schwarz,JJ, Bucana,C, Olson,EN. Control of mouse cardiac morphogenesis and myogenesis by the myogenic transcription factor MEF2C. Science 276:1404-1407.
13. Bruneau,BG, Bao,ZZ, Tanaka,M, Schott,JJ, Izumo,S, Cepko,CL, Seidman,JG, Seidman,CE. Cardiac expression of the ventricle-specific homeobox gene Irx4 is modulated by Nkx2-5 and dHand. Dev.Biol. 217:266-277.
14. Bao,ZZ, Bruneau,BG, Seidman,JG, Seidman,CE, Cepko,CL. regulation of chamber-specific gene expression in the developing heart by Irx4. Science 283:1161-1164.
15. Biben,C, Harvey,RP. Homeodomain factor Nkx2-5 controls left/right asymmetric expression of bHLH gene eHand during murine heart development. Genes Dev. 11:1357-1369.
16. Tanaka,M, Chen,Z, Bartunkova,S, Yamasaki,N, Izumo,S. The cardiac homeobox gene Csx/Nkx2.5 lies genetically upstream of multiple genes essential for heart development. Development 126:1269-1280.
17. Srivastava,D, Cserjesi,P, Olson,EN. A Subclass of bHLH Proteins Required for Cardiac Morphogenesis. Science 270:1995-1999.
18. Durocher,D, Chen,CY, Ardati,A, Schwartz,RJ, Nemer,M. The atrial natriuretic factor promoter is a downstream target for Nkx- 2.5 in the myocardium. Mol.Cell Biol. 16:4648-4655.

19. Chen,CY, Schwartz,RJ. Recruitment of the tinman homolog Nkx-2.5 by serum response factor activates cardiac alpha-actin gene transcription. Mol.Cell Biol. 16:6372-6384.
20. Harvey,RP, Biben,C, Elliott,DA. Transcriptional control and pattern formation in the developing vertebrate heart: studies on Nk-2 class homeodomain factors. In: Heart Development. Harvey,RP, Rosenthal,N, eds. 1999. Academic Press, San Diego.
21. Chen,J-N, Fishman,MC. Zebrafish tinman homolog demarcates the heart field and initiates myocardial differentiation. Development 122:3809-3816.
22. Cleaver,OB, Patterson,KD, Krieg,PA. Overexpression of the tinman-related genes XNkx-2.5 and XNkx-2.3 in Xenopus embryos results in myocardial hyperplasia. Development 122:3549-3556.
23. Fu,Y, Izumo,S. Cardiac myogenesis: overexpression of XCsx2 or XMEF2A in whole Xenopus embryos induces the precocious expression of XMHCα gene. Roux's Arch Dev Biol 205:198-202.
24. Schott,JJ, Benson,DW, Basson,CT, Pease,W, Silberbach,GM, Moak,JP, Maron,BJ, Seidman,CE, Seidman,JG. Congenital heart disease caused by mutations in the transcription factor NKX2-5. Science 281:108-111.
25. Biben,C, Weber,R, Kesteven,S, Stanley,E, McDonald,L, Elliott,DA, Barnett,L, Koentgen,F, Robb,L, Feneley,M, Harvey,RP. Cardiac Septal and Valvular Dysmorphogenesis in Mice Heterozygous for Mutations in the Homeobox Gene Nkx2-5. Circulation Research 87:888-895.
26. Ihle,JN. The challenges of translating knockout phenotypes into gene function. Cell 102:131-134.
27. Chen,J, Kubalak,SW, Chien,KR. Ventricular muscle-restricted targeting of the RXRalpha gene reveals a non-cell-autonomous requirement in cardiac chamber morphogenesis. Development 125:1943-1949.
28. Mark,M, Ghyselinck,NB, Wendling,O, Dupe,V, Mascrez,B, Kastner,P, Chambon,P. A genetic dissection of the retinoid signalling pathway in the mouse. Proc Nutr.Soc. 58:609-613.

6
ISOLATED MYOCYTE MECHANICS AND CALCIUM TRANSIENTS

Zhi Su, M.D. Ph.D. and William H. Barry, M.D.
University of Utah Health Sciences Center, Salt Lake City, UT 84132, USA

INTRODUCTION

Studies of contraction and relaxation of myocardium in isolated heart and cardiac muscle strips can be complicated by coronary circulation (whole heart), chamber geometry (whole heart), extracellular matrix, and neurohumoral factors. Measurement of $[Ca^{2+}]_i$ transients in intact myocardium preparations can be complicated because fluorescence measurements can be affected by dye loading into cells other than myocytes, motion artifact, inner filter effects, and unstable background fluorescence (1,2). Dissociated cardiac myocytes provide an experimental model that is devoid of the complicating factors mentioned above. In addition, isolated myocytes allow application of single cell voltage-clamp technique to study cardiac ion channels and transporters and of confocal microscopy. The study of isolated myocyte mechanics and $[Ca^{2+}]_i$ transients has proved to be an especially important tool in the investigation of cardiac physiology and pathophysiology in genetically engineered mice (3-11) and this chapter will focus on the methods that are employed.

DISSOCIATION OF ADULT MOUSE VENTRICULAR MYOCYTES

Preparing healthy and Ca^{2+} tolerant cells is a prerequisite for successful studies of isolated cardiac myocytes. Dissociation of cardiac myocytes from adult mice is much more difficult than from larger animals because the mouse heart is very small, and mouse myocytes are easily overloaded with Ca^{2+} due to high $[Na^+]_i$ plus high Na/Ca exchange density (4,12). Ca^{2+} overload leads to cell damage and death, thus reducing the yield of rod-shaped myocytes. Because of the rapidly growing number of genetically engineered mouse models used in the studies of cardiac physiology and pathophysiology, considerable effort has been devoted to develop isolation methods for murine myocytes. At present, mouse myocyte isolation with slight modifications or variations has been successful in a number of laboratories (3-5,11-15). We have previously described a protocol which has been successfully applied in our laboratory to isolate viable adult mouse ventricular myocytes since 1996 (4,6,12,16,17).

It is very important to note the following guidelines in preparing mouse cardiac myocytes: 1) it is necessary to test multiple batches of collagenase, until one is identified which gives cells with satisfactory quality and yield; 2) the duration of collagenase perfusion and perfusion pressure are very important; 3) keep the time required for cannulation as short as possible; 4) the age of mice is important; higher yields are obtained with younger animals; 5) the addition of BDM to the isolation solution reduces Ca^{2+} overload injury and increases yield; 6) the addition of serum or albumin to the cell suspension increases the viability of cells; 7) a key factor for success is to have a dedicated person who routinely performs the isolations.

MYOCYTE MECHANICS

Mechanical properties of isolated cardiac myocytes can be monitored by a wide variety of techniques (18, 19) including: 1) force measurement; 2) analysis of sarcomere length; 3) detection of cell motion. Because of its relative simplicity and low instrumentation cost, the video edge detection method of measuring cell shortening has become most popular. Recently, this technique has been widely employed in the studies of mechanical responses of single cardiac myocytes isolated from genetically engineered mice (5,9,10,14,15,20). The video edge detection system has been previously described in a greater detail (21). A brief description is given below.

A video image is produced by focusing a myocyte with an inverted microscope on the light-sensitive element of a charge-coupled device (CCD) camera which rapidly scans the image area with 256 parallel horizontal lines from top to bottom. The voltage along each scan line varies in response to changes in light intensity on the element. The edge detection system described by Steadman et al. (21) relies on these voltage changes to identify each end of the cell or to identify specific sites on the cell surface. The vertical sweep rate of the CCD camera is typically every 16.7 ms for a standard camera with an interlace defeat, and as short as 4.2 ms with high speed cameras. High speed CCD cameras are highly recommended for study of murine myocytes, because of the relative rapidity of the shortening and lengthening of these cells. The horizontal position of the detected edge is continuously updated at this frequency. If both ends of the cell are monitored the total extent of cell shortening can be measured, and combined with determination of end diastolic length, fractional shortening is calculated.

Myocyte shortening is dependent on the $[Ca^{2+}]_i$ transient. To simultaneously measure the $[Ca^{2+}]_i$ and shortening (Figure 1), it is necessary to add an edge detector to the $[Ca^{2+}]_i$ measurement system. This is achieved by placing a long short pass dichroic mirror (650 nm) in the side port of the microscope, and illuminating the cells with red (750 nm) light from a 12 V DC light source. This red light produces an image which is detected with the intensified CCD camera and projected on a TV monitor for analysis of contraction and relaxation with a motion detector (22).

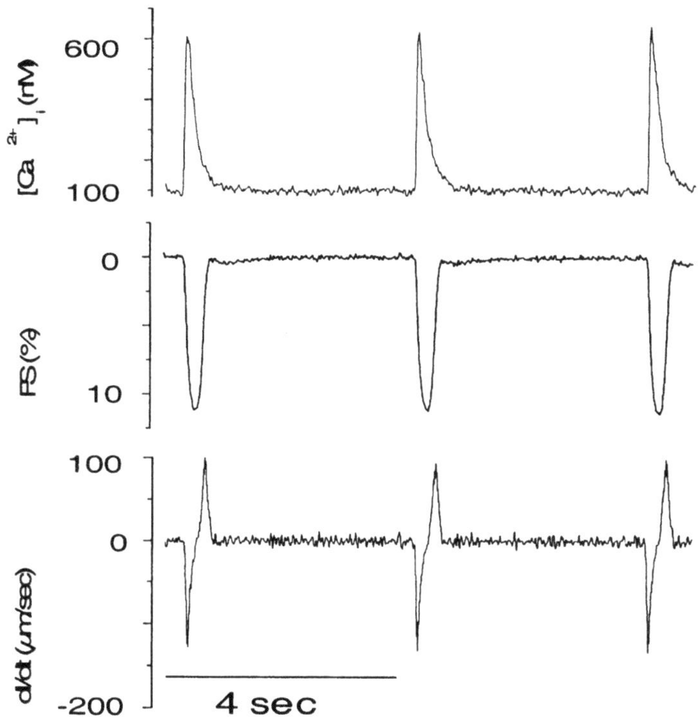

Figure 1. Simultaneous recordings of $[Ca^{2+}]_i$ transients and shortening of mouse ventricular myocytes. Cells were field stimulated at 0.25 Hz. Top panel: stable $[Ca^{2+}]_i$ transients measured by fluo-3 fluorescence. The fluorescence transients were calibrated to $[Ca^{2+}]_i$ transients by pseudo-calibration method. Middle panel: the simultaneously recorded cell fractional shortening (FS) by video edge detector. Bottom panel: the corresponding derivatives (dl/dt, μm/sec) of the cell shortening.

MYOCYTE CALCIUM TRANSIENTS

The calcium transient in cardiac myocytes is initiated by Ca^{2+} influx via L-type Ca^{2+} channels and possibly the reverse mode Na/Ca exchange, the release of Ca^{2+} from the sarcoplasmic reticulum (SR) via Ca^{2+} release channels (ryanodine receptors, RyR) in the SR membrane, with subsequent extrusion of $[Ca^{2+}]_i$ via forward mode Na/Ca exchange. This in turn is coordinated with reuptake of Ca^{2+} into the SR by the SR Ca^{2+}-

ATPase (SERCA). A fundamental principle of this process is that to maintain steady-state calcium homeostasis, the amount of Ca^{2+} entering the cell with each contraction must be extruded before the subsequent contraction. The force developed by myofilaments in cardiac myocytes is modulated by variations in the magnitude of the $[Ca^{2+}]_i$ transient. Factors that modify calcium homeostasis and/or Ca^{2+} sensitivity of myofilaments may significantly alter the force of contraction of an individual myocyte and thus of the intact heart. The current approaches to examining calcium homeostasis are to measure global $[Ca^{2+}]_i$ transients, localized $[Ca^{2+}]_i$ transients (calcium sparks), and the kinetics of $[Ca^{2+}]_i$ transients.

Global $[Ca^{2+}]_i$ Transient

The most commonly used method in measuring $[Ca^{2+}]_i$ transients in cardiac myocytes is by means of calcium sensitive fluorescent indicators. There are a number of commercially available fluorescent indicators and this number is continuously growing (24,25). Fluo-3, Fura-2, and Indo-1 are the most common fluorescent indicators employed at the present time (3-6,9-11,14,15). It should be noted that although fura-2 and indo-1 are still commonly used, intracellular Ca^{2+} buffering and dye saturation can present a problem with these relatively high-affinity Ca^{2+} indicators, which may lead to underestimates of the peak $[Ca^{2+}]_i$ (25,26). In this section we detail the experimental procedure of $[Ca^{2+}]_i$ transient measurement using Fluo-3, which has been successfully applied in our laboratory and others to measure the $[Ca^{2+}]_i$ transients in wild-type and transgenic mice (4,6,11,27,28). The general protocols and many of the cautions described below are similar for most of the fluorescent indicators.

Measurement of Global $[Ca^{2+}]_i$ Transient

To measure the global $[Ca^{2+}]_i$ transient, the following experimental equipment is required: 1) an inverted microscope equipped with an epifluorescence attachment; 2) a photomultiplier tube or an intensified CCD camera and a video monitor; 3) a xenon or mercury lamp; 4) a video cassette recorder and a computer and/or chart recorder; 5) an electric stimulator; 6) a computer workstation for data analysis.

Isolated myocytes are usually used for experiments within 6 hours after dissociation. Myocytes are attached to laminin-coated glass coverslips and then incubated in the loading solution which contains 1-2 μM fluo-3/AM (Molecular Probes, USA). Loading is usually carried out at 30°C in the dark for 30 min. Dye loading is also commonly accomplished in cell suspensions. Fluo-3/AM stock solution (100 μM) is prepared in 0.45% pluronic F-127 (Molecular probes), 10% dimethyl sulfoxide, and 90% heat-inactivated fetal calf serum. Aliquots of fluo-3/AM stock solution are kept at deep freezer

(-70°C). Fluo-3/AM, the acetoxymethyl ester form of fluo-3, diffuses freely across the cell membrane and is hydrolyzed inside the cell by cytoplasmic esterases, regenerating fluo-3 itself, which is cell membrane impermeant and therefore remains trapped in the cytosol. HEPES solution consists of (mM): NaCl 126, KCl 4.4, $MgCl_2$ 1.0, $CaCl_2$ 1.08, HEPES 24, Glucose 11, NaOH 13, and probenecid 0.5 (pH 7.4). Probenecid inhibits the anion transporter which removes fluo-3 from the cytoplasm. Fluo-3-loaded myocytes on the coverslip are then washed twice with dye-free HEPES solution and placed in a flow-through chamber on the microscope. The dye loaded myocytes are excited by a mercury-arc lamp system at a 485 nm wavelength through an epifluorescence attachment (505 nm dichroic mirror, Omega) and a x40 fluor oil objective lens (Nikon). Fluorescence (530 nm, DF30, Omega) is detected with a photomultiplier tube (Hitachi). The intensity of the fluorescence at 530 nm increases with an increase in $[Ca^{2+}]_i$. Fluorescence transients in cardiac myocytes are triggered by either field-stimulating with platinum electrodes, or applying voltage clamp pulses, or exposing the cell to caffeine. The fluorescence transients are recorded on video tapes or stored on computer disks for later analysis.

Calibration of the fluorescence transients to $[Ca^{2+}]_i$ transients is performed with a modification of the method of Kao et al. (29). After the fluorescence transients are recorded, the myocyte is superfused with HEPES solution containing 10 μM ionomycin and 30 mM BDM. In 5 min, the fluorescence intensity increases as Ca^{2+} begins to enter the cell via the ionomycin channels. Then the myocyte is perfused with 10 mM $MnCl_2$, 10 μM ionomycin, and 30 mM BDM in HEPES solution. Mn^{2+} quenches fluo-3 in the cytosol, yielding an intensity of fluorescence (F_{Mn}), which is 1/5 F_{max}. After F_{Mn} is recorded, the intensity of the fluorescence from the field (F_{bkg}) is measured by blowing the myocytes away from the filed with a pipette. The mean value of the autofluorescence of 10 unloaded myocytes in the same day is used as an autofluorescence (F_{auto}) value for each myocyte. The F_{auto} of cardiac myocytes is very low under our experimental conditions and the variation of F_{auto} is also very small. F_{max} is calculated with the following formula: $F_{max}=(F_{Mn}-F_{bkg}-F_{auto})x5$. F_{min} is 1/40 of F_{max}. Therefore, $[Ca^{2+}]_i$ can be calculated with the following formula: $[Ca^{2+}]_i = Kd[(F-F_{bkg}-F_{auto})-F_{min}]/[F_{max}-(F-F_{bkg}-F_{auto})]$. In this formula, F is the measured fluorescence intensity of the dye-loaded myocytes, and Kd is the dissociation constant. The Kd of fluo-3 is known to be temperature dependent and is reported to be 400 and 864 nM at 22°C and 37°C, respectively (29). We use 493 nM as the Kd at 25°C, assuming a linear relationship between Kd and temperature. The fluorescence intensity of fluo-3 also can be converted to $[Ca^{2+}]_i$ by a pseudo-calibration method (30,31).

Two major factors that influence this measurement are quality of the isolated cells and variation of dye loading. Major variables in dye loading are: 1) loading methods (we can load the cell with fluo-3 via incubation with fluo-3/AM as described above or deliver the fluo-3 into the cell via diffusion from the patch-clamp pipettes that are filled with fluo-3 salt containing solution); 2) fluo-3/AM concentration; 3) incubation time and temperature; 4) addition of probenecid to inhibit or suppress loss of dye from the cell; 5) the extent to which AM forms of the dye are sequestered in intracellular organelles, such as mitochondria.

Analysis of Global $[Ca^{2+}]_i$ Transients and Its Determinants

Quantitative analysis of the characteristics of the $[Ca^{2+}]_i$ transient usually includes diastolic $[Ca^{2+}]_i$, peak systolic $[Ca^{2+}]_i$, amplitude of the $[Ca^{2+}]_i$ transient (equal to the peak systolic $[Ca^{2+}]_i$ minus the diastolic $[Ca^{2+}]_i$), time to the peak, and the time constant (τ) of the decline of the $[Ca^{2+}]_i$ transient (4,6,10,11,14,32).

The amplitude of $[Ca^{2+}]_i$ transient in cardiac myocytes is determined by several key factors (23,33): 1) amount of triggering Ca^{2+} influx through L-type Ca^{2+} channel and possibly reverse mode Na^+-Ca^{2+} exchange; 2) the efficiency of E-C coupling gain ($\Delta[Ca^{2+}]_i$ $/I_{Ca}$); 3) releasable SR $[Ca^{2+}]_i$ content; 4) intracellular Ca^{2+} buffers. Several mechanisms contribute to the decline of the $[Ca^{2+}]_i$ transient (23,32,33): 1) Ca^{2+} reuptake into SR by SR Ca^{2+}-ATPase; 2) Ca^{2+} extrusion from the cell by the forward mode Na^+-Ca^{2+} exchange; 3) Ca^{2+} extrusion from the cell by SL Ca^{2+}-ATPase; 4) Ca^{2+} accumulation by mitochondria; 5) Ca^{2+} binding to intracellular buffers (including fluorescent indicators). The most important of these mechanisms are 1 and 2, with SR Ca^{2+}-ATPase mediated Ca^{2+} uptake the dominant mechanism in mouse myocytes.

Calcium Sparks

Calcium sparks are the elementary SR Ca^{2+} release events underlying excitation-contraction coupling in heart muscle (30). Calcium sparks occurring in resting myocytes are called spontaneous Ca^{2+} sparks (30). The probability of the spontaneous Ca^{2+} sparks depends on three major factors (30,34): 1) cytosolic $[Ca^{2+}]_i$, 2) SR Ca^{2+} content, and 3) time after previous activity (i.e., recovery from the adapted or inactivated state of the SR ryanodine receptor (RyR)). The $[Ca^{2+}]_i$ sparks can also be triggered by Ca^{2+} entry via L-type Ca^{2+} channels (31,35). Those localized $[Ca^{2+}]_i$ transients or Ca^{2+} sparks represent a localized CICR mechanisms in cardiac myocytes (31,35). The triggered local release arises from the openings of a cluster of channels (30,36-38). Although the exact nature and origin of Ca^{2+} sparks are not fully understood, the prevailing view is that the whole-cell $[Ca^{2+}]_i$

transient is produced by the temporal and spatial summation of a large number of Ca^{2+} sparks (17,28,31,35).

The amplitude of Ca^{2+} sparks is determined by several factors (11,31,39): 1) SR Ca^{2+} loading conditions, 2) intrinsic gating properties of the RyRs, 3) SR Ca^{2+}-ATPase. Factors that influence the decline of global $[Ca^{2+}]_i$ transients also contribute to the decline of local $[Ca^{2+}]_i$ during a Ca^{2+} spark. Ca^{2+} diffusion from the source of Ca^{2+} release also contributes to local $[Ca^{2+}]_i$ decline during the Ca^{2+} spark (27).

Visualization of Ca^{2+} Sparks

To observe Ca^{2+} sparks, the basic instrument requirement is a laser confocal microscope. Currently there are several line scan confocal microscopy systems commercially available, such as Bio-Rad MRC600 (Cambridge, MA, USA), Zeiss LSM410 (Carl Zeiss, Inc., Germany), Olympus LSM-GB200 (Olympus, Japan), which use photomultiplier tubes as detectors. A Nipkow disc confocal system (Yokogawa, Tokyo, Japan) is also available, which uses an intensified CCD camera as a detector. Line scan devices (LSCM) give higher temporal resolution, but are quite expensive. The Nipkow device has a temporal resolution limited by the frame rate of the detector camera, but is simpler to use and less expensive. Both types of confocal systems use an argon laser (488 nm) as a light source. Therefore, it has been difficult to use UV excitation required for confocal imaging of other Ca^{2+} indicators such as indo-1. This is gradually becoming less of a problem as UV-based confocal systems and multi-photon approaches are allowing use of UV-excitation probes in confocal situations, but they are still very expensive. Recently, a new LSCM system (Radiance 2000 MP) from Bio-Rad is able to provide limited variations (457-514 nm) in the excitation wavelength of argon laser.

Fluo-3 is one of the most suitable Ca^{2+} indicator used in combination with LSCM to visualize Ca^{2+} sparks (11,17,30,31,35,38) because it can be excited with a currently available argon ion laser. Recently, a low-affinity, fast Ca^{2+} indicator, Oregon Green 488 BAPTA-5N (Kd $\approx$ 30 μM) was also used to optimize the detection of localized high $[Ca^{2+}]_i$ in cardiac myocytes (40). Fluo-3 can be introduced into the cell either by using Fluo-3/AM loading method as described above or by diffusion from the patch pipette filled with Fluo-3 salt containing solution (7,8,11). Fluorescence images can be acquired by using the line scan mode (30,31), or two-dimensional (full frame) scan mode (8,17). To increase temporal resolution it has been customary to use the line scan mode of the confocal microscopes.

Image processing and analysis are carried out by using special software such as IDL software (Research System Inc. Boulder. CO, USA) and NIH Image (NIH, Bethesda, MD, USA). Water-fall display and three-dimension surface plot of the line scan images are widely used to present the Ca^{2+} spark images (Figure 2). While analyzing the spark images, the following parameters are usually measured (11,17,30,36): 1) frequency of Ca^{2+} spark occurrence; 2) Ca^{2+} spark size; 3) amplitude of Ca^{2+} sparks; 4) time course of Ca^{2+} sparks; 5) time constant of decay of Ca^{2+} sparks; 6) location of Ca^{2+} sparks.

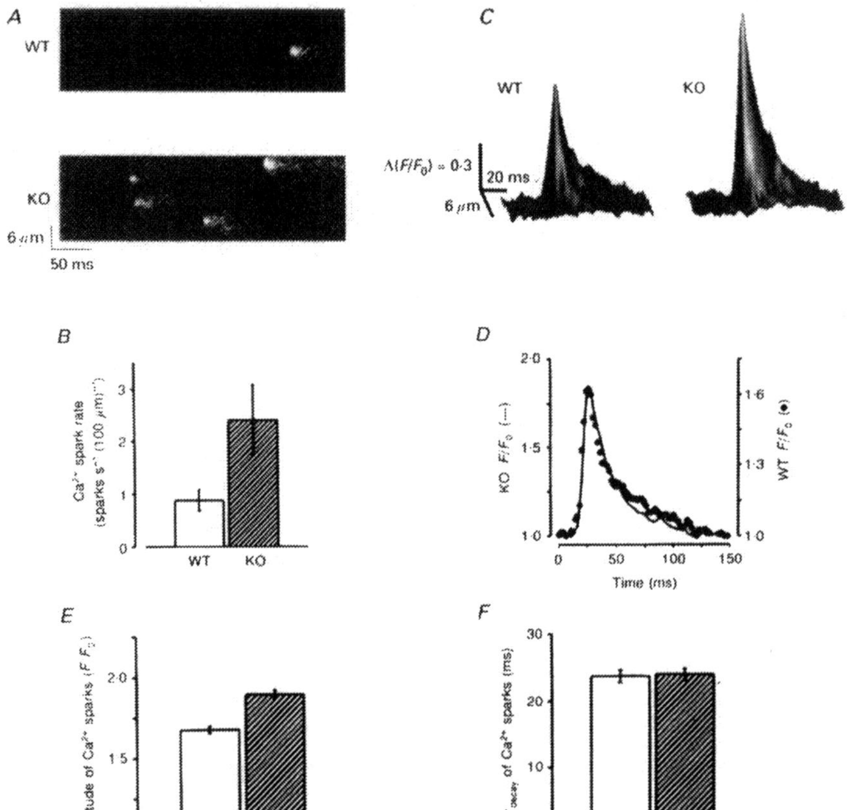

Figure 2. Examples of spontaneous Ca^{2+} sparks in WT and phospholamban knock-out (KO) ventricular myocytes. A, representative line-scan images from WT (top) and KO (bottom) ventricular myocytes. These line-scan images were collected at a holding potential of -80 mV. B, rate of spontaneous Ca^{2+} spark occurrence in WT and KO ventricular cells. C, surface plot of averaged Ca^{2+} sparks from WT and KO myocytes. The vertical scale bar in C represents a fluorescence change of 0.3 in F/F_0, where baseline F/F_0 =1. D, time course of the averaged WT (•) and KO (continuous line) Ca^{2+} sparks shown in C, E, amplitude of spontaneous Ca^{2+} sparks in WT and KO ventricular myocytes. F, time constant of decay (τ_{decay}) of Ca^{2+} sparks in WT and KO cells. Reprinted with permission from Santana et al. (11).

FUNCTIONAL ASSESSMENT OF INDIVIDUAL Ca^{2+} REGULATORY MECHANISMS IN INTACT MYOCYTES FROM GENETICALLY ENGINEERED MICE

The amplitude and kinetics of $[Ca^{2+}]_i$ transients in cardiac myocytes are mainly controlled by the interplay between several calcium-handling components including L-type Ca^{2+} channel, Na^+/Ca^{2+} exchanger, SR RyR/Ca^{2+} release channel, and the SR Ca^{2+}-ATPase. Techniques are available now which permit the quantitation of the function of these different components, by combining $[Ca^{2+}]_i$ measurements, cell shortening, rapid solution switcher, and voltage clamp techniques. We will now discuss how function of these individual Ca^{2+} control systems can be assessed in intact single ventricular murine myocytes, and give examples of their application in genetically engineered animals.

Assessment of SR function

Ca^{2+}-ATPase

The most common approach in the past has been to assess the rate of SR Ca^{2+}-ATPase dependent resequestration of Ca^{2+} in the SR by measuring the first derivative of relengthening (-dL/dt), when motion measurements are made, or the time constant (τ) of decline of the $[Ca^{2+}]_i$ transient. There are several problems with this approach. First, the rate of relengthening of isolated myocytes does not exactly track the rate of decline of the $[Ca^{2+}]_i$ transient. In addition, the τ of decline in the $[Ca^{2+}]_i$ transient is influenced by function of the Na/Ca exchanger which is extruding Ca^{2+} from the cell in parallel to the reuptake of Ca^{2+} by the SR Ca^{2+}-ATPase. In mouse myocytes, however, Ca^{2+} extrusion by the Na/Ca exchanger contributes only slightly to the τ of the $[Ca^{2+}]_i$ decline (6). A more serious problem is that the time constant of decline of the $[Ca^{2+}]_i$ transient is related to the peak of the $[Ca^{2+}]_i$ achieved during the transient (Fig. 3) (32).

These considerations have led to a number of different approaches to assess the function of the SR Ca^{2+}-ATPase in an intact single cell. Bers et al. have removed extracellular Na^+ and Ca^{2+}, then resupplied extracellular Ca^{2+}allowing the observation of transients induced by normal EC coupling, in the absence of extracellular Na^+, so that the Na/Ca exchanger function is eliminated (41-43). A rapid solution switcher device can be used to inhibit Na/Ca exchange abruptly by removing extracellular Na^+ simultaneous with production of depolarization (44), or by a combination of removal of extracellular Na^+ and exposure to Ni^{2+}, an agent which inhibits Na/Ca exchange (4, 6). We and others have used these techniques to examine SR Ca^{2+}-ATPase function in myocytes from mice overexpressing the SR Ca^{2+}-ATPase (6) or with phospholamban (PLB) ablated (14). While helpful in eliminating the contribution of the Na/Ca exchanger, these approaches are still complicated by the fact that the τ of decline of the transient is dependent on the peak $[Ca^{2+}]_i$ (Fig. 3). Another approach which we have employed is to use the rapid solution switcher to abruptly expose isolated myocytes to 10 mM caffeine for 100 ms, in the absence of extracellular Na^+ and Ca^{2+}. As shown in Figure 4, this induces a transient which produced a near maximal peak $[Ca^{2+}]_i$, and the decline of the $[Ca^{2+}]_i$ then occurs in the absence of caffeine which is washed away by the rapid solution switcher. We have used this technique to assess SR Ca^{2+}-ATPase function in myocytes from animals overexpressing the Na/Ca exchanger (4) and to compare SR Ca^{2+}-ATPase function in myocytes from different species (45).

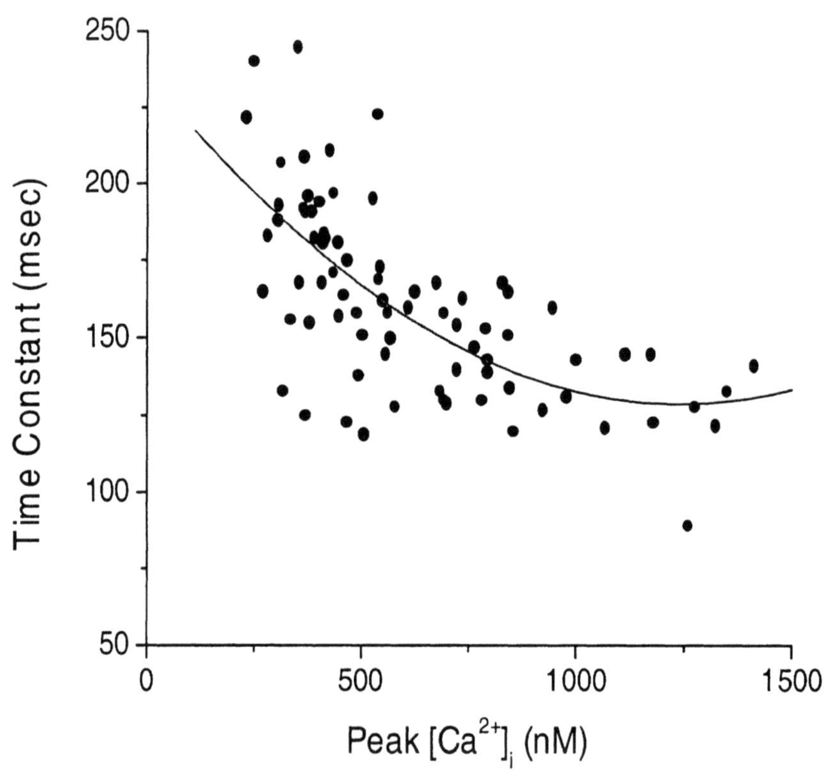

Figure 3. The time constant (τ) of the decline of each $[Ca^{2+}]_i$ transient is plotted as a function of the peak $[Ca^{2+}]_i$ in mouse ventricular myocytes field stimulated at 0.25 Hz. The τ is shorter at higher peak $[Ca^{2+}]_i$.

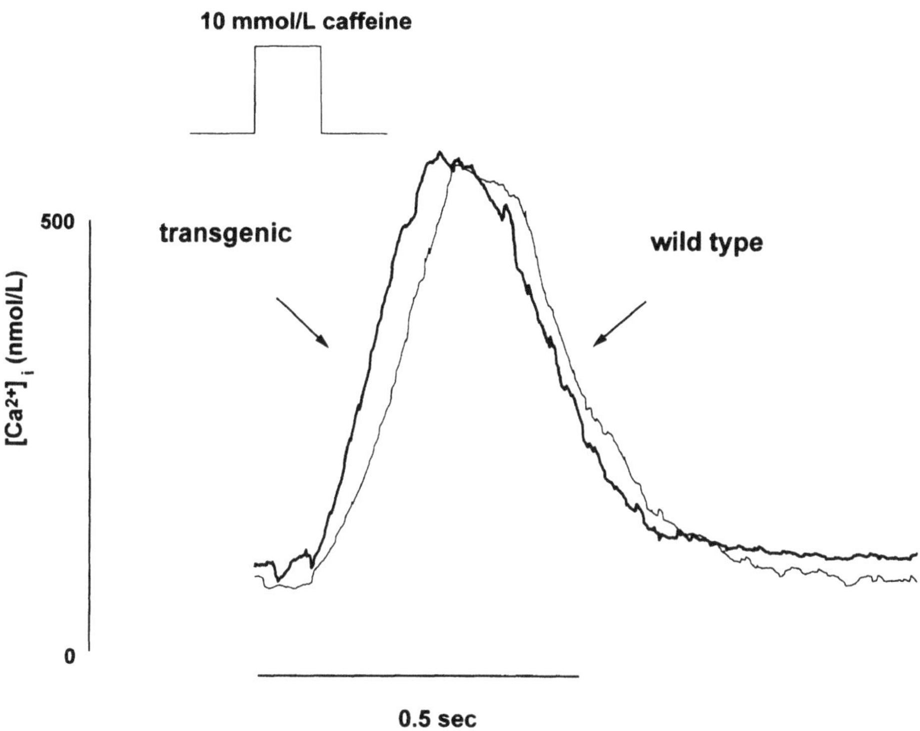

Figure 4. Examples of short-pulse caffeine transients in a WT and a TG myocyte. A $[Ca^{2+}]_i$ transient was induced by a 100-ms exposure to 10 mM caffeine in 0 Na^+/0 Ca^{2+} solution in a WT (thin line) and a TG (thick line) myocyte. Both transients have an identical peak and resting $[Ca^{2+}]_i$. The rapid decline of both transients, mediated by SR Ca^{2+}-ATPase, is similar. Reprinted with permission from Yao et al. (4).

SR Ca^{2+} Content

Frequently it is of interest to know if SR Ca^{2+} content is altered in a given pathologic or transgenic model, or in studying drug effects. There are several approaches that have been used to assess SR Ca^{2+} content, almost all involving the abrupt exposure to caffeine which activates the ryanodine receptor, and releases Ca^{2+} from the SR. The amplitude of contracture (shortening) induced by rapid exposure to caffeine in myocytes can be used as an index of SR Ca^{2+} content. A difficulty with this approach is that unloaded shortening may be near maximal and not reflect accurately variations in the amount of Ca^{2+}

released. In addition, in some situations an action potential can be triggered by abrupt exposure to caffeine due to forward Na/Ca exchange and the resulting inward current which can cause the membrane potential to reach threshold (46). A better approach is to expose myocytes to caffeine abruptly and measure the peak $[Ca^{2+}]_i$. This provides a more accurate measurement of SR Ca^{2+} content than shortening, but still can be influenced by the Na/Ca exchanger, which may extrude Ca^{2+} and limit the peak Ca^{2+} level achieved. Nevertheless, this technique has been used to study SR Ca^{2+} loading in calsequestrin-overexpressing, PLB knock-out, and hypertrophied murine ventricular myocytes (7, 11, 47).

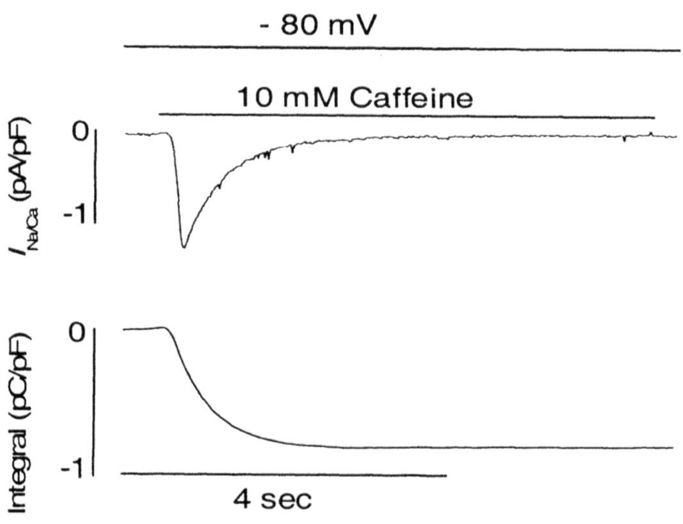

Figure 5. Assessment of SR Ca^{2+} content. During exposure of voltage clamped (-80 mV) myocytes to caffeine (10 mM) with a rapid solution switcher, Ca^{2+} is released from SR. This results in forward mode Na/Ca exchange, in which one Ca^{2+} is extruded by the Na/Ca exchanger in exchange for 3 Na^+, which generates an inward current. The integral of the inward current is used as an indicator of SR Ca^{2+} content.

Currently the most accurate technique to measure SR Ca^{2+} content involves myocyte exposure to caffeine with a rapid solution switcher, and quantitation of the integral of the inward Na/Ca exchange current resulting from exchanger mediated extrusion of Ca^{2+} (4, 6, 11, 15). This technique is illustrated in Figure 5. In all these approaches, several principles are important. First, it is important that the caffeine exposure be very abrupt. If caffeine exposure is gradual, Ca^{2+} release will not be rapid enough to allow accurate assessment of SR Ca^{2+} content. Abrupt exposure to caffeine can be best accomplished with a rapid solution switcher. Secondly, the caffeine exposure must continue during the entire measurement, to inhibit resequestration of Ca^{2+} by the SR. Finally, the measurements should be made at a steady state of pacing so that SR loading via Ca^{2+} influx through the L-type Ca^{2+} channel can be maintained.

RyR Function

Ryanodine receptor function, which is responsible for Ca^{2+}-induced Ca^{2+} release and thus EC coupling in cardiac myocytes, can be assessed by measurement of the global $[Ca^{2+}]_i$ transient, and by measurement of Ca^{2+} sparks. Assessment of RyR function using the global $[Ca^{2+}]_i$ transient is usually carried out by measuring EC coupling "gain", which is the magnitude and/or rate of rise of the $[Ca^{2+}]_i$ transient, divided by the amplitude of the L-type Ca^{2+} current (7, 11). This approach allows assessment of the efficiency by which L-type Ca^{2+} channel current induces Ca^{2+} release from the SR. However, it is influenced by SR Ca^{2+} loading, in that the greater the SR Ca^{2+} load, the greater the EC coupling gain (11). Thus in experiments designed to determine EC coupling gain, it is very important to ascertain if SR Ca^{2+} loading is comparable under the conditions which are being compared.

As discussed previously, it is now possible to measure the elementary Ca^{2+} release events, Ca^{2+} sparks, by confocal microscopy. The effects of interventions on ryanodine receptor function can therefore be now characterized by alterations of Ca^{2+} spark characteristics. This approach has been used in the study of several genetically engineered mouse models such as PLB knock-out mice and calsequestrin-overexpressing mice (8, 11), and effects of FK506 (48), and cardiac hypertrophy (49). Ca^{2+} spark characteristics which are usually measured are the probability of spontaneous spark occurrence, the amplitude of sparks, the size of sparks, and spark kinetics (Fig. 2). In addition, Santana et al have related the probability of spark production to the sarcolemmal L-type Ca^{2+} current, in an attempt to provide another layer of accuracy to the measurement of EC coupling gain (50). Assessment of RyR function from measurement of Ca^{2+} sparks and L-type channel mediated current is limited by the fact that the number of RyRs that are involved in producing a single spark is not well defined, and the number of L-type Ca^{2+} channels that relate to a given cluster of RyRs producing a single spark is not clear. It is also not technically possible at the present time to measure the current via L-type Ca^{2+} channels in a localized area simultaneous with the occurrence of local sparks in the same area, preventing exact determination of the magnitude of the current required to generate a local spark.

Na/Ca Exchanger Function

Function of the Na/Ca exchanger in intact myocytes has been usually assessed by measuring the rate of relaxation, and/or the rate of decline of the $[Ca^{2+}]_i$, during a caffeine-induced contraction or $[Ca^{2+}]_i$ transient. An example of this type of approach is shown in Figure 6.

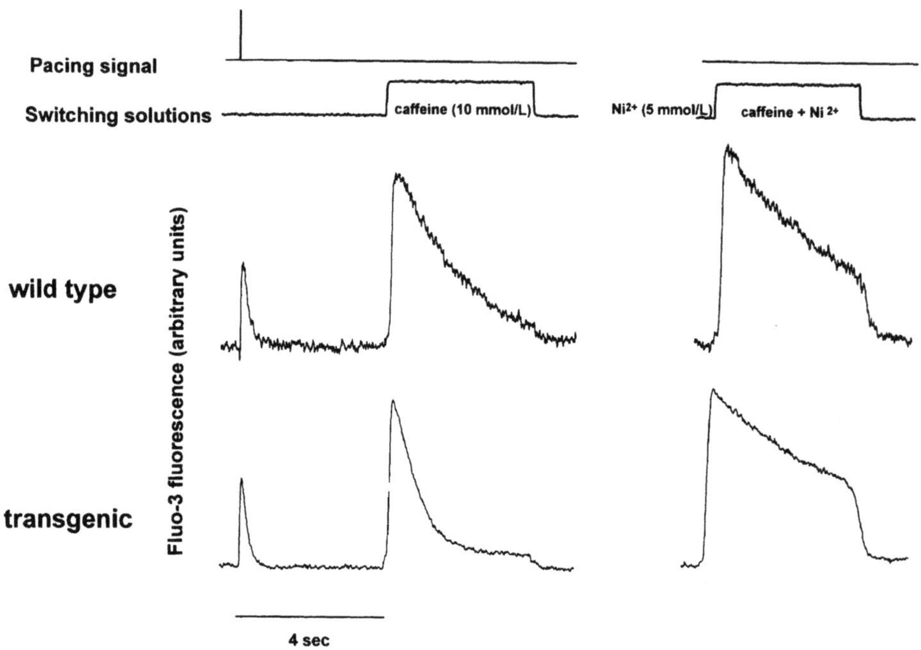

Figure 6. Representative tracings of Fluo-3 fluorescence transients triggered by field stimulation or caffeine in mouse ventricular myocytes. After the last electrically stimulated beat, a caffeine-induced $[Ca^{2+}]_i$ transient (left tracings) was induced by a 4-s exposure to caffeine in a wild-type (WT) myocyte and a transgenic (TG) myocyte overexpressing Na/Ca exchanger. Note that the rate of decline of $[Ca^{2+}]_i$ in the WT myocyte when the SR is disabled by caffeine is ≈1/10 that of a normal electrically stimulated transient. The decline of the caffeine induced $[Ca^{2+}]_i$ transient was markedly accelerated in the TG myocyte. There is no apparent difference in the decline of the caffeine-induced transients (right tracings) between WT and TG myocytes in the presence of 5 mM Ni^{2+}. Reprinted with permission from Yao et al. (4).

The theory here is that the reuptake of Ca^{2+} released by caffeine is prevented by the continued presence of caffeine, and the decline of the $[Ca^{2+}]_i$ and associated relaxation are due to extrusion of Ca^{2+} from the cell by Na/Ca exchange. Thus the rate of this extrusion can be taken as an index of the function of the exchanger. While useful, this approach is complicated by the fact that the function of the exchanger is very dependent on the transarcolemmal Na^+ concentration gradient, and thus intracellular Na^+ concentration, and by membrane potential. Unless the cell is voltage clamped to control membrane potential, and the cell dialyzed with a specific Na^+ concentration solution, these factors cannot be adequately controlled to allow precise assessment of exchanger function and density.

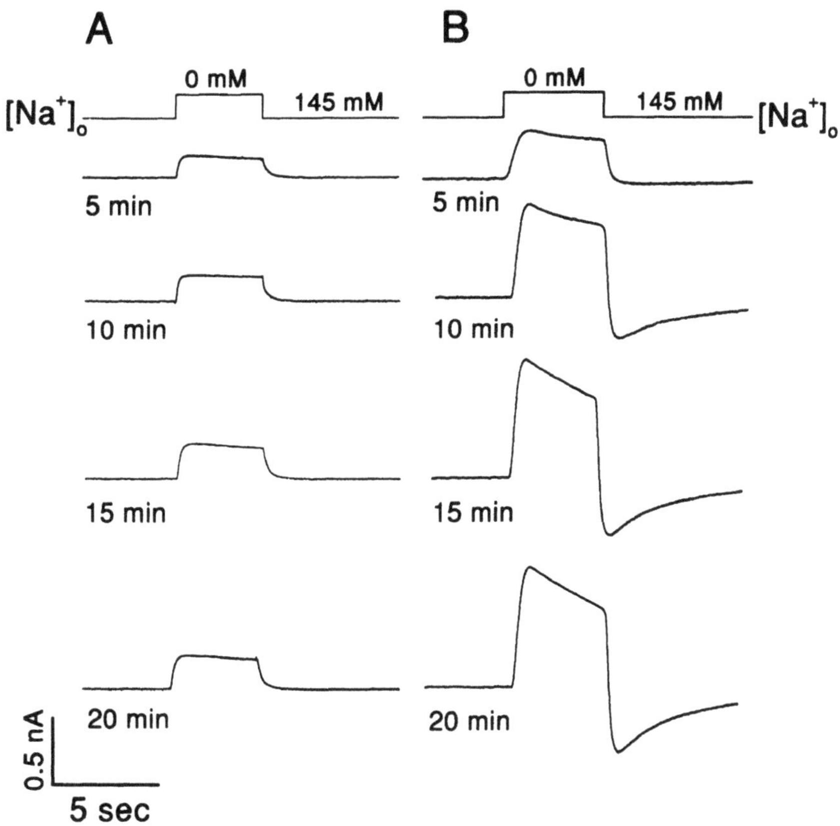

Figure 7. Examples of Na/Ca exchange current in mouse ventricular myocytes. A, recordings in a WT mouse ventricular myocyte, obtained at 5, 10, 15, and 20 min after formation of a whole-cell patch. In each panel, the current was elicited by abrupt removal of extracellular Na^+ and replacement with Li^+, by means of a rapid switcher device. The timing of the switcher pulse is shown in the top tracings. B, similar recordings in a transgenic myocyte overexpressing the Na/Ca exchanger. Current density in transgenic cells is 2.5 times that in wild-type cells. Currents increased with time and plateaued within 10-15 min. Reprinted with permission from Su et al. (12).

An alternative approach we have used is to combine voltage clamp measurements and use of a rapid solution switcher to measure the outward current produced by the exchanger when extracellular Na^+ is abruptly removed. An example of this approach in transgenic mice overexpressing the Na/Ca exchanger is shown in Figure 7. In these experiments, membrane potential is maintained constant, the cell is dialyzed with a known Na^+ concentration containing solution via the pipette, and the Na^+ pump is completely inhibited by exposure to zero extracellular K^+, to make sure that the Na^+ concentration in the subsarcolemmal space ("fuzzy space") is adequately controlled by the pipette Na^+ concentration (12, 51). We have used this approach to compare the density of the Na/Ca exchanger in murine myocytes with that in other myocytes from other species including humans (12).

L-type Ca^{2+} Channel Function

L-type Ca^{2+} channel function can be assessed in murine ventricular myocytes by conventional voltage clamp techniques, and these are described in Chapter 7 of this book.

Other Ca^{2+} Homeostatic Systems

It is possible to quantitate Ca^{2+} extrusion from the cell via the sarcolemmal Ca^{2+}-ATPase and Ca^{2+} uptake by mitochondria (41-43), and to estimate intracellular Ca^{2+} buffering (52). The rate of Ca^{2+} extrusion from the cell by the sarcolemmal Ca^{2+}-ATPase, and uptake by mitochondria is 1 to 2 orders of magnitude slower than Ca^{2+} transport by the Na/Ca exchanger, and the sarcoplasmic reticulum Ca^{2+}-ATPase (41). Therefore these systems are thought to play less of an important role in the dynamic regulation of $[Ca^{2+}]_i$ within cardiac myocytes, and less attention has been devoted to their study. However, they are obviously important systems and further definition of their role in murine cardiac myocyte Ca^{2+} homeostasis will be necessary in the future.

Intracellular Ca^{2+} buffering is very important in influencing the magnitude and kinetics of the $[Ca^{2+}]_i$ transient. This is readily apparent from the consideration that only approximately 1-2% of the total Ca^{2+} entering the cell and released from the SR appears as the free Ca^{2+} transient (52, 53). Also, overexpression of the principal Ca^{2+} buffer in the SR, calsequestrin, has dramatic effects on myocyte Ca^{2+} homeostasis and cardiac function (7, 8). Quantitation of Ca^{2+} buffering requires detailed knowledge of the Ca^{2+} binding characteristics, as well as the concentrations of the various Ca^{2+} buffers within the cell. This is a very complex issue and at present a lack of precise information regarding these buffers has currently limited the approaches to semiquantitative ones.

CONCLUSIONS

The development of techniques to modify expression of specific genes in mice has markedly enhanced our ability to investigate the molecular mechanisms of cardiac pathophysiology. An important component of this work entails quantitation of these molecular processes at the cellular and subcellular level in the intact living cell. Much progress has been made in this regard with respect to Ca^{2+} homeostasis and contractile function, but clearly much remains to be accomplished. Further improvement in fluorescent probes, in confocal microscopic techniques, and in patch and whole cell voltage clamp methods, as well as development of techniques to measure force development by intact single cells, will undoubtedly advance these efforts in the study of transgenic mouse cardiac myocytes.

REFERENCES

1. Field WL, Azzawi A, Styles P, Henderson C, Seymour A-M L, Radda GK. Intracellular Ca^{2+} transients in isolated perfused rat heart: measurement using the fluorescent indicator fura-2/AM. Cell Calcium 1994;16:87-100.
2. Del Nido PJ, Glynn P, Buenaventura P, Salama G, Koretsky AP. Fluorescence measurement of calcium transients in perfused rabbit heart using rhod 2. Am J Physiol 1998;274:H728-H741.
3. Adachi-Akahane S, Lu L, Li Z, Frank JS, Philipson KD, Morad M. Calcium signaling in transgenic mice overexpressing cardiac Na^+-Ca^{2+} exchanger. J Gen Physiol 1997;109:717-729.
4. Yao A, Su Z, Nonaka A, Zubair I, Lu L, Philipson KD, Bridge JHB, Barry WH. Effects of overexpression of the Na^+-Ca^{2+} exchanger on $[Ca^{2+}]_i$ transients in murine ventricular myocytes. Circ Res 1998;82:657-665.
5. He H, Giordano FJ, Hilal-Dandan R, Choi D, Rockman HA, McDonough PM, Bluhm W, Meyer M, Sayen MR, Swanson E, Dillmann WH. Overexpression of the rat sarcoplasmic reticulum Ca^{2+} ATPase gene in the heart of transgenic mice accelerates calcium transients and cardiac relaxation. J Clin Invest 1997;100:380-389.
6. Yao A, Su Z, Dillmann WH, Barry WH. Sarcoplasmic reticulum function in murine ventricular myocytes overexpressing SR CaATPase. J Mol Cell Cardiol 1998;31:1327-1336.
7. Jones LR, Suzuki YJ, Wang W, Kobayashi YM, Ramesh V, Franzini-Armstrong C, Cleemann L, Morad M. Regulation of Ca^{2+} signaling in transgenic mouse cardiac myocytes overexpressing calsequestrin. J Clin Invest 1998;101:1385-1393.
8. Wang W, Cleemann L, Jones LR, Morad M. Modulation of focal and global Ca^{2+} release in calsequestrin-overexpressing mouse cardiomyocytes. J Physiol 2000;524:399-414.
9. Wolska BM, Stojanovic MO, Luo W, Kranias EG, Solaro RJ. Effect of ablation of phospholamban on dynamics of cardiac myocyte contraction and intracellular Ca^{2+}. Am J Physiol 1996(a);271:C391-C397.
10. Kadambi VJ, Ponniah S, Harrer JM, Hoit BD, Dorn II GW, Walsh RA, Kranias EG. Cardiac-specific overexpression of phospholamban alters calcium kinetics and resultant cardiomyocyte mechanics in transgenic mice. J Clin Invest 1996;97:533-539.

11. Santana LF, Kranias EG, Lederer WJ. Calcium sparks and excitation-contraction coupling in phospholamban-deficient mouse ventricular myocytes. J Physiol 1997;503:21-29.
12. Su Z, Bridge JHB, Philipson KD, Spitzer KW, Barry WH. Quantitation of Na/Ca exchanger function in single ventricular myocytes. J Mol Cell Cardiol 1999;31:1125-1135.
13. Wolska B, Solaro RJ. Method for isolation of adult mouse cardiac myocytes for studies of contraction and microfluorimetry. Am J Physiol 1996(b); 271:H1250-H1255.
14. Li L, Chu G, Kranias EG, Bers DM. Cardiac myocyte calcium transport in phospholamban knockout mouse: relaxation and endogenous CaMKII effects. Am J Physiol 1998;274:H1335-H1347.
15. Terracciano CMN, De Souza AI, Philipson KD, MacLeod KT. Na^+-Ca^{2+} exchange and sarcoplasmic reticular Ca^{2+} regulation in ventricular myocytes from transgenic mice overexpressing the Na^+-Ca^{2+} exchanger. J Physiol 1998;512:651-667.
16. Wagoner LE, Zhao L, Bishop K, Chan S, Xu s, Barry WH. Lysis of adult ventricular myocytes by cells infiltrating rejecting murine cardiac allografts. Circulation 1996;93:111-119.
17. Ritter M, Su Z, Spitzer KW, Ishida H, Barry WH. Caffeine-induced Ca^{2+} sparks in mouse ventricular myocytes. Am J Physiol 2000;278:H666-H669.
18. Brady AJ. Mechanical properties of Isolated Cardiac Myocytes. Physiol Rev 1991;71:413-428478.
19. Palmer RE, Brady AJ, Roos KP. Mechanical measurements from isolated cardiac myocytes using a pipette attachment system. Am J Physiol 1996;270:C697-C704.
20. Chu G, DornII GW, Luo W, Harrer JM, Kadambi VJ, Walsh RA, Kranias EG. Monomeric phospholamban overexpression in transgenic mouse hearts. Circ Res 1997;81:485-492.
21. Steadman BW, Moor KB, Spitzer KW, Bridge JHB. A video system for measuring motion in contracting heart cells. IEEE Trans Biomed Eng 1988; 35:264-272.
22. Peeters GA, Hlady V, Bridge JHB, Barry WH. Simultaneous measurement of calcium transients and motion in culture heart cells. Am J Physiol 1987;253:H1400-H1408.
23. Barry WH & Bridge JHB. Intracellular calcium homeostasis in cardiac myocytes. Circulation 1993;87:1806-1815.
24. Takahashi A, Camacho P, Lechleiter AD, Herman B. Measurement of intracellular calcium. Physiol Rev 1999;79:1090-1125.
25. Simpson AWM. Fluorescent measurement of $[Ca^{2+}]_c$. Methods Mol Biol 1999;114:3-30.
26. Berlin J & Konishi M. Ca^{2+} Transients in cardiac myocytes measured with high and low affinity Ca^{2+} indicators. Biophys J 1993;65:1632-1647.
27. Gomez AM, Cheng H, Lederer WJ, Bers DM. Ca^{2+} diffusion and sarcoplasmic reticulum transport both contribute to $[Ca^{2+}]_i$ decline during Ca^{2+} sparks in rat ventricular myocytes. J Physiol 1996;496:575-581.
28. Cannell MB, Cheng H, Lederer WJ. Spatial non-unuiformities in $[Ca^{2+}]_i$ during excitation-contraction coupling in cardiac myocytes. Biophys J 1994; 67:1942-1956.
29. Kao JPY, Harootunian AT, Tsien RY. Photochemically generated cytosolic calcium pulses and their detection by fluo-3. J Biol Chem 1989;264:8171-8178.
30. Cheng H, Lederer WJ, Cannell MB. Calcium sparks: elementary events underlying excitation-contraction coupling in heart muscle. Science 1993; 262:740-744.
31. Cannell MB, Cheng H, Lederer WJ. The control of calcium release in heart muscle. Science 1995;268:1045-1049.
32. Bers DM & Berlin JR. Kinetics of $[Ca]_i$ decline in cardiac myocytes depend on peak $[Ca]_i$. Am J Physiol 1995;268:C271-C277.
33. Bers DM. Excitation-contraction coupling and cardiac contractile force. Kluwer Academic Publisher. Dordrecht, The Netherlands. 1991.

34. Bers DM, Li L, Satoh H, McCall E. Factors that control sarcoplasmic reticulum calcium release in intact ventricular myocytes. Ann N Y Acad Sci 1998;853:157-177.
35. Lopez-Lopez JR, Shacklock PS, Balke CW, Wier WG. Local calcium transients triggered by single L-type calcium channel currents in cardiac cells. Science 1995;268:1042-1045.
36. Cheng H, Lederer MR, Xiao RP, Zhou YY, Ziman B, Spurgeon H, Lakatta EG, Lederer WJ. Excitation-contraction coupling in heart: new insights from Ca^{2+} sparks. Cell Calcium 1996;20:129-140.
37. Parker I, Zang WJ, Wier WG. Ca^{2+} sparks involving multiple Ca^{2+} release sites along Z-lines in rat heart cells. J Physiol 1996;497:31-8.
38. Bridge JHB, Ershler PR, Cannell MB. Properties of Ca^{2+} sparks evoked by action potentials in mouse ventricular myocytes. J Physiol 1999;518:469-478.
39. Lukyanenko V, Wiesner TF, Gyorke S. Termination of Ca^{2+} release during Ca^{2+} sparks in rat ventricular myocytes. J Physiol 1998; 507:667-677.
40. Song L, Sham JSK, Stern MD, Lakatta EG, Cheng H. Direct measurement of SR release flux by tracking Ca^{2+} 'spikes' in rat cardiac myocytes. J Physiol 1998;512:677-691.
41. Bassani RA, Bassani JW, Bers DM. Mitochondrial and sarcolemmal Ca^{2+} transport reduce $[Ca^{2+}]_i$ during caffeine contractures in rabbit cardiac myocytes. J Physiol 1992;453:591-608.
42. Bassani JW, Bassani RA, Bers DM. Relaxation in rabbit and rat cardiac cells: species-dependent differences in cellular mechanisms. J Physiol 1994;476:279-293.
43. Bers DM, Bassani JW, Bassani RA. Na-Ca exchange and Ca fluxes during contraction and relaxation in mammalian ventricular muscle. Ann N Y Acad Sci 1996;779:430-442.
44. Yao A, Matui H, Spitzer KW, Bridge JH, Barry WH. Sarcoplasmic reticulum and Na/Ca exchanger function during early and late relaxation in ventricular myocytes. Am J Physiol 1997;23:H2765-H2773.
45. Su Z, Zubair I, Barry WH. Sarcoplasmic reticulum Ca^{2+}-ATPase function in human, rabbit and mouse ventricular myocytes. Circulation 1999;100 (suppl I):I842.
46. Zaniboni M, Yao A, Barry WH, Musso E, Spitzer KW. Complications associated with rapid caffeine application to cardiac myocytes that are not voltage clamped. J Mol Cell Cardiol 1998;30:2229-2235.
47. Ito K, Yan X, Tajima M, Su Z, Barry WH, Lorell BH. Contractile reserve and intracellular calcium regulation in mouse myocytes from normal and hypertrophied failing hearts. Circ Res 2000;87:588-595.
48. McCall E, Li L, Satoh H, Shannon TR, Blatter LA, Bers DM. Effects of FK-506 on contraction and Ca^{2+} transients in rat cardiac myocytes. Circ Res 1996;79:1110-1121.
49. Gomez AM, Valdivia HH, Cheng H, Lederer MR, Santana LF, Cannell MB, McCune SA, Altschuld RA, Lederer WJ. Defective excitation-contration coupling in experimental cardiac hypertrophy and heart failure. Science 1997;276:800-805.
50. Santana LF, Cheng H, Gomez AM, Cannell MB, Lederer WJ. Relation between the sarcolemmal Ca^{2+} current and Ca^{2+} sparks and local control theories for cardiac excitation-contraction coupling. Circ Res 1996;78:166-171.
51. Su Z, Zou A, Nonaka A, Zubair I, Sanguinetti MC, Barry WH. Influence of prior Na^+ pump activity on pump and Na^+/ Ca^{2+} exchange currents in mouse ventricular myocytes. Am J Physiol 1998;275:H1808-H1817.
52. Berlin JR, Bassani JWM, Bers DM. Intrinsic cytosolic calcium buffering properties of single rat cardiac myocytes. Biophys J 199;67:1775-1787.
53. Sipido and Wier. Flux of Ca^{2+} across the sarcoplasmic reticulum of guinea-pig cardiac cells during excitation-contraction coupling. J Physiol 1991;435:605-630.

7
SINGLE CELL PATCH-CLAMP ANALYSIS OF MOUSE CARDIAC MYOCYTES

Nipavan Chiamvimonvat, M.D.
University of California, Davis, Davis, CA 95616 USA
M. Jane Lalli, Ph.D.
Atsuko Yatani, Ph.D.
Hackensack University Medical Center, Hackensack, NJ, 07601 USA

The last few years have seen an explosion in the number of transgenic mouse models being generated. Exciting new data have been obtained. In particular, in the field of cardiac electrophysiology, the use of single-cell patch-clamp techniques on isolated cardiomyocytes offers a powerful method with which we can utilize to analyze the effects of selective genetic manipulation on the electrophysiological process of the cells.

SINGLE CELL PATCH-CLAMP TECHNIQUES

General overview of the patch-clamp techniques

By the early 1970s, it had become clear that discrete integral membrane proteins underlie the electrical signal of nerve and muscle. The extracellular patch-clamp technique has allowed currents passing through single ionic channels to be observed for the first time (1). In this technique, a small fire-polished glass pipette is pressed against the cell membrane forming an electrical seal with a resistance of the order of 50-100 MΩ. Using such configuration, most of the currents originating in a small patch of membrane flow into the pipette. However, the major drawbacks of measurements with low-resistance seals included limited resolution and excessive level of background noise. All these problems were greatly improved when it was found that the application of slight suction within a freshly prepared pipette readily induced the phenomenon of 'giga-seal' formation (2). The physical basis of the giga-seal is not clear, but it certainly reduces the background noise of the recording by several orders of magnitudes and allows a patch of membrane to be voltage-clamped without the use of conventional microelectrodes.

Four different patch configurations are possible following the formation of a giga-seal. In addition, perforated patch recording (3) and other minor variants (4) are also described. In the cell-attached patch recording, currents through ion channels trapped within the pipette orifice can be recorded with subpicoampere resolution following seal formation. It was recognized that the giga-seal provided mechanical stability, so that patches could be 'excised' from the parent cell and studied in isolation. Such configuration is referred to as inside-out excised patches. Alternatively, following initial seal formation, it is also possible to rupture the membrane patch trapped within the pipette by application of strong suction keeping the pipette cell-attached. This provides a direct low-resistance access to the cell interior thereby allowing potential recording and voltage clamping of small cells. One particular advantage of the whole-cell patch clamp technique is that one can dialyze the cell interior through the relatively large electrode tip. This technique has been utilized to test the effects of several large molecular weight proteins on various types of ion channels. Finally, if the pipette is withdrawn from the cell during whole-cell recording, a tether of membrane is drawn away from the cell, which normally breaks off and reseals on the pipette as an outside-out patch.

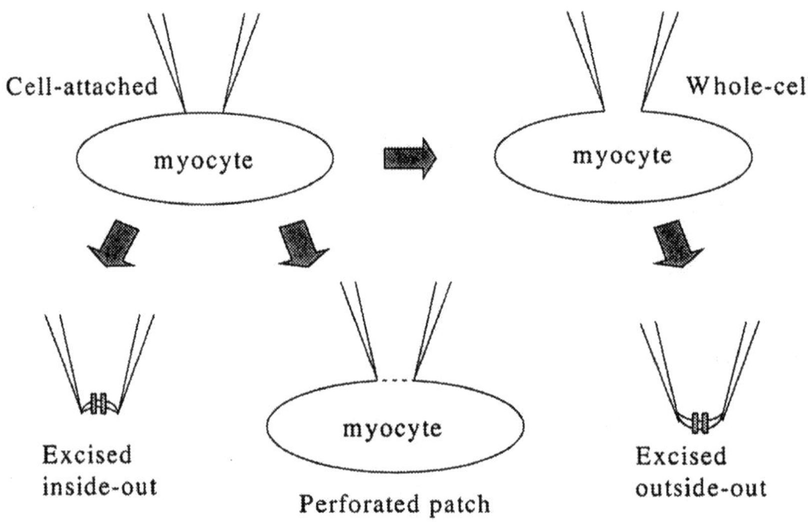

Figure 1. Different configurations of patch-clamp techniques. All methods start with a clean patch pipette forming a gigaohm seal against an intact cell. Channel activities underneath the patch pipette can be recorded in this cell-attached mode as minute currents passing between the pipette solution and the cytoplasm. Further suction results in rupture of the membrane patch underneath the pipette forming a whole-cell configuration. Alternatively, excised patches can be formed by isolating the small patch of membrane trapped at the tip of the pipette. Perforated patch configuration results from the inclusion of nystatin (or amphotericin) in the pipette solution (Adapted from Hille, 1992) (5).

Theory

Voltage-clamp techniques allow us to directly measure the ionic currents by eliminating the capacitative currents during the voltage-clamp step, as follows:

$$I_t = I_c + I_i; \text{ and } I_c = C_M(\partial V/\partial t)$$

where I_t, I_c and I_i are the total current through the membrane, the capacity current and the ionic current, respectively, C_M is the membrane capacitance per unit area and t is time (6). During voltage-clamp step, $\partial V/\partial t = 0$, so that the ionic current can be obtained directly from the experimental records.

A given class of ion channel discriminate among the many types of ions in the physiological milieu, allowing only those of a certain size and charge to cross the membrane, a property known as the 'selectivity' of the channel proteins. Apart from being highly selective, the channel must also allow the passage of ions at a very high throughput rate, up to a million ions per second. Furthermore, these ion channel proteins are exquisitely sensitive to changes in membrane voltages or specific ligands, the fundamental property known as the 'gating' function of the channels. Whole-cell patch-clamp techniques allow us to record currents passing through a population of channels. In contrast, single-channel recording offer us the opportunity to separate the permeation and gating properties of the channel. For a homogeneous population of statistically independent channels, the mean macroscopic current (I) is defined as: $I = Nip_{open}$, where N is the number of channels, i is the single channel current amplitude and p_{open} represents the open probability of the channel (5). As such, the two fundamental properties of the channel can be determined directly using single-channel recordings (Figure 2).

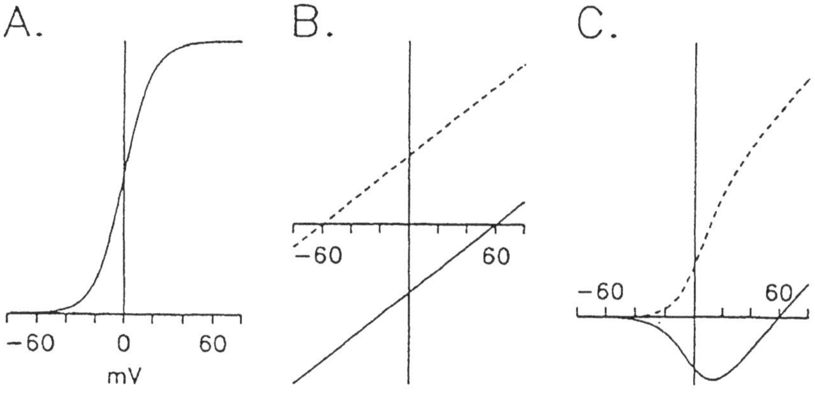

Figure 2. Current-voltage curves for hypothetical voltage-dependent currents. The vertical scales are arbitrary. A, Open probability (p_{open}) for a voltage-dependent channel, assuming the channel open probability changes e-fold for 5 mV, with half of the channels open at 0 mV. B, Current-voltage relations for single-channel opening. An Ohmic behavior is observed. C, In contrast, whole-cell current-voltage relations reflect both the gating and permeation properties of a population of channels with gating as in A and ion selectivity as in B. (Adapted from Jones, 1990) (7).

Limitations of the patch-clamp technique

Since only ionic currents are being directly measured during voltage-clamp experiments, two major impediments during whole-cell patch-clamp technique are the series resistance and space clamp artifacts. In fact, without proper compensation for series resistance and a demonstration of a good space clamp, all measurements of membrane currents are only qualitative.

Heterogenous expression of the transgene has been documented in different transgenic animals. The exact mechanisms of the variation from cell to cell in the level of expression of the transgene is not known. Therefore, caution should be exercised when using single-cell analysis. Electrophysiologic recordings need to be obtained from a relatively large number of isolated myocytes. The data will need to be averaged and statistical significant differences between the transgenic and nontransgenic animals will need to be sought. Statistical differences obtained from functional studies can be further crosschecked using biochemical and molecular biological techniques.

One other disadvantage of the patch-clamp technique is the phenomenon known as 'run-down' of the channels during the recordings whereby the activities of the channels decrease over time. For example, it is well known that whole-cell L-type Ca^{2+} channel currents rapidly run down during whole-cell patch-clamp recordings. One of the reasons for the run-down is presumably due to the equilibration of the intracellular milieu with the pipette solution and the depletion of the requisite intracellular regulatory proteins. To prevent progressive run-down, some of the cellular constituents such as Mg-ATP should be included in the intracellular solution.

THE USE OF GENETICALLY ENGINEERED MICE TO STUDY ION CHANNELS IN THE HEART

In this section, we will concentrate on three exciting areas of transgenic mouse models: β-adrenergic and G-protein signaling, Ca^{2+} cycling proteins, and genetic manipulation of cardiac ion channels. We will provide an overview of these models, and compare and contrast them with particular emphasis on new insights into ion channels modulation learned from these models.

β-adrenergic and G-protein signaling

Neurohormonal regulation of beat-to-beat cardiac function is vital to maintain a healthy heart. Indeed, one of the hallmarks of heart failure is the inability to adequately respond to sympathetic drive. With this as motivation, several laboratories have generated transgenic mouse models in which a component of the β-adrenergic receptor (β-AR) pathway has been altered (8,9,10,11,12,13,14).

Adrenergic receptor (AR) stimulation by catecholamines provides the most important regulatory mechanism for cardiovascular performance. The ARs were classified as α (for excitatory) and β (for inhibitory) (15). β-ARs could be further categorized into two distinct subtypes based on their relative potencies for the ligands (16). The cloning of cDNA for the β-ARs indicates that the receptors are members of the G protein–coupled receptor superfamily, which shares a common feature of seven-transmembrane-spanning domains (17,18).

A classical notion of cardiac β-AR signal transduction is that the agonist-bound β-AR selectively interacts with the stimulatory G protein (G_s), which activates adenylyl cyclase (AC), catalyzing cAMP formation. Activation of cAMP-dependent protein kinase A (PKA) leads to phosphorylation of regulatory proteins involved in cardiac excitation-contraction (EC) coupling and energy metabolism, including L-type Ca^{2+} channels, the sarcoplasmic reticulum (SR) protein phospholamban (PLB), myofilament proteins, and glycogen phosphorylase kinase. PKA also phosphorylates and activates an endogenous protein phosphatase inhibitor I (19,20). In addition to the acute effects, chronic β-AR stimulation affects multiple cellular functions, including gene transcription, cell growth, and death.

Although several β-AR subtypes have been cloned and pharmacologically characterized, those expressed in cardiomyocytes were initially thought to be exclusively the β_1-AR subtype. However, recent studies have provided compelling evidence that at least both β_1-AR and β_2-AR functionally coexist in cardiomyocytes of many species, including humans (21,22). Indeed, it has been shown that β-AR subtypes elicit distinct cardiac responses through different signaling pathways. Increasing evidence has demonstrated striking qualitative and quantitative differences in the functions and signaling mechanisms of the cardiac β_1-AR and β_2-AR subtypes (21).

To better understand the importance of various effectors in the pathway, transgenic mouse models with overexpression or ablation of key players in the β-AR pathway have been generated (8,9,10,11,12,13,14). A landmark study by Lefkowitz and co-workers revealed that β_2-AR overexpressing mice had increased basal adenylyl cyclase activity and increased *in vivo* cardiac function (8). Initial studies showed that Ca^{2+} cycling, as measured by SR Ca^{2+} release and uptake using fura-2 was greatly increased in isolated myocytes, consistent with increased phospholamban activation by increased β_2-AR signaling (23). In addition, this same study showed that Ca^{2+} spark frequency (the elementary Ca^{2+} release event) was increased; however, SR Ca^{2+} content and Ca^{2+} channel activity were unchanged (24). However, a second transgenic mouse model in which β_2-AR receptor was overexpressed at various levels (60, 100, 150, 350 times background) showed that at high levels of overexpression, basal L-type Ca^{2+} current is greatly downregulated in transgenic compared to non-transgenic myocytes (Figure 3) (25).

The importance of this work is born out by the exciting gene-therapy studies in rats receiving transplanted hearts (26). Donor hearts received β_2-AR recombinant adenovirus before transplantation; five days after transplantation, basal left ventricular pressure was significantly higher in treated *versus* non-treated animals (peak +dP/dt 6250 *versus* 3152 mmHg). Thus, information obtained from these β_2-AR transgenic mouse models has provided new insights into pathological processes and exciting promises for new therapy.

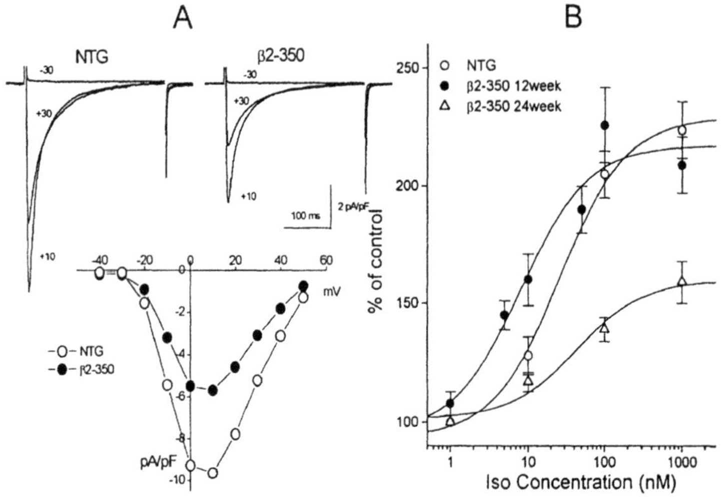

Figure 3. ß$_2$-350 cardiomyocyte electrophysiological properties. A, Representative whole-cell Ca^{2+} current recorded in nontransgenic (NTG) and ß$_2$-350–overexpressing myocytes at 24 weeks of age. B, Concentration-dependent effects of isoproterenol (Iso) on Ca^{2+} current for NTG and ß$_2$AR-overexpressing myocytes. EC_{50}s were 26.5±6.9, 8.2±4.7, and 43.0±30 nmol/L for NTG, ß$_2$-350–overexpressing mice at 12 weeks, and the same at 24 weeks, respectively. Data are mean ± S.E.M. from 6 to 40 cells (Reproduced by permission from Liggett SB, et al. *Circulation.* 2000;101:1707-1714).

Altered G-protein coupled signaling in transgenic mice

β-adrenergic signaling is far more interactive than simply activating receptors, and cascading through a defined compliment of proteins. Rather, the signaling between β_1-AR and β_2-AR is highly dynamic, with proteins in the cascade sensing and responding to a variety of inputs. While β_1-AR signaling couples exclusively to stimulatory G-proteins (G_s), β_2-AR couples to both G_s and inhibitory G-protein, G_i. Thus, transgenic animals that have altered G-protein signaling have proven to be essential in helping to parse the role of G_s and G_i in β-signaling. These studies have demonstrated that the cardiac β_1-AR and β_2-AR are markedly different not only with regard to G protein coupling, but also, cAMP handling, target protein phosphorylation, and modulation of cardiac excitation-contraction coupling (21). Many of these differences can be explained by the additional coupling of ß$_2$-AR to G_i proteins, which, in some species and developmental stages, functionally localizes the ß$_2$-AR/G_s–mediated cAMP/PKA signaling to the subsarcolemmal microdomain. This

β_2-AR/G_i pathway contributes to the distinct phenotypes of overexpression of cardiac β_2-ARs *versus* β_1-ARs in transgenic mouse models. Thus, it is necessary and important to distinguish β_2-ARs from β_1-ARs with regard to their physiological and pathophysiological roles (21).

G proteins are heterotrimers consisting of a guanine nucleotide-binding α subunit and an associated dimer of β and γ subunits (22,27,28). G proteins are classified according to the identity of their α subunits, which are broadly grouped into major classes; $G\alpha_s$, $G\alpha_i$, $G\alpha_q$, and $G\alpha_o$. Members of the $G\alpha_i$ and $G\alpha_o$ are substrates for ADP ribosylation by pertussis toxin. In contrast, $G\alpha_s$ stimulates AC and $G\alpha_q$ stimulates phospholipase C. $G\alpha_s$ and $G\alpha_i$ figure most prominently in ß-AR subtype signaling, as they couple to the stimulation or inhibition of AC enzyme activity, respectively. Historically, G protein function was attributed entirely to the actions of the freed α subunit. For $G\alpha_i$, this was accomplished largely with pertussis toxin, which catalyzes the ADP ribosylation of $G\alpha_i$, a covalent modification that blocks $G\alpha_i$-receptor interactions and thereby functionally inactivates signal transduction pathways mediated by α_i subunits. However, more recent studies establish that G protein activation is a bifurcating process, with the liberated $\beta\gamma$ dimers also controlling various effector functions (either alone or in concert with the α subunits). Additional functions of the $\beta\gamma$ dimer include directing the fidelity of G protein-coupled receptor-effector interactions and facilitating the agonist-dependent β-AR phosphorylation/desensitization process by recruiting the G protein-coupled receptor kinase (GRK or βARK) to the plasma membrane.

$G\alpha_s$

Stimulatory GTP-binding protein ($G\alpha_s$) is critical for increasing cardiac isotropy and lusitropy. To better understand which components of the β-AR signaling pathway were due to activation of this stimulatory pathway, transgenic mice which overexpress $G\alpha_s$ were generated (29). In addition to the predictable response of increased cardiac contractility, the most striking effect $G\alpha_s$ overexpression was on Ca^{2+} current, which was ~50% higher in transgenic *versus* non-transgenic myocytes. Equally noteworthy was the finding that intracellular dialysis of non-transgenic myocytes with GTP-γs also results in dramatically increased Ca^{2+} currents. This effect was independent of PKA activation, as Ca^{2+} currents can be enhanced by GTP-γs even in the presence of PKA inhibitors. The data indicate that $G\alpha_s$ overexpression is associated with a constitutive enhancement of Ca^{2+} current, which is independent of the cAMP pathway and activation of endogenous AC. Since chronic $G\alpha_s$ overexpression has been shown to lead to both hypertrophy (which progresses to cardiomyopathy) and / or apoptosis (30), detailed understanding of this pathway will elucidate many of the important proteins in these diseases.

The importance of $G\alpha_s$ was further confirmed in studies in which mice overexpressing $G\alpha_s$ were administered chronic β-adrenergic receptor blockade. Propranolol treatment was started after cardiac function was depressed and cardiomyopathy had developed in Gα_s mice. In the treated mice, propranolol treatment prevented further hypertrophy and fibrosis, while apoptosis was eliminated and premature mortality was abolished. Thus, it is clear that chronic Gα_s activation is deleterious *via* its effect on the β-AR signaling pathway (31).

Gα_o

The Gα_o protein is the most abundant G subtype in the nervous system, but it is also found in the heart. Its function is not completely known, although it is required for regulation of N-type Ca^{2+} channels. To directly analyze the function of Gα_o, transgenic mice lacking both isoforms of Gα_o by homologous recombination were generated (32). Surprisingly, the nervous system is grossly intact, despite the fact that Gα_o makes up 0.2-0.5% of brain particulate protein. In contrast, Gα_o -/- mice have a clear and specific defect in ion channel regulation in the heart. Normal muscarinic regulation of L-type Ca^{2+} channels in ventricular myocytes is absent in the mutant mice. The L-type Ca^{2+} channel responds normally to isoproterenol, but there is a lack of muscarinic inhibition. In contrast, muscarinic regulation of atrial K^+ channels is normal. These studies show that muscarinic inhibition of L-type Ca^{2+} channels requires coupling of the muscarinic receptor to Gα_o. In this instance, transgenic techniques proved to be a powerful tool in determining its role. Therefore, this model is an excellent example of the cogent use of transgenic / ablation technique to elucidate protein function.

Gα_q

GTP-binding protein Gα_q plays an important role in cardiac G-protein signaling. To determine its exact function, transgenic mice that overexpressed Gα_q were generated. Four-fold overexpression of this protein resulted in cardiac hypertrophy, increased fetal-gene expression, and decreased β-adrenergic signaling sensitivity; markers of hypertrophy / heart failure (33,34,35,36). The electrophysiological basis of the altered myocardial function was examined in left ventricular myocytes isolated from transgenic Gα_q mice (35,36). Action potential duration was significantly prolonged in Gα_q compared with nontransgenic (NTG) myocytes (Figure 4). The densities of inward rectifier K^+ currents, transient outward K^+ currents (I_{to}), and Na^+/Ca^{2+} exchange currents were reduced in Gα_q myocytes. Consistent with functional measurements, Na^+/Ca^{2+} exchanger gene expression was reduced in Gα_q hearts. Kinetics or sensitivity of I_{to} to 4-aminopyridine was unchanged, but 4-aminopyridine prolonged the action potential more in Gα_q myocytes. Isoproterenol increased L-type Ca^{2+} currents (I_{Ca}) in both groups, with a similar EC_{50}, but the maximal response in Gα_q myocytes was ~24% of that in NTG myocytes. In NTG myocytes, the maximal increase of I_{Ca} with isoproterenol or forskolin was similar. In Gα_q myocytes, forskolin was more effective and enhanced I_{Ca} up to ~55% of that in NTG myocytes. These results indicate that changes in ionic currents and multiple defects in the β-AR/Ca^{2+} channel signaling pathway contribute to altered ventricular function in this model of cardiac hypertrophy.

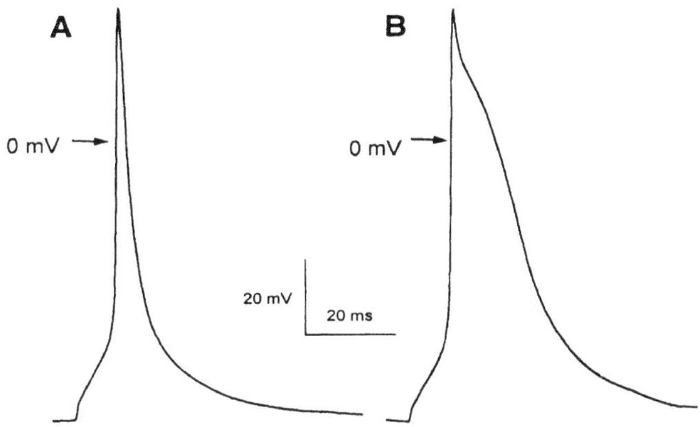

Figure 4. Typical action potentials recorded in ventricular myocytes isolated from hearts of nontransgenic (NTG, A) and Gα_q (B) mice. Membrane potential was recorded in the current-clamp mode with using a patch electrode filled with K^+-rich internal solution. The external solution was normal Tyrode solution. Myocytes were stimulated at 0.2 Hz *via* the patch pipette (Reproduced by permission from Mitarai S, et al. *Am J Physiol.* 2000;279: H139-H148).

Alterations in Ca^{2+} handling proteins

Na^+-Ca^{2+}-Exchanger

Na^+-Ca^{2+} exchanger (NCX) can work to remove Ca^{2+} from the cell (forward mode) or bring Ca^{2+} into the cell (reverse mode). However, the role of the NCX remains controversial, especially on a beat-to-beat basis during various pathological conditions. Transgenic mouse models that overexpress NCX (NCX-OE) have helped to elucidate the role of NCX in contributing to the Ca^{2+} transient and determining SR Ca^{2+} load (37,38). Surprisingly, 9-fold overexpression of NCX did not result in altered twitch Ca^{2+} transient or SR Ca^{2+} stores, as determined by application of caffeine. This suggests that neither increased forward or reverse mode NCX activity effects beat-to-beat Ca^{2+} handling or long-term Ca^{2+} storage. Further, it was shown using NCX-OE mouse model that, in the presence of nifedipine to inactivate the L-type Ca^{2+} channel, Ca^{2+} influx via NCX is sufficient to trigger SR Ca^{2+} release (39). The exact role of NCX on cardiac action potential profile was studied using NCX-OE model showing an increase in action potential plateau and prolonged action potential duration in mice overexpressing NCX compared to wild-type mice (39).

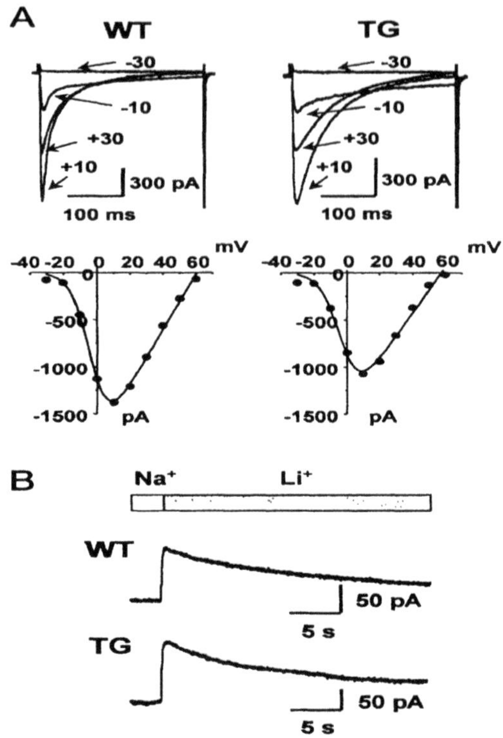

Figure 5. Ca^{2+} current and Na^+-Ca^{2+} exchanger current in wild-type and calsequestrin-overexpressing transgenic ventricular myocytes. A: upper panels, typical whole-cell L-type Ca^{2+} currents recorded in wild-type (WT) and transgenic (TG) cells. Shown are traces of currents that were recorded from a holding potential of -50 mV to the indicated test potentials. Lower panels, the peak current-voltage relationships obtained from these cells. The extracellular Ca^{2+} concentration was 2 mM. B: Na^+-Ca^{2+} exchanger currents measured from wild-type and transgenic cells. The Na^+-Ca^{2+} exchanger current was estimated as the outward current when the external solution was changed from 150 mM Na^+ to 150 mM Li^+. The traces show continuous recording at a holding potential of -40 mV. Extracellular and intracellular Ca^{2+} concentrations were adjusted to 1 and 67 nM, respectively. (Reproduced by permission from Sato Y, et al., *J Biol Chem* 1998;273:28470-28477).

Calsequestrin

Calsequestrin is a high-capacity, high affinity, Ca^{2+} binding protein that functions to chelate and store Ca^{2+} inside the SR. It forms a complex with junctin, triadin, and the ryanodine receptor. Transgenic mice that overexpress calsequestrin show a significant increase in SR Ca^{2+} storage capacity, but a marked decrease in excitation-contraction coupling gain (amount of SR Ca^{2+} release for Ca^{2+} influx *via* the L-type channel) than wild-type mice (40,41,42,43). Interestingly, proteins involved in Ca^{2+}-release, ryanodine receptor, junctin, and triadin, were downregulated, whereas those involved in Ca^{2+}-uptake, SERCA and phospholamban, were unchanged or slightly increased. Compensation among Ca^{2+} handling proteins is incomplete, as these mice develop dilated cardiomyopathy and die prematurely. Specifically, overexpression of calsequestrin was associated with cardiac hypertrophy and induction of a fetal gene expression program. Isolated transgenic cardiomyocytes exhibited diminished shortening fraction, shortening rate, and relengthening rate. The Ca^{2+} transient amplitude was also depressed, although the SR Ca^{2+} storage capacity was augmented, as suggested by caffeine application studies. These alterations were associated with a decrease in L-type Ca^{2+} current density. A significant prolongation of the inactivation kinetics can be appreciated and is likely due to a decrease in the Ca^{2+}-dependent inactivation. There were no detectable changes in Na^+-Ca^{2+} exchanger current density (Figure 5). Single-channel analysis showed no detectable changes in single-channel conductances (Figure 6). Left ventricular function analysis in Langendorff perfused hearts and closed-chest anesthetized mice also indicated depressed rates of contraction and relaxation of transgenic hearts. These findings suggest that calsequestrin overexpression is associated with increases in SR Ca^{2+} capacity, but a decrease in Ca^{2+}-induced SR Ca^{2+} release, leading to depressed contractility in the mammalian heart.

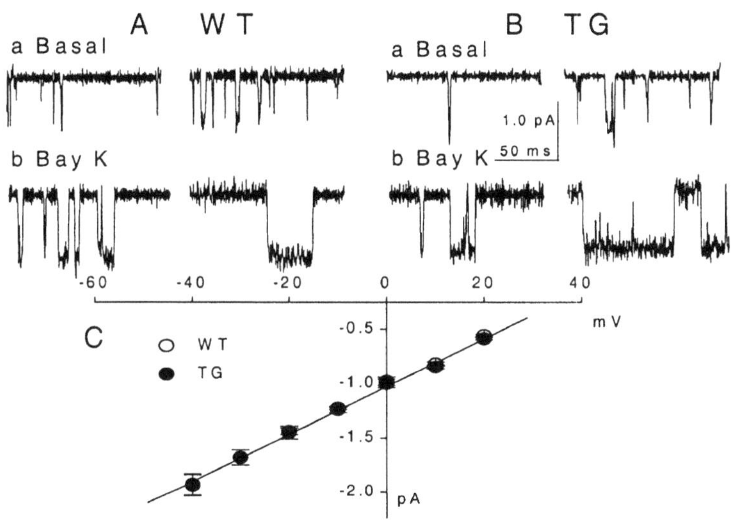

Figure 6. Single-channel currents recorded from cell-attached patches from myocytes isolated from wild-type (WT) *versus* calsequestrin overexpressed (TG) mice. A and B: Typical single-channel Ba^{2+} currents elicited by depolarizing pulses to 0 mV from a holding potential of -80 mV in the absence (a) and in the presence of

1μM Bay K 8644 (b). C: Current-voltage relationships of the single-channel currents. The solid line is a linear regression plot with the slope conductance of 23 pS. Data represent mean ± SE from 5-6 experiments with 200-3000 events.

Phospholamban

Phospholamban is a critical regulator of Ca^{2+}-reuptake, by virtue of its regulatory interaction with SERCA pump, as well as a key effector in the β-adrenergic signaling pathway. Indeed, much has been learned from both overexpression and ablation of this protein (44,45). Phospholamban (PLB) ablation in transgenic mice results in hypercontractility, increased Ca^{2+} transients, increased SR Ca^{2+} stores, and decreased RyR levels. These mice are healthy and reproduce well. Whole-cell Ca^{2+} channel density is unchanged, however, time course of inactivation is hastened due to an increase in the Ca^{2+}-dependent inactivation (46). The results further demonstrate the feedback interaction between SR and L-type Ca^{2+} channel. This enhanced contractility has been used to improve cardiac function, decrease pathology, and improve longevity when cross-bred with other mouse models of cardiac hypertrophy (47).

Sarcoplasmic reticulum Ca^{2+} ATPase (SERCA)

SERCA is a major component of the beat-to-beat Ca^{2+} cycling proteins in the heart, and a critical regulator of Ca^{2+} removal from the cytosol. There are two predominant isoforms that function in cells in which large amounts of Ca^{2+} are transported: SERCA2a, the endogenous cardiac isoform, or SERCA1a, a pump with faster kinetics than 2a. Both pumps have been overexpressed in the heart with similar phenotypes: enhanced Ca^{2+} transport with a concomitant boost in contractility (48,49,50). Similarly, adenoviral-mediated gene transfer into adult myocytes demonstrated an enhanced rate of shortening and relaxation, with an increased rate of Ca^{2+} uptake and release (51,52,53). Because SERCA1a has faster kinetics than 2a, and it can be expressed at higher levels than 2a, it is an attractive candidate for possible gene therapy. Recent study of SERCA1a overexpressing mice showed expression and function of other Ca^{2+} handling proteins were also altered. SR Ca^{2+} content was increased 2-fold, while RyR levels were decreased ~40%. Most striking was the finding that the L-type Ca^{2+} current amplitude was decreased over 50%, while the time course of inactivation remained unchanged (unpublished observation). This was an unexpected result, as decreased current amplitude is normally associated with prolonged channel inactivation time. This is because Ca^{2+} itself inactivates the channel. It is not currently known what role SR Ca^{2+} plays in channel inactivation, but it is believed that Ca^{2+} that enters *via* the channel and Ca^{2+} that is released from the SR both contribute. Thus, this mouse model provides an exciting paradigm in which to study altered steady-state levels of several Ca^{2+} handling proteins, and the result on excitation-contraction coupling.

Compensatory Changes in Genetically Manipulated Animal Models

Much information was gained in recent years using different transgenic mouse models. One common finding among different models was compensatory changes in the animals in response to the targeted genetic manipulation. One such example was recently demonstrated in a mouse model carrying a null mutation in one copy of the SERCA2 isoform gene, in which SERCA2 protein levels are reduced by ~35% (54). When compared with wild-type controls, SR Ca^{2+} stores and Ca^{2+} release in myocytes of SERCA2 heterozygous mice were decreased by ~40-60% and ~30-40%, respectively, and the rate of myocyte shortening and relengthening were each decreased by ~40%. However, the rate of Ca^{2+} transient decline was not altered significantly, suggesting that compensation was occurring in the removal of Ca^{2+} from the cytosol. Phospholamban, which inhibits SERCA2, was decreased by ~40% in heterozygous hearts, and basal phosphorylation of Ser-16 and Thr-17, which relieves the inhibition, was increased ~2 fold. These results indicate that reduced expression and increased phosphorylation of phospholamban provides compensation for decreased SERCA2 protein levels in heterozygous heart. Furthermore, both expression and current density of the sarcolemmal Na^{+}-Ca^{2+} exchanger were up-regulated (Figure 7). These results demonstrate that a decrease in SERCA2 levels can directly modify intracellular Ca^{2+} homeostasis and myocyte contractility. However, the resulting deficit is partially compensated by alterations in phospholamban/SERCA2 interactions and by up-regulation of the Na^{+}-Ca^{2+} exchanger. In fact, in many transgenic models, a specific or even a global remodeling process can be accompanied by the upregulation or downregulation of a gene or set of genes. Although these compensatory changes might be beneficial for the animal's survival, they introduce complications in understanding their phenotypes. Thus, caution should be exercised when drawing a direct mechanistic link between genetic manipulations and phenotypes.

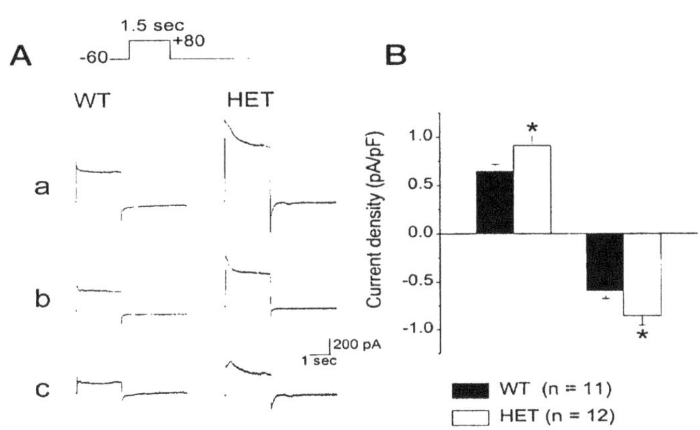

Figure 7. Na^{+}-Ca^{2+} exchanger currents in wild-type and heterozygous mice. A, representative tracings of Na^{+}-Ca^{2+} exchange currents from myocytes during 1.5-s depolarizing pulses from a holding potential of -60 mV to +80 mV. The Ni^{2+}-sensitive currents (c) were obtained by subtracting current records obtained during exposure to 5 mM Ni^{2+} (b) from the control record (a). B, Na^{+}-Ca^{2+} exchanger current density (normalized to the cell capacitance) shows a significant increase in

the sustained outward current and peak tail current density of the Ni^{2+}-sensitive current from six heterozygous (HET) mice as compared with five wild-type (WT) mice. One to three cells were selected from each heart, "n" in parentheses represents the numbers of cells in wild-type and heterozygous hearts. Data represent mean ± S.E. *, $p < 0.05$ (Reproduced by permission from Ji Y, et al. *J Biol Chem.* 2000;275: 38073-38080).

Dominant Negative Functional Knockout of Cardiac Ion Channels

Ionic fluxes across the membrane are the foundation of the action and membrane potentials. Changes in the ionic flux balance can lead to alterations in the cardiac action potential profile and resulting pathology. Indeed, mutations in a variety of ion channels have now been linked to hereditary diseases in humans including congenital long QT syndrome (55,56,57,58,59), Brugada syndrome (60,61,62,63) and sudden cardiac death. In addition, it is often difficult to determine which gene(s) underlies a particular ionic current in the heart. This is especially the case with K^+ currents, as a plethora of channel families generates a multitude of currents. K^+ channels consist of four identical pore-forming subunits (α subunits) that coassemble to form a functional channel together with smaller β subunits; this is in contrast to the Na^+ and Ca^{2+} channels, which are composed of one pore-forming subunit (α subunit) that has a tetrameric-repeat structure associated with smaller β subunits (64,65). This unique four pore-forming subunit design for K^+ channels can be utilized to generate a functional knockout of various K^+ channels, without using gene ablation. Subunits can be generated that contain mutations in their pore forming region that render them incompetent to conduct current. The resulting subunit (dominant negative), when coassembles with wild-type subunits, results in non-functional channels. Thus, overexpression of dominant negative subunits results in a functional knockout of K^+ current. This functional knockout strategy can be utilized to study the function of a particular ion channel gene in the generation of the native currents. Thus, the combined techniques of genetic manipulation of targeted genes and electrophysiologic study offer a powerful tool for determining the molecular correlates of different underlying ionic currents; information that is crucial in the design of antiarrhythmic drugs as well as the understanding and diagnosis of different congenital and acquired cardiac arrhythmias.

K^+ Channels

Myocardial K^+ currents function to control resting membrane potentials, the heights and durations of action potentials, and refractoriness and automaticity. They are important targets for the actions of transmitters, hormones or drugs known, or postulated, to modulate cardiac function. A variety of K^+ currents that subserve these functions have now been identified in myocardial cells isolated from different species, as well as in cells isolated from different regions of the heart (66,67,68). These currents include the various voltage-gated K^+ channels, such as the transient outward (I_{to}) and delayed rectifier (I_{Kr}, I_{Ks}) currents, as well as the inwardly rectifying currents, I_{K1}, $I_{K(ACh)}$, and $I_{K(ATP)}$. Advances have been made in cloning K^+ channel subunits of the various voltage-gated (K_v, eag, I_{sK}) and inward rectifier (K_{ir}) families (69). The relationships between the various cloned K^+ channel subunits and the functional K^+ channels characterized electrophysiologically in myocardial cells have recently been explored using transgenic mouse models (70,71,72,73).

The electrophysiology and biophysical properties of the ionic currents expressed in the mouse heart are not fully characterized. The outward current of adult mouse ventricular myocytes has been noted to have an incomplete inactivation phase characterized by a sum of a pedestal and two exponentials with a fast time constant similar to that of classic I_{to} and a slow time constant. Detailed analysis of the outward currents in mouse ventricular myocytes revealed at least four distinct K^+ current components: a rapidly inactivating, transient outward K^+ current, $I_{to,f}$; a slowly inactivating, transient outward K^+ current, $I_{to,s}$; a slowly inactivating delayed rectifier K^+ current, $I_{K,slow}$; and a non inactivating, steady state current, I_{ss} (74). Furthermore, significant regional heterogeneity in the expression has been documented. For example, it was observed that all cells isolated from the apex express $I_{to,f}$; $I_{to,s}$ is not detected in these cells. In contrast, all of the cells isolated from septum express $I_{to,s}$ and in the majority of cells, $I_{to,f}$ is also present. The density of $I_{to,f}$ in septum cells, however, is significantly lower than that in cells isolated from the apex. Moreover, $I_{to,f}$, $I_{to,s}$, and $I_{K,slow}$ display differential sensitivities to 4-AP, tetraethylammonium (TEA), and Heteropoda toxin-3 (74). $I_{K,slow}$ can be blocked selectively by low (10-50 μM) concentrations of 4-AP and by TEA. Although both $I_{to,f}$ and $I_{to,s}$ are blocked by high concentrations of 4-AP (> 100 μM) and are relatively insensitive to TEA, $I_{to,f}$ can be selectively blocked by nanomolar concentrations of Heteropoda toxin-3, and $I_{to,s}$ (as well as $I_{K,slow}$ and I_{ss}) is unaffected. I_{ss} is partially blocked by high concentrations of 4-AP or TEA.

K_v1

Transgenic mouse model has been utilized to study the molecular correlate of $I_{K,slow}$. A transgenic mouse model with a long-QT phenotype (LQT mouse) by overexpressing a truncated K_v1.1 channel gene that contains the N-terminus and first transmembrane segment of K_v1.1, has recently been generated (71). Electrophysiological study of the myocytes isolated from LQT mice, compared with controls, revealed prolonged APD due to a significant reduction in the density of a rapidly activating, slowly inactivating, and highly 4-AP-sensitive outward K^+ current ($I_{K,slow}$). These studies provided new evidence that the slowly inactivating component of the outward current of mouse myocytes is likely encoded by K_v1 family.

K_v2

Using similar *in vivo* experimental strategy, a mutant K_v2.1 subunit that functions as a dominant negative was exploited in studies focused on exploring the role of members of the K_v2 subfamily of pore-forming subunits in the generation of functional voltage-gated K^+ channels (72). A mutant K_v2.1 subunit (K_v2.1N216) was designed to produce a truncated protein containing the intracellular N terminus, the S1 membrane-spanning domain, and a portion of the S1/S2 loop. Using the myosin heavy chain promoter to direct cardiac specific expression of the transgene, K_v2.1N216-expressing transgenic mice were generated. Electrophysiological recordings from ventricular myocytes isolated from these animals revealed that $I_{K,slow}$

is selectively reduced. The attenuation of $I_{K,slow}$ is accompanied by marked action potential prolongation, and, occasionally, spontaneous triggered activity was observed. In addition, unlike $I_{K,slow}$ in wild-type cells, the component of $I_{K,slow}$ remaining in the mutant-expressing cells was insensitive to 25 mM TEA. *In vivo* telemetric recordings also reveal marked QT prolongation, consistent with a defect in ventricular repolarization in mutant-expressing transgenic mice. Taken together, these observations suggest that there are two distinct components of $I_{K,slow}$ in mouse ventricular myocytes and that K_v2 subunits underlie the more slowly inactivating, TEA-sensitive component of $I_{K,slow}$.

HERG

Two K^+ channel genes, K_vLQT and human ether-a-go-go or HERG have been identified as loci of mutations in familial long-QT syndromes (55,56,57). HERG underlies I_{Kr} (the rapid I_K), whereas K_vLQT1 contributes to I_{Ks} (the slow I_K). Mutations in the HERG gene are linked to the LQT2 form of the inherited long-QT syndrome.

Using information obtained from these hereditary diseases, transgenic mice were generated expressing high myocardial levels of a particularly severe form of LQT2-associated HERG mutation (G628S) (73). Hearts from G628S mice appeared normal except for a modest enlargement seen only in females. Ventricular myocytes isolated from adult wild-type hearts consistently exhibited an inwardly rectifying E-4031–sensitive K^+ current resembling the rapidly activating cardiac delayed rectifier K^+ current (I_{Kr}) in its time and voltage dependence; this current was not found in cells isolated from G628S mice. Action potential duration was significantly prolonged in single myocytes from G628S ventricle. ECG intervals, including QT duration, were unchanged, although minor aberrancies were noted in the G628S mice, primarily involving the QRS complex and, more rarely, T-wave morphology. The aberrations were more commonly observed in females than males. These results establish the presence of I_{Kr} in the adult mouse ventricle and demonstrate the ability of the G628S mutation to exert a dominant negative effect on endogenous I_{Kr} *in vivo*, leading to the LQT2 phenotype of prolonged repolarization at the single cell level but not QT prolongation in the intact animal.

SUMMARY

The application of single cell patch-clamp techniques to genetically engineered mice provides a powerful combination of molecular biology and cardiovascular ion channel physiology and biophysics. Creating animals in which modified genes or the cognate of cDNAs are placed into the genome and the encoded proteins expressed in an organ-specific manner offers a unique opportunity to study the *in vivo* structure-function relation of a particular ion channel gene of interest.

REFERENCES

1. Neher E, Sakmann B. Single channel current recorded from membrane of denervated frog muscle fibres. *Nature.* 1976;260:799-802.
2. Hamill OP, Marty A, Neher E, Sakmann B, Sigworth FJ. Improved patch clamp techniques for high-resolution current recording from cells and cell-free membrane patches. *Pflügers Archiv.* 1981;391:85-100.
3. Korn SJ, Marty A, Connor JA, Horn R. Perforated patch recording. *Methods in Neurosci.* 1991;4:264-373.
4. Hilgemann DW. Giant excised cardiac sarcolemmal membrane patches: sodium and sodium-calcium exchange currents. *Pflügers Archiv.* 1989;415:247-249.
5. Hille B. *Ionic channels of excitable membrane.* Sinauer Associates Inc., MA, 1992.
6. Hodgkin AL, Huxley AF, Katz B. Measurement of current-voltage relations in the membrane of the giant axon of Loligo. J Physiol. 1952;116:424-448.
7. Jones SW. Whole-cell and microelectrode voltage clamp. From Neuromethods, Vol. 14: *Neurophysiological Techniques: Basic Methods and Concepts.* Edited by: Boulton AA, Baker GB, Vanderwolf CH. The humana Press Inc., NJ, 1990.
8. Milano CA, Allen LF, Rockman HA, Dolber PC, McMinn TR, Chien KR, Johnson TD, Bond RA, Lefkowitz RJ. Enhanced myocardial function in transgenic mice overexpressing the β_2-adrenergic receptor. *Science.* 1994;264:582–586.
9. Koch WJ, Rockman HA, Samama P, Hamilton R, Bond RA, Milano CA, Lefkowitz RJ. Cardiac function in mice overexpressing the ß-adrenergic receptor kinase or a ßARK inhibitor. *Science.* 1995;268:1350–1353.
10. Bond RA, Leff P, Johnson TD, Milano CA, Rockman HA, McMinn TR, Apparsundaram S, Hyek MF, Kenakin TP, Allen LF. Physiological effects of inverse agonists in transgenic mice with myocardial overexpression of the β_2-adrenoceptor. *Nature.* 1995;374:272–276.
11. Engelhardt S, Hein L, Wiesmann F, Lohse MJ. Progressive hypertrophy and heart failure in ß1-adrenergic receptor transgenic mice. *Proc Natl Acad Sci U S A.* 1999;96:7059–7064.
12. Rohrer DK, Desai KH, Jasper JR, Stevens ME, Regula DP, Barsh GS, Bernstein D, Kobilka BK. Targeted disruption of the mouse β_1-adrenergic receptor gene: developmental and cardiovascular effects. *Proc Natl Acad Sci U S A.* 1996;93:7375–7380.
13. Chruscinski AJ, Rohrer DK, Schauble E, Desai KH, Bernstein D, Kobilka BK. Targeted disruption of the β_2-adrenergic receptor gene. *J Biol Chem.* 1999;274:16694–16700.
14. Rohrer DK, Chruscinski AJ, Schauble E, Bernstein D, Kobilka BK. Cardiovascular and metabolic alterations in mice lacking both β_1- and β_2-adrenergic receptors. *J Biol Chem.* 1999;274:16701–16708.
15. Ahlquist RP. A study of the adrenotropic receptors. *Am J Physiol.* 1948;153:586–600.

16. Lands AM, Arnold A, Mcauliff JP, Luduena FP, Brown TG. Differentiation of receptor systems activated by sympathomimetic amines. *Nature.* 1967;214:596–598.
17. Dixon RA, Kobilka BK, Strader DJ, Benovic JL, Dohlman HG, Frielle T, Bolanowski MA, Bennett CD, Rands E, Diehl RE. Cloning of the gene and cDNA for mammalian ß-adrenergic receptor and homology with rhodopsin. *Nature.* 1986;321:75–79.
18. Yarden Y, Rodriguez H, Wong SK, Brandt DR, May DC, Burnier J, Harkins RN, Chen EY, Ramachandran J, Ullrich A. The avian ß-adrenergic receptor: primary structure and membrane topology. *Proc Natl Acad Sci U S A.* 1986;83:6795–6799.
19. Gupta RC, Neumann J, Watanabe AM, Lesch M, Sabbah HN. Evidence for presence and hormonal regulation of protein phosphatase inhibitor-1 in ventricular cardiomyocytes. *Am J Physiol.* 1996;270:H1159—H1164.
20. Neumann J, Gupta RC, Schmitz W, Scholz H, Nairn AC, Watanabe AM. Evidence for isoproterenol induced phosphorylation of phosphatase inhibitor-1 in the intact heart. *Circ Res.* 1991;69:1450–1457.
21. Xiao RP, Cheng HP, Zhou YY, Kuschel M, Lakatta EG. Recent Advances in Cardiac ß2-Adrenergic Signal Transduction. *Circ Res.* 1999;85:1092-1100.
22. Steinberg SF. The Molecular Basis for Distinct ß-Adrenergic Receptor Subtype Actions in Cardiomyocytes. *Circ Res.* 1999;85:1101-1111.
23. Zhou YY, Song LS, Lakatta EG, Xiao RP, Cheng H. Constitutive beta-adrenergic signaling enhances sarcoplamsic reticulum Ca cycling to augment contraction in mouse heart. *J Physiol.* 1999;521:351-361.
24. Zhou YY, Cheng H, Song LS, Lakata WD, Xiao RP. Spontaneous β(2)-adrenergic signaling fails to modulate L-type Ca^{2+} current in mouse ventricular myocytes. *Mol Pharmacol.* 1999;56:485-493.
25. Liggett SB, Tepe NM, Lorenz JN, Canning AM, Jantz TD, Mitarai S, Yatani A, Dorn, II GW. Early and Delayed Consequences of ß2-Adrenergic Receptor Overexpression in Mouse Hearts: Critical Role for Expression Level. *Circulation.* 2000;101:1707-1714.
26. Kypson A, Hendrickson S, Akhter S, Wilson K, McDonald P, Lilly R, Dolber P, Glower D, Lefkowitz R, Koch W. Adenovirus-mediated gene transfer of the beta2-adrenergic receptor to donor hearts enhances cardiac function. *Gene Ther.* 1999;6:1298-1304.
27. Hamm HE. The many faces of G protein signaling. *J Biol Chem.* 1998;273:669–672.
28. Neer EJ. Heterotrimeric G proteins: organizers of transmembrane signaling. *Cell.* 1995;80:249–257.
29. Lader AS, Xiao YF, Ishikawa Y, Cui Y, Vatner DE, Vatner SF, Homcy CJ, Cantiello HF. Cardiac $G\alpha_s$ overexpression enhances L-type calcium channels through an adenylyl cyclase independent pathway. *Proc Natl Acad Sci U S A.* 1998;95:9669-9674.

30. Geng YJ, Ishikawa Y, Vatner DE, Wagner TE, Bishop SP, Vatner SF, Homcy CJ. Apoptosis of cardiac myocytes in Gsalpha transgenic mice. *Circ Res.* 1999;84:34-42.
31. Asai K, Yang GP, Geng YJ, Takagi G, Bishop S, Ishikawa Y, Shannon RP, Wagner TE, Vatner DE, Homcy CJ, Vatner SF. Beta-adrevergic receptor blockade arrests myocyte damage and preserves cardiac function in the transgenic G(salpha) mouse. *J Clin Invest.* 1999;104:551-558.
32. Balenzuela D, Han X, Mende U, Fankhauser C, Mashimo H, Huang P, feffer J, Neer EJ, Fishman MC. Galpha-o is necessary for muscarinic regulation of Ca channels in mouse heart. *Proc Natl Acad Sci U S A.* 1997;94:1727-1732.
33. D'Angelo DD, Sakata Y, Lorenz JN, Boivin GP, Walsh RA, Liggett SB, Dorn GW. Transgenic $G\alpha_q$ overexpression induces cardiac contractile failure in mice. *Proc Natl Acad Sci U S A.* 1997;94:8121-8126.
34. Adams JW, Sakata Y, Davis MG, Sah VP, Wang Y, Liggett SB, Chien KR, Brown JH, Dorn II GW. Enhanced $G\alpha_q$ signaling: A common pathway mediates cardiac hypertrophy and apoptotic heart failure. *Proc Natl Acad Sci U S A.* 1998;95:10140-10145.
35. Yatani A, Frank K, Sako H, Kranias EG, Dorn II GW. Cardiac-specific overexpression of $G\alpha_q$ alters excitation-contraction coupling in isolated cardiac myocytes. *J Mol Cell Cardiol* 1999;31:1327-1336.
36. Mitari S, Reed TD, Yatani A. Changes in ionic currents and beta-adrenergic receptor singaling in hypertrophied myocytes overexpressing $G\alpha_q$. *Am J Physiol:Heart.* 2000;279:H139-148.
37. Adaci-Akahane S, Lu L, Li Z, Frank JS, Philipson KD, Morad M. Calcium signaling in transgenic mice overexpressing cardiac Na^+-Ca^{2+} exchanger. *J Gen Physiol.* 1997;109:717-729.
38. Terracciano CM, Souza AI, Philipson KD, MacLeod KT. Na^+-Ca^{2+} exchange and sarcoplasmic reticular Ca^{2+} regulation in ventricular myocytes from transgenic mice overexpressing the Na^+-Ca^{2+} exchanger. *J Physiol.* 1998;512:651-667.
39. Yao A, Su Z, Nonaka A, Zubair I, Lu L, Philipson KD, Bridge JHB, Barry WH. Effects of overexpression of the Na^+-Ca^{2+} Exchanger on $[Ca^{2+}]_i$ transients in murine vantricular myocytes. *Circ Res.* 1998;82:657-65.
40. Jones LR, Suzuki YJ, Wang W, Kobayashi YM, Ramesh V, Franzini-Armstrong C, Cleemann L, Morad M. Regulation of Ca^{2+} signaling in transgenic mouse cardiac myocytes overexpressing calsequestrin. *J Clin Invest.* 1998;101:1385-93.
41. Cho MC, Rapacciuolo A, Koch WJ, Kobayashi Y, Jones LR, Rockman HA. Defective beta-adrenergic receptor signaling precedes the development of dilated carediomyopathy in transgenic mice with calsequestrin overexpression. *J Biol Chem.* 1999;274:22251-22256.
42. Wang W, Cleemann L, Jones LR, Morad M. Modulation of focal and global Ca^{2+} release in calsequestrin overexpressing mouse cardiomyocytes. *J Physiol.* 2000;524:399-414.

43. Sato Y, Ferguson DG, Sako H, Dorn GW, Kadambi VJ, Yatani A, Hoit BD, Walsh RA, Kranias EG. Cardiac-specific overexpression of mouse cardiac calsequestrin is associated with depressed cardiovascular function and hypertrophy in transgenic mice. *J Biol Chem* 1998;273:28470-28477.
44. Luo W, Grupp IL, Harrer J, Ponniah S, Grupp G, Duffy JJ, Doetschman T, Kranias EG. Targeted ablation of the phospholamban gene is associated with markedly enhanced myocardial contractility and loss of β-agonist stimulation. *Circ Res.* 1994;75:401-409.
45. Kadambi VJ, Ponniah S, Harrer JM, Hoit BD, Dorn GW, Walsh RA, Kranias EG. Cardiac-specific overexpression of phospholamban alters calcium kinetics and resultant cardiomyocyte mechanics in transgenic mice. *J Clin Invest.* 1996;97:533-539.
46. Masaki H, Satao Y, Luo W, Kranias EG, Yatani A. Phospholamban defieiency alters inactivation kinectics of L-type Ca^{2+} channel in mouse ventricular myocytes. *Am J Physiol.* 1997;272:H606-612.
47. Minamisawa S, Hoshijima M, Chu G, Ward CA, Frank K, Gu Y, Martone ME, Wang Y, Ross J Jr., Kranias EG, Giles WR, Chien KR. Chronic phospholamban-sarcoplasmic reticulum calcium ATPase interaction is the critical calcium cycling defect in dilated cardiomyopathy. *Cell.* 1999;99:313-322.
48. Baker DL, Hashimoto K, Grupp IL, Ji Y, Reed T, Loukianov E, Grupp G, Bhagwhat A, Hoit B, Walsh R, Marban E, Periasamy M. Targeted overexpression of the sarcoplasmic reticulum Ca^{2+}-ATPase increases cardiac contractility in transgenic mouse hearts. *Circ Res.* 1998;83:1205-1214.
49. Loukianov E, Ji Y, Grupp IL, Kirkpatrick DL, Baker DL, Loukianova T, Grupp G, Lytton J, Walsh RA, Periasamy M. Enhanced myocardial contractility and increased Ca^{2+} transport function in transgenic hearts expressing the fast-twitch skeletal muscle sarcoplasmic reticulum Ca^{2+}-ATPase. *Circ Res.* 1998;83:889-897.
50. Ji Y, Loukianov E, Loukianova T, Jones LR, Periasamy M. SERCA1a can functionally substitute for SERCA2a in the heart. *Am J Physiol.* 1999;276:H89-97.
51. Inesi G, Wade R, Rogers T. The sarcoplasmic reticulum Ca^{2+} pump: inhibition by thapsigargin and enhancement by adenovirus-mediated gene transfer. *Ann N Y Acad Sci.* 1998;853:195-206.
52. Schmidt U, del Monte F, Miyamoto MI, Matsui T, Gwathmey JK, Rosenzweig A, Hajjar RJ. Restoration of diastolic function in senescent rat hearts through adenoviral gene transfer of sarcoplasmic reticulum Ca(2+)-ATPase. *Circulation.* 2000;101:790-796.
53. Miyamoto MI, del Monte F, Schmidt U, DiSalvo TS, Kang ZB, Matsui T, Guerrero JL, Gwathmey JK, Rosenzweig A, Hajjar RJ. Adenoviral gene transfer of SERCA2a improves left-ventricular function in aortic-banded rats in transition to heart failure. *Proc Natl Acad Sci U S A.* 2000;97:793-798.
54. Ji Y, Lalli MJ, Babu GJ, Xu Y, Kirkpatrick DL, Liu LH, Chiamvimonvat N, Walsh RA, Shull GE, Periasamy M. Disruption of a Single Copy of the SERCA2

Gene Results in Altered Ca^{2+} Homeostasis and Cardiomyocyte Function *J Biol Chem.* 2000;275: 38073-38080.
55. Wang Q, Curran ME, Splawski I, Burn TC, Millholland JM, VanRaay TJ, Shen J, Timothy KW, Vincent GM, de Jager T, Schwartz PJ, Towbin JA, Moss AJ, Atkinson DL, Landes GM, Connors TD, Keating MT. Positional cloning of a novel potassium channel gene: *KVLQT1* mutations cause cardiac arrhythmias. *Nature Gen.* 1996;12:17-23.
56. Curran ME, Splawski I, Timothy KW, Green ED, Vincent GM, Keating MT: A molecular basis for cardiac arrhythmia: HERG mutations cause long QT syndrome. *Cell.* 1995;80:795-803.
57. Sanguinetti MC, Jiang C, Curran ME, Keating MT. A mechanistic link between an inherited and an acquired cardiac arrhythmia: HERG encodes the I_{Kr} potassium channel. *Cell.* 1995;81:299-307.
58. Wang Q, Shen J, Splawski I, Atkinson D, Li Z, Robinson JL, Moss AJ, Towbin JA, Keating MT. *SCN5A* mutations associated with an inherited cardiac arrhythmia, long QT syndrome. *Cell.* 1995;80:805-811.
59. Abbott GW, Sesti F, Splawski I, Buck ME, Lehmann MH, Timothy KW, Keating MT, Goldstein SAN. MiRP1 Forms IKr Potassium Channels with HERG and Is Associated with Cardiac Arrhythmia. *Cell.* 1999;97:175-187.
60. Veldkamp MW, Viswanathan PC, Bezzina C, Baartscheer A, Wilde AA, Balser JR. Two distinct congenital arrhythmias evoked by a multidysfunctional Na^+ channel. *Circ Res.* 2000;86:E91-E97.
61. Deschenes I, Baroudi G, Berthet M, Barde I, Chalvidan T, Denjoy I, Guicheney P, Chahine M Electrophysiological characterization of SCN5A mutations causing long QT (E1784K) and Brugada (R1512W and R1432G) syndromes. *Cardiovasc Res.* 2000;46:55-65.
62. Rook MB, Alshinawi CB, Groenewegen WA, van Gelder IC, van Ginneken AC, Jongsma HJ, Mannens MM, Wilde AA. Human SCN5A gene mutations alter cardiac sodium channel kinetics and are associated with the Brugada syndrome. *Cardiovasc Res* 1999;44:507-517.
63. Bezzina C, Veldkamp MW, van Den Berg MP, Postma AV, Rook MB, Viersma JW, van Langen IM, Tan-Sindhunata G, Bink-Boelkens MT, van Der Hout AH, Mannens MM, Wilde AA. A single Na^+ channel mutation causing both long-QT and Brugada syndromes. *Circ Res.* 1999;85:1206-1213.
64. Armstrong CM, Hille B. Voltage-gated ion channels and electrical excitability. *Neuron.* 1998;20:371-380.
65. Catterall WA. From ionic currents to molecular mechanisms: the structure and function of voltage-gated sodium channels. *Neuron.* 2000;26:13-25.
66. Barry DM, Nerbonne JM. Myocardial potassium channels: electrophysiological and molecular diversity. *Annu Rev Physiol* 1996;58:363-594.
67. Nerbonne JM. Molecular basis of functional voltage-gated K^+ channel diversity in the mammalian myocardium. *J Physiol.* 2000;525:285-298.

68. Anumonowo JMB, Freeman LC, Kwok WM, Kass RS. Potassium channels in the heart: electrophysiology and pharmacological regulation. *Cardiovasc Drug Rev.* 1991;9:299-316.
69. Pongs O. Molecular biology of voltage-dependent potassium channels. *Physiol Rev.* 1992;72:S69-S88.
70. Barry DM, Xu H, Schuessler RB, Nerbonne JM. Functional knockout of the transient outward current, long-QT syndrome, and cardiac remodeling in mice expressing a dominant-negative Kv4 alpha subunit. *Circ Res.* 1998;83:560-567.
71. London B, Jeron A, Zhou J, Buckett P, Han X, Mitchell GF, Koren G. Long QT and ventricular arrhythmias in transgenic mice expressing the N terminus and first transmembrane segment of a voltage-gated potassium channel. *Proc Natl Acad Sci USA.* 95:2926-2931, 1998.
72. Xu H, Barry DM, Li H, Brunet S, Guo W, Nerbonne JM. Attenuation of the slow component of delayed rectification, action potential prolongation, and triggered activity in mice expressin a dominant-negative K_v2-alpha subunit. *Circ Res.* 1999;85:623-633.
73. Babij P, Askew GR, Nieuwenhuijsen B, Su CM, Bridal TR, Jow B, Argentieri TM, Kulik J, DeGennaro LJ, Spinelli W, Colatsky TJ. Inhibition of cardiac delayed rectifier K^+ current by overexpression of the long-QT syndrome HERG G628S mutation in transgenic mice. *Circ Res.* 1998;83:668-678.
74. Xu H, Guo W, Nerbonne JM. Four Kinetically Distinct Depolarization-activated K^+ Currents in Adult Mouse Ventricular Myocytes. *J Gen Physiol.* 1999;113:661-678.

8
THE ISOLATED ISOVOLUMIC (LANGENDORFF) HEART PREPARATION

Martin M. LeWinter, M.D.
University of Vermont, Burlington, VT 05401 USA

INTRODUCTION

Isolated, Langendorff-perfused, isovolumic preparations have been employed for many years to study cardiac function. In order to achieve isovolumic conditions a compliant balloon filled with a known volume of fluid is inserted into the left ventricle (LV). These preparations have a number of advantages over the in situ heart for the study of cardiac mechanics. First, hemodynamic and mechanical variables can be well controlled. The balloon allows accurate measurement and independent control of ventricular preload (end-diastolic volume), while isovolumic contraction results in a constant afterload. Heart rate can also be maintained constant by electronic pacing. Second, coronary perfusion pressure or flow, which independently influence both systolic and diastolic function, is independently maintained at desired levels. Moreover, *left ventricular* coronary flow and oxygen consumption and coronary vascular resistance can be estimated. Third, independent perfusion eliminates variable concentrations of circulating catecholamines and other substances, including anesthetic agents, which can influence cardiac performance. The fact that the heart is denervated also eliminates the effects of variations in direct autonomic nervous system stimulation.

MEASUREMENTS OF VENTRICULAR PERFORMANCE IN THE ISOVOLUMIC PREPARATION

The advantages discussed above allow assessment of ventricular function under much more controlled conditions than are possible in the in situ heart. Thus, the influence of preload on contractile function (the Frank-Starling relation) as well as the fully relaxed or end-diastolic pressure-volume relation can be assessed without confounding, uncontrolled changes in neurohumoral activation or coronary perfusion that inevitably occur in in situ preparations. A straightforward way to assess ventricular function in the context of the Frank-Starling relationship is to relate end-diastolic volume to developed LV pressure or peak positive dP/dt (1).

Isolated, perfused, isovolumically contracting preparations are especially well-suited for delineation of end-systolic elastance, a commonly used, relatively load-independent index of ventricular function. This index dates to the early 1970s, when Suga, Sagawa and colleagues (2,3) developed the time varying elastance model of ventricular contraction. The model was based on the observation that as ventricular volume and loading are varied under constant contractility conditions in both isovolumic and ejecting cross-perfused canine hearts, at any time *t* following the onset of contraction the relation between pressure and volume is linear and becomes progressively steeper until end-systole, when it reaches a maximum. Thus, the ventricle behaves like a spring whose elastance relationship (or stiffness) increases during contraction and decreases during relaxation. *End-systolic* elastance, or Emax, the slope of the end-systolic pressure-volume relationship (ESPVR), was observed to increase or decrease as a function of acute changes in contractility without a change in the volume axis intercept (Vd), and did not seem to vary as a function of loading conditions (Figure1). Hence, Emax behaves as a "load-independent" index of ventricular contractility.

The time-varying elastance model has subsequently undergone modification in several ways. Emax does have a modest degree of load dependence, most likely due to the load dependence of activation. Moreover, the end-systolic elastance relation is not as linear as originally thought. As contractility is altered above or below typical baseline conditions in large mammals, significant curvilinearity often becomes evident. Curvilinearity is also more apparent when the relation is obtained in in situ preparations. In rodents, the baseline relation obtained in isolated, perfused preparations is normally curvilinear. As a result, rather than quantifying the elastance relationship by fitting the ESPVR to a straight line, the slope of the relationship is estimated using curvilinear regression statistics (see below) and is termed E'max.

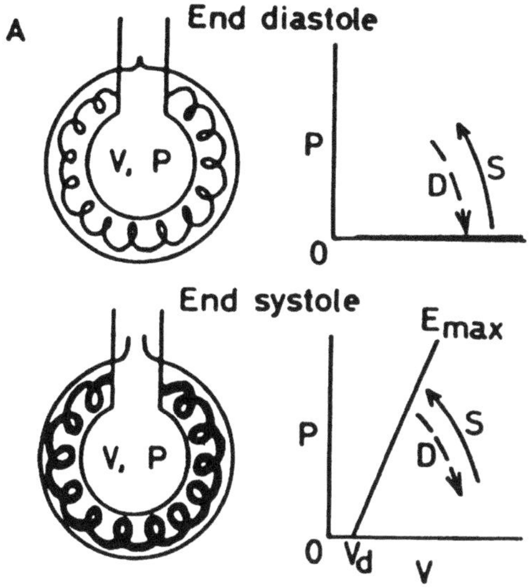

Figure 1. The time varying elastance concept. At left, the ventricle represented as an elastic "spring" whose stiffness increases during systole and decreases during diastole. At right, the relation between pressure and volume at any time t during variably loaded contractions under constant contractility conditions is linear and reaches a maximum value at end-systole (Emax). V, volume; P, pressure; Vd, dead volume; S, systole; D, diastole. Reprinted with permission from Suga, Am J Physiol, 1979.

Suga extended the elastance concept to the study of ventricular mechanoenergetics by proposing LV *pressure-volume area* (PVA) as a measure of the *total* mechanical energy of LV contraction (4,5). The PVA concept is shown schematically in Figure 2, top. The total mechanical energy of contraction can be considered to consist of two components, external work, in pressure-volume terms the area enclosed within the pressure-volume loop, and potential energy. The latter is largely converted to heat during relaxation. The potential energy stored in the ventricular "spring" at end-systole is theoretically equal to the area under its elastance relationship to the left of the end-systolic pressure and above its relaxed or end-diastolic pressure-volume relationship. A good way to conceive of the two components of mechanical energy is to consider the two extremes of possible loading conditions during contraction. With unloaded conditions, all mechanical energy is converted to external work. In contrast, during isovolumic conditions no external work is performed, yet it is intuitively obvious that mechanical energy has been generated, in this case all of it in the form of potential energy. Contractions between these two extremes have variable ratios of external work to potential energy.

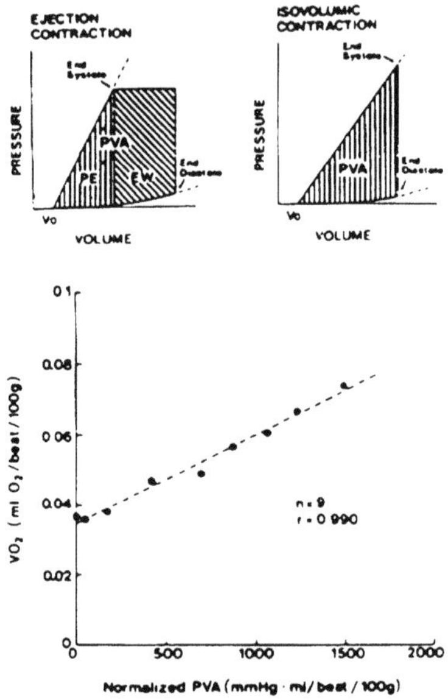

Figure 2. Top, schematic representation of pressure-volume area (PVA) for ejection and isovolumic contractions. Bottom, relation between VO2 and PVA in the isovolumic rabbit left ventricle. (See text.) Reprinted with permission from Goto et al, Am J Physiol, 1988.

The unique feature of the PVA concept is its ability to quantify the potential energy component of mechanical energy and thus estimate *total* mechanical energy by addition of this component to external work. The accuracy of PVA as a measure of total mechanical energy is borne out by the fact that there is a remarkably tight linear correlation (r values routinely > .98) between LV VO2 per beat and PVA per beat over a wide range of experimental and contraction conditions in several different species (Figure 2, bottom) (5-7). The VO2-PVA relation is obtained during the same load change protocols used to delineate the ESPVR by combining pressure-volume data with measurements of LV VO2 made at each steady-state.

The VO2-PVA approach to mechanoenergetics has proven to be remarkably robust, and a source of mechanistic insights into the behavior of both the contractile machinery and the excitation-contraction coupling system (5,8). To understand this it is useful to dissect the components of the VO2-PVA relationship, as shown in Figure 3. The VO2 intercept of the VO2-PVA relationship is the unloaded VO2 (also termed PVA-independent VO2). In an isovolumically contracting preparation, this corresponds to a point at which LV volume is reduced such that the LV peak pressure is 0 mmHg. Theoretically, at this point there is no mechanical energy production and therefore no energy expenditure for crossbridge cycling (in fact, there is some low level of crossbridge cycling under these conditions because the heart changes its shape with each beat). The VO2 under unloaded conditions should largely reflect two processes, excitation-contraction coupling (mainly energy expended for pumping of calcium ions) and basal metabolism, and indeed there is experimental evidence that this is the case. Since it is possible to measure basal metabolism by arresting the heart (e.g., with KCl) while maintaining perfusion, the energy expenditure for excitation-contraction coupling can be quantified. Both acute and chronic changes in excitation-contraction coupling energy consumption have been detected as shifts in the unloaded VO2 (1,4).

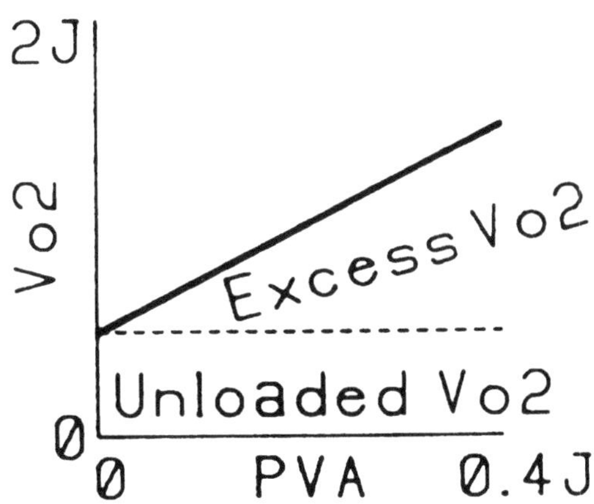

Figure 3. The VO2-PVA relation divided into unloaded or PVA-independent and "excess" or PVA-dependent components. (See text.) Reprinted with permission from Suga et al, Circ Res, 1988.

Oxygen consumption over and above the unloaded value ("excess" or PVA-dependent VO2) is used by the contractile machinery for crossbridge cycling. As PVA increases, PVA-dependent VO2 increases in a linear fashion, consistent with the fixed stoichiometry of energy utilization for crossbridge cycling considered present under most or all conditions (i.e., one ATP split per cycle). Since PVA-dependent VO2 can be converted to terms of energy (joules) and constitutes the energy input of the contractile machinery and PVA can be considered its total energy output, the ratio of PVA per beat to PVA-dependent VO2 per beat, or simply the inverse slope of the VO2-PVA relationship, is a dimensionless measure of the thermodynamic efficiency of the contractile machinery (3,4). This formulation contrasts with the more conventional use of external work divided by total VO2 as an expression of efficiency. The latter is strongly load-dependent, whereas efficiency in terms of the VO2-PVA relationship is virtually insensitive to load, as would be expected if PVA truly represents total mechanical energy output. Thus, the slope of the VO2-PVA relation should vary with manipulations that influence the efficiency of energy utilization for crossbridge cycling. This is borne out by experimental observations in which efficiency has been shown to change as an inverse function of myofibrillar ATPase activity (1,9). It is important to keep in my mind that efficiency reflected by the slope of the VO2-PVA relationship consists of two sequential steps, conversion of oxygen to high energy phosphate bonds followed by conversion of chemical energy liberated by hydrolysis of high energy phosphates to mechanical energy. Thus, metabolic alterations that influence the relative utilization of glycolytic versus aerobic pathways can also influence the VO2-PVA relationship.

The isolated, perfused, isovolumic preparation is ideally-suited for elastance based measurements of contractility and mechanoenergetics because Emax can be assessed accurately and, most important, LV VO2 can be measured under changing experimental conditions. The latter is problematic in intact, in situ preparations because it requires cannulation of the coronary sinus to measure coronary arteriovenous O2 content difference and it is very difficult to measure the LV venous O2 content because of contamination by venous blood coming from other cardiac chambers.

PERFUSION OF THE ISOVOLUMIC PREPARATION: GENERAL CONSIDERATIONS

In small mammals, perfusate is delivered by a roller pump via cannulation of the aortic root after rapid removal of the heart from the chest. It is critically important to minimize the time the heart is not perfused. However, in larger animals perfusion can be accomplished with whole blood via cross-perfusion from an intact donor animal (2,3,5,7). This is obviously the most physiologic perfusate, but this method exposes the heart to variable levels of catecholamines and anesthetic agents from the donor animal. The technical difficulty of cross-perfusion is considerable in small hearts, and this technique has not been reported to date in animals smaller than rabbits. In smaller animals, perfusion is usually accomplished by delivering oxygenated, buffered electrolyte solutions containing glucose as the metabolic substrate to the coronary circulation (1,7). (It is possible to include fatty acids in the perfusate but this in turn necessitates addition of albumin to eliminate foaming. In our experience in rabbit and rat hearts, inclusion of fatty acids and albumin does not improve performance or preparation stability sufficiently to warrant their routine use.) Perfusion can also be accomplished with red blood cells suspended in buffer solutions (10). In our experience, these preparations function quite well but use of red blood cell suspensions is frequently limited by increases in coronary resistance that typically appear after 90-120 minutes of perfusion. Moreover, in order to avoid recirculation and associated hemolysis, large volumes of red blood cells are required. This consideration mandates the use of red cells obtained from large animals such as cows (cow red blood cells have the advantage of being relatively small in volume).

Using the roller pump, either coronary perfusion pressure or flow can be manipulated, but the investigator must pick one of these as the independent and the other as the dependent variable. Most commonly, pressure is maintained constant at some predetermined, physiologic value (e.g., 80-90 mmHg) and flow is allowed to vary during an experimental intervention. It is always important to keep in mind the possibility of induction of ischemia if flow decreases. For some protocols during which changes in coronary vascular resistance are anticipated, it may be more sensible to maintain coronary flow constant by altering the perfusion pressure. Obviously, the isovolumically contracting, independently perfused heart is an excellent preparation for studies of the effects of changes in global perfusion pressure or flow under controlled conditions.

DISADVANTAGES OF THE ISOVOLUMIC PREPARATION

As with any choice of an experimental preparation, there are disadvantages that must be kept in mind. Most important is the fact that the trauma of removal of the heart from the chest and manipulation required to attach it to the perfusion apparatus as well as the use of non-phyiologic perfusate undoubtedly results in intrinsic depression of cardiac performance and metabolic stress. It is very difficult to quantify the extent to which intrinsic contractile performance is actually depressed because a certain amount of this effect is related to the absence of neurohumoral stimulation. Second, even in the most expert hands, these preparations are time-limited and typically show significant degeneration after two to three hours of perfusion. Third, it is obvious from the foregoing that these preparations are not suitable for assessing responses that are critically dependent on an intact peripheral circulation and neurohumoral axes.

METHODOLOGIC CONSIDERATIONS IN THE MOUSE HEART

Isolated, perfused, isovolumically contracting preparations have previously been employed in animals ranging in size from the dog to the rat (1,5,6). As with virtually all methods used to study function of the mouse heart, its small size represents a technical challenge that mandates some variations from methods developed in larger species. However, the modifications required for the mouse heart are relatively straightforward and have been described in detail (7). At the same time, it must be stated that successful preparations demand considerable manual dexterity. The preparation is well-suited to assess the end-diastolic pressure-volume relationship, the influence of preload on ventricular performance, the ESPVR, and mechanoenergetics using the VO2-PVA relation.

Isolated heart preparation

Mice are anesthetized with a 1:5 diluted mixture of ketamine hydrochloride (100mg/kg) and xylazine (5 mg/kg i.p.). Heparin (5,000 IU/kg i.p.) is administered at the same time. A tracheotomy is then performed, the lungs are ventilated with a small animal respirator, and the chest is opened at the midline of the sternum. The pulmonary hili and superior vena cava are then ligated simultaneously and the heart is removed from the chest and immediately submerged in oxygenated, 2,3-

butanedione monoxime (BDM) (30mM)-Ringer solution at room temperature. Under a microscope, the severed end of the aorta is cannulated and perfused with warmed perfusate (35-37°C, see below for composition) via a 20-gauge needle. The right atrium (RA) is widely opened and a flexible tube inserted into the right ventricle (RV) via the RA. RA and inferior vena cava are then ligated and the left atrium is opened widely. A collapsed, thin, high-density polyethylene balloon mounted on a 22-gauge polyethylene (see below for details) catheter is placed in the LV via the mitral orifice. The pressure-volume relationship of the balloon is measured before each experiment. We require that the pressure in the balloon be zero up to a volume ≥0.03 ml. A 2.5-Fr micromanometer catheter (Millar Instruments, Houston, TX) is introduced just above the mitral orifice via a sideport and the balloon is connected to a 100-μl graduated syringe (Hamilton, Reno, NV) another port. Pacing electrodes connected to an electronic stimulator (model S9, Grass Instruments, Quincy, MA) are attached to the LV apex. The RV is maintained collapsed by continuous hydrostatic drainage to minimize RV VO2. The heart is then placed in a chamber with a heating jacket and temperature is maintained at 35-37C°. Coronary flow is measured by timed collection of the RV drainage. We measure coronary arteriovenous O2 content difference continuously with a dual channel platinum electrode system (Instech, Plymouth Meeting, PA). The electrode is zeroed at the beginning of each experiment using sodium dithionate, which extracts O2 from solution. The gain of the electrode is calibrated using perfusate saturated with 100% O2. O2 content is calculated from the known value of O2 solubility (2.3 vol/100ml) in the perfusate solution.

Perfusate preparation

Standard perfusate is composed of (in mM) 108.0 NaCl, 4.0 KCl, 1.4 KH_2PO_4, 25.0 $NaHCO_3$, 11.0 dextrose, 10.0 sodium pyruvate, and 2.5 $CaCl_2$. The solution is equilibrated with 95% O_2-5% CO_2, warmed to 37°C, and pH is adjusted to 7.35-7.45 by changing CO_2. A variable flow pump is used to transport perfusate to the perfusion tubing. Perfusion pressure is controlled by a pressurized arterial reservoir connected to a pressure regulator and compressed air. Perfusate temperature is maintained at 35-37°C with water jackets around the container and the pressurized arterial reservoir by constant-temperature circulators. Perfusate is not recirculated.

Experimental protocols

For most protocols, hearts are stimulated electrically at 240 beats/min by LV pacing and 30 minutes are allowed for stabilization and washout of BDM once perfusion has been established. For delineation of the relation between end-diastolic volume and variables of interest as well as the ESPVR and VO2-PVA relationships, measurements of LV pressure, coronary perfusion pressure and flow and coronary arteriovenous O2 content difference are made during steady-state conditions at various balloon volumes from 0 ml to a maximum of 0.03-0.04 ml in 0.002 ml steps using a manual micrometer syringe driver (Stoelting, Wood Dale, IL). Two minutes are allowed to pass after each volume change to allow conditions to stabilize. We term this a volume run.

Because of the size of the 20-gauge perfusion cannula there is a small pressure drop across it. The actual pressure drop (<10 mmHg) depends on coronary flow and is directly measured and subtracted from the proximal perfusion pressure to obtain the actual coronary perfusion pressure. RV and LV weight are measured at the end of each experiment.

LV balloon and frequency response of the system

The small size of the mouse heart necessitates special considerations pertaining to the LV balloon. We employ high density polyethylene obtained from commonly available plastic shopping bags. A small piece of this material is stretched with the round tip of a small centrifuge tube and gently mounted on a 22-gauge polyethylene intravenous catheter (SERFLO, Terumo, Japan). The length of the balloon is ~5 mm. The volume of the balloon wall plus the tip of the tubing within the balloon is determined by water displacement after loading a known volume of fluid within the balloon. The balloon is checked for airtightness by measuring the volume change required to overinflate it to ~150 mmHg. After two hours of overinflation, removal of the same volume that was instilled indicates an airtight system.

An important consideration in measurement of LV volume in balloon in ventricle preparations is the residual volume between the balloon and the internal surface of the chamber. It is usually assumed that this is small, but it is critical to verify that this is the case. Accordingly, we have determined the residual volume between the balloon and the LV inner surface in formalin fixed mouse hearts at various intraventricular pressures using methods similar to those employed previously for canine and rabbit hearts (7). Pressures (mmHg) and corresponding average residual volumes as a % of total volume were as follows: 90 mmHg, 4.5%; 40 mmHg, 8.5%; 15 mmHg, 19.5%; 5 mmHg, 35.5%. These errors are large compared to larger hearts. However, they have relatively modest average effects on the ESPVR (14% underestimate of E'max, 16% underestimate of Vd) and small effects on the VO2-PVA relation (4% slope overestimate, 0.2% VO2 intercept overestimate).

The 2.5-Fr pressure transducer is too large to pass through the 22-gauge transducer introduced into the LV balloon. Therefore, the sensing tip is positioned ~1 cm from the balloon. The frequency response of the system configured in this way has been tested using previously published methods. The damping coefficient is .619, near the optimal value of .64, and the natural frequency is high (65.3 Hz), resulting in a flat frequency response to 57.5 Hz. As a result, the system is applicable for measurement of pressures up to heart rates of 5.8 Hz (assuming the physiologically significant signal is contained within the first 10 harmonics of the original wave) and for dP/dt up to heart rates of 2.8 Hz (up to the first 20 harmonics of the original wave). Thus, it must be kept in mind that dP/dt measurements have borderline accuracy at heart rates of 240 beats/min.

QUANTIFICATION OF END-SYSTOLIC PRESSURE-VOLUME AND VO2-PVA RELATIONS IN THE ISOVOLUMIC MOUSE HEART PREPARATION

Examples of LV pressure measured at varying LV volumes are shown in Figure 4. End-systolic pressure is taken as peak systolic pressure. LV max dP/dt averages ~1700 mmHg/s in this preparation under baseline conditions. This value is considerably lower than that reported in intact mouse hearts (7). This is likely related to some combination of intrinsically depressed contractility, absence of adrenergic stimulation and lower heart rate. In addition, as described earlier, the frequency response of the system is borderline for dP/dt measurements.

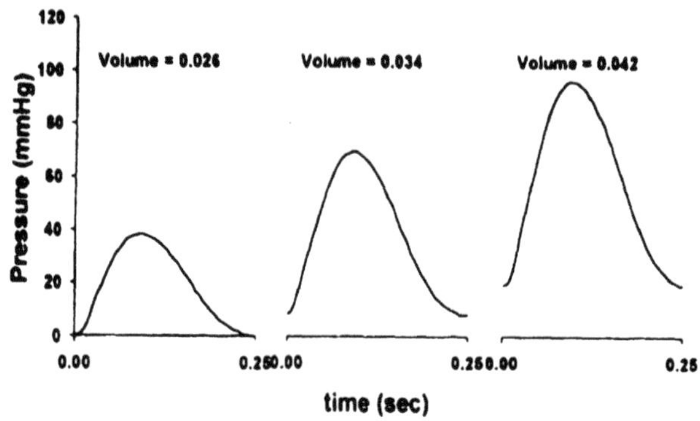

Figure 4. Pressure recordings at three different volumes (ml) in the isovolumic mouse heart (see text). Reprinted with permission from Kameyama et al, Am J Physiol, 1998.

Examples of end-systolic (closed circles) and end-diastolic (open triangles) pressure-volume relationships obtained during volume runs are shown in Figure 5, left panel. Also shown is the pressure-volume relationship of the balloon itself (crosses). As expected, the end-diastolic relation is curvilinear. It is characteristically much steeper than that observed in large mammals, consistent with recent data demonstrating that passive stiffness of rodent myocardium is much larger than that of large mammals. The baseline ESPVR is also almost invariably significantly curvilinear. Accordingly, its slope is quantified using a nonlinear regression analysis. However, decreases in contractility usually decrease the curvilinearity of the ESPVR.

For an isovolumic preparation, PVA is calculated for each steady-state condition as the area circumscribed by the ESPVR, the end-diastolic pressure-volume relation and the systolic pressure-volume trajectory. RV VO2 is calculated by subtracting the RV fraction of VO2 present at zero PVA (based on RV/LV weight). Since the RV is vented at all times, this value is subtracted from total VO2 to calculate LV VO2 for all conditions. PVA is then plotted as a function of LV VO2 per beat. As shown in Figure 5, right panel, the relation is highly linear and of the form VO2 = a × PVA + b, where a represents the oxygen cost of PVA and b is the VO2 intercept (unloaded VO2). Contractile efficiency for conversion of VO2 to PVA is the reciprocal of the slope (1/a). The value of the VO2 intercept in the mouse heart (corrected for LV weight) is greater than that reported in

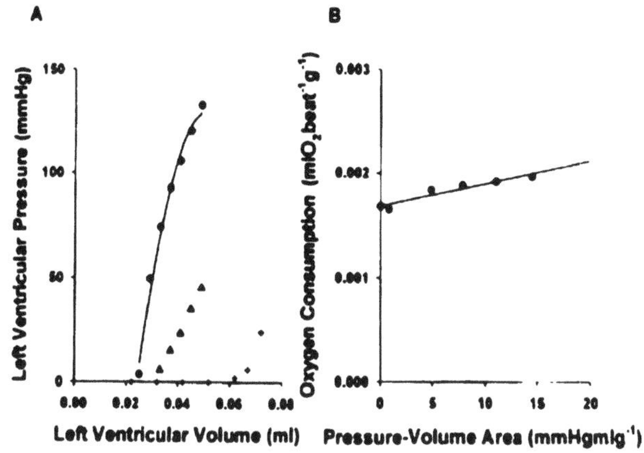

Figure 5. Left panel, end-systolic (filled circles) and end-diastolic (open triangles) pressure-volume relations in mouse heart. Crosses indicate balloon pressure-volume relation. Right panel, VO2-PVA relation. (See text.) Reprinted with permission from Kameyama et al, Am J Physiol, 1998.

larger hearts. We speculate that this may be due to the relatively low heart rate employed in this preparation *in relation to the* normal *rate of the mouse heart.* Since VO2 is estimated per beat, the contribution of basal metabolism per beat should be larger compared with other species that have lower normal rates. Contractile efficiency of the mouse heart averages ~.21, lower than the range of .30-.40 typically observed in rats, rabbits and dogs. This may be related to intrinsic differences in contractile proteins and crossbridge cycling kinetics in the mouse heart

An example of changes in the ESPVR and VO2-PVA relation occurring in response to decreasing perfusate calcium concentration from the normal value of 2.5 mM to 1.5 mM is shown in Figure 6. As expected, the ESPVR (left panel) is shifted rightward and is less curvilinear at the lower calcium concentration, and the VO2-PVA relation demonstrates a parallel downward shift (right panel), the latter consistent with decreased O2 consumption for excitation-contraction coupling. We have also shown that chronic changes in contractility and mechanoenergetics induced by hypothyroidism can be detected using this preparation.

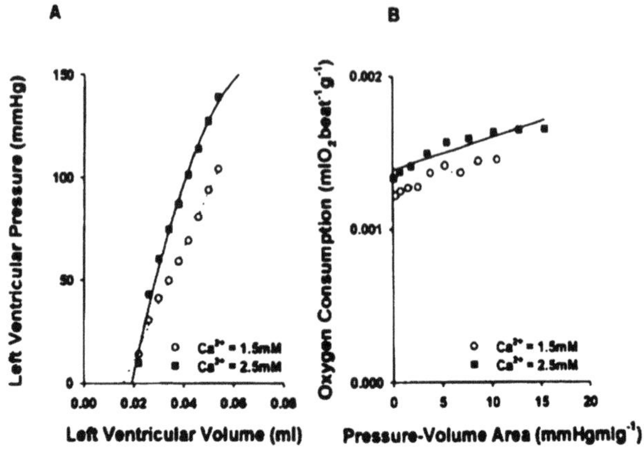

Figure 6. Left, the ESPVR of the mouse heart at normal (solid squares) and reduced (open circles) perfusate calcium concentration. Right, the VO2-PVA relation at normal and reduced perfusate calcium concentration. (See text.) Reprinted with permission from Kameyama et al, Am J Physiol, 1998.

CONCLUSIONS

Although the isovolumic, perfused mouse heart requires considerable dexterity to achieve success because of its small size, with careful attention to detail it is possible to make all of the measurements of ventricular function reported in larger isovolumic preparations. Assessment of mechanoenergetics using the VO2-PVA framework is a particular strength because of the possibility of delineating changes in energy consumption for excitation-contraction coupling separately from changes in crossbridge cycling energy consumption. Thus, estimation of energy used for excitation-contraction coupling should provide insights in models in which calcium handling is altered while estimation of contractile efficiency may be of particular interest in genetically engineered animals simulating hypertrophic cardiomyopathy or situations in which the kinetics of crossbridge cycling may be altered, for example, changes in protein kinase C activity.

REFERENCES

1. Kameyama T, Chen Z, Bell SP, Maughan D, Van Buren P, LeWinter, MM. Mechanoenergetic alterations during the transition from cardiac hypertrophy to failure in Dahl salt sensitive rats. Circulation 1998; 98:2919-2929.
2. Sagawa K. The end-systolic pressure-volume relationship of the ventricle: definition, modification and clinical use. Circulation 1981; 63:1223-1227.
3. Kass D, Maughan DL. From "Emax" to pressure-volume relations: a broader view. Circulation 1988; 77:1203-1212.
4. Suga H. Total mechanical energy of a ventricle model and cardiac oxygen Consumption. Am J Physiol 1979; 236:H498-H505.
5. Suga H. Ventricular mechanoenergetics. Physiol Rev 1990; 70:247-277.
6. Goto Y, Slinker BK, LeWinter MM: Similar normalized Emax and O consumption-pressure volume area relation in rabbit and dog. Am J Physiol 1988; 255:H366-H374.
7. Kameyama T, Chen Z, Bell SP, Fabian J, LeWinter MM: Mechanoenergetic studies in isolated mouse hearts. Am J Physiol 1998; 274:H366-H374.
8. Suga H, Goto Y, Igarashi Y, Yasumura Y, Nozawa T, Futaki S, Tanaka N. Cardiac cooling increases Emax without affecting relation between O2 consumption and systolic pressure-volume area in dog left ventricle. Circ Res 1988; 63:61-71.
9. Goto Y, Slinker BK, LeWinter MM: Decreased contractile efficiency and increased nonmechanical energy cost in hyperthyroid rabbit heart. Circ Res 1990; 66:999-1011.
10. Yaku H, Slinker BK, Myhre ESP, Watkins MW, LeWinter MM: Stability of myocardial O consumption-pressure-volume area relation in red cell-perfused rabbit heart. Am J Physiol 1991; 61:H1630-H1635.

9

THE ISOLATED WORK-PERFORMING AND EJECTING MOUSE HEART PREPARATION

COMPARISON AND QUANTIFICATION OF CARDIAC PERFORMANCE IN TRANSGENIC AND WILD-TYPE MICE

Ingrid L. Grupp, M.D., Jo El Schultz, Ph.D., George Sfyris, and Gunter Grupp, M.D. Ph.D.

Departments of Pharmacology and Cell Biophysics and Molecular and Cellular Physiology, College of Medicine University of Cincinnati

1. INTRODUCTION

Over the past ten years an abundance of transgenic mouse models (gene ablations, overexpressions, or mutations) with variable consequences on heart function have become available. Consequently, quantitative methods had to be developed to analyze cardiac function in the mouse heart. Since the mouse heart is so small (heart weight approximately 125-180 mg) new types of instrumentation for the heart produce new challenges for the cardiovascular and cardiac physiologist. We developed the buffer perfused work-performing mouse heart preparation to analyze myocardial function independent from autonomic or hemodynamic feedback or other compensation. The resultant preparation permits exact control of preload and afterload of the heart. The ejecting heart allows the evaluation of all phases of the intraventricular pressure curve, (systolic, diastolic, and end-diastolic pressure). Furthermore, in order to cover the widest possible range of contractile parameters, ±dP/dt, time to peak pressure (TPP), and ½ time to relaxation (RT ½) must be recorded. In order to indeed provide reliable quantitative myocardial contractile comparison non-transgenic (NTG) and transgenically (TG) altered littermates have to be directly compared in every case under exactly identical loading conditions. The ideal comparison should be made by the construction of cardiac function curves based upon this pressure and volume loading.

2. DESCRIPTION OF THE PERFUSION APPARATUS

The basic perfusion system of the isolated preloaded heart was originally described by Neely *et.al* (1). We modified it for hearts of rabbits, guinea pigs, hamsters and rats (2,3,4,5), and now it has also been adapted to the mouse heart (6,7,8,9) (Figure 1). Others have recently described similar methods adopted to the mouse heart physiology (10,11,12).

2.1. Two Major Components of the System:

2.1.1 The Langendorff or Retrograde Component

This part is only used in the beginning of the preparation for the work-performing heart studies. Here the hearts are perfused in retrograde fashion for about 5 minutes to flush the coronary system and to adjust the temperature to 37.5°C. It is also used to provide a recovery period when the heart has been overloaded and begins to fail. The Langendorff preparation is extensively covered in chapter 8.

2.1.2 Work-Performing Component

In the working mode the heart can be preloaded by varying venous return and afterloaded by varying aortic pressure. Analysis of the normal, ejecting heart allows for evaluation of all components of intraventricular pressure development during minimal and maximal loading. Oxygen consumption can also be measured at all loading points (using for instance the Ciba-Corning 248 Blood gas analyzer). The design of the downsized apparatus for the mouse heart is shown in Figure 1.

2.2. Preparation of the Perfusion System

The perfusion system consists of three cylinders, each containing 1000 ml of Krebs-Henseleit buffer. All cylinders and tubings are water-jacketed to achieve and maintain a temperature of 37.5° in the heart. The buffer is filtered through a 45 micron Millipore filtering system and gassed with a mixture of 95% O_2 and 5% CO_2 one hour before the experiment. This mixture is continuously circulated within the cylinders by a Harvard Apparatus peristaltic perfusion pump. Temperature is monitored and recorded with a Physitemp thermometer (Model TH/8, Probe type MT-23/3) located just before the inflow cannulas on both the Langendorff and the venous return sides. It is important to keep the buffer at these constant conditions throughout the experiment. The top cylinder is mounted 78cm above the heart producing a retrograde perfusion pressure of 55mmHg. The second and third cylinders are reservoirs that continuously fill the top cylinder. The top cylinder, in turn, has an overflow line leading back to these reservoirs in order to create and maintain a constant perfusion pressure. A bubble trap with an additional overflow in the downstream line assures faster turnover of the fluids. The extra overflow is important because cardiac output of the mouse heart only ranges between 3 and 10 ml/min.

2.3. Controlling Drug Infusions, Venous Return, and Preload

The inflow line is branched off for retrograde and anteriograde perfusion ("T" on Figure 1). After this division there are two receptacles for drug infusion, one for the Langendorff mode and the other for the work-performing mode. Each receptacle can accommodate up to six infusion lines. There are two micro-infusion pumps (Harvard, Model 55-1111) for either layover of concentrations of one infusion, or for different infusions at the same time (i.e. agonist and antagonist) (see also section 7.3). Adding the infusion agents to the perfusate and not to the cylinders helps to avoid contamination of the entire buffer system.
The venous return line contains a custom made micrometer that allows precise manipulation of the inflow that produces the preload condition of the method. There is also an inline flow probe (TM1N) for a Transonic flowmeter (Dual Channel Flowmeter, Transonic Systems Inc.) to continuously monitor venous return.

2.4. Cannula Assembly

The suspension of the mouse heart on the mounting cannula assembly is shown in Figure 2. The two cannulas are fixed in a Teflon disc and can be easily moved to accommodate the exact relationship of the openings (aorta and pulmonary vein). The aorta is connected to the 20-gauge cannula of the assembly. There is a groove in the cannula 1mm away from the opening to facilitate fixation of the vessels to it. The aortic side of the cannula has a side port, which is connected to the aortic pressure transducer. The intraventricular catheter (Figure 3C) is then inserted. The catheter itself is a 7cm piece of PE 50 (Intermedic, Clay Adams #7411). The proximal end is flanged in order to anchor it against the ventricular wall (Figures 3D and 4). The distal end is connected to a longer (10 cm), more compliant silastic catheter (I.D. 1.47mm) (Figure 3B), which connects to the transducer (Figure 3A). The length and diameter of the catheters are important for the fidelity of pressure recordings (Figure 3). The intraventricular pressures compare favorably with the measurements of high fidelity Millar catheters. The cannula for the venous return side is inserted into the left pulmonary vein of the heart (Figure 4). It also has a side port to measure left atrial pressure. All pressures are measured with COBE (CDX^{TM}III) pressure transducers. Figure 4 describes a complete schematic overview of the heart in its relationship to the working preparation.

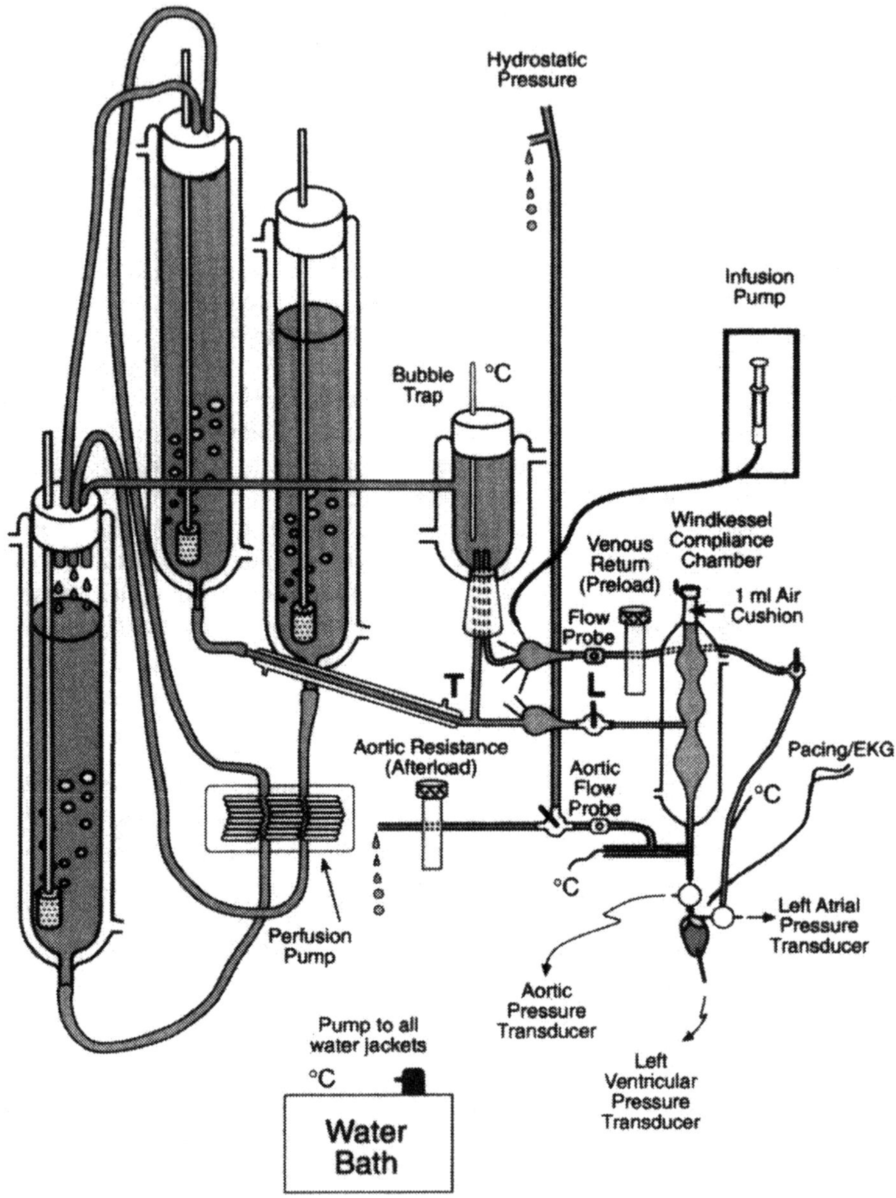

Figure 1. Work-performing mouse heart apparatus

2.5. Controlling the Afterload

In the working mode the left ventricle of the suspended heart ejects fluid through the aorta against a custom-made "windkessel" or elastic recoil (see Figure 1), which contains 1ml of air at the top that serves as a pressure cushion. This chamber is important for the maintenance of cardiac function for the duration of the experiment. It serves as a buffer against the "stiff" tubing and facilitates coronary flow, which only takes place in the work-performing heart during diastole. A sidearm of this chamber permits the aortic flow to exit through a tubing that contains the micrometer for afterload regulation and a second Transonic flow probe to measure aortic flow. It also can be switched to a constant pressure ("hydrostatic") column instead of the mechanical resistor (see Figure 1) (11). Coronary flow is either collected from the pulmonary artery and weighed or computer calculated from venous return (=cardiac output) minus aortic flow (both measured with a Transonic flowmeter).

2.6. The Electrocardiogram and Pacing

The electrocardiogram of the hanging heart is monitored continuously via 2 electrodes (silver or platinum), which are either connected to the aortic and venous return cannulas (Figure 4) or placed directly on the atrium and right ventricle. These leads are connected to an ECG preamplifier of a 6 channel Grass P7 Polygraph. The R-R intervals are simultaneously computed as heart rate. Pacing of the heart, if required, is accomplished through these electrodes using a Grass S48 Stimulator connected to a Grass SIU5 Stimulus Isolation Unit.

3. PREPARATION OF BUFFER SOLUTIONS

There are a variety of balanced buffer solutions (BBS) that can be used in the preparations (2,13). We prefer the Krebs-Henseleit solution (KHS). KHS is composed of (in mM) 118 NaCl, 4.7 KCl, 2.5 $CaCl_2$, 1.2 $MgSO_4$, 1.2 KH_2PO_4, 0.5 Na-EDTA, 25 $NaHCO_3$, and 11 glucose. Ion compositions of the BBS may vary from laboratory to laboratory. It is of importance, however, to remain consistent within studies and laboratories. Some labs may want to add pyruvate, lactate, β-hydroxybutyrate, and/or insulin in addition to glucose.
The buffer can be prepared the day before the experiment with the exception of glucose, which should be added the same day just prior to the experiment. The buffer needs to be stored overnight in the refrigerator. It is of great importance to use the purest, deionized water (i.e. Millipore: Milli-Q Biocel). All of the chemicals should be added one at a time and dissolved in the premeasured water before adding the next. $CaCl_2$ should be added last in order to prevent precipitation. Na-EDTA (0.5mM) is added to chelate any heavy metals, therefore adjusting the effective $CaCl_2$ concentration to 2mM in all of our experiments. There are several suggestions regarding the concentration of glucose; we prefer 11mM for the mouse heart.

Figure 2. (right) Cannula assembly with heart hanging.

Figure 3. (far right)
A-Pressure Transducer (COBE CDXIII).
B, C, D-Intraventricular pressure recording catheter.
B-Silastic catheter.
C-7.5cm PE-50 catheter.
D-Magnified view of PE-50 catheter flange to seal the IV catheter to the ventricular wall

Figure 4. (below) Diagram of cannulated heart

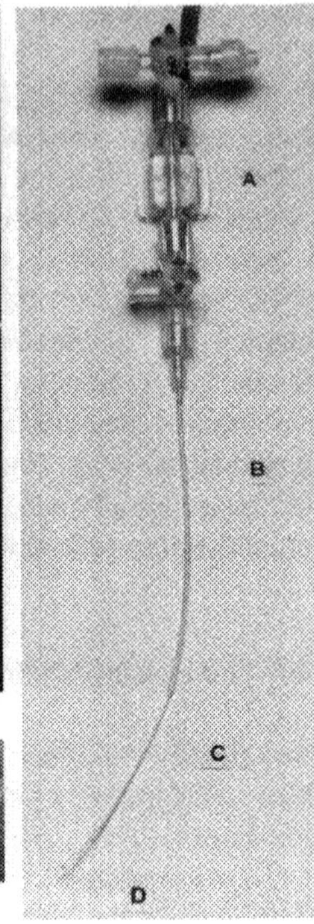

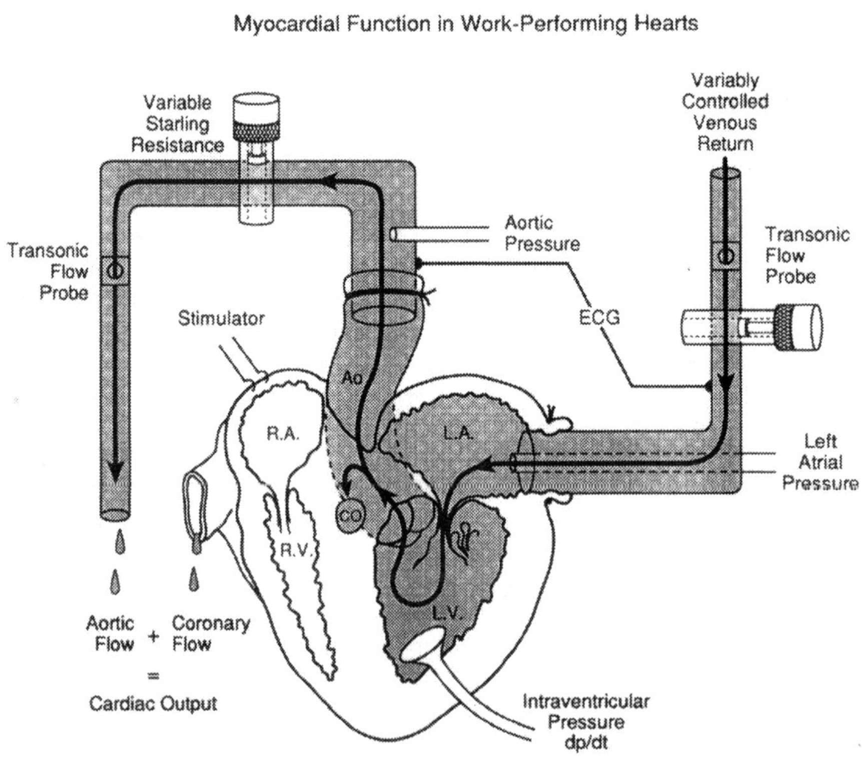

4. PREPARATION OF THE ANIMAL

Before the scheduled experiment systolic blood pressures and heart rate are measured via the tail cuff method (Natsume KN-210) (Figure 5A) in unanesthetized transgenic and wild type mice. The mice can easily and quietly be restrained in specific containers (Figure 5B). The tail cuff method compares very well with the intra-femoral determination of an indwelling catheter (bottom tracing) (Figure 5A). Comparisons of TG and NTG mice are made on the same day to avoid daily fluctuations. Prior to the working heart experiment the mice are anesthetized with 30-40 mg/kg sodium pentobarbital i.p. and heparinized (5000U/kg) to protect the heart against microthrombi. Other anesthetics like Avertin (Aldrich) can also be used. Heart rate and all 12 leads of the electrocardiogram are recorded (Figure 6). The mice are respired artificially via tracheotomy with a respirator (Harvard Small Rodent ventilator #683). The chest is opened at the sternum and the heart is quickly removed and placed in a separate preparatory tissue bath (Radnotti) of warm, heparinized, and oxygenated Krebs-Henseleit solution, where the aorta is cleaned of the surrounding tissue. In contrast to the rat heart, stunning of the mouse heart with cold fluid severely affects function.

4.1. Cannulation and Suspension of the Heart

The technical part of the apparatus is discussed in section 2. The sequence of suspension is as follows. The aorta is connected to the cannula, (see section 2.5) (Figure 2), preserving the aortic valve and the coronary artery ostia. The aortic pressure transducer is now connected to the side port of this cannula to record the mean aortic pressure. Retrograde perfusion (Langendorff) is started immediately. This shortens the time to begin perfusion and flushes the coronary system. While the heart is perfused and beating, the distal end of the IVP-pressure catheter (Figure 3C) is introduced through the left pulmonary vein, through the left atrium, and into the left ventricle. Great care has to be taken to pass the catheter during the opening of the mitral valve to prevent injury. The catheter is pulled through the left ventricle at the apex. The flanged part of the catheter (Figure 3D) rests tightly against the apical wall. The distal end is then connected to the larger, more compliant catheter (Figure 3B). This catheter is connected to the pressure transducer (Figure 3A) and records intraventricular pressure (IVP): [systolic, diastolic, and end-diastolic pressures]. After the placement of the catheter, the venous return cannula is tied into the pulmonary vein opening of the left atrium while carefully preserving both the atrium and the atrial septum. There are usually two pulmonary vein roots; the intervening tissue between them is cut to make the opening larger for cannulation. At this point the side port of this cannula can be connected to the left atrial pressure transducer to begin measurement. Flow is then switched from retrograde to anteriograde mode (see stopcock at position "L" in Figure 1) thus terminating the Langendorff mode of perfusion. The heart now begins the normal operation of pumping fluid through the left atrium and ventricle. The pulmonary artery drains into the open, a small cut in the pulmonary artery proximal to the outflow tract allows sampling of the coronary sinus effluent with a capillary tube to determine oxygen consumption by the heart (Figure 4). O_2 consumption is calculated with the following formula (14):

$$\mathbf{MVO_2=(PaO_2\text{-}PVO_2) \bullet Cor.\ Flow\ (ml/min) \bullet (C/760) \bullet 1000 \bullet Heart\ Wt\ (g)^{-1}}$$

5. EXPERIMENTAL PROTOCOL

We developed the following protocol, which is applied to all hearts, both transgenic and non transgenic, based on the consideration that the methodology can determine the functional limits of cardiac performance as defined by either cardiac minute work or cardiac power.

1. The hearts are mounted and stabilized in the Langendorff mode for about 5 minutes before switching to the work-performing mode.
2. After conversion to "work", the hearts are equilibrated for 30 min. at an average basal workload of 5ml/min venous return and 50mmHg aortic pressure for adult hearts (150-180mg). This results in a cardiac minute work of 250 ml/min•mmHg (venous return • mean aortic pressure). For smaller hearts this workload is adjusted to lower heart weights.
3. Changes in afterload are accomplished by sequential increases and decreases of aortic pressure in increments of 10 mmHg from 30mmHg to 90mmHg.
4. 30 min re-stabilization at basal loading condition (5ml-50mmHg).
5. Changes in preload are accomplished by sequential increases and decreases of venous return, in increments of 1ml. This can be performed as a classical Frank-Starling loading by keeping the aortic resistance constant, or by keeping the aortic pressure constant by utilizing a hydrostatic column of a given pressure (i.e.50mmHg). The second method mentioned may be beneficial under certain circumstances of experimental design.
6. After this procedure and following a 30 min. re-stabilization at basal loading, relevant pharmacological agents are infused.

6. DATA ACQUSITION

In addition to the computer acquisition of all data online, we also monitor and record all-important parameters on a Grass Model 7B Polygraph throughout the experiment. This enables us to see a complete overview of all trends at any time and, furthermore alerts us the insight of sudden events such as irregularities of the hearts. An example of part of such tracings is seen in figures 7 and 8; the aortic pressure, the IVP, and +/- dP/dt are shown. Other recorded parameters (not shown in the figure) are ECG, heart rate, atrial pressure, and venous return (which equals cardiac output). The pressure curve (see Figure 7) is used to calculate peak pressure, time to peak pressure (TPP), and TPP/mmHg of the IVP in order to normalize contractility. TPP is the time it

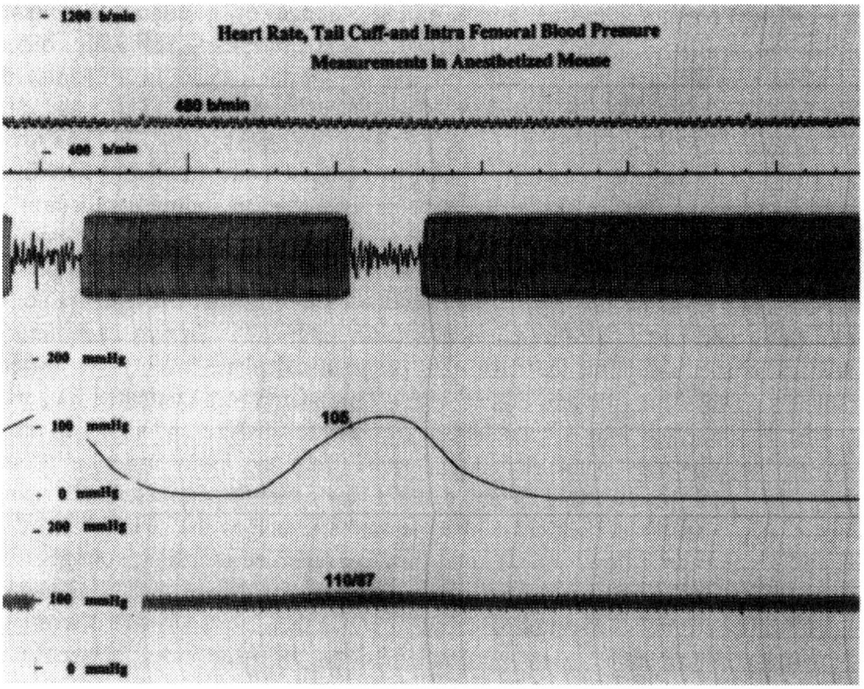

Figure 5A. (above) Systolic blood pressure recordings

Figure 5B. (above) Mouse capsule

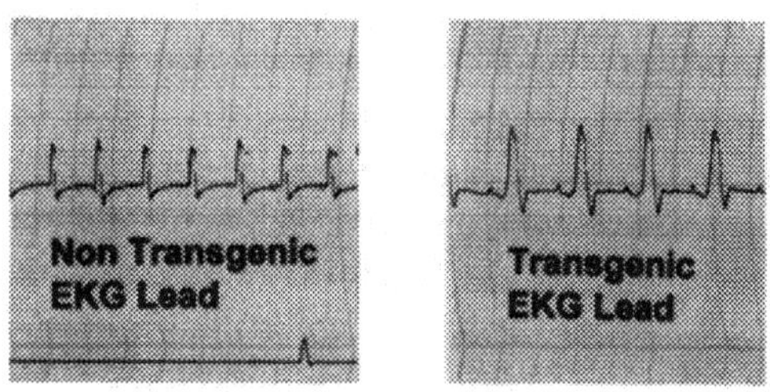

Figure 6. Electrocardiogram recordings

takes to reach peak pressure, it is calculated as the time from EDP (end diastolic pressure) to SP (systolic pressure). Additionally, ½ Relaxation Time, [RT ½ =time from SP to DP /2], and RT ½ /mmHg of IVP can be obtained to normalize relaxation (7). The rate of pressure development and relaxation (first derivative of IVP) determines +dP/dt and -dP/dt (Figure 7).

The cardiac analysis system consists of a hardware component and a custom-made software component, which performs the online data analysis and recording. The hardware contains a signal conditioner with high precision instrumentation amplifiers for the intraventricular, aortic, and atrial pressure sensors. It also includes an amplifier for conditioning the small output voltage of a digital thermometer (Physitemp), and two buffer amplifiers for the dual flowmeter (Transonic) outputs. The signal conditioner is connected to a Lab Master (Axon Instruments) data acquisition system, which is installed in a 486 PC computer. The Lab Master contains three digital Input/Output ports, five external clocks, two Digital to Analog converters and a sixteen channel Analog to Digital converter with twelve bits resolution (error < 1%). The software component is an online acquisition and analysis program, which samples, computes, records, and displays several parameters of the heart during the experiment. The program simultaneously samples the intraventricular pressure, the aortic pressure, the atrial pressure, the temperature, the aortic flow, and the total flow in ten second intervals. During each interval, immediately after sampling the data, this program calculates twenty-two different parameters (Table1) and records them with a time stamp in a computer file, which can later be imported into a spreadsheet program to analyze and plot the results. Many of the important parameters are derived from the intraventricular pressure signal. The peak positive and negative dP/dt, for example, are calculated using a standard algorithm for derivatives. Other parameters such as the atrial pressure, the temperature, the aortic flow, and the total flow are recorded as analog voltage signals from their specific measuring instruments. These signals are then digitized one hundred times at one kHz sampling frequency and then averaged and converted to their corresponding units. The program displays the heart rate, intraventricular pressure (systolic, diastolic, end diastolic), intraventricular maximum and minimum dP/dt, TPP, and RT ½. All twenty-two instantaneous parameters are printed at any important moment during the experiment (Table 1).

7. APPLICATIONS

This section details the ways in which we use the methodology to compare cardiac performance between complimentary hearts (transgenic and/or treated against their control counterparts). The ability of each heart or group of hearts dictates, in part, the best way to make these comparisons. Comparisons of basal loading, for example, are very important in hearts that have an impaired ability to respond to the normal preload and afterload regimen. The cardiac function curves, based on the Frank-Starling principle, are still the best overall method of direct physiological comparison. In addition, pharmacological dose response curves are also an effective method of comparison. The guiding principle behind these applications is that all require the accurate observation and measurement of all core parameters of contraction and relaxation (IVP and its derivatives, see sections 6 and 7.1).

7.1. Basal Loading of the Hearts

The first step in the protocol is to equilibrate the hearts under basal loading conditions (see sections 5 & 6). Once equilibrium is reached several computer

Online Computer Analysis (Contractility)

Left Ventricular Pressure	
1. Systolic	**87 mmHg**
2. Diastolic	**-9 mmHg**
3. End Diastolic	**+8 mmHg**
Left Atrial Pressure	**+11 mmHg**

Contraction

+dP/dt = +2750 mmHg/sec

Peak Press (PP): 1 - 3 = 79 mmHg
TPP 41 msec
TPP/PP 0.519 msec/mmHg

Relaxation

-dP/dt = -2700 mmHg/sec

½ Relax. Press: (1-2)/2 = 48 mmHg
RT½ 31 msec
RT½ / ½ RP 0.646 msec/mmHg

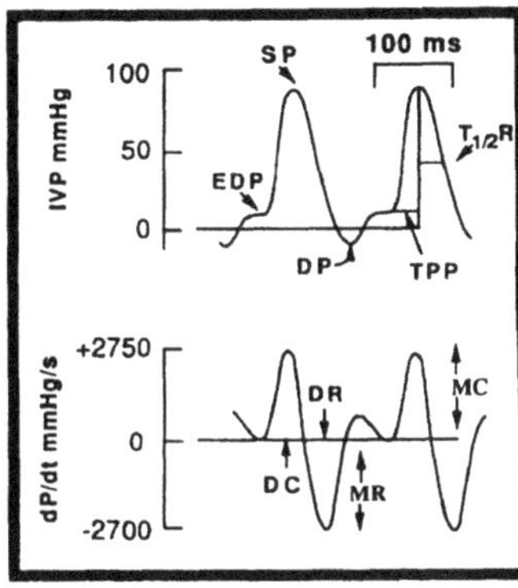

Figure 7. Computation of selected parameters

Table 1. Example of a realtime experimental printout

Heart Rate...	**323 bpm**
Mean Aortic Presure...................................	**50 mmHg**
Systolic Pressure.......................................	**92 mmHg**
Diastolic Pressure.....................................	**-5 mmHg**
End Diastolic Pressure.............................	**+2 mmHg**
Systolic - End Diastolic pressure....................	90 mmHg
Systolic - Diastolic Pressure..........................	97 mmHg
½ Relaxation..	48.5 mmHg
Time to Peak Pressure.................................	33 msec
Duration of Relaxation................................	50 msec
Time ½ of Relaxation..................................	20 msec
Time to Peak/ Max Pressure.......................	**.371 msec/mmHg**
T ½ / ½ Relaxation Pressure.......................	**.412 msec/mmHg**
Positive dP/dt...	**+4248 mmHg/sec**
Negative dP/dt...	**-3720 mmHg/sec**
Left Atrial Pressure..................................	**6 mmHg**
Cardiac Output...	**5.2 ml/min**
Work...(MAP·CO).....................................	**258 mmHg·ml/min**
Stroke Volume	**16 µl**
Aortic Flow...	**1.1 ml/min**
Coronary Flow..	**4.1 ml/min**
Temperature...	**37.6 C°**

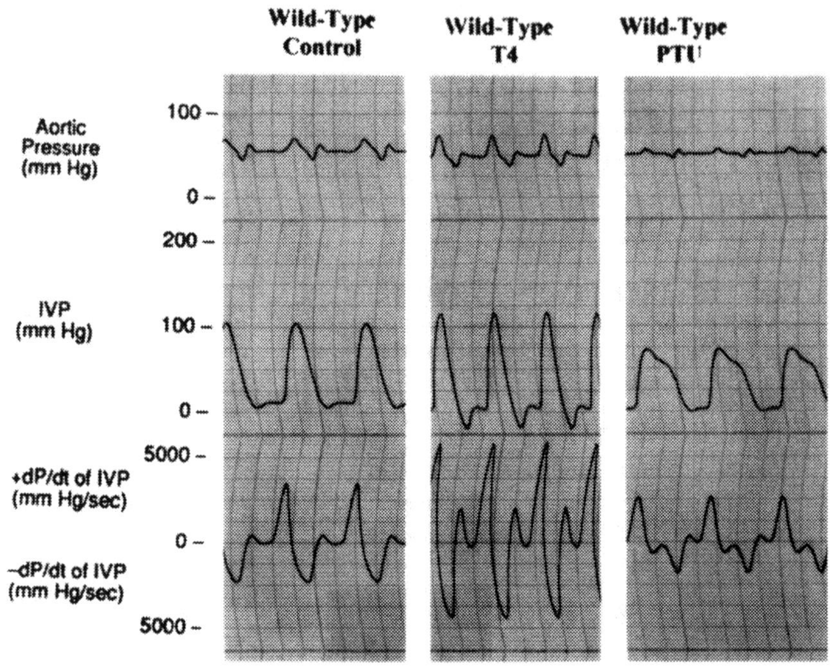

Figure 8. Key parameters of a polygraph recording

readings are acquired for later comparisons. An example of the key parameters of a polygraph recording is shown in figure 8 (7). The three hearts are from age matched control, hyperthyroid (daily injection of l-L-thyroxine), and hypothyroid mice (0.15% 5-propyl-2 thiouracil (PTU) per mouse for 4wks). All heart weights were similar. When weights fall outside of normal range (150-180mg), it becomes necessary to normalize measured parameters by their individual heart weights. All three hearts were compared under exactly identical loading conditions: venous return (preload) = 5ml/min and aortic pressure (afterload) = 50mmHg and cardiac minute work of 250 ml/min • mmHg: The three equally loaded ventricles show completely different contractile parameters. Compared with the control, the hyperthyroid shows a significant increase in contractility whereas, under the same load, the hypothyroid is severely reduced and slowed (hearts are paced at the same rate as controls) (Figure 8). Molecular analysis of these three groups shows a direct correlation of the α (fast) and β (slow) myosin heavy chain (MHC) transcripts with the contractile state of the hearts. A slight elevation of α-MHC with no β-MHC expression was observed in hyperthyroid ventricles; whereas, β-MHC was significantly elevated with no detectable α-MHC in hypothyroid ventricles (7).

7.2. Cardiac Function Curves (Pressure and Volume Loading)

Tables 2A, 2B, and 2C contain the spreadsheets of three different loading experiments in the same heart. Table 2A contains pressure loading data, 2B volume loading at fixed mechanical resistance, and 2C volume loading with fixed aortic pressure (hydrostatic pressure column, see Figure 1). The +dP/dt data of the three tables were separately evaluated using linear regression analysis. The resulting cardiac function curves (Starling curves) were compared in pairs in Figure 9. Figure 9.1 compares pressure and volume loading of table 2A and 2B; Figure 9.2 compares pressure and hydrostatic volume loading of tables 2A and 2C.
To test whether the different volume loads affect the normal mouse heart differently, we also compared in Figure 9.3 volume loading with a fixed resistance against volume loading with fixed aortic pressure (hydrostatic pressure). Below is a list of the slopes, y intercepts of +dP/dt, and correlation coefficients.

	Slope	Intercept	Correlation Coefficient
Pressure load:	7.1	2366	.96
Volume load w/ const. Resistance:	3.0	3000	.89
Volume load w/ const. Pressure:	2.6	3400	.81

The data show that the pressure and volume slopes in Figure 9.1 and Figure 9.2 are about 2:1, i.e. pressure load requires about double the +dP/dt per work unit than during volume load. Furthermore, the comparison of the two different types of volume load is not statistically significantly different (P=.0703), (i.e. volume load compared to hydrostatic volume load). Linear regression plots have also been obtained of ±dP/dt, TPP, and RT ½ , aortic and coronary flows using the same spreadsheets. All of these parameters show the same relationship.

7.3 Dose Response Relationships (Infusion of Pharmacological Agents)

The infusion of cardiovascular agents, of their inhibitors, and of agents affecting contractile and regulating proteins is of great analytical use. Section 2.3 describes how these agents can be added to the perfusates without contaminating the buffer containing cylinders and tubing. Figure 10 depicts examples of Isoproterenol (Iso) dose response curves that compare transgenic phospholamban (PLB-a phosphoprotein regulator of calcium homeostasis) ablated and nontransgenic wild-type (littermates) mouse hearts. The concentration infused is calculated on the basis of cardiac output. In this example 0.8 nM to 0.8 μM of Iso was infused in increasing doses. The plotted parameters show the effect of Iso on + and -dP/dt, TPP, RT1/2, and heart rate. The data represent six nontransgenic and six transgenic hearts each. The rates of pressure development and relaxation are greatly increased, and the time to peak (TPP) and half time of relaxation (RT ½) are markedly reduced in the wild-type hearts. However, Iso does not affect these parameters in the PLB-ablated hearts

Table 2. Spreadsheets of cardiovascular parameters during pressure and volume loads in the same heart

Afterload Changes (30-81 mmHg) at Constant Preload (5.15 ml/min Venous Return).

A	HR bpm	MAP mmHg	VR=CO ml/min	Work mmHg· ml/min	SV µl	+dP/dt mmHg/ sec	TPP ms/ mmHg	-dP/dt mmHg/ sec	RT1/2 ms/ mmHg	Aortic Flow ml/min	Cor. Flow ml/min
Control	331	51	5.2	263	16	4416	.362	3576	.396	1.4	3.8
High Press	335	60	5.1	304	15	4728	.343	3600	.404	1.7	3.4
High Press	331	59	5.2	305	16	4464	.354	3312	.432	1.5	3.7
High Press	341	69	5.1	350	15	4752	.336	3456	.433	1.1	4.0
High Press	331	81	5.2	419	16	5232	.308	3696	.433	0.4	4.8
Low Press	324	39	5.1	198	16	3864	.397	3624	.405	1.5	3.6
Low Press	324	30	5.1	198	16	3720	.408	3288	.434	1.5	3.6

Preload Changes (3.5-8.1 ml/min Venous Return) at Constant Aortic Resistance (Aortic Pressure = 40-59 mmHg).

B	HR bpm	MAP mmHg	VR=CO ml/min	Work mmHg· ml/min	SV µl	+dP/dt mmHg/ sec	TPP ms/ mmHg	-dP/dt mmHg/ sec	RT1/2 ms/ mmHg	Aortic Flow ml/min	Cor. Flow ml/min
Control	284	50	5	250	15	3816	.433	-2808	.500	2.1	2.9
High Flow	311	55	6	330	17	4104	.409	-3120	.474	2.2	3.9
High Flow	308	58	7.1	412	20	4248	.394	-3216	.458	2.4	4.7
High Flow	308	59	8.1	478	23	4248	.394	-3216	.450	2.4	5.7
Low Flow	267	44	4.1	180	12	3600	.469	-2592	.370	1.9	2.2
Low Flow	271	44	4.1	180	12	3600	.469	-2640	.516	1.9	2.1
Low Flow	268	40	3.5	140	10	3192	.521	-2328	.585	1.7	1.9

Preload Changes (3.1-8.9 ml/min) at Constant Aortic Pressure

C	HR bpm	MAP mmHg	VR=CO ml/min	Work mmHg· ml/min	SV µl	+dP/dt mmHg/ sec	TPP ms/ mmHg	-dP/dt mmHg/ sec	RT1/2 ms/ mmHg	Aortic Flow ml/min	Cor. Flow ml/min
Control	345	50	4.9	245	12	3936	.424	-3360	.426	2	2.9
High Flow	343	51	6	306	15	4224	.400	-3456	.442	2.9	3.0
High Flow	333	51	7	357	18	4392	.390	-3768	.417	3.7	3.3
High Flow	321	51	8	408	22	4392	.396	-3816	.439	4	3.9
High Flow	324	51	8.9	454	25	4512	.388	-3912	.416	4.4	4.6
Control	294	51	5	255	14	4272	.388	-3312	.442	2.1	2.9
Low Flow	288	50	4.1	205	12	4008	.407	-2928	.466	1.3	2.8
Low Flow	317	50	3.1	155	7	3888	.429	-2688	.505	0.6	2.5
Control	283	50	4.9	245	14	4032	.413	-2952	.481	1.9	3.1

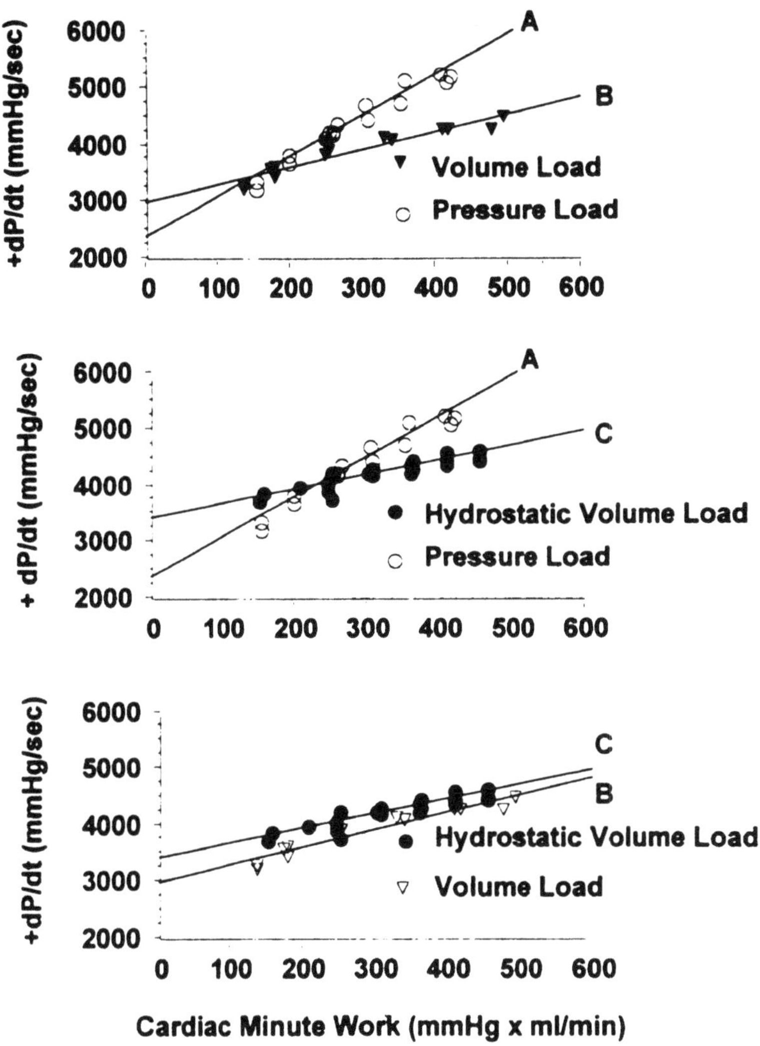

Figure 9. Cardiac function curves: +dP/dt vs. cardiac work

that are normally hyperdynamic. This demonstrates the regulatory importance of PLB on both relaxation and contraction (15). It also shows that pacemaker activity is not regulated by PLB, since the effect of Iso on heart rate remains the same in both wild type and PLB ablated mouse hearts.

7.4. Ischemia-Reperfusion

A number of genetically modified mouse models have been generated to study receptors and signaling pathways involved in ischemia-reperfusion injury. As a consequence, special *in vivo* and *ex vivo* techniques evolved, including the work-

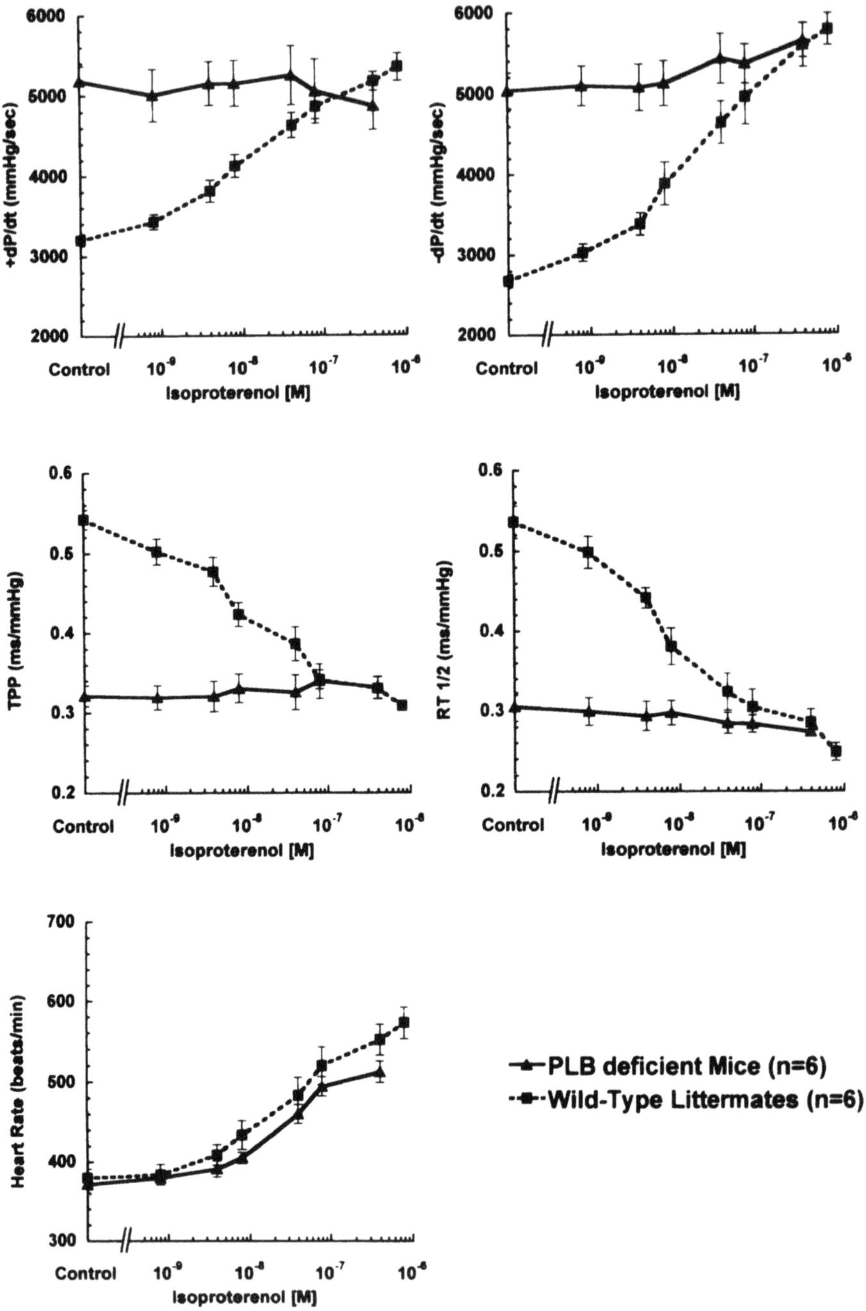

Figure 10. Examples of isoproterenol dose response curves. Effects of isoproterenol on ± dP/dt, TPP, RT ½ , and Heart Rate. PLB deficient mice hearts show a lack of stimulatory effects.

performing heart preparation, to elicit ischemic insults. In the mouse, these techniques to induce myocardial ischemia include different procedures such as coronary artery ligation and other no-flow methods (16,17,18). Myocardial ischemia occurs when the reduction of coronary blood flow is so severe that the oxygen supply is inadequate to the oxygen demands of the heart. This oxygen imbalance results in poor oxygen delivery to tissue, accumulation of cytotoxic metabolites, alterations in high-energy phosphate levels or availability, and mechanical dysfunction. Restitution of coronary flow leads to reperfusion arrhythmias, microvascular damage, formation of free radical, calcium overload, and inflammatory responses, all of which can further damage the myocardium and lead to cell death (19). The endpoints of ischemia-reperfusion injury can be represented as cell necrosis and/or apoptosis, arrhythmias, and contractile dysfunction, depending on the scope of the study in question.

A vast range of experimental models for the ischemic heart exists, from *in vivo* to isolated organ and myocyte preparations, and has been of tremendous importance in characterizing the physiology, biochemistry and genetics of ischemia-reperfusion. The advantages and disadvantages of each experimental design, species, and endpoint have been well described (20,21). Importantly, it has been suggested that induction of low-flow global and regional ischemia is better controlled in isolated perfused hearts (both Langendorff and work-performing preparations) of small mammalian animal models (20). Previous *ex vivo* murine studies of myocardial ischemia have been performed on the Langendorff isolated heart model; however, the work-performing murine heart preparation is a load-dependent (preload and after load) model and has the advantage of measuring work-related indices of whole heart function during normal and ischemic episodes. In the working mode, the ischemic heart is a loaded system, which results in a supply-demand imbalance of oxygen and an accumulation of cytotoxic metabolites; whereas, the Langendorff isolated heart preparation is an unloaded, quiescent system with no significant oxygen imbalance during ischemia. In the work-performing heart model, regional ischemia (coronary artery occlusion) or global ischemia (no-flow and low-flow) can be induced and many key hemodynamic, biochemical, and histological parameters can be measured in a more physiologically relevant setting, including percent recovery of function, left ventricular developed pressure, rhythmic and dysrhythmic patterns, coronary flow changes, high energy phosphate levels, creatine kinase (CK) or lactate dehydrogenase (LDH) release, cardiac morphology and vascular anatomy, and infarct size. It should be noted that the isolated work-performing heart preparation is the model of choice for characterizing post-ischemic stunning (21).

Low-flow ischemia was utilized in our model and the protocol, based upon preliminary results in our lab and in previous murine studies (18), is as follows. After a 30-minute equilibration period at a basal workload of 5 mL/min venous return and 50 mm Hg aortic pressure, the venous return is reduced stepwise to a flow of 1 mL/min for 30 minutes. At each stepwise decrease in flow and during designated time points of ischemia, hemodynamic measurements, blood gases, and coronary flow effluent are obtained. The heart rate is maintained at baseline levels during ischemia by pacing. Following 30 minutes of low-flow ischemia, venous return is increased stepwise to a flow of 5 mL/min and reperfusion occurs for 30 minutes. Again, hemodynamic parameters, blood gases, and coronary flow effluent are obtained at each stepwise increase in flow and during designated time points of

reperfusion. It is extremely important to maintain strict control of physiological temperature, as slight alterations in temperature have been shown to influence ischemia-reperfusion injury (22,23).
Figure 11 depicts the percent recovery of contractile function of wild type mice, mice deficient of fibroblast growth factor-2 (*Fgf2* KO), a peptide implicated in protecting the myocardium from ischemic injury (24), and mice with a cardiac-specific over expression of FGF2 (*Fgf2* Tg). Following 30 minutes of low-flow ischemia and 30

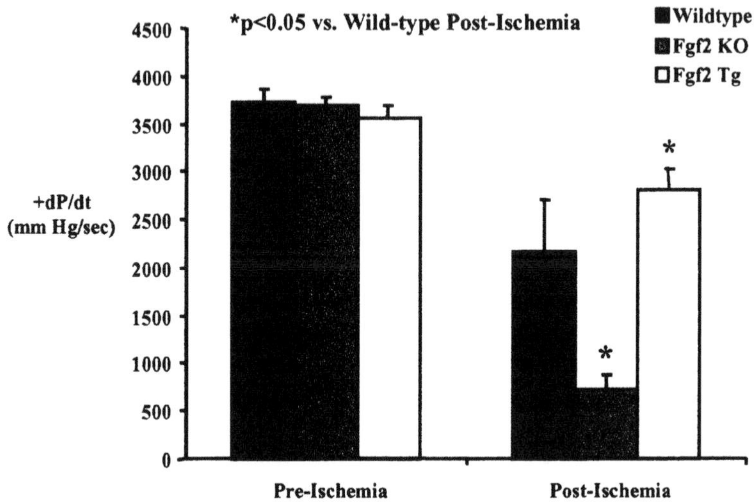

Figure 11. Graph depicting the percent recovery of contractile function measured as +dP/dt, for wild-type mice (black bar), FGF2 deficient mice (Fgf2 KO, gray bar), and cardiac-specific FGF2 over-expressing mice (Fgf2 Tg, white bar)

minutes of reperfusion, the wild type mice had a 58% recovery of contractile performance (+dP/dt), whereas, the *Fgf2* KO mice recovered to only 19% of their contractile function ($p<0.05$). Alternatively, over expressing FGF2 in the heart resulted in a significant increase in the percent recovery of function (80%) compared to wild type mice ($p<0.05$). No significant change in coronary flow or vascular anatomy was detected in any of the groups, implying that the protective effect of FGF2 is not mediated by coronary perfusion changes. Therefore, we have demonstrated, with our experimental model, a direct role of FGF2 in protecting the heart from ischemia-reperfusion injury.

7.5. Hypoxia-Reoxygenation

To establish hypoxia-reoxygenation injury in genetically engineered mice the work-performing heart is an excellent tool (25,26). Hypoxia is induced after a stabilization period of 30 min. by switching to a low PO_2 buffer balanced with NO_2. PO_2 is decreased from 660 mmHg to 161 mmHg. PH and PCO_2 remain the same. Hypoxia is maintained for 30 min., followed by reperfusion with high PO_2 buffer (660 mmHg) for 30 min. Table 3 and Figure 12 show an example of protection against hypoxia-reoxygenation in the absence of poly (ADP-ribose) synthesase (PARS). The absence of PARS provided an almost full protection against the hypoxia-reoxygenation induced systolic and diastolic dysfunction.

Table 3 & Figure 12. (below) Comparison of hypoxia-reperfusion induced injury of wildtype and PARS ablated mouse hearts

	WT Before Hypoxia (n=5)	KO Before Hypoxia (n=5)	WT After Hypoxia (n=5)	KO After Hypoxia (n=5)
Intraventricular Pressure				
Systolic	+90.0±1.11	+95.2±1.67*	###+78.3±1.98	#+89.9±0.87***
Diastolic	-3.2±1.00	-6.0±0.91	##+1.6±1.08	##-1.7±0.90*
End Diastolic	+12.4±0.99	+10.3±0.82	#+15.5±0.83	+12.7±1.30
Heart Rate (Bpm)	367.5±9.85	357.6±6.50	352.3±7.68	347.8±5.56
Venous Return (ml/min) **(Preload)**	5.1±0.02	5.1±0.02	5.1±0.02	5.0±0.02
LV Minute Work (mmHg•ml/min)	253.9±1.19	250.8±1.42	247.0±2.87	249.5±1.09
Stroke Volume (μl)	14.3±0.68	14.1±0.63	14.5±0.70	14.7±1.05
LA Pressure (mmHg)	11.48±0.87	11.45±1.59	13.29±1.23	13.24±1.19
PaO_2 (mmHg)	662.7±1.53	663.1±3.22	660.9±3.90	660.6±2.53
PvO_2 (mmHg)	210.5±17.17	263.4±35.00	300.7±50.16	334.0±29.92
MVO_2 ($\mu L\ O_2 \cdot min^{-1} \cdot g^{-1}$)	174.0±28.84	170.0±28.17	246.8±39.93	153.02±23.74

$$MVO_2(\mu L\ O_2 \cdot min^{-1} \cdot g^{-1}) = \frac{(PaO_2 - PvO_2) \cdot \text{Coronary Flow(mL/min)} \cdot (C/760) \cdot 1000}{\text{Heart Wet Weight (g)}}$$

C=0.0239 Bunsen solubility coefficient
Values are mean±SEM
*p<.05; **p<.01; ***p<.001 Compares **WT Before vs KO Before** and **WT After vs KO After**
#p<.05; ##p<.01; ###p<.001 Compares **WT Before vs WT After** and **KO Before vs KO After**

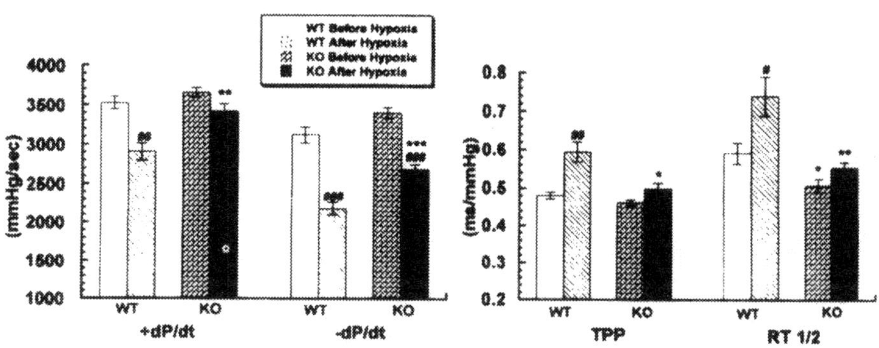

8. TISSUE PREPARATIONS

We have frequently encountered newborn hearts in transgenic mice, which did not develop beyond 10-14 days after birth. In these cases isolated tissues of the transgenic offspring have to be studied either before birth or shortly after, often requiring even smaller physiological probes. These hearts can be studied in the isolated tissue preparation apparatus. In our lab we can compare eight ventricular preparations (papillary muscles or right ventricular outflow tract strips) and/or eight atrial preparations at the same time. The methods have been referenced for human trabeculae (27), for rat ventricle and atria (28,29), for newborn mouse ventricles (30), and for adult mouse atria and ventricles (31,32,33). The same methodology has also been applied to normal and transgenic skeletal muscles of mice (34,35).

9. SUMMARY

The working-performing buffer-perfused heart preparation offers a completely controlled realistic picture of left ventricular function of the mouse heart without modulation by hemodynamic or autonomic compensation. Contraction and relaxation are separately observable. The advantage of the work-performing methodology is the definition of the complete range of performance of each heart under both pressure and volume load and its comparison to a selected counterpart, i.e. transgenic versus nontransgenic, or control heart versus treated heart. Cardiac function curves allow single heart, and heart to heart comparisons of volume and/or pressure loading of the mouse ventricle. Low flow ischemia and hypoxia was utilized successfully in our model to compare reperfusion injury in transgenic mouse hearts.

10. FUTURE DIRECTIONS

It is clear from our studies on work performing mouse hearts of transgenic and non-transgenic mice that the Frank-Starling Relationship occupies a central position in the quantitation and understanding of contraction and relaxation of these hearts. We are now focusing on ischemia- and hypoxia-reperfusion injury and direct protection of genetically altered mice against oxidant and free radical mediated cell death. We are also concentrating on the aging heart and it's protecting mechanism in various diseased models (36,37).

ACKNOWLEDGMENTS

We thank Gilbert Newman for his inventive technical skills, Jason P. Straus, James Donovan, and Julia Sillis for their invaluable assistance with the manuscript and figures, Anne and Dr. Michael Behbehani for the earliest version of the computer acquisition, Farad Bahremand for assisting with the computer programming, Gwen Kraft for the art work of Figure 1, and Jay P. Slack for editing and proofing contributions.

This work was supported in parts by NIH grants PO1 HL46826, PO1 HL22619 and PO1HL41496

REFERENCES

1. Neely JR, Liebermeister H, Batterby H, Morgan H. Effect of pressure development on O_2 consumption by isolated rat hearts. Am J Phys 1967; 212: 804-814.
2. Grupp IL, Grupp G. "Isolated heart preparation perfused or superfused with balanced salt solutions." In *Methods in Pharmacology*. New York: Plenum Publishing, 5: 111-128, 1984.
3. Grupp G. "Derived Indices of Myocardial Function in Isolated Work-Performing Hearts." In *Methods in Pharmacology*, New York: Plenum Publishing, 5: 129-139, 1984.
4. Szymanska G, Grupp IL, Slack JP, Harrer JM, Kranias EG. Alterations in sarcoplasmic reticulum calcium uptake, relaxation parameters and their responses to β-adrenergic agonists in the developing rabbit heart. J Mol Cell Cardiol 1995; 27: 1819-1829.
5. Grupp G, Grupp IL, Kojima M, Sperelakis N, Tsuchiya Y, Hosokawa T, Schwartz A. Hemodynamic and electrophysiological effects of a novel positive inotropic drug, OPC-8212, in normal and "failing" guinea pig heart preparations. J Cardiovasc Pharm 1986; 8:428-440.
6. Ng WA, Grupp IL, Subramaniam A, and Robbins J. Cardiac myosin heavy chain mRNA expression and myocardial function in the mouse heart. Circ Res 1991; 69:1742-1750.
7. Grupp IL, Subramaniam A, Hewett T, Robbins J, Grupp G. Comparison of normal, hypo, and hyperdynamic mouse hearts using isolated work performing heart preparations. Am J Phys 1993; 265 (Heart Circ Phys 34): H1401-1410.
8. Hewett TE, Grupp IL, Grupp G, Robbins J. α-Skeletal actin is associated with increased contractility in the mouse heart. Circ Res 1994; 74:740-746.
9. Grupp IL, Kranias EG, Kiss E, Koss K, Harrer JM, Slack J, Edes I, Low W, Grupp G. The contribution of phospholamban to myocardial contractility. Heart Fail 1995; 11:48-61.
10. Chen EP, Bittner HB, Petersen DS, Van Trigt P. A Work-performing heart preparation for myocardial performance. J Surg Res 1996; 64: 57-62.
11. Gauthier NS, Matherne GP, Morrison RR, Headrick, JP. Assessment of contractile function in working mouse hearts. FASEB J 1997; 11: A67.
12. Desai KH, Sato R, Schauble E, Baosli GS, Kobilka BK, Bernstein D. Cardiovascular indexes in the mouse at rest and with exercise: new tools to study models of cardiac disease. Am J Phys 1997; 272: H1053-1061.
13. Zack, Radovan. *Basic Facts for Basic Science*. New York: Raven Press; 235-236, 1990.
14. Chu G, Luo W, Slack JP, Kranias EG, Grupp IL, et. al. Compensatory mechanisms associated with the hyperdynamic function of phopholamban deficient mouse hearts. Circ Res 1996; 79: 1064-1076.
15. Luo W, Grupp IL, Harrer J, Ponniah S, Grupp G, Duffy JJ, Doetschmann T, Kranias, E. Targeted ablation of the mouse phospholamban gene results in altered myocardial contractility. Circ Res 1994; 75: 401-409.
16. Michael LH, Entman ML, Hartley CJ, Youker KA, Zhu J, Hall SR, Hawkins HK, Berens K, Ballantyne CM. Myocardial ischemia and reperfusion: a murine model. *Am J Physiol* 1995;269:H2147- H2154.
17. Brooks WW, Garibaldi BA, Conrad CH. Myocardial injury in the mouse induced by transthoracic cauterization. *Lab Animal Science* 1998;48:374-378.
18. Sumeray MS, Yellon DM. Characterisation and validation of a murine model of global ischaemia-reperfusion injury. *Mol Cell Biochem* 1998;186:61-68.
19. Opie LH. The heart: physiology, from cell to circulation, 3rd ed. Philadelphia:Lippincott-Raven Publishers, 1998.
20. Hearse DJ, Sutherland FJ. Experimental models for the study of cardiovascular function and disease. *Pharmacol Res* 2000;41:597-603.
21. Ytrehus K. The ischemic heart-experimental models. *Pharmacol Res* 2000;42:193-203.
22. Tveita T, Mortensen E, Hevroy O, Refsum H, Ytrehus K. Experimental hypothermia: effects of core cooling and rewarming on hemodynamics, coronary blood flow, and myocardial metabolism in dogs. *Anesth Analg* 1994;79:212-218.
23. Broderick TL, Currie RW, Paulson DJ. Heat stress induces rapid recovery of mechanical function of ischemic fatty acid perfused hearts by stimulating glucose oxidation during reperfusion. *Can J Physiol Pharmacol* 1997;75:1273-1279.
24. Padua RR, Sethi R, Dhalla NS, Kardami E. Basic fibroblast growth factor is cardioprotective in ischemia-reperfusion injury. *Mol Cell Biochem* 1995;143:129-135.
25. Dhar P, GrupplL, Schwartz A, Grupp G, Matlib MA; Reduction of Carnitine Content by Inhibition of Its Biosynthesis Results in Protection of Isolated Guinea Pig Hearts Against Hypoxia Damage. *J Cardiovasc Pharmacol Therapeut* 1996; I(3):235-242
26. Grupp IL, Jackson TM, Hake P, Grupp G, Szabo C; Protction against Hypoxia-reoxygenation in the Absence of Poly (ADP-ribose) Synthetase in Isolated Working Hearts. *J Mol Cell* Cardiol 1999; 31:297-303

27. Shamraj OI, Grupp IL, Grupp G, Melvin D, Gradoux N, Kremers W, Lingrel JB, De Pover A. Characterisation of Na/K-ATPase, its isoforms, and the inotropic response to ouabain in isolated failing human hearts. Cardiovasc Res 1993; 27: 2229-2237.
28. Grupp I, Im WB, Lee CO, Lee SW. Relation of sodium pump inhibition to positive inotropy at low concentrations of ouabain in rat heart muscle. J Physiol 1985; 360: 149-160.
29. Grupp IL, Grupp G, Schwartz A. Digitalis receptor desensitization in rat ventricle: ouabain produces two inotropic effects. Life Sci 1981; 29:2789-2794.
30. Shull MM, Diebold RJ, Eis M, Boivin G, Grupp IL, Doetschman T. Homologous recombination and growth factors in molecular genetics and gene therapy of cardiovascular diseases, Stephen C. Mockrin, ed. 26: 14, 371-403, 1996, Marcel Dekker, New York.
31. Koss KL, Ponniah S, Jones WK, Grupp IL, Kranias EG. Differential phospholamban gene expression in murine cardiac compartments. Circ Res 1995; 77: 342-353.
32. Koss KL, Grupp IL, Kranias EG. The relative phospholamban and SERCA2 ratio: a critical determinant of myocardial contractility. Basic Res Cardiol 1997; 92 suppl. 1: 17-24.
33. Askew GR, Lingrel JB, Grupp IL, Grupp G. Direct correlation of Na^+/K^+-ATPase isoform abundance and myocardial contractility in mouse heart. Biol Chem 1993; 374: 545-624.
34. Slack JP, Grupp, IL, Luo W, Kranias EG. Phospholabman ablation enhances relaxation in the murine soleus. Am J Phys 1997; 273: C1-C6.
35. Slack JP, Grupp IL, Ferguson DG, Rosenthal N, Kranias EG. Ectopic expression of phospholamban in fast-twitch skeletal muscle alters sarcoplasmic reticulum Ca^{2+} Transport and muscle relaxation. J Biol Chem 1997; 272: 18862-18868.
36. Grupp IL, Slack JP, Tilgmann C, Luo W, Tamura T, Gerdes AM, Kranias EG. Effects of Age on the Hyperdynamic Cardiac Function of Phospholamban Knockout Mice. AHA 70th Scientific Sessions 1997; 249.
37. Slack JP, Grupp IL, Holder D, Schmidt A, Gerst MJ, Tamura T, Tilgmann C, James PF, Johnson R, Gerdes AM, Kranias EG. The enhanced contractility of the phospholamban-deficient mouse heart persists with aging. J Biol Chem. Accepted 2001.

10

^{31}P NMR SPECTROSCOPY OF THE MOUSE HEART

Joanne S. Ingwall, Ph.D., Maryam M. Javadpour, Ph.D. and Wenfeng Miao, M.D., Ph.D.
Brigham & Women's Hospital, Boston MA 02115 USA

INTRODUCTION

NMR spectroscopy of tissues and organs yields two kind of information: the steady state tissue content of abundant NMR-observable metabolites and ions and the kinetics of certain enzymes and metabolic pathways. Because these measurements are non-invasive, they can be made repetitively throughout a protocol. They can also be made simultaneously with assessment of physiologic function. In the case for ^{31}P NMR spectroscopy of the isolated heart, and for the heart *in situ*, it is possible to make measurements of ATP and phosphocreatine (PCr) contents and turnover rates while simultaneously assessing systolic and diastolic performance. Such information is unique and provides a valuable solution to the perennial conflict between biochemistry and physiology. The biochemist dares to extrapolate results obtained from dilute solutions to the protein-rich cytosol, hoping that the results have functional meaning. The physiologist is so convinced of the importance of studying the intact heart, that he/she treats the heart as a black-box about which no molecular information can be known. The biophysicist has the best of both worlds: using NMR spectroscopy, one can learn about biochemical events within the cell while respecting the integrity of the physiological system.

With the application of molecular and genetic tools to the cardiovascular system, we now have the opportunity to apply NMR spectroscopy to the study of intact hearts in which specific gene products are over expressed, ablated or mutated. NMR spectroscopy is one of the tools important for defining the functional genomics (formerly called biology) of bioengineered hearts. In this chapter, we describe our experience adapting the NMR experiment to the isolated perfused heart of both adult and very young (17-23 days old) mice. Our examples will focus on ^{31}P NMR spectroscopy, but spectra based on other NMR-observable nuclei such as ^{19}F, ^{13}C and ^{23}Na have also be obtained. To illustrate the wide diversity of the scientific problems being studied in this way, we also present a sampling of the rapidly growing literature in this field.

BASIC PRINCIPLES OF NMR SPECTROSCOPY

Excellent texts describing the NMR experiment in detail (1) and reviews describing the basics of ^{31}P NMR spectroscopy of the isolated rat heart (2) and of the human heart (3) are available. Some of the basic principles will be reviewed here.

The main components of the NMR spectrometer (Figure 1) are a superconducting magnet, a radiofrequency (rf) transmitter and receiver, and a computer to run the experiment and process the signal. The sample sits within the magnet in a probe designed to transmit and receive rf energy. At a given field strength, the frequency range for each type of NMR-observable nucleus (P, Na, H, etc.) is distinct and can be easily accessed, much like dialing in your favorite FM station on a radio. The experiment begins by applying a rf pulse to the sample thereby exciting the nuclei. The NMR signal is the rf energy emitted as excited nuclei return to the ground state. The magnitude of the rf signal is small, and must be amplified to be detected. NMR is inherently an insensitive technique. Generally, many individual rf transmit/rf receive experiments must be summed to obtain enough signal to form an interpretable spectrum. For a given number of acquisitions, the signal is directly proportional to the number of nuclei present and hence the number of molecules or ions present in the NMR-sensitive volume[1]. More sample yields more signal. Increasing the number of acquisitions yields more signal, but the time of the experiment increases linearly. Higher magnetic field strengths also yield more signal. Note that signal is not equivalent to concentration (amount/volume); to convert amount of metabolite into concentration, one must obtain estimates for volume independently.

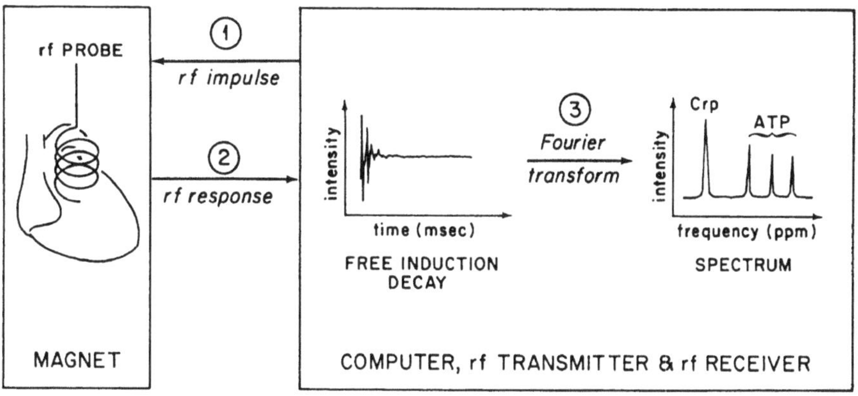

Figure 1. Diagram of a nuclear magnetic resonance (NMR) spectrometer. The NMR experiment requires a magnet, radiofrequency (rf) transmitter and receiver, and computer. The sample is placed in a magnet. A rf probe either surrounds (for cells or isolated organs) or is placed on the region of interest. To initiate the experiment, a rf pulse is transmitted (*step 1*). After a set delay, the rf receiver is turned on to measure rf signal emitted as nuclei return to their relaxed or ground state (*step 2*). Energy (intensity) as a function of time is transformed into an interpretable spectrum relating intensity and frequency (*step 3*). Redrawn from Ingwall (1982).

The kinds of information obtained in a typical ^{31}P NMR spectrum from a mammalian heart are illustrated in the Figures 2-4. ^{31}P NMR spectra contain five major resonances or peaks. The peaks report the presence of phosphorus nuclei corresponding to (from left to right)

---

[1] The relationship between signal to noise ratio (S/N) and the number of acquisitions is not linear: to double S/N requires increasing the number of acquisitions by 4-times.

inorganic phosphate (Pi), PCr and the three phosphates of ATP, the γ, α and β-phosphates. The shoulder to the right of the α-phosphate peak arises from the family of NAD molecules present. It is rare to observe other phosphate-containing molecules present in heart either because they are present in the cytosol in low concentrations (such as free ADP or cAMP) or because they are part of large molecular weight molecules such as proteins, nucleic acids or phospholipids, which tumble so slowly that their phosphate moieties con-tribute very broad peaks lost in the baseline. Under some circum-stances, glycolytic inter-mediates, di- and monophosphate esters and even AMP are observable.

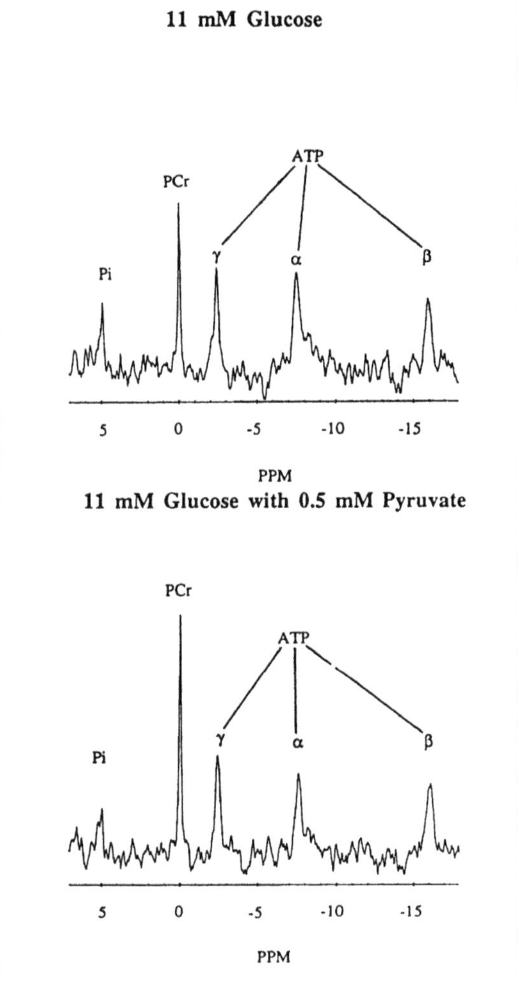

Figure 2. ^{31}P NMR spectra of an isolated perfused beating mouse heart.

Because the area of the peak is proportional to the amount of phosphate present in the sample, we can draw conclusions about relative amounts of phosphate - containing metabolites in the heart. The three phosphates of ATP are equal in amount chemically, so the areas corresponding to these phosphates must also be equal. The β-phosphate peak is broader than the γ- and α-phosphate areas; this is why peak areas are more reliable indicators of amount than peak heights. The PCr to ATP and PCr to Pi ratios are usually greater than 1. The position (called chemical shift) of the Pi resonance reports intracellular pH (4); the relative positions of the three ATP resonances provide a rough measure of intra-cellular [Mg^{2+}] (5). Because ^{31}P NMR spectroscopy provides information about ATP, PCr and H^+ concentrations, it provides the only practical method available to calculate free ADP concentrations in intact hearts. The method uses the creatine kinase equilibrium expression: [ADP] = [ATP][free creatine]/[PCr][H^+] x $1.66x10^9$ and requires input of total creatine concentration determined from other experiments.

Using magnetization transfer techniques in which one of the phosphates is magnetically tagged by selectively applying a rf pulse to that region of the spectrum, one can measure the unidirectional reaction velocities of the creatine kinase reaction (PCr + ADP + H^+ $\longleftrightarrow$ ATP + creatine) and the sum of all ATP synthesis reactions (Pi $\rightarrow$ ATP) (6,7).

THE CHALLENGE OF THE ISOLATED MOUSE HEART

The challenge of performing an NMR experiment using a mouse heart may be best explained by comparing it to the isolated perfused rat heart preparation. The adult rat heart typically studied by NMR spectroscopy weighs ~1 - 1.5 g. The heart can be perfused with buffers such as Krebs-Henseleit buffer containing a variety of oxidizable substrates, red-blood cell enriched buffers, or even blood. Hearts can be prepared either in the isovolumic mode or Neely-Morgan ejecting heart mode, either of which is easily accommodated by a 20 mm NMR tube which fits into a 20 mm NMR probe. The cytosolic volume of a 1-g rat heart is ~ 0.5 ml/g wet weight. The NMR-sensitive volume of a 20 mm probe is ~5 ml; so the ratio of cytosolic volume to total NMR-sensitive volume is ~0.1. The total amount of ATP, PCr and Pi in the cytosol of such a heart is ~5, 10 and 2 μmoles, respectively. In a commonly used 300-400 MHz spectrometer, good signal is easily obtained with 4 min of signal averaging. Even 30 s spectra can provide usable information about peak position. Since an isolated perfused rat heart preparation is stable, both physiologically and biochemically, for ~90-120 mins, many measurements for even complex protocols can be made.

The adult mouse heart weighs 90-140 mg, or about one-tenth the size of a rat heart. The corresponding amounts of abundant phosphate-containing metabolites in the cytosol of only 50 μl are 0.5, 1 and 0.2 μmoles. Because the NMR signal is proportional to the number of nuclei in the sample, it would take 10-times longer, or 40 mins, to achieve the same signal for the mouse heart as compared to 4 mins of signal averaging for the rat heart when placed in the same 20 mm NMR probe, if all others factors are equal [2]. It goes without saying that this does not allow for many repetitive measures. The challenge is to increase the ratio of cytosolic volume (region of interest) to total NMR-sensitive volume.

THE SOLUTION FOR THE ADULT MOUSE HEART

The obvious answer to this problem is to use a smaller probe. But the physical limitations imposed by the plumbing of the isolated heart are formidable. The ejecting heart preparation so elegantly developed by Ingrid Grupp at the University of Cincinnati uses a cannula placed in the left ventricle (LV) to record systolic and diastolic pressures. This is crucial information that cannot be sacrificed. The smallest diameter NMR probe accommodating both the ejecting mouse heart and the cannula is 16 mm, not much better than 20 mm. To solve this problem, we sought to develop the isovolumic mouse heart preparation. This preparation uses a balloon placed in the LV attached to a pressure transducer, thereby reducing the bulk of the physiologic preparation so that it will fit into a smaller diameter probe, namely 10 mm. The manipulation of isolating and cannulating the mouse heart seemed formidable; placing a suitably thin balloon into the LV seemed impossible. However, Kurt Saupe and Matthias Spindler found that using polyvinyl chloride instead of latex as the balloon material provided the needed flexibility of the balloon so that

[2] Because S/N varies as the square root of the number of acquisitions, it would take 100 times longer to achieve the same S/N.

the heart is not injured (first described in 8). The typical 110-140 mg adult mouse heart perfused in the isovolumic mode fits well into a 10 mm probe. Good signal can be obtained in 8 mins in the wide-bore GE 400 MHz spectrometer in our laboratory. Use of more advanced electronics reducing noise in the spectrometer and higher magnetic field strengths would substantially reduce this time.

Figure 2 shows two ^{31}P NMR spectra for a mouse heart perfused in the isovolumic mode (perfusion pressure 75 mmHg). The heart weighed ~120 mg. It was paced at 420 beats/min and its systolic and diastolic pressures were stable at 90-100 mmHg and 8-10 mmHg during the course of the experiment. Spectra were obtained at 9.4 Tesla by averaging signal obtained from 208 rf pulses of 15 μs duration (60 degree flip angle) and delays of 2.1 s. This required 8 mins.

The spectrum on the top was obtained for the heart perfused with 11 mM glucose as the sole exogenous substrate; the one on the bottom using 11 mM glucose and 0.5 mM pyruvate as the exogenous substrates. The major difference between the two spectra is the ratio of PCr to ATP resonance areas. Just as observed for the rat heart, the pyruvate-perfused mouse heart has higher PCr to ATP and PCr to Pi ratios than for the glucose-perfused heart. Since the ATP concentration measured in freeze-clamped mouse hearts is ~30 μmoles/g Lowry protein regardless of substrate, the different ratios report different PCr contents. This is generally interpreted as showing the increased rate of high-energy phosphate production in pyruvate-supplied hearts.

Figure 3 illustrates the second type of information obtained from ^{31}P NMR spectroscopy measurements of hearts, namely phosphoryl turnover rate. We have performed magnetization transfer experiments for mouse hearts, where the resonance corresponding to the phosphoryl group of the terminal phosphate of ATP is magnetically labeled, and its rate of exchange between PCr and ATP is measured. Theoretical and experimental details of the method for measuring the unidirectional reaction velocity for creatine kinase have been described (6,7).

The spectra shown here were obtained for a 112 mg mouse heart perfused with 11 mM glucose and 0.5 mM pyruvate. Each spectrum is the sum of 128 scans (90 degree flip angle) signal averaged over 15 mins. In the spectrum on the top, the γ-phosphate of ATP was selectively irradiated with a low power pulse for 4.8 s, making it NMR-invisible. The spectrum on the bottom was obtained in an interleaved fashion with the saturation pulse applied for 0 s. Because of chemical exchange via the creatine kinase reaction, the now-NMR invisible phosphoryl group contributes to the PCr resonance. The chemical amount of PCr is constant throughout this experiment, but part of the signal is NMR-invisible. The ratio of the PCr resonance areas with and without saturation is directly related to the unidirectional rate constant for the creatine kinase reaction. By multiplying this rate constant by the PCr concentration, the unidirectional velocity is obtained. In this example, the rate constant is 0.55/s and the velocity is 7.5 mM/s.

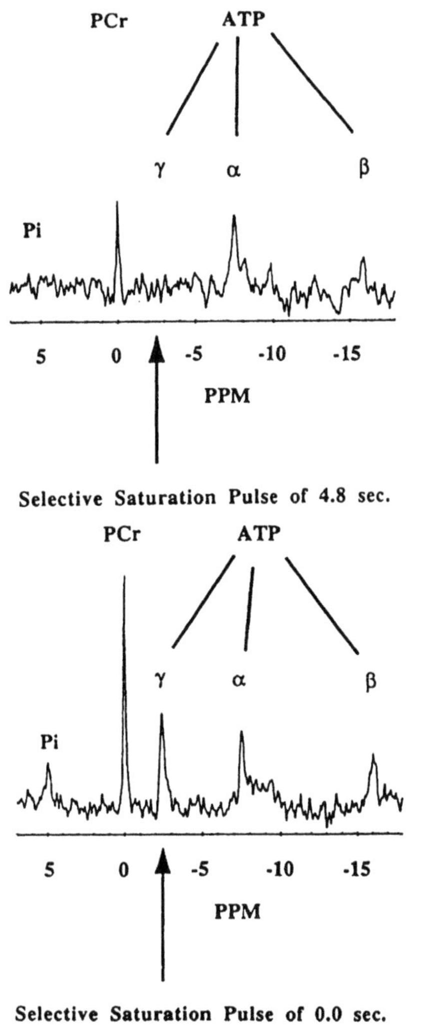

Figure 3. ^{31}P magnetization transfer: the forward creatine kinase reaction

THE SOLUTION FOR THE IMMATURE MOUSE HEART

Some problems in the biology and chemistry of the bioengineered mouse heart require measures of contractile performance and energetics in neonatal and very young mouse hearts. One of us (WM) set out to master the isovolumic heart preparation in 15 to 17-day-old mouse hearts, a time when some gene products are transiently expressed. These hearts weighed only ~40 mg. By careful scaling down of the cannula, using very, very thin plastic (stretched saran wrap or trash bags) to make the balloon and having great hands, WM was the first person in our laboratory (and likely the world) to succeed in obtaining repetitive measures of contractile performance and also oxygen consumption in such tiny hearts.

Another co-author (MMJ) has succeeded in using this approach to obtain the first ^{31}P NMR spectra from a 23-day-old mouse heart. The spectrum shown in Figure 4A was obtained using a 61 mg heart (hybrid of C57BL6 and ICR[3]). 780 acquisitions were signal averaged over 30 mins. As expected for an immature heart, the PCr/ATP was ~1.1. The ATP concentration was the same as for adult mouse (and all mammalian) hearts. These hearts contain less PCr, because the size of the creatine pool is developmentally regulated. Figure 4B is the physiologic record obtained for this mouse heart.

These results show that it is now possible to perform longitudinal studies of contractile performance and energetics of bioengineered mouse hearts beginning as early as a few weeks postpartum.

---

[3] Hearts of the ICR strain are larger than usual strains used, making this experiment using such young hearts feasible.

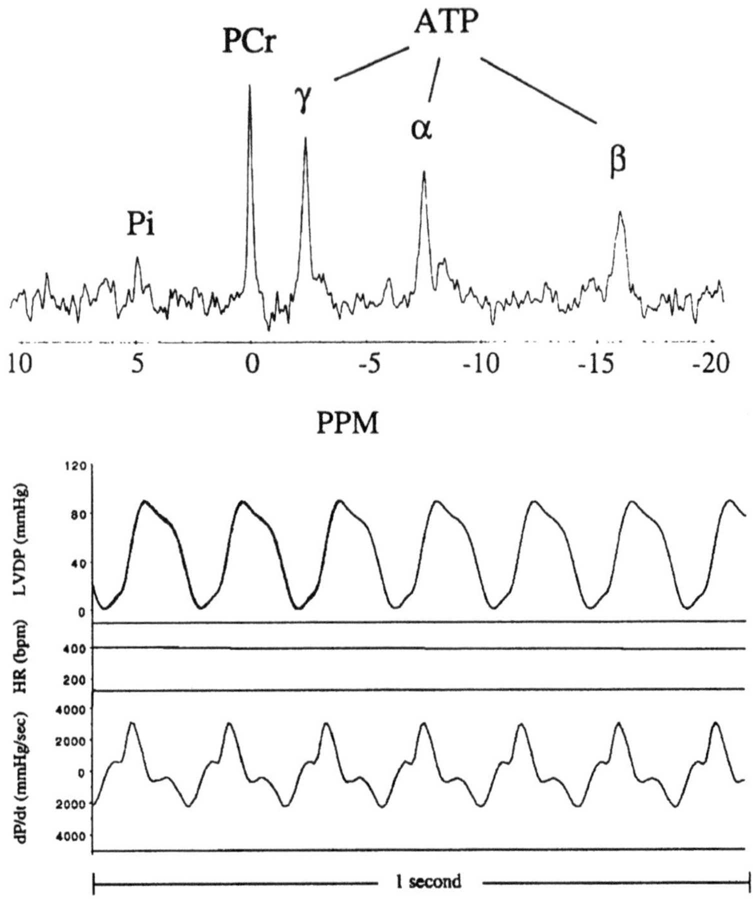

Figure 4. (Top) ^{31}P NMR spectrum of a 61 mg mouse heart. (Bottom) Tracing obtained simultaneously with the NMR experiment.

APPLICATIONS OF NMR SPECTROSCOPY TO BIO-ENGINEERED MOUSE HEARTS: A SAMPLING

The illustrations given here show that NMR spectroscopy, in combination with measurements of isovolumic contractile performance, can be used for the study of the biology and chemistry of the mouse heart. Our laboratory and now many laboratories around the world have applied this tool to mouse hearts bioengineered by ablating, replacing or increasing copy number of many specific genes. A sampling of these is presented in the table.

Genetic Modification	Physiology	Reference
70 k-Da HSP overexpression	ejecting	Radford et al, PNAS, 1996 (9)
Phospholamban knockout	isovolumic	Chu et al, Circ Res, 1996 (8)
Creatine kinase isozyme knockouts	isovolumic	Saupe et al, Circ Res, 1998 (10) Van Dorsten et al, AJP, 1998 (11) Saupe et al, JBC, 2000 (12)
Arg403Gln mutation in myosin heavy chain	isovolumic	Spindler et al, JCI, 1998 (13)
Protein kinase C overexpression	isovolumic	Tian et al, PNAS, 1999 (14)
eNOS knockout	isovolumic	Flogel et al, JMCC, 1999 (15)
Myoglobin knockout	isovolumic	Godecke et al, PNAS, 1999 (16)
β-ARK-1 overexpression	isovolumic	Cross et al, Circ Res 1999 (17)
GLUT4 knockout	isovolumic	Abel et al, JCI, 1999 (18)

ACKNOWLEDGMENTS

This work was supported by grants from the National Institutes of Health 5P50HL52320 to JSI and the National Research Science Award 1F32HL10435 to MMJ.

REFERENCES

1. Gadian DG. ed: NMR and its Applications to Living Systems, second edition. 1995; New York, Oxford University Press, Inc.
2. Ingwall JS. Phosphorus nuclear magnetic resonance spectroscopy of cardiac and skeletal muscle. Am J Physiol. 1982; 242:H729-H744.
3. Ingwall JS, Weiss RG. ^{31}P NMR Spectroscopy. The noninvasive tool for the study of the biochemistry of the cardiovascular system. Trends Cardiovasc Med. 1993; 3:29-37.
4. Moon RB, Richards JH. Determination of intracellular pH by ^{31}P NMR. J Biol Chem. 1973; 248:7276-7278.
5. Gupta RK. On the state of magnesium ion in intact cells as observed by non-invasive P-31 NMR spectroscopy. Int J Quantum Chem. 1980; 7:67-73.

6. Bittl JA, Ingwall JS. Reaction rates of creatine kinase and ATP synthesis in the isolated rat heart. J Biol Chem. 1985; 260(6):3512-3517.
7. Freidrich J, Nascimben L, Liao R, Ingwall JS. Phosphocreatine T_1 measurements with and without exchange in the heart. Magn Reson Med. 1993; 30:45-50.
8. Chu G, Luo W, Slack JP, Tilgmann C, Sweet WE, Spindler M, Saupe KW, Boivin GP, Moravec CS, Matlib MA, Grupp IL, Ingwall JS, Kranias EG. Compensatory mechanisms associated with the hyperdynamic function of phospholamban-deficient mouse hearts. Circ Res. 1996; 78:1064-1076.
9. Radford NB, Fina M, Benjamin IJ, Moreadith RW, Graves KH, Zhao P, Gavva S, Wiethoff A, Sherry AD, Malloy CR, Williams RS. Cardioprotective effects of 70-kDa heat shock protein in transgenic mice. Proc Natl Acad Sci U S A. 1996; 93(6):2339-42.
10. Saupe KW, Spindler M, Tian R, Ingwall JS. Impaired cardiac energetics in mice lacking muscle-specific isoenzymes of creatine kinase. Circ Res. 1998; 82:898-907.
11. Van Dorsten FA, Nederhoff MG, Nicolay K, Van Echteld CJ. 31P NMR studies of creatine kinase flux in M-creatine kinase-deficient mouseheart. Am J Physiol. 1998; 275(4 Pt 2):H1191-9.
12. Saupe KW, Spindler M, Hopkins JC, Shen W, Ingwall JS. Kinetic, thermodynamic, and developmental consequences of deleting creatine kinase isoenzymes from the heart: reaction kinetics of the creatine kinase isoenzymes in the intact heart. J Biol Chem. 2000; 275(26):19742-19746.
13. Spindler M, Saupe KW, Christe ME, Sweeny HL, Seidman CE, Seidman JG, Ingwall JS. Diastolic dysfunction and altered energetics in the □MHC $^{403/+}$ mouse model of familial hypertrophic cardiomyopathy. J Clin Invest. 1998; 101:1775-1783.
14. Tian R, Miao W, Spindler M, Javadpour MM, McKinney R, Bowman JC, Buttrick PM, Ingwall JS. Long-term expression of protein kinase C in adult mouse hearts improves postischemic recovery. Proc National Acad Sci. 1999; 96(23):13536-13541.
15. Flogel U, Decking UK, Godecke A, Schrader J. Contribution of NO to ischemia-reperfusion injury in the saline-perfused heart: a study in endothelial NO synthase knockout mice. J Mol Cell Cardiol. 1999; (4):827-36.
16. Godecke A, Flogel U, Zanger K, Ding Z, Hirchenhain J, Decking UK, Schrader J. Disruption of myoglobin in mice induces multiple compensatory mechanisms. Proc Natl Acad Sci USA. 1999; 96(18):10495-500.
17. Cross HR, Steenbergen C, Lefkowitz RJ, Koch WJ, Murphy E. Overexpression of the cardiac beta(2)-adrenergic receptor and expression of a beta-adrenergic receptor kinase 1 (betaARK1) inhibitor both increase myocardial contractility but have differential effects on susceptibility to ischemic injury. Circ Res. 1999; 85(11):1077-84.
18. Abel ED, Kaulbach HC, Tian R, Hopkins JCA, Duffy J, Doetschman T, Minnemann T, Boers M-E, Hadro E, Oberste-Berghaus C, Quist W, Lowell BB, Ingwall JS, Kahn BB. Cardiac hypertrophy with preserved contractile

function after deletion of GLUT4 from the heart. J Clin Invest. 1999; 104(12):1703-1714.

11
ELECTROPHYSIOLOGICAL CHARACTERIZATION OF THE MOUSE HEART USING VOLTAGE-SENSITIVE DYES

Benjamin C. Eloff, M.S. and David S. Rosenbaum, M.D.
The Heart and Vascular Research Center, MetroHealth Campus, Case Western Reserve University, Cleveland, Ohio 2500 Metrohealth Drive, Cleveland OH, 44109

ASSESSING ELECTROPHYSIOLOGICAL PHENOTYPE IN THE MOUSE HEART

The recent explosion of interest in the molecular basis of cardiac arrhythmias was spawned by discoveries of mutations of sarcolemmal ion channels which underlie the Long QT Syndrome in humans.[1-4] For instance, it is now recognized that a mutation in the delayed rectifier potassium channel (HERG) can produce loss of function resulting in prolongation of cellular repolarization and the QT interval, and can potentially explain the electrophysiological basis for arrhythmias under such circumstances. Genetic disorders involving mutations in ion channels and intracellular calcium handling proteins may underlie a number of cardiac arrhythmias such as those seen in the Brugada Syndrome,[5,6] catecholaminergic ventricular tachycardia,[7] and some forms of atrial fibrillation.[8] Importantly, investigators have evaluated the molecular basis for cardiac arrhythmias by selectively underexpressing sarcolemmal currents,[9-11] connexin channels which transmit electrical activity between cells,[12-14] or sarcoplasmic reticulum (SR) calcium handling proteins[15] using genetically engineered intact mouse hearts.

Several standard methods have been used to investigate electrophysiological phenotypes in the intact mouse heart. One choice is to impale a cell with glass microelectrodes, or record monophasic action potentials from a few sites on the heart. Another choice is to record extracellular electrograms, which does not directly measure transmembrane potential. Both of these techniques require rather large areas of tissue in which to place the electrodes, not to mention the difficulty maintaining electrode position on a viable contracting mass of myocardium. Berul et. al.[16] described an elegant technique in mouse for performing a full electrophysiological study in vivo, comparable to electrophysiological evaluation performed with intracardiac catheters in humans. The aforementioned techniques are limited by the fact that the small dimensions of the mouse heart hamper multi-site simultaneous recordings required to measure conduction pathways, conduction velocity, and action potential characteristics in the intact heart. Therefore, conventional electrode recording techniques are substantially limited in their capabilities for performing detailed electrophysiological evaluation in the intact transgenic mouse heart.

In contrast to direct electrode recordings, fluorescent imaging techniques provide an attractive alternative for performing electrophysiological characterization of small preparations such as the mouse heart. Neurobiologists first utilized fluorescent mapping techniques in the 1960's.[17] While these studies measured intrinsic changes in optical properties of excitable tissue during the action potential, the signal to noise ratio (SNR) was low, making it difficult to measure the responses. Further studies were performed where dye fluorescent light intensity changes due to transmembrane potential were measured from cardiac tissue loaded with different potentiometric dyes. The earliest dyes tested were merocyanine compounds which possessed several advantageous physical properties; e.g., they were easy to load into the preparation, they were resistant to washout, they bound the cell membrane with high affinity, and they exhibited changes in fluorescence that were proportional to transmembrane potential.[18] Since mechanical electrodes are not required (i.e. action potentials are imaged), optical mapping with voltage-sensitive florescent dyes allowed for the possibility of recording hundreds to thousands of action potentials from very small hearts with essentially unlimited spatial resolution. Consequently, with optical mapping it is possible to assess; 1. cell function in multi-cellular preparations (e.g. the intact mouse heart), 2. investigate cell to cell interactions 3. map propagation and arrhythmias, 4. record membrane responses free

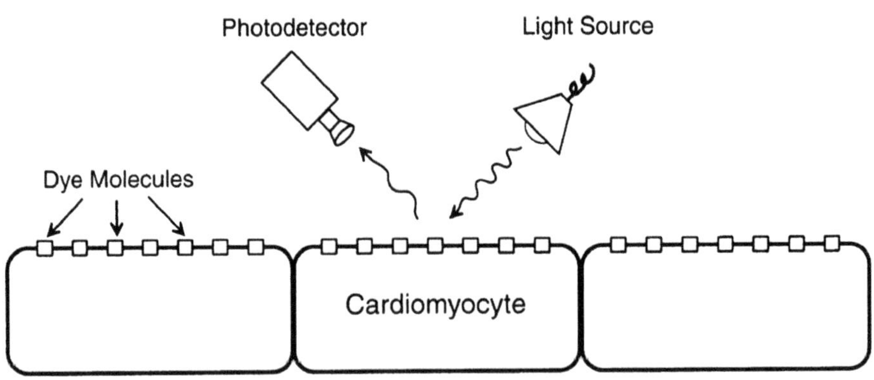

Figure 1. Basic principles of optical mapping. Voltage-sensitive dye (squares) binds plasma membrane of cardiomyocytes with high affinity. Dye molecules are excited by a specific wavelength of light, and fluoresce light in proportion to transmembrane potential. Fluoresced light emanating from multiple small aggregates of myocytes on the heart's surface is detected by an array of photodetectors. Note that the wavelength of fluoresced light is longer than that of excitation light (i.e. Stokes shift) permitting one to selectively discriminate voltage-sensitive fluorescence from the excitation light source.

of electrical artifacts produced by pacing stimuli, and 5. also evaluate other relevant physiological parameters that can be measured with florescent probes (e.g. intracellular calcium, NADH, pH).

However, there are important differences between the characteristics of action potentials recorded optically, and those recorded with conventional microelectrodes that have been reviewed in detail previously.[19]

OPTICAL MAPPING PRINCIPLES

The basic principles underlying optical mapping are outlined schematically in Figure 1. The experiment consists of loading a preparation with voltage-sensitive dye, exciting florescence from the membrane bound dye with light, and measuring fluoresced light with a photodetector. An optical mapping system consists of three major components: 1. The experimental preparation which has been loaded with a voltage sensitive dye, 2. A system of lenses and filters through which fluoresced light from the dye-stained preparation is collected and focused onto a detector, and 3. A photodetector and data acquisition system that measures the fluoresced light intensity from up to thousands of sites on the heart's surface and digitizes this information using a high-speed data acquisition computer. Each of these components is considered separately below. A schematic diagram of a typical system is shown in Figure 2.

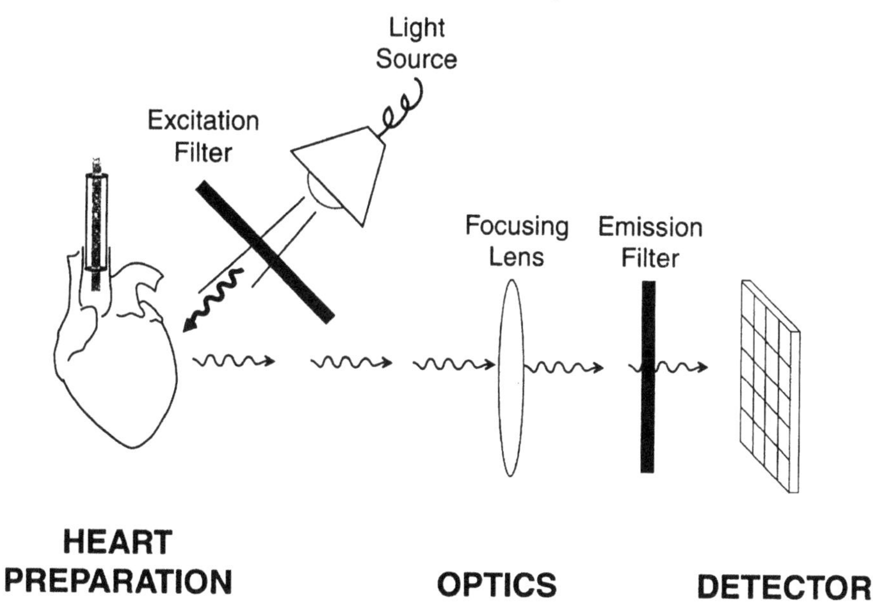

Figure 2. Typical optical mapping system. Voltage-sensitive dye is excited with light from the light source that is filtered to correspond with the peak of the dye's absorption spectrum. Fluoresced light is transmitted through the system of optics and focused onto the photodetector. Fluoresced light incident on the photodetector is filtered to remove excitation light (i.e. light reflected from the light source) and to isolate the emission spectrum.

Voltage-Sensitive Dyes

Optical mapping requires the use of a potentiometric dye that can respond to a change in transmembrane potential. Di-4-ANEPPS is the most commonly used dye for optical mapping experiments in cardiac tissue. The dye's structure is shown in Figure 3. Notice the hydrophobic structure, and the negatively charged SO_3^- group localized to one end of the molecule. These features allow the dye molecule to bind strongly to the cell membrane, while preventing diffusion into the cell. The localization of the dye to the cell membrane exposes the molecule to large potential differences. When an excited dye molecule is subjected to a change in its local electrical field, the emission spectrum of the dye shifts, which results in a change in fluoresced light intensity. The change in fluoresced light intensity is linearly related to the change in electrical potential, allowing the accurate measurement of dynamic changes in transmembrane potential. Dyes used for optical mapping purposes are required to have a response time that is as fast or faster than the time for an electrophysiological variable to change. For example, di-4-ANEPPS, has a response time of approximately 10^{-6} msec, while the most rapid change in transmembrane potential is on the order of 1 msec.

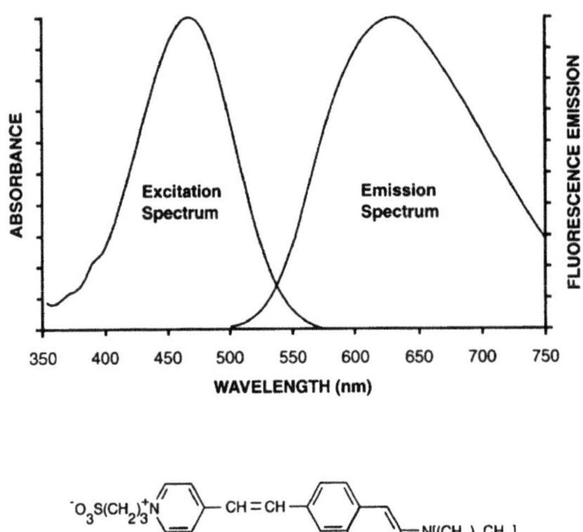

di-4-ANEPPS

Figure 3. Characteristics of the voltage-sensitive dye, di-4-ANEPPS. The dye exhibits two distinct spectra for absorbance(excitation) and emission when bound to phospholipid bilayer membranes (Top). The hydrophobic structure and negatively charged SO_3^- group on one end of the molecule (Bottom) combine to make the dye bind to the cell membrane with high affinity, and to prevent diffusion into the intercellular space. The localization of the dye to the cell membrane exposes it to large electrical fields, so maximum fluorescence can be stimulated. Modified from Molecular Probes Inc. http://www.molecularprobes.com

To measure fluorescence from di-4-ANEPPS, the investigator must select optical filters that correspond to the excitation and fluorescence spectra of the dye. Figure 3 shows the absorbance and emission spectra for di-4-ANEPPS when bound to an artificial lipid bilayer. Peak excitation is at approximately 475 nm, while peak emission occurs at around 640 nm. The shift to longer wavelengths between excitation and emission of dye is referred to as the Stokes shift, and is an important physical property of voltage-sensitive dyes which allows one to selectively identify voltage-dependent florescent signal from the voltage-independent excitation signal.

Fluorescent dyes are well suited for assessing the time course of the action potential and the timing of depolarization and repolarization between cells. However, it is difficult to calibrate the fluorescent intensity to absolute transmembrane potential since small variations in excitation light intensity and dye concentration can lead to large changes in fluoresced light intensity. However, these problems can potentially be overcome by ratiometric methods, which can permit calibration of fluorescent signal in some circumstances.[20]

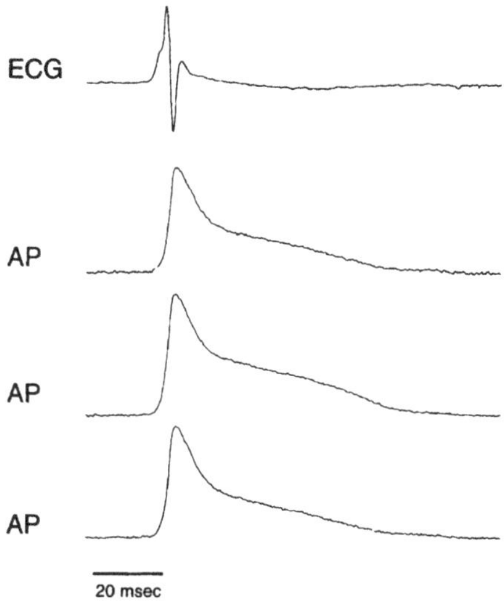

Figure 4. Electrocardiogram (top) and optical action potentials (AP) recorded from three representative sites on the epicardial surface of the mouse ventricle. Optical action potentials delineated the time course of membrane potential changes in mouse ventricle with high signal fidelity. Note the difficulty of identifying the time course of repolarization from the ECG as compared to the AP.

Figure 4 shows representative optically recorded ventricular action potentials and an ECG from a Langendorff-perfused mouse heart loaded with di-4-ANEPPS. The depolarization phase of the action potential corresponds with the QRS complex, as expected. However, the large and early phase 1 repolarization causes repolarization (the start of the t-wave) to encroach on depolarization (i.e. the QRS complex), making it very difficult to distinguish final depolarization from early repolarization. Therefore, it is difficult to completely distinguish propagation from repolarization from the mouse ECG. In contrast to the ECG, the optically recorded action potentials allow detailed and unambiguous measurements of activation and repolarization simultaneously from hundreds of sites on the ventricular epicardium.

Optics

Optical mapping systems require a specialized system of optics (i.e. filters, lenses, light sources) that is used to excite the dye and transmit the fluoresced light to the photodetector. There are many different imaging modalities that have been used which vary in complexity. For the purposes of illustration, a simple single-lens setup is depicted in figure 2.

A quartz tungsten halogen light source is used to deliver excitation to the voltage sensitive dye. Other light sources can be used (e.g. laser or Mercury arc lamp), but these provide less stable illumination intensity. The output of the light source is band-pass filtered to correspond with the excitation band of the dye (500 nm for di-4-ANEPPS). This light is directed to a flat imaging window against which the heart preparation is placed. Fluoresced light emanating from the heart is collected by the lens, and focused onto the photodetector. A filter placed between the lens and the photodetector is used to isolate the emission wavelengths of the dye. The optical magnification of the system is controlled by varying the distances of the recording chamber and the lens from the photodetector.

There are many different types of optical systems that can be used to collect and transmit light fluoresced by voltage sensitive dyes. For example, photographic lenses arranged coaxially face-to-face (tandem-lens system) can be used to increase light throughput efficiency.[21,22] A standard fluorescence microscope with a photodetector attached to its camera port can also be used, particularly when high magnifications are required. However, the small working distances associated with microscopes usually precludes use in the intact heart. Optical fibers can be used in place of lenses to collect and transmit light directly from a cultured monolayer to the photodetector.[23,24] Another system of optics is the optrode, which uses an optical fiber inserted into a micropipette for mechanical stability.[25] When combined with a voltage-sensitive dye, the optrode is capable of recording local transmembrane potential in tissue with minimal motion artifacts.

Photodetectors

There are two principal types of photodetectors, photodiode arrays and charge coupled device (CCD) video cameras. Both types of detectors are formed on silicon chips and consist of many separate pixels where incident photons are converted into electrical charge. Photodiodes allow for continuous simultaneous recording from every pixel in the array. CCD cameras collect light for a set period of time, pixels are then read sequentially from the chip. CCD detectors are analogous to film cameras, where a shutter is opened for a set amount of time, light is collected onto the film (chip), then the shutter is closed, and the film is advanced to the next frame. While this entire process is performed mechanically in a standard camera, CCD cameras use electronics to perform the same task. In order to increase signal intensity, one can increase the amount of time that the "shutter" is open, which increases the amount of incident light on the chip. There is a tradeoff, however, that increasing collection time decreases the frame (sampling) rate of the detector. In contrast, when using a photodiode array voltage resolution and sensititivy does not diminish as sampling rate increases, making this type of detector preferable for recording details of the time course of transmembrane potential.

The data acquisition systems are usually capable of recording signals from hundreds to thousands of sites simultaneously at high speeds (>1 kHz) with high (8-12 bit) precision. While both primary forms of photodetectors share many characteristics, there are very important differences that must be taken into account when designing an experimental system.

Both photodiode arrays and video cameras allow the investigator to record optical signals from many sites simultaneously. While photodiode arrays are limited usually to 128 or 256 recording sites (pixels), video cameras allow for far greater spatial resolution, by including many thousands of pixels. The tradeoff is that photodiodes are capable of recording signals with greater signal to noise ratio at faster sampling rates than video cameras. For example, the video camera is well suited for recording global propagation across the heart, and examining the nature of wavefronts and their complex interactions. In contrast, a photodiode array more readily measures the time course of the action potential. Importantly, the rapid sampling rates offered by the photodiode detector is potentially an important advantage for investigating the mouse heart, where propagation spreads rapidly over very small distances.

PREVIOUS OPTICAL MAPPING STUDIES IN MICE

Vaidya, et. al.[26] used a video detector to record epicardial propagation during ventricular fibrillation in Langendorff-perfused mouse hearts. The detailed measurements of electrical activity allowed for measurement of wavelength of reentrant circuits. By demonstrating ventricular fibrillation in very small hearts such as the mouse, this study challenged the critical mass hypothesis of ventricular fibrillation proposed originally by Garrey.[27]

Another recent study used a photodiode array to examine the electrophysiological effects of the elimination of the slowly inactivating K^+ current in mouse.[28] The loss of current was due to an overexpression of an N-terminal fragment of the K^+ channel Kv1.1.[29] This fragment causes trapping of Kv1.x subunits in the sarcoplasmic reticulum, thus inhibiting currents carried by these channels.[30] Using optical mapping, it was shown that the loss of K^+ current caused twofold increases in action potential duration and an increase in the slope of restitution. Because of these changes, hearts were more susceptible to initiation of reentrant ventricular tachycardia by the addition of a premature beat. One limitation to studies of mouse repolarization is that the repolarization currents differ significantly between mice and humans, making it difficult to apply this knowledge to clinical arrhythmias.

Several optical mapping studies have investigated the role of reduced expression of the connexin family of proteins on ventricular conduction. Connexins are proteins which comprise the principal connections between cells, gap junctions, which permit cell to cell conduction and communication. Each gap junction hemichannel is composed of six connexin proteins. The two cardiac connexin proteins that have been successfully manipulated and studied using optical mapping are Connexin43 (Cx43), the most abundant ventricular gap junction protein, and Connexin40 (Cx40), which is found primarily in the atria and the specialized conduction system of the heart. There have been two types of manipulation of the Cx43 protein, a conventional knockout that targets all Cx43 protein expression in the entire genome,[31] and more recently, a conditional knockout, where only Cx43 expression in the heart is manipulated.[32] The removal of Cx43 expression from the entire genome of the mouse results in a right ventricular outflow tract obstruction that causes Cx43-/- mice to die shortly after birth, The conditional knockout prevents this, allowing the animals to survive infancy and make them candidates for electrophysiological study.

Morley et. al.[12] used optical mapping to investigate conduction patterns in mice expressing the heterozygous mutation for Cx43. These mice produce roughly 50% of the normal quantity of connexin protein,[14] and, therefore, are expected to exhibit conduction slowing when compared to animals with normal levels of Cx43. A CCD video camera was used to measure propagation, and a velocity vector algorithm was used to quantify conduction velocity at each of multiple sites recorded from the epicardial surface. Defining the fast and slow axes of conduction by the maximum and minimum magnitudes of conduction velocity, a population of mice was compared across genotype to determine if there was indeed any conduction slowing in the Cx43+/- mouse. Surprisingly, there was no clear conduction abnormalities identified in Cx43+/- mice.

More recently, a photodiode array system was used by Eloff et. al.[33] to record propagation in Cx43+/- mice. Action potentials were recorded from 256 sites on the surface of Langendorff-perfused mouse hearts. Using this system, action potentials were recorded simultaneously from 256 ventricular sites in Langendorff –perfused mouse hearts with spatial (367 μm) voltage (1mV) and temporal (0.33 msec) resolutions that

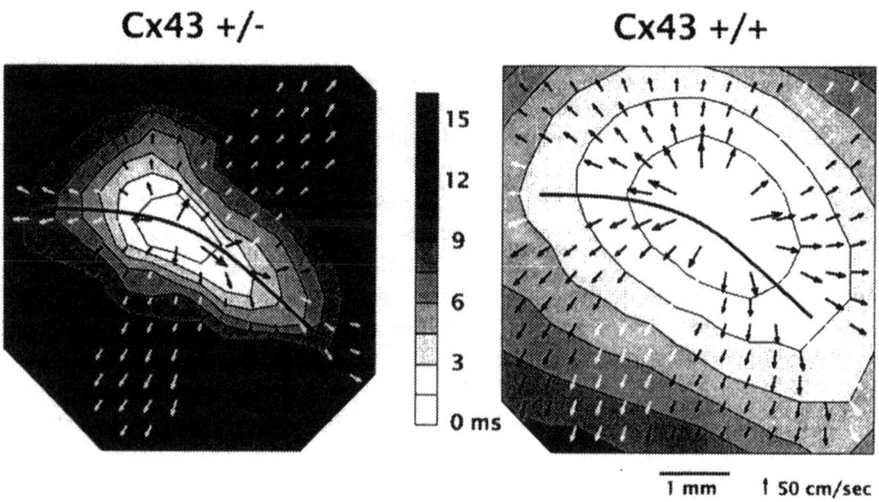

Figure 5. Contour plots (1.5ms isochrone lines) demonstrating the sequence and pattern of propagation from the point of stimulation in a representative Cx43+/- heterozygous (left) and Cx3+/+ wildtype (control) counterpart shown to the right. In each case, the impulse spreads anisotropically according to fiber orientation. Although the patterns of conduction are similar, the Cx43+/- heart is associated with substantially slower conduction velocity (crowding of isochrones). Method used to calculate conduction velocity was based on calculation of local velocity vectors (arrows) at each recording site. A computer algorithm was used to group vectors along any selected line of propagation. Shown in this example are vectors selected for propagation along lines of conduction longitudinal and transverse to fibers (white arrows), in addition to vectors determined to fall outside these lines (black arrows) by the algorithm. Conduction velocity along any path was calculated from the average of all vectors selected along a given conduction path. (Scales shown for distance and velocity vector magnitude)

were sufficient to measure the time course of membrane potential change in great detail (examples in Figure 4). Figure 5 shows representative maps of propagation from a point stimulus taken from 4X4 mm areas on the epicardial surface of Cx43+/- and wildtype mice. Notice that the entire array is depolarized within 9 ms in control, whereas the Cx43+/- mouse heart depolarizes fully in 15 ms. These conduction delays combined with the relative crowding of isochrones indicate conduction velocity slowing in the Cx43+/- mouse. The velocity vectors (arrows) allow for quantification of this result in the directions of propagation both longitudinal and transverse to cardiac fiber orientation. Conduction velocities were found to be significantly slowed (by 23% to 35%) in both axes of propagation for Cx43+/- mice when compared to control. As compared to controls (Cx43+/+), Cx43+/- mice exhibited no significant difference in PR (67 ± 20 ms versus 51 ± 9 ms) and QT (26 ± 10 versus 19 ± 9) intervals of the ECG. There was a trend toward QT prolongation in the Cx43+/- mice, and since action potential duration was unchanged(Cx43+/- 19.0 ± 3.6 ms, Cx43+/+ 20.9 ± 3.4 ms, NS), any QT prolongation could only be explained by prolongation of the QRS complex. This underscores the limitations of using the ECG to measure subtle electrophysiological phenotypes in the mouse.

The conditional Cx43 knockout mouse is unlike the conventional knockout in that it targets only the Cx43 protein expression in cardiomyocytes, leaving the rest of the animal's expression levels as normal. Because of this targeted deletion, the right ventricular outflow tract obstruction was eliminated, allowing the Cx43-/- mice to survive infancy.[31] Interestingly, the Cx43-/- mice were highly arrhythmogenic and experienced 100% mortality within two months.[34] The arrhythmogenic substrate was caused by the profound slowing in ventricular conduction and the extreme structural and electrical remodeling found in these mice, all of which were caused by elimination of cardiac gap junctions.

Cx40 is instead primarily found in the atria and specialized conduction system of the heart. The optical mapping study of the Cx40 mouse revealed no reduction in ventricular epicardial conduction velocity, as expected.[35] However, the patterns of epicardial breakthrough were found to be consistent with bundle branch block. This study demonstrated that optical mapping is not only useful for the measurement of epicardial conduction, but can be used to infer more complex conduction patterns.

CHALLENGES ASSOCIATED WITH OPTICAL MAPPING IN MICE

Tissue movement due to contraction causes a change in fluorescence which is included in the optical signal as a motion artifact. This artifact obscures the repolarization phase of the action potential, making it difficult to measure action potential duration. To reduce motion artifact, electromechanical uncoupling agents like diacetyl monoxime (DAM) and cytocholasin-D are often used, but they can be associated with unwanted electrophysiological side effects.[28] Another commonly used approach is to apply gentle pressure to the preparation with a moveable piston to mechanically stabilize the heart against the imaging window.[19]

The small size of the mouse hearts requires that a relatively high optical magnification be used because the imaging area is so small. Furthermore, the rapid conduction seen in the mouse heart requires high spatial resolution in order to measure conduction delays occurring over small distances. Ninety-five percent of fluoresced light emanates from the most epicardial 500mm of the preparation.[19] Therefore, fluoresced light intensity is directly related to the pixel area, and falls as the square of magnification. It is because of this that one must devise a method where fluoresced light can be maximized. The fluorescent signal is dependent on many factors: excitation light intensity, dye concentration, pixel size, and numerical aperture. An increase in the dye concentration could be toxic. An increase in excitation light intensity can cause photobleaching or even phototoxicity. Therefore, the remaining factor to control is the optical throughput of the system. For this purpose, it may be necessary to use fluorescence microscopes, or the tandem lens optical system mentioned above. Both systems have many advantages and drawbacks, and the optimal choice must be guided by experimental need.

The relatively rapid time course of the murine cardiac action potential introduces complexities into the design of the data acquisition system. The depolarization time varies with magnification due to the blurring of the upstroke because of propagation across a recording pixel, but is generally in the order of 1 – 0.5 ms. In order to interpret the details of the action potential, one must be capable of recording signals so that the fastest phase (depolarization) of the action potential can be accurately registered. Therefore, it is necessary to record optical action potentials with sufficiently fast sampling rates. Typically, sampling rates of 2-4 thousand samples/sec/recording site are adequate.

In conclusion, optical mapping is a powerful technique for electrophysiological characterization of mouse hearts. Thus far, the technique has been used successfully in measuring conduction, repolarization, and arrhythmias from the surface of the heart. The technique offers clear advantages over electrocardiographic assessment of mouse phenotypes, since the electrocardiogram conveys only global electrical activity, and is neither sensitive or specific. As is the case with microelectrode impalements, optical mapping conveys important information on the time course of membrane potential change. However, optical mapping is the only technique that allows for monitoring of cellular activity from hundreds to thousands of closely placed recording sites in the intact and beating heart. Clearly, this approach will be used increasingly to investigate cell to cell interactions, propagation, and arrhythmias in transgenic animals.

REFERENCES:

1. Curran ME, Splawski I, Timothy KW, Vincent GM, Green ED, Keating MT. A molecular basis for cardiac arrhythmia: *HERG* mutations cause long QT syndrome. *Cell.* 1995;80:795-803.
2. Sanguinetti MC, Jiang C, Curran ME, Keating MT. A mechanistic link between an inherited and an acquired cardiac arrhythmia: *HERG* encodes the I_{Kr} potassium channel. *Cell.* 1995;81:299-307.
3. Sanguinetti MC, Curran ME, Spector PS, Keating MT. Spectrum of HERG K^+-channel dysfunction in an inherited cardiac arrhythmia. *Proc.Natl.Acad.Sci.USA.* 1996;93:2208-2212.
4. Wang Q, Curran ME, Splawski I, Burn TC, Millholland JM, VanRaay TJ, Shen J, Timothy KW, Vincent GM, De Jager T, Schwartz PJ, Towbin JA, Moss AJ, Atkinson DL, Landes GM, Connors TD, Keating MT. Positional cloning of a novel potassium channel gene: *KVLQT1* mutations cause cardiac arrhythmias. *Nature Genet.* 1996;12:17-23.
5. Chen QY, Kirsch GE, Zhang DM, Brugada R, Brugada J, Brugada P, Potenza D, Moya A, Borggrefe M, Breithardt G, Ortiz-Lopez R, Wang Z, Antzelevitch C, O'Brien RE, Schulze-Bahr E, Keating MT, Towbin JA, Wang Q. Genetic basis and molecular mechanism for idiopathic: ventricular fibrillation. *Nature.* 1998;392:293-296.
6. Gussak I, Antzelevitch C, Bjerregaard P, Towbin JA, Chaitman BR. The Brugada syndrome: Clinical, electrophysiologic and genetic aspects. *J.Am.Coll.Cardiol.* Jan 1999;33:5-15.
7. Priori SG, Napolitano C, Tiso N, Memmi M, Vignati G, Bloise R, Sorrentino V, Danieli GA. Mutations in the cardiac ryanodine receptor gene (*hRyR2*) underlie catecholaminergic polymorphic ventricular tachycardia. *Circulation.* 2001;103:196-200.
8. Brugada R, Tapscott T, Czernuszewicz GZ, Marian AJ, Iglesias A, Mont L, Brugada J, Girona J, Domingo A, Bachinski LL, Roberts R. Identification of a genetic locus for familial atrial fibrillation. *N Engl J Med.* 1997;336:905-.
9. London B, Jeron A, Zhou J, Buckett P, Han XG, Mitchell GF, Koren G. Long QT and ventricular arrhythmias in transgenic mice expressing the N terminus and first transmembrane segment of a voltage-gated potassium channel. *Proc.Natl.Acad.Sci.USA.* 1998;95:2926-2931.
10. Barry DM, Xu HD, Schuessler RB, Nerbonne JM. Functional knockout of the transient outward current, long-QT syndrome, and cardiac remodeling in mice expressing a dominant-negative Kv4 α subunit. *Circ.Res.* 1998;83:560-567.
11. Xu HD, Li HL, Nerbonne JM. Elimination of the transient outward current and action potential prolongation in mouse atrial myocytes expressing a dominant negative Kv4 α subunit. *J.Physiol.(Lond.).* 1999;519:11-21.
12. Morley GE, Vaidya D, Samie FH, Lo C, Delmar M, Jalife J. Characterization of conduction in the ventricles of normal and heterozygous Cx43 knockout mice using optical mapping. *J.Cardiovasc.Electrophysiol.* 1999;10:1361-1375.
13. Guerrero PA, Schuessler RB, Davis LM, Beyer EC, Johnson CM, Yamada KA, Saffitz JE. Slow ventricular conduction in mice heterozygous for a connexin43 null mutation. *J.Clin.Invest.* 1997;99:1991-1998.
14. Thomas SA, Schuessler RB, Berul CI, Beardslee MA, Beyer EC, Mendelsohn ME, Saffitz JE. Disparate effects of deficient expression of connexin43 on atrial and ventricular conduction - Evidence for chamber-specific molecular determinants of conduction. *Circulation.* 1998;97:686-691.
15. Masaki H, Sako H, Kadambi VJ, Sato Y, Kranias EG, Yatani A. Overexpression of phospholamban alters inactivation kinetics of L-type Ca^{2+} channel currents in mouse atrial myocytes. *J.Mol.Cell.Cardiol.* 1998;30:317-325.

16. Berul CI, Aronovitz MJ, Wang PJ, Mendelsohn ME. In vivo cardiac electrophysiology studies in the mouse. *Circulation.* 1996;94:2641-2648.
17. Cohen LB, Keynes RD, Hille B. Light scattering and birefringence changes during nerve activity. *Nature (London).* 1968;218:438-441.
18. Salama G. Optical measurements of transmembrane potentials in heart. In: Spectroscopic Probes of Membraine Potential.1988. CRC Uniscience Publications.
19. Girouard SD, Laurita KR, Rosenbaum DS. Unique properties of cardiac action potentials recorded with voltage-sensitive dyes. *J.Cardiovasc.Electrophysiol.* 1996;7:1024-1038.
20. Knisley SB, Justice RK, Kong W, Johnson PL. Ratiometry of transmembrane voltage-sensitive fluorescent dye emission in hearts. *Am.J.Physiol.Heart Circ.Physiol.* 2000;279:H1421-H1433.
21. Libbus I., Eloff B. C., Girouard S. D., Rosenbaum D. S.Advantages and limitations of a tandem lens design in high resolution optical imaging. *Ann Biomed Eng* 98;26:S-19. Abstract.
22. Ratzlaff EH, Grinvald A. A tandem-lens epifluorescence macroscope: hundred-fold brightness advantage for wide-field imaging. *J Neurosci Methods.* 1991;36:127-137.
23. Thomas SP, Bircher-Lehmann L, Thomas SA, Zhuang JP, Saffitz JE, Kléber AG. Synthetic strands of neonatal mouse cardiac myocytes - Structural and electrophysiological properties. *Circ.Res.* 2000;87:467-473.
24. Fast VG, Kléber AG. Anisotropic conduction in monolayers of neonatal rat heart cells cultured on collagen substrate. *Circ.Res.* 1994;75:591-595.
25. Neunlist M, Zou S, Tung L. Design and use of an "optrode" for optical recordings of cardiac action potentials. *Pflugers Arch.* 1992;420:611-6171.
26. Vaidya D, Morley GE, Samie FH, Jalife J. Reentry and fibrillation in the mouse heart - A challenge to the critical mass hypothesis. *Circ.Res.* Jul 23 1999;85:174-181.
27. Garrey W. The nature of fibrillary contraction of the heart: Its relation to tissue mass and form . *Am.J.Physiol.* 1914;33:397-414.
28. Baker LC, London B, Choi BR, Koren G, Salama G. Enhanced dispersion of repolarization and refractoriness in transgenic mouse hearts promotes reentrant ventricular tachycardia. *Circ.Res.* 2000;86:396-407.
29. Zhou J, Jeron A, London B, Han XQ, Koren G. Characterization of a slowly inactivating outward current in adult mouse ventricular myocytes. *Circ.Res.* 1998;83:806-814.
30. Folco E, Mathur R, Mori Y, Buckett P, Koren G. A cellular model for long QT syndrome - Trapping of heteromultimeric complexes consisting of truncated Kv1.1 potassium channel polypeptides and native Kv1.4 and Kv1.5 channels in the endoplasmic reticulum. *J.Biol.Chem.* 1997;272:26505-26510.
31. Reaume AG, de Sousa PA, Kulkarni S, Langille BL, Zhu D, Davies TC, Juneja SC, Kidder GM, Rossant J. Cardiac malformation in neonatal mice lacking connexin43. *Science.* 1995;267:1831-1834.
32. Gutstein DE, Morley GE, Tamaddon H, Vaidya D, Schneider MD, Chen J, Chien KR, Stuhlmann H, Fishman GI. Conduction slowing and sudden arrhythmic death in mice with cardiac-restricted inactivation of connexin43. *Circ Res.* 2001;88(3):333-9.
33. Eloff B. C., Lerner D. L., Yamada K. A., Schuessler R. B., Saffitz J. E., Rosenbaum D. S.High resolution optical mapping reveals conduction slowing in connexin43 deficient mice. *Circulation* 2000;102:II-3. Abstract.
34. Gutstein D. E., Morley G. E., Tamaddon H. S, Vaidya D. M., Schneider M. D., Chen J., Chien K. R., Stuhlmann H., Fishman G. I.Genetic manipulation of connexin43 expression in the heart: Establishing a role for gap junction remodeling in arrhythmogenesis and ventricular dysfunction. *Circulation* 2000;102:II-15. Abstract.
35. Tamaddon HS, Vaidya D, Simon AM, Paul DL, Jalife J, Morley GE. High-resolution optical mapping of the right bundle branch in connexin40 knockout mice reveals slow conduction in the specialized conduction system. *Circ.Res.* 2000;87:929-936.

12
ECHOCARDIOGRAPHIC ASSESSMENT OF THE MOUSE HEART AND AORTA

Brian D. Hoit, M.D.
Case Western Reserve University and University Hospitals of Cleveland 11100 Euclid Avenue, Cleveland OH, 44106-5038

INTRODUCTION

Two-dimensionally (2D) directed echocardiography is arguably the leading method for imaging the cardiovascular system in small animals because it is noninvasive, versatile, readily available, and well suited for serial studies over a wide range of ages. M-mode and Doppler echocardiography have been used to phenotype the cardiovascular system in a variety of genetic mouse models by numerous laboratories over the past five years. More recently, reports of 2D and 3D echocardiography in mice have appeared. These techniques have proved useful in determining the interplay between altered cardiovascular gene expression and compensatory physiologic cardiovascular regulation in the genetically engineered mouse. Although the rapid beating frequency and small size of the mouse heart place constraints on an echocardiographic assessment, the use of increasingly sophisticated transducers operating at a high frequency results in highly reliable and reproducible image quality. The major advantages of the *in vivo* echocardiographic approach include the ability to assess age- or disease-related cardiovascular changes serially over time and to determine the effects of genetic, surgical, and pharmacological interventions in the intact animal. Thus, the structural and functional data that are derived from echocardiography complement and extend information obtained from other *in vitro and in vivo* physiological studies.

The objective of this chapter is to describe the imaging and Doppler echocardiographic methods that are currently used to evaluate cardiovascular function in genetically engineered mice.

Theoretical Background

Ultrasound is a mechanical radiation (and therefore requires a medium for propagation) that can be directed in a beam, conforms to the laws of reflection and refraction, and is reflected by small objects [1]. The major limitation of ultrasound is that it travels poorly through air and highly dense substances (i.e. lung and bone, respectively). When ultrasound travelling in one medium reaches a medium with different acoustic properties (acoustic impedance, or density), a small percent of ultrasound energy is reflected. The intensity of the reflected energy is directly related to the differences in acoustic impedances of the two media and is used by the scan converter on the ultrasonograph to form an image. Reflection is also related to the thickness of the target and the angle of incidence of the ultrasound beam. An important physical characteristic of ultrasound is that it is attenuated as it travels through a medium, and this loss of acoustic energy is directly related to the frequency of ultrasound. However, attenuation is generally not an issue in mice because of the shallow imaging depths that are employed.

Either a single crystal or an array of elements that is either mechanically or electronically steered produces a three-dimensional ultrasound beam. The beam intensity is optimal in the focal zone (compared to the near and far fields); focusing is accomplished either acoustically or electronically.

The most important factors that influence resolution (i.e. the ability to separate two targets as distinct echos) are the frequency of the ultrasound beam and the beam width; axial resolution increases directly with transmission frequency and lateral resolution varies inversely with beam width. To the extent that system gain directly affects beam width, gain is a determinant of resolution. Therefore, imaging with high frequency transducers and appropriate use of the instrument's gain controls are critically important for obtaining accurate ultrasound images.

The Doppler equation relates the velocity (V) of a moving target (e.g. red blood cells) to the shift in frequency (Δf) produced by the moving target, the speed of sound (c), the transducer frequency (f_t), and the intercept angle (θ) between the ultrasound beam and the direction of the target:

$$V = \frac{\Delta f \times c}{2f_t (\cos \theta)} \quad (1)$$

This relation dictates that for recording accurate velocities, the angle between the ultrasound beam and the direction of blood flow should be $< 20^\circ$ (i.e. cos θ is near 1), and that the transducer frequency is inversely related to the ability to record high velocities. Thus, the optimal recording conditions for Doppler (which resolves best with a parallel beam and low frequencies) are diametrically opposed to those for imaging (which resolves best with a perpendicular beam and high frequencies). Consequently, optimal imaging may not provide optimal Doppler, and vice versa.

METHODOLOGY

In our laboratory, cardiac ultrasound studies are performed with an ultrasonograph (Acuson Sequoia™) that uses a 15 MHz linear array transducer (theoretical axial resolution of ~ 0.2 mm). Although other imaging systems using as low as a 7.0 MHz phased array transducer have been used in the past, higher frequency transducers (12-15 MHz) are now routinely employed. Centrifuged acoustic coupling gel (which eliminates air bubbles) is placed on a thin layer of gel applied to the cleanly shaved left hemithorax. We continue to favor the use of tribromoethanol (formerly, Avertin) 2.5% (0.01 ml/g I.P.) anesthesia because of its relative ease-of-use, short half-life, and minimal cardiodepressant properties [2].

Imaging Echocardiography

2D-directed M-mode echocardiography is an attractive modality because of its ability to noninvasively and repeatedly assess ventricular dimensions and shortening with both high temporal and axial resolution (Figure 1). M-mode echo provides excellent temporal resolution owing to its high sampling rate (1,000/sec), but is limited in that a unidimensional, "ice pick" view of the heart is produced.

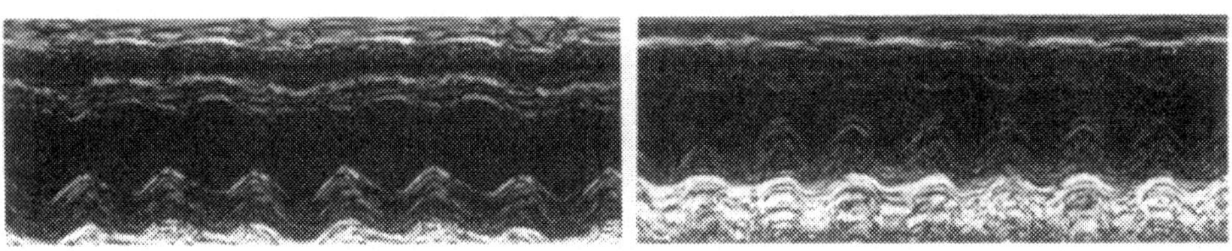

Figure 1. Representative M-mode tracing from wild type control mice. Left panel : C57Bl/6J. Right panel: A/J. IVS = interventricular septum; pw = posterior wall.

Although M-mode echo can be used to assess left ventricular morphology, this technique is optimally suited only for ventricles with uniform geometry and wall motion. Newer generation ultrasonographs with high frame rate two-dimensional echo potentially overcomes problems due to regional differences; however, the limited acoustic windows and imaging planes available in the mouse remain significant constraints. Nevertheless, M-mode echocardiography has been shown to provide reasonably accurate estimates of left ventricular mass, dimensions and percentage fractional shortening [3-6]. Correlations between echocardiographic and gravimetric LV weights have been quite good, although absolute values are usually not attainable [7]. The large standard deviation and interpretative variability for wall thickness measurements preclude detection of subtle differences (e.g. < 25%) in LV mass between two groups. Thus, it is essential that studies are performed and measurements made by blinded observers, and that transgenic animals are compared with their wildtype littermate controls.

Examples of M-mode and Doppler in Genetically Altered Mice

Echocardiography has been used to study several well-characterized mouse genetic models of heart disease [3-6, 8-10]. In our initial studies, we reported echocardiographic findings in mice with either phospholamban overexpression (PLBOE) or deficiency (PLBKO). Phospholamban reversibly regulates the activity of the SR Ca^{2+} ATPase, and thereby modulates basal and catecholamine-stimulated contractility of the heart. Compared to controls, PLBKO mice displayed enhanced LV systolic function, indicated by velocity-dependent indices of contraction, such as Vcf [3]. In contrast, PLBOE mice exhibited depressed LV fractional shortening and Vcf [4]. Furthermore, in both studies the response of LV function parameters to isoproterenol stimulation (2.0 μg I.P.) was directly related to the amount of expressed phospholamban.

A particular strength of echocardiography is the ability to examine serially and repetitively the cardiac effects of an intervention. This is illustrated by the following study in which the transverse aorta of PLBKO and isogenic wild type mice were banded and studied with echocardiographically at 2.5, 5 and 10 weeks [11]. The magnitude of the aortic constriction was also determined noninvasively using continuous wave Doppler (see below). The time course and magnitude of hypertrophy, and the development of pulmonary congestion (necropsy lung weight) were similar in PLBKO and wild types (7 of 13 and 5 of 16, respectively). Thus, these studies suggest that despite their hyperdynamic function and increased energy demand, PLBKO mice are able to tolerate long-term aortic banding.

PKC isoforms initiate important intracellular signaling pathways; echocardiography helped elucidate the *in vivo* functional consequence of isoform-specific activation. For example, in mice with transgenic overexpression of the PKCβ2 isoform, impairment of LV systolic function was associated with reduced LV chamber dimension and increased LV mass [8]. Moreover, administration of a specific PKCβ isoform inhibitor normalized left ventricular hypertrophy and markedly improved LV fractional shortening. The relevance of this transgenic model is underscored by the finding that this PKC isoform is activated in end stage failure human hearts and in streptozotocin-induced diabetic rats. In contrast, transgenic mice with cardiac-specific overexpression of the constituitively active PKCε isoform demonstrated concentric hypertrophy with normal *in vivo* cardiac function [12]. These data suggest that distinct PKC isoforms play differential roles in cell signaling pathways in cardiac hypertrophy and failure.

As a final example, concentric hypertrophy (increased M-mode LV mass, increased h/D ratio) with increased LV systolic function was produced in transgenic mice overexpressing the sarcoplasmic reticulum calcium storage protein, calsequestrin (Figure 2). These examples are only illustrative; other models illustrating the utility of ultrasound in genetically altered mice may be found throughout this monograph.

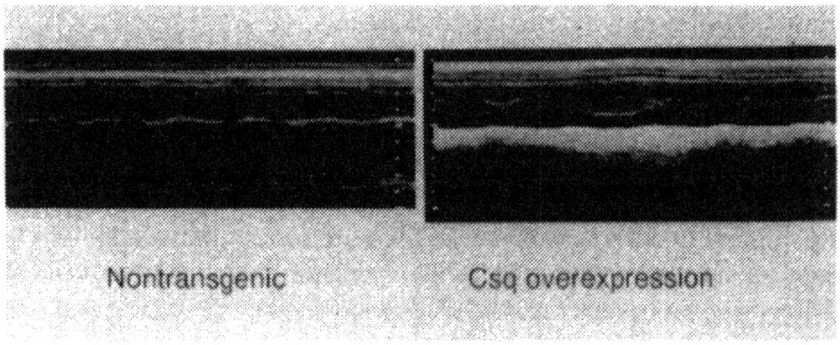

Figure 2. M-mode echocardiograms in control FVB/N (left) and calsequestrin overexpressing mice (right).

Two-Dimensional Echocardiography

Newly available linear array, broadband transducers operating at 12-15 MHz enhance spatial, contrast, and temporal resolution and near field definition. These transducers obviate the need for acoustic standoffs (simplifying and speeding image acquisition), and facilitate imaging cardiac chambers other than the left ventricle (Figure 3).

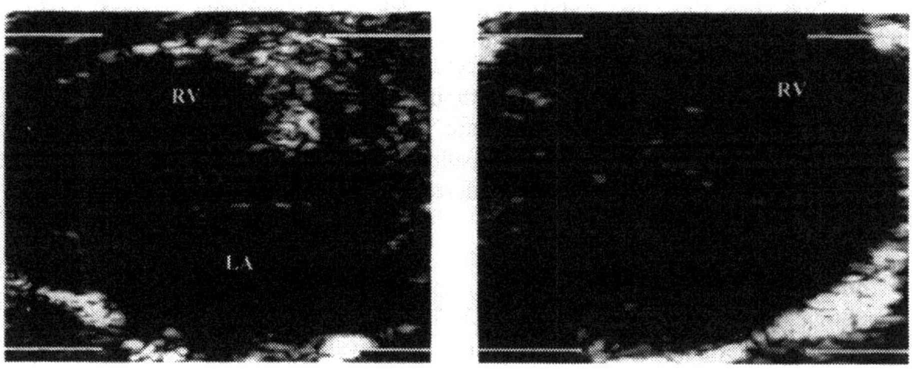

Figure 3. Long (left) and short axes (right) from mice exposed to chronic hypoxia. Note the prominent right ventricle (RV) and the left atrium (LA).

Newer generation scanners with high frame rates permit high-resolution real time tomographic imaging in multiple planes, making it possible to calculate two-dimensional LV volume and mass using standard geometric formulae. Accurate, serial assessment of LV mass after aortic banding was initially established with M-mode [11, 13], and was recently demonstrated with 2-D area-length methods [14] Biplanar area-length measurements of LV mass are feasible and were shown to be superior to M-mode estimates when compared to necropsy LV mass (r=.91, SEE=5.9 mg, slope =.795 vs r= .78, SEE= 8.8 mg, slope = .577, 2-D vs M-mode echo, respectively) [15].

In ventricles with uniform geometry and wall motion, we found only a minimally improved correlation with necropsy LV weight using the 2D area-length than M-mode echo method (r=0.87 vs 0.85); however, the slope of the regression more closely approached unity (slope=0.87 vs 0.71) and there was a better degree of agreement (10.2± vs 13.1±%) [16].

Importantly, precise two-dimensional echo measurements eliminate the M-mode echo requirements for uniform ventricular geometry. Thus, 2-D echo has been shown to accurately quantitate global and regional changes in ventricular volume that occur in the post-infarction model [17]. Using this model, LV volumes, ejection fraction, and echo-determined cardiac output were generated by three-dimensional reconstruction of four to five tomographic 2-D short axis sections using Simpson's rule algorithm, and were validated against aortic flow probe flow [r=0.8, slope = 1.05]).

Assessment of right ventricular structure and function is especially challenging because of the unusual geometry of the right ventricle, and the limited access to that chamber. Right ventricular volume and flow estimates were derived from transesophageal images in mice using a 3.5F, 30MHz intravascular catheter and were validated with postmortem casts and flow probe-determined flow [18]. The echo-derived free wall mass and necropsy RV free wall mass correlated well (r=. 81), although the transesophageal echo resulted in systematic overestimation of flow (slope =. 53). The same group used these methods to detect and quantify the hypertrophic effects of chronic hypoxia and congenital NOS3 deficiency on the right ventricle [18, 19]. Although this innovative method is potentially useful, frame rates with this catheter are slow (8 Hz), and the transesophageal echocardiogram necessitates tracheal intubation, thereby increasing complexity, risk of animal loss, cost, and length of the procedure.

Doppler Echocardiography

Aortic outflow and diastolic transmitral LV inflow velocities are interrogated from angulated parasternal long axis views using pulsed wave Doppler (5 - 7.5 MHz). The smallest sample volume available should be used. Echoes from the mitral valve annulus and aortic root are readily defined, and we have found that color flow mapping Doppler assists sample volume placement. In contrast, Hartley and colleagues blindly record ascending aortic flow with a custom-designed, 20 MHz instrument coupled to a 0.5 mm diameter transducer that is mounted on a blunted 20 G needle [20].

Doppler interrogation of transvalvular flows permits assessment of global left ventricular systolic and diastolic function without regard for ventricular geometry. Thus, aortic Doppler waveform analysis can estimate LV stroke volume and systolic performance. The area under the aortic velocity curve during systole (Ao_{vti}) represents the distance blood is ejected each beat and when multiplied by a measurement of aortic cross-sectional area (CSA) derived from M-mode at the aortic annulus yields stroke volume (SV):

$$CSA = \pi(AoD/2)^2 \qquad (2)$$

$$SV = (Ao_{vti})(CSA) \quad (3)$$

The potential sources of error in estimating cardiac output by Doppler include the angle between the ultrasound beam and flow, and the measurement accuracy of the ascending aortic diameter. Irrespective of the instrument and technique used, it is important to align the ultrasound beam as parallel as possible to the flow of interest and record the highest obtainable velocity. The peak aortic velocities we reported are comparable to those obtained by others using similar equipment, whereas Hartley et al recorded higher peak velocities using a custom-built transducer (90±11 (SD) cm/s); these higher velocities probably reflect a more parallel alignment between the sound beam and aortic flow [20]. These authors noninvasively estimated stroke volume (33 µl) and cardiac output (15.6 ml/min) in mice from the aortic Doppler waveform and an average cast-derived aortic diameter of 1.2 mm. Fentzke and coworkers reported measurements of aortic diameter only slightly smaller than those reported by Hartley et al; nevertheless, small errors in diameter measurements result in significant errors in the estimate of cross-sectional area [21]. As a result, the accuracy of Doppler cardiac output measurements typically range from ±10 to ±25%. Imaging aortic flow with a narrow intercept angle is often impossible and estimates of LV stroke volume and aortic flow are therefore limited with Doppler. Accordingly, these methods are better suited for serial (relative) measurements of flow velocity, rather than absolute volumetric flow.

Pressure gradients (high velocity), such as those across a surgically-produced transverse aortic constriction , are unambiguously detected with continuous wave CW Doppler. The pressure gradient (ΔP) is derived from the simplified Bernoulli formula using the maximal velocity (V) across a stenosis:

$$\Delta P=4V^2 \quad (4)$$

Errors are due primarily to an inability to direct the ultrasound beam parallel to flow (i.e. a large intercept angle, θ). Viscous and inertial losses are usually negligible.

In contrast to velocity estimates, measurements of ejection time (ET in ms) are independent of the orientation of the ultrasound beam, and when combined with LV fractional shortening (FS), result in the mean velocity of circumferential shortening (V_{cf}, in sec $^{-1}$):

$$V_{cf} = FS/ET \quad (5)$$

Peak aortic velocity and mean acceleration of aortic flow velocity [(peak aortic velocity)/(time from onset to peak velocity)] correlate with maximal LV dP/dt, peak flow and maximum flow rate; however, these parameters may be difficult to measure accurately in the mouse. In addition, like all ejection phase indices, they are influenced by loading conditions and are therefore unable to distinguish between the mechanical effects of abnormal loads and intrinsic alterations in inotropic (contractile) state (14).

Transmitral Doppler waveform analysis provides insight into the temporal distribution of left ventricular filling (Figure 4). Specifically, ventricular filling patterns (e.g. E/A ratios) characteristic of impaired LV relaxation and reduced ventricular compliance are described in species larger than the mouse [22]. Although the transmitral flow profile may identify abnormalities of LV diastolic function [23], these indices alone are relatively insensitive. Several "new" indices of diastolic function are potentially useful in this regard, but are each technically challenging. For example, in larger animals, the pattern of pulmonary venous flow is coupled with transmitral profiles to more accurately diagnose diastolic dysfunction [24], and the isovolumic relaxation time (IVRT) is employed as an index of LV pressure decay. However, these methods have serious limitations. Specifically, pulmonary venous flow will vary with changes in atrial pressure and function, and the IVRT varies not only with LV relaxation rate, but with the level of aortic diastolic and left atrial pressures (i.e. the loading conditions), which are often unknown. In addition, heart rate, age, and technical variables such as the sample volume site influence noninvasive indices of left ventricular diastolic function [25].

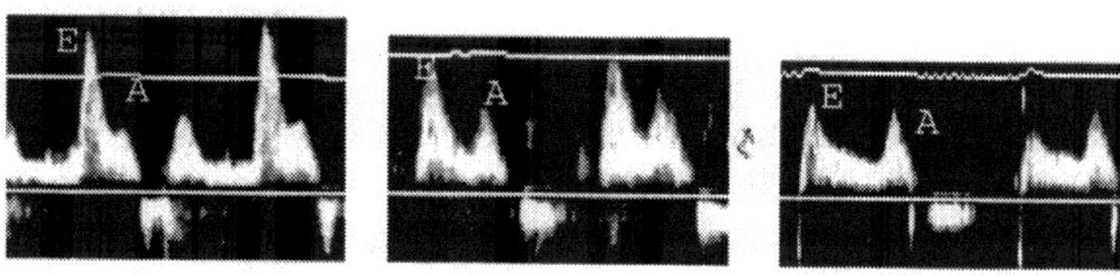

a. b. c.

Figure 4. Transmitral flow profiles: a. Increased E/A ratio (restrictive pattern) suggesting a noncompliant ventricle or increased left atrial pressures; b. normal or pseudonormal pattern with a normal E/A ratio; c. Reduced E/A ratio (impaired relaxation pattern). E= early diastolic flow; A= late (atrial systolic) diastolic flow.

A relatively load-independent measure of LV diastolic relaxation, the propagation velocity (Vp) of transmitral diastolic flow[26], is derived from color M-mode flow (Figure 5); this technique provides the spatiotemporal distribution of velocities at many points along a single scan line (in this case, bisecting mitral inflow from annulus to apex). Preliminary work in our laboratory has demonstrated the ability to identify diastolic dysfunction using these Doppler indices from transgenic mice overexpressing a missense mutation of the α tropomyosin molecule (TMGlu180Gly) that results in hypertrophic cardiomyopathy [23]. In addition, we found that a "superinhibiting" phospholamban mutant (Asparagine[27] to Alanine, PLB-N27A), exhibits impaired diastolic function as determined by reduced transmitral E/A ratio, prolonged IVRT, and reduced velocity of diastolic transmitral flow propagation [27].

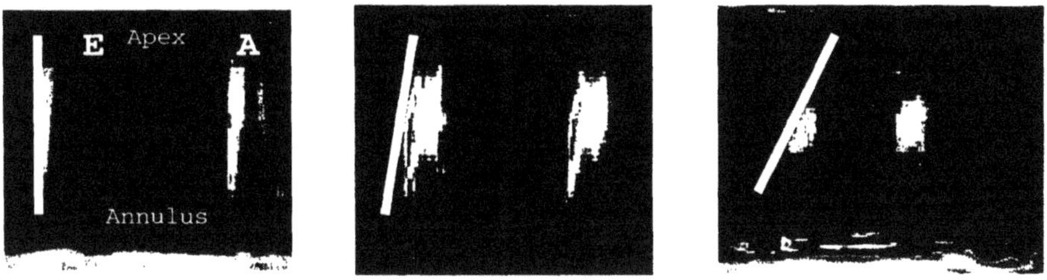

Figure 5. Color M mode of diastolic flow. The M-mode cursor bisects the mitral annulus and apex. The slope of early diastolic flow (E) represents the propagation velocity, Vp.

LV Stress-Shortening Relation

The LV end-systolic stress-shortening relation is a useful index of contractility which is relatively load-independent, and has been used in a variety of experimental preparations and animal species [28]. It should be recognized that LV fractional shortening may be normal in a ventricle with abnormal myocardial contractility if end-systolic stress is low and conversely, that LV fractional shortening may be reduced in an otherwise normal ventricle if the afterload is excessive. Furthermore, an increase in LV hypertrophy reduces end-systolic stress and may either mask intrinsic contractile dysfunction or mimic a hypercontractile state. Stress-shortening relations overcome these limitations, and can be used to compare mice of different weights and sizes because end-systolic wall stress (σ) incorporates LV chamber diameter (D), pressure (P) and ventricular wall thickness (h):

$$\sigma \alpha \frac{D \cdot P}{h} \qquad (5)$$

Left ventricular wall stresses are oriented in the circumferential, meridional, and radial directions. Although circumferential stress has the theoretical advantage of being directed in the same plane as the majority of LV fibers, the requisite LV long axis cannot be determined readily from M-mode echo. However, the LV diameter allows measurement of meridional wall stress (σ_{mes}), which is related to LV performance [29]. The latter can be computed at end-systole (es) as:

$$\sigma_{mes} = \frac{[(P_{es})(D_{es})(1.35)]}{4(h_{es})(1+ h_{es}/D_{es})} \qquad (6)$$

We previously showed that linear meridional stress-shortening relations can be determined in the adult mouse with the use of methoxamine-induced changes in systemic arterial pressure (measured with a fluid-filled catheter in the femoral artery), and that these relations are capable of detecting load independent changes in myocardial contractility [30]. Global LV function was analyzed in mice both before and after dobutamine- and verapamil-altered inotropic state using the relation between the M-mode echocardiographically-derived LV end-systolic meridional stress (using the peak systolic pressure) and LV shortening fraction (Figure 6). Of note, single stress-shortening coordinates could identify intrinsic myocardial contractile dysfunction in hearts from transgenic mice overexpressing phospholamban. Subsequently, Fentzke and coworkers reported a similar inverse linear V_{cf}-end systolic meridional wall stress relation using a high fidelity catheter in the ascending aorta [21].

Calsequestrin overexpression mice (CSQ-OE) and their isogenic controls (WT) were studied in an integrative approach using transthoracic echocardiography, stress-shortening relations, and invasive hemodynamics in intact closed-chest mice [31]. M-mode echocardiography indicated that calsequestrin overexpression resulted in concentric hypertrophy (+ 52 %) and an increase in LV ejection phase indices. However, mean end-systolic stress-shortening coordinates revealed that at matched end-systolic wall-stress, fractional shortening was depressed in CSQ-OE mice. This was confirmed by depressed indices of LV isovolumic contraction and relaxation in CSQ-OE vs. WT mice.

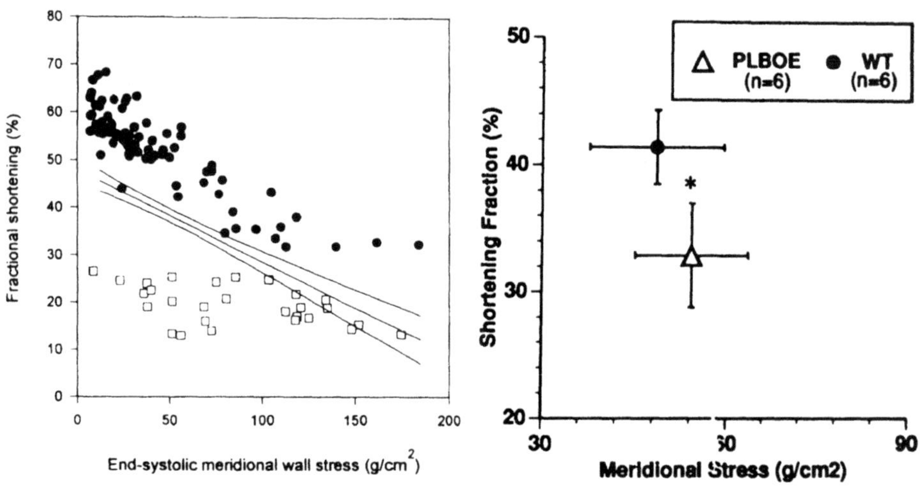

Figure 6. a) Relationship between LV end systolic meridional wall stress and percentage fractional shortening (SF%) produced by pharmacologically altered loading with methoxamine. Data points after dobutamine (filled circles) and after verapamil (open squares) are shown respectively above and below the mean ±95% confidence intervals for the regression line generated from the composite data at baseline. b) Mean coordinates from 6 transgenic mice overexpressing PLB (Δ) and 6 controls (•) at matched end-systolic stress. *p<0.01, ±SD. Reproduced with permission from Hoit et al: In vivo determination of left ventricular wall stress-shortening relationship in normal mice. Am J Physiol 1997;272:H1049 and H1050.

Conscious vs Anesthetized Echo

Echocardiographic studies are typically performed on anesthetized mice; however, the influences of anesthesia on cardiovascular parameters are often tacitly ignored. Recently, it was shown that pentobarbital and ketamine-xylazine anesthesia markedly depress left ventricular fractional shortening and echocardiographically determined cardiac output in mice [32]. We found that compared to conscious mice, the short acting, alcohol- based anesthetic, avertin (tribromoethanol), produces only a modest (12%)

reduction in left ventricular fractional shortening, a greater slowing of heart rate (29%) accompanied by left ventricular enlargement (20%), without a change in echo-calculated cardiac output [2]. If these changes are unacceptable for a particular investigation, conscious echo is a feasible alternative. However, echocardiography in conscious mice requires several training sessions, increases operator demands, necessitates strict attention to decorum in the examining area, and often eventuates in a less than thorough examination.

Arterial Hemodynamics

Ventriculovascular coupling represents the interaction between pulsatile left ventricular ejection (stroke volume) and the arterial forces and loads that resist ejection. These forces may be either external (e.g. aortic input impedance) or internal to the left ventricle (e.g. end systolic stress). The input impedence of the arterial system is an important determinant of the stroke work (stroke volume times pressure) and power output of the left ventricle, and is comprised of the systemic vascular resistance, total arterial compliance, viscosity of blood, and wave reflections from the arterial tree [33]. Quantitation of aortic input impedance requires determination of instantaneous aortic pressure and flow with a temporal resolution that cannot be achieved in commercially available ultrasound instruments. In contrast, measures of aortic compliance (change in volume for a given change in pressure) are both feasible and physiologically relevant. A decrease in aortic compliance represents an increase in the resistance to left ventricular ejection and therefore is a component of the ventricular afterload. As such, altered aortic compliance influences the function of the normal and failing heart.

Fentzke and coworkers quantified global arterial function with echo Doppler and high fidelity aortic pressure [21]. Left ventricular stroke volume was determined from M-mode echocardiographic measurements of the diameter of the proximal ascending aorta (A_oD) and integrated ascending aortic blood velocity (Ao_{vti}). Left ventricular ejection time was determined from the digitized aortic pressure tracings taken from the onset of the systolic rise in aortic pressure to the dicrotic notch and was normalized by the square root of the preceding electrocardiographic R-R interval. Cardiac output (CO) was calculated as the produce of stroke volume and heart rate, and total vascular resistance (TVR) was determined from simultaneous invasive pressure and Doppler echo measurements. TVR was calculated as:

$$TVR = \frac{(MAP) \times 0.080}{CO} \quad (7)$$

where TVR is in dyne-s-cm^{-5}, MAP is mean arterial pressure in mmHg, and CO is cardiac output in ml/min.

Arterial compliance was determined by the area method of Liu and coworkers [34]. This method assumes that vasculature can be modelled as a two element Windkessel, and that the pressure-volume relation is linear:

$$C = \frac{A_d}{(TVR)(P_{es} - P_d)} \quad (8)$$

where C is global arterial compliance in cm^3/mmHg, A_d is the area under the diastolic

portion of the arterial pressure wave, P_{es} is the end-ejection pressure, and P_d is the diastolic pressure.
Isoproterenol increased stroke volume by 32% and cardiac output by 49% without a change in mean aortic pressure, and decreased calculated total vascular resistance by 38%. In contrast, methoxamine increased total vascular resistance and mean aortic pressure two-fold while decreasing cardiac output by 21%. Global arterial compliance was 7.38±0.73 (SE) ml/mmHg x 10^4, was unchanged significantly with isoproterenol (12%) but fell a significant 33% with methoxamine. Thus, these studies establish the feasibility of quantifying global arterial properties in the mouse, which should prove useful for transgenic and gene targetted studies of cardiovascular function.

Future Directions

Novel applications of cardiac ultrasound in mice include the use of ultrasonic contrast agents to opacify the LV cavity and myocardium [35, 36]. Intravenous bolus injections of albumin-coated perfluorocarbon (Optison™) and lipid-coated perfluoropropane gas (Definity™) have the potential to enhance endocardial definition and quantify ischemic risk area, respectively. Despite encouraging preliminary reports, the number of available acoustic windows and imaging planes in the mouse, and the need for venous access remain important limitations of murine contrast echocardiography.

Higher frequency transducers with improved spatial resolution are limited by economic, rather than engineering considerations; the transesophageal approach using high frequency intravascular transducers is one potential solution. Transducer footprints more compatible with the small intercostal spaces and size of the mouse, improved focusing capabilities, and very high frame rate scanning are realistic goals. Refinement of techniques that enhance the ability to assess regional function, such as color kinesis and tissue Doppler imaging, would be useful additions.

REFERENCES:

1. Feigenbaum H. Echocardiography 5th ed. In: (ed) FH, ed. Instrumentation. Malvern, PA: Lea and Febiger, 1994.
2. Kiatchoosakun S, Kirkpatrick D, Walsh R, Hoit B. The effect of tribromoethanol anesthesia on the echocardiographic assessment of left ventricular function. Comparative Medicine 2001; 51:27-31.
3. Hoit BD, Khoury SF, Kranias EG, Ball N, Walsh RA. In vivo echocardiographic detection of enhanced left ventricular function in gene-targeted mice with phospholamban deficiency. Circ Res 1995; 77:632-7.
4. Kadambi VJ, Ponniah S, Harrer JM, et al. Cardiac-specific overexpression of phospholamban alters calcium kinetics and resultant cardiomyocyte mechanics in transgenic mice. J Clin Invest 1996; 97:533-9.
5. Iwase M, Bishop SP, Uechi M, et al. Adverse effects of chronic endogenous sympagthetic drive induced by cardiac Gsα overexpression. Circ Res 1996; 78:517-524.
6. Tanaka N, Dalton N, Mao L, et al. Transthoracic echocardiography in models of cardiac disease in the mouse. Circulation 1996; 94:1109-1117.
7. Gardin JM, Siri FM, Kitsis RN, Edwards JG, Leinwand L. Echocardiographic assessment of left ventricular mass and systolic function in mice. Circ Res 1995; 76:907-914.
8. Wakasaki H, Koya D, Schoen FJ, et al. Targeted overexpression of protein kinase C beta2 isoform in myocardium causes cardiomyopathy. Proc Natl Acad Sci U S A 1997; 94:9320-5.
9. Sato Y, Ferguson D, Sako H, et al. Cardiac-specific overexpression of mouse cardiac calsequestrin is associated with depressed cardiovascular function and hypertrophy in transgenic mice. J Biol Chem 1998; 273:28470-28477.

10. Arber S, Hunter JJ, J Ross J, et al. MLP-deficient mice exhibit a disruption of cardiac cytoarchitectural organization, dilated cardiomyopathy, and heart failure. Cell 1997; 88:393-403.
11. Kiriazis H, Kadambi V, Hoit B, Sato Y, Schmidt A, Kranias E. Long-term transverse aortic banding is associated with similar functional alterations in phospholamban knockout and wild-type mice. J Mol Cell Cardiol 2000; 32:A75.
12. Takeishi Y, Ping P, Bolli R, DL K, Hoit B, RA W. Transgenic overexpression of consitituitively active protein kinase Cε causes cardiac concentric hypertrophy. Circ Res 2000; 86:1218-1223.
13. Sakata Y, Hoit BD, Liggett SB, Walsh RA, Dorn GWI. Decompensation of pressure-overload hypertrophy in G alpha q- overexpressing mice [see comments]. Circulation 1998; 97:1488-95.
14. Fard A, Wang C, Takuma S, et al. Noninvasive assessment and necropsy validation of changes in left ventricular mass in ascending aortic banded mice. J Am Soc Echocardiogr 2000; 13:582-587.
15. Youn H, Rokosh G, Lester S, Simpson P, Schiller N, Foster E. Two-dimensional echocardiography with a 15-MHz transducer is a promising alternative for in vivo measurement of left ventricular mass in mice. J Am Soc Echocardiogr 1999; 12:70-75.
16. Kiatchoosakun S, Restivo J, Kirkpatrick D, Hoit B. Assessment of left ventricular mass in mice: comparison between two-dimensional and M-mode echocardiography. Echocardiography In Press.
17. Scherrer-Crosbie M, Steudel W, Hunziker P, et al. Three-dimensional echocardiographicassessment of left ventricular wall motion abnormalities in mouse myocardial infarction. J Am Soc Echocardiogr 1999; 12:834-840.
18. Scherrer-Crosbie M, Steudel W, Hunziker P, et al. Determination of right ventricular structure and function in normoxic and hypoxic mice: a transesophageal echocardiographic study. Circulation 1998; 98:1015-1021.
19. Steudel W, Scherrer-Crosbie M, Bloch K, et al. Sustained pulmonary hypertension and right ventricular hypertrophy after chronic hypoxia in mice with congenital deficiency of nitric oxide synthase 3. J Clin Invest 1998; 101:2468-2477.
20. Hartley C, LH M, Entman M. Noninvasive measurement of ascending aortic blood velocity in mice. Am J Physiol 1995; 268:H499-H505.
21. Fentzke RC, Korcarz CE, Shroff SG, et al. Evaluation of ventricular and arterial hemodynamics in anesthetized closed-chest mice. J Am Soc Echocardiogr 1997; 10:915-925.
22. Lenihan DJ, Gerson MC, Hoit BD, Walsh RA. Mechanisms, diagnosis, and treatment of diastolic heart failure. Am Heart J 1995; 130:153-66.
23. Prabhakar R, Boivin GP, Hoit B, Wieczorek DF. Rescue of high expression beta-tropomyosin transgenic mice by PTU: regulating the alpha-myosin heavy chain promoter. J Bio Chem 1999; 274:29558-29563.
24. Hoit BD, Shao Y, Gabel M, Walsh RA. Influence of loading conditions and contractile state on pulmonary venous flow. Validation of Doppler velocimetry. Circulation 1992; 86:651-9.
25. Lenihan DJ, Gerson MC, Hoit BD, Walsh RA. Mechanisms, diagnosis, and treatment of diastolic heart failure. Am Heart J 1995; 130:153-166.
26. Garcia MJ, Thomas JD, Klein AL. New Doppler echocardiographic applications for the study of diastolic function. J Am Coll Cardiol 1998; 32:865-75.
27. Schmidt A, Zhai J, Gerst M, Kranias E, Hoit B. Echocardiographic assessment of LV diastolic function in genetically altered mice. J Mol Cell Cardiol 2000; 32:A36.
28. Borow K, Green L, Grossman W, Braunwald E. Left ventricular end-systolic stress shortening and stress length relations in humans. Am J Cardiol 1982; 50:1301-1308.
29. Grossman W, Jones D, McLauren LP. Wall stress and patterns of hypertrophy in the human left ventricle. J Clin Invest 1975; 56:56-64.
30. Hoit BD, Khan ZU, Pawloski-Dahm CM, Walsh RA. In vivo determination of left ventricular wall stress-shortening relationship in normal mice. Am J Physiol 1997; 272:H1047-52.
31. Schmidt A, Kadambi V, Ball N, et al. Cardiac specific overexpression of calsequestrin results in left ventricular hypertrophy, depressed force frequency relation and pulsus alternans in vivo. J Mol Cell Cardiol 2000; 37:1735-1744.
32. Yang X, Liu Y, Rhaleb N, Kurihara N, Kim H, Carretero O. Echocardiographic assessment of cardiac function in conscious and anesthetized mice. Am J Physiol 1999; 277:H1967-H1974.

33. Milnor W. Arterial impedence as ventricular afterload. Circ Res 1975; 36:565-570.
34. Liu Z, Brin K, Yin F. Estimation of total arterial compliance: an improved method and evaluation of current methods. Am J Physiol 1986; 251:H588-H600.
35. Mor-avi V, Korcarz C, Fentzke R, Lin H, Leiden J, Lang R. Quantitative evaluation of left ventricular function in a transgenic mouse model of dilated cardiomyopathy with 2-dimensional contrast echocardiography. J Am Soc Echocardiogr 1999; 12:209-214.
36. Scherrer-Crosbie M, Steudel W, Ullrich R, et al. Echocardiographic determination of risk area size in a murine model of myocardial ischemia. Am J Physiol 1999; 277:H986-992.

13
IN VIVO INVASIVE HEMODYNAMIC STUDIES IN MICE

Brian D. Hoit, M.D.
Case Western Reserve University and University Hospitals of Cleveland
11100 Euclid Avenue, Cleveland OH, 44106-5038

INTRODUCTION

Although noninvasive techniques are used routinely in the mouse, a more complete and complimentary characterization of the cardiovascular phenotype requires invasive catheter-based approaches. Ultrasonic methods provide the means to quantitate the heart's ability to shorten and generate flow, whereas catheter/transducer systems enable the determination of the heart's ability to reversibly generate tension. However, pressure measurements from fluid-filled catheters are particularly problematic in mice because of the temporal delay produced by the transmission of pressure waves in the fluid (especially important at rapid heart rates in the mouse) and the low frequency response of the recording system. The latter is due in part to the small caliber of the catheter and tubing that results in a system with a low natural frequency. Furthermore, fluid-filled pressure measurements are often excessively damped, further reducing the frequency response. These limitations are overcome by the use of high fidelity micromanometers, which consist of miniature strain gauge elements bonded to the tip of a catheter. These catheters are available at sizes as small as 1.4F and readily allow retrograde LV catheterization via the carotid artery; the LV may also be approach from apical or transmitral routes, but then opening the chest is required. It should be recognized that these catheters are delicate, expensive, temperature sensitive, and may have electrical instability (i.e. drift); importantly, they cannot be mechanically calibrated *in vivo*, and matching pressures to properly zeroed fluid-filled pressures is impractical. In addition, this method is highly invasive, requires microsurgical skills, precludes serial study, often produces "nonphysiological" hemodynamics, and may be traumatic. Despite these limitations, high fidelity catheter based methods are critical for assessing the cardiovascular phenotype *in vivo*. Accordingly, this chapter will review indices of myocardial contraction, relaxation, and ventricular performance derived from intracardiac pressures that are used extensively in cardiovascular physiologic studies in larger mammals and that have been adapted and miniaturized for study in mice.

THEORETICAL CONSIDERATIONS

Selection of an Assay

There are several issues that should be considered when selecting an *in vivo* assay: 1) critical assay attributes such as accuracy, reproducibility and diagnostic discrimination may not be similar among laboratories; this is particularly true for several of the invasive techniques that have been studied by only a few groups; 2) anesthesia and surgery may independently and variably influence the specific physiologic variable under study; 3) the suitability of a phenotypic assay is influenced by the developmental stage/age of the mouse. For example, technical considerations preclude reliable catheterization of closed chest juvenile animals; 4) the particular strain of mouse may significantly and selectively affect parameters of cardiovascular performance; and 5) local expertise, technical difficulty of the procedure, and cost are often paramount considerations.

Determinants of Left Ventricular Function

Left ventricular performance may be assessed by evaluating the mechanics of the constituent myocyte or muscle fibers, but a more integrated (and realistic) analysis considers the left ventricle as a muscle pump coupled to the systemic and venous circulations. The principle determinants of left ventricular function include the loading conditions (preload and afterload), the inotropic (contractile) state, and the heart rate. Measures of overall ventricular performance typically include cardiac output (the quantity of blood delivered to the circulation), stroke volume (quantity of blood ejected/beat), and stroke work (the product of pressure and stroke volume).

The influence of preload on measures of ventricular performance defines the left ventricular function curve, known as the Frank Starling relationship (Figure 1a). Because a representative fiber length (i.e. preload) is difficult to determine in the left ventricle, changes in the myocardial fiber length are estimated from changes in either the left ventricular end-diastolic volume or left ventricular end-diastolic pressure. Not surprisingly under certain circumstances, ventricular pressure may inaccurately reflect the ventricular volume. Moreover, changes in ventricular volume may erroneously be inferred from changes in cardiac pressures, which may result only from alterations in ventricular compliance (dV/dP). For example, chronic volume overload may shift the ventricular diastolic pressure-relation so that volume is increased at a normal end-diastolic pressure.

Afterload in the intact heart may be considered as the tension that resists ventricular ejection (wall stress during systole) or as the arterial input impedance. Input impedance is a complex function of arterial pressures, elasticity, vessel dimension, and blood viscosity, which requires measurement of instantaneous aortic pressure and flow (See Chapter 9). Although forces within the ventricular wall are difficult to measure, initial estimates of systolic wall stress can be derived from application of the Laplace relationship in which

$$\text{wall tension} = (P \bullet r)/2h, \qquad (1)$$

where P refers to pressure, r to ventricular radius, and h to wall thickness. More complex derivations based on various geometric assumptions are used to calculate end-systolic wall stress (Chapter 9). Owing to its simplicity, aortic pressure is often used as a surrogate for afterload.

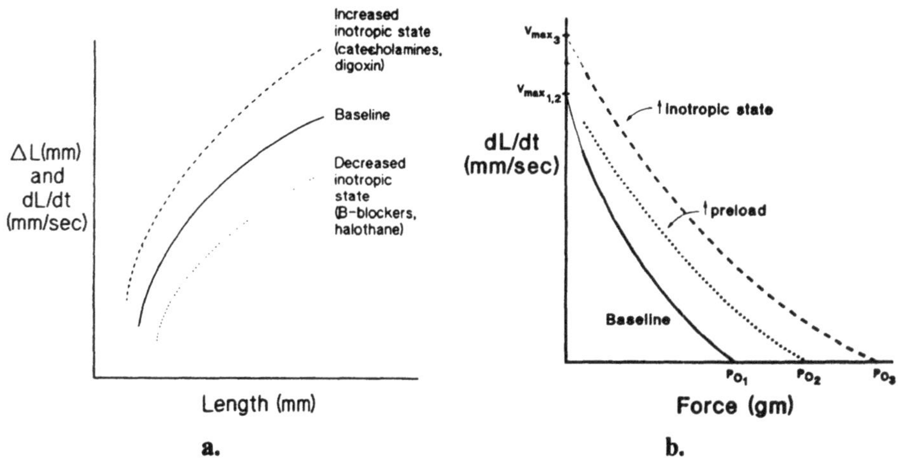

Figure 1a. Length-shortening and length-velocity relationships from isotonic contractions at a constant afterload. As muscle length (preload) increases, shortening (ΔL) and velocity of shortening (dL/dt) increase. An increased inotropic state shifts the curve upward and to the left; conversely, a decreased inotropic state shifts the curve downward and to the right. **b.** Force-velocity relationship from variably afterloaded contractions. Increased preload causes an increase in the maximum isometric tension (P_{O1} to P_{O2}) without a change in the extrapolated velocity of an unloaded contraction (V_{max}). An increase in the inotropic state increases P_o and V_{max}. (dL/dt = velocity of contraction.) Reproduced with permission from Hoit et al: Determinants of left ventricular performance and cardiac output. *In* Essentials of Basic Science: Physiology, Second Edition, Edited by Sperelakis and Banks. Boston: Little, Brown and Company, Figure 23-2 and 23-3, p 271, 1996.

Changes in inotropic state are usually defined operationally by shifts of the various ventricular function curves; by definition, these effects are independent of loading conditions (Figure 1). For example, a drug with positive inotropic activity (e.g. dobutamine) shifts the Frank-Starling relationship (analogous to the length-shortening curve in papillary muscle preparations) upward and to the left, and changes in the stress-shortening relationship (analogous to the force- velocity curve) upward and to the right. Another inotropic stimulus, dealt with in detail below, is heart rate (treppe or force-frequency relationship).

Indices of Contractility

The rate of pressure development in the left ventricle during isovolumic systole (dP/dt) is used frequently as an index of the inotropic state. Although LV $+dP/dt_{max}$ provides a measure of the rate of tension development and of myocardial contractility, this index is preload dependent, due in part to length-dependent changes in myofilament Ca^{2+} sensitivity [1]. However, $+dP/dt_{max}$ is largely independent of afterload, provided that the maximum rate of rise occurs before aortic valve opening. Although changes in the maximal rate of rise of ventricular pressure are highly sensitive to acute changes in contractility and are useful to assess directional changes in inotropic state, absolute dP/dt_{max} is not as useful for assessment of basal contractility as are the ejection phase indices, LV fractional shortening and Vcf (see Chapter 9) [2]. Moreover, dP/dt_{max} cannot be corrected for changes in muscle mass produced by left ventricular hypertrophy, in which case it is best to compare peak stress ($d\varepsilon/dt$). Because of the direct influence of preload on dP/dt, dP/dt at a common developed pressure (LV systolic minus diastolic pressure) and the slope of the dP/dt-end-diastolic volume curve have been proposed as preload independent indices of the inotropic state [3]. End systolic elastance (E_{es}) is a relatively load independent measure of contractility, but its determination requires linear regression of end systolic pressure and volume points over a range of preload [3] (see Chapter 11).

Heart rate dependence of cardiac contractility reflects basic cycling kinetics of calcium and is critically dependent on function of the sarcoplasmic reticulum (SR). Manifestations of the force-interval relation (force-frequency, mechanical restitution, post-extrasystolic potentiation) have been studied extensively in a variety of preparations [4,5] and changes in these parameters are observed in congestive heart failure [6]. Processes related to force-interval behavior are important insofar as 1) they are fundamental physiological control mechanisms; 2) they are utilized as indices of myocardial function; and 3) they play a role in the response to exercise and the development and maintenance of heart failure [7,8].

Recent data from our laboratory and others demonstrate that the force-frequency relationship in the anesthetized mouse is biphasic (i.e. there is both an ascending and a descending limb), and the heart rate at which maximal rate of contraction and relaxation ($\pm dP/dt_{max}$) reaches a maximum (i.e. the critical heart rate) is a sensitive measure of LV function [9]. Mechanical restitution represents the restoration of contractile force in premature beats having progressively longer extrasystolic intervals (ESI), and is related to intracellular release of activator Ca^{2+} by the ryanodine receptor. An increased force of contraction characterizes the beat that follows an extrastimulus. This phenomenon, called postextrasystolic potentiation (PESP), is also a manifestation of myocyte calcium handling, and has been attributed to an increase in the amount of calcium released by the sarcoplasmic reticulum (SR) during the postextrasystolic beat [10-14] Postextrasystolic potentiation contributes to the beat to beat variability of the pulse in atrial fibrillation [15], and has been suggested as both a sensitive indicator of viability and inotropic reserve, and as a framework for treating severe heart failure.

Indices of Relaxation

Diastolic properties of the ventricle are complex, related to the speed and synchrony of myocardial relaxation, loading conditions, viscoelasticity, heart rate, atrial function and ventricular interaction [16]. The pressure-volume relation during early diastole reflects the lusitropic state (relaxation) of the heart, analogous to the inotropic state (contraction) measured during systole. The rate of left ventricular relaxation may be estimated from the maximal rate of pressure decay ($-dP/dt_{max}$) and indices (e.g. $RT_{1/2}$) that are related to the time necessary for ventricular relaxation, but these measurements are highly dependent on the prevailing load of the intact circulation. In contrast, Tau, the time constant of left ventricular relaxation during isovolumic diastole provides a more accurate, less load-dependent measure of relaxation [17]; Tau is shortened by beta-adrenergic stimulation (cyclase-dependent phosphorylation of phospholamban and Troponin I) and prolonged with beta-adrenergic antagonists. Although several mathematical models of the exponential decay of left ventricular pressure exists, we employ a simple monoexponential model that declines to zero:

$$P_{(t)} = P_0 e^{-t/T} \qquad (2)$$

where P(t) is the left ventricular pressure at any time, t, Tau (T) is the relaxation constant, P_o is the left ventricular pressure at the onset of relaxation and e is the base of the natural logarithm [18]. The natural logarithmic transformation of both sides of the equation yields:

$$\ln P = -1/T + \ln P_o. \qquad (3)$$

Tau is derived by obtaining the negative of the reciprocal of the slope of ln P (t) vs. time, t from aortic valve closure to mitral valve opening (isovolumic diastole). High fidelity catheter tip micromanometers are necessary for measurement of (-) dP/dt_{max} and Tau.

Another index, relaxation restitution, represents the time for restoration of relaxation in premature beats having progressively longer ESIs, and is thought to be related to uptake of Ca^{2+} by the sarcoplasmic reticulum.

METHODOLOGY

We anesthetize mice with a mixture of ketamine (100 mg/kg), xylazine (5 mg/kg), and morphine (2.5 mg/kg), 0.25 cc given intraperitoneally . The trachea is exposed, ligated with 6-0 silk, and an airway is established by puncturing between the tracheal rings with a 24G Jelco catheter; the sheath is advanced 6-7 mm and secured with a 6-0 silk ligature. Instrumentation and analysis have been used in both the open and closed chest, but for reasons detailed below [19], the closed chest preparation is preferable and most often employed.

In the open chest preparation, the chest is widely opened via a subxiphoid incision and bilateral thoracotomies and the sternum is reflected and secured with a hemostat. Animals are ventilated with a tidal volume of 500 μl at 100 cycles/min using a Harvard Respirator and room air. A 1.4 or 1.8F Millar catheter is placed in the LV via the apical dimple and secured with a purse string suture. (An alternative approach used by others is pulmonary vein and transmitral LV catheterization.) Insulated platinum wires (0.127 mm diameter) are attached to the right and left atrial appendages for pacing. The jugular vein is cannulated with flame-stretched Nalgene PE tubing for intravenous access.

In the closed chest preparation, the animal is allowed to breathe room air spontaneously. The right jugular vein is catheterized and a custom made 1F bipolar pacemaker (Numed) is positioned in the right atrium via the right internal jugular vein; this avoids the need to pre-instrument the animal with pacing wires. The right carotid artery is longitudinally incised, and a 1.4F Millar high fidelity catheter introduced into the vessel is advanced past the distal tie with a single, quick continuous movement into the left ventricle.

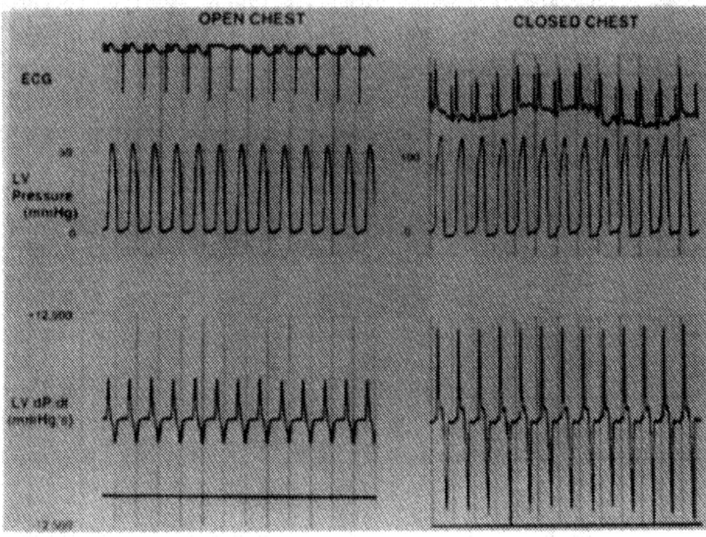

Figure 2. Hemodynamic tracings from open chest (left) and closed chest (right) mice. Atria are paced at 300 bpm.

In both cases the micromanometer is electronically calibrated *in vitro* by submersing the tip of the catheter in warmed saline with the reference zero level taken at mid-chest. The analog LV dP/dt signal is obtained on-line by electronic differentiation of the high fidelity LV pressure signal. A constant infusion minipump is used for volume infusion and a stimulator is used for atrial pacing.

Protocols

A programmable stimulator is used for our restitution and postextrasystolic potentiation protocols [20,21]. Pacing thresholds are determined and stimulation is performed with a 1.0-millisecond pulse width at twice the capture threshold.

Force-frequency Protocol

Pacing is initiated just above the intrinsic heart rate (slowed by 0.1 cc I.V of the sinus node I_f channel inhibitor, DKAH 0269) to avoid competing rhythms and is increased at 12 beats/min (0.2 Hz) increments until the LV dP/dtmax is visually decreased or until atrioventricular block unresponsive to 0.04 to 0.08 ng/g atropine supervenes.

Figure 3. Single extrastimulus protocol for mechanical and relaxation restitution. Three different extrastimuli are shown. S_1 is the basic cycle length and S_2 is the variable extrastimulus.

Restitution Protocol

The right atrium is stimulated at a basic cycle length of 200 milliseconds (300 bpm). After a series of at least 15 beats, an atrial extrastimulus is introduced in the refractory period of the AV node, and extrastimuli intervals (ESI) are increased at 10 milliseconds intervals (producing beats with progressively increased cycle length) until the basic cycle length is achieved (Figure 3). In order to analyze mechanical restitution, dP/dt_{max} is normalized to the dP/dt from beats having the basic cycle length, and expressed as a percent [4,6].

$$\text{Normalized } dP/dt = Y_{max}\,(1\text{-EXP }[ESI_0\text{-}ESI]/TC)$$

where Y_{max} *is the maximal or plateau response,* ESI_0 *is the largest ESI that fails to produce a response and TC is the time constant.*

Relaxation restitution is calculated by normalizing peak -dP/dt of the extrastimulus beat to the peak -dP/dt of the basic cycle length, and fitting the relation between 1/normalized dP/dt and the ESI to a monoexponential function [6,22].

Postextrasystolic Potentiation Protocol

The right atrium is paced at a basic cycle length of 200 milliseconds (S1) and after a fixed extrastimulus (S2, 120 ms), a second extrastimulus (S3) is introduced over a range of postextrasystolic intervals (PESI) of 10-100 ms intervals until 1000 ms (producing beats with progressively increased cycle length), or an intrinsic sinus beat captures the ventricle (Figure 4). In order to analyze PESP, dP/dt_{max} of the potentiated beat (S3) was normalized to the dP/dt_{max} from beats having the basic cycle length (S1), and was expressed as a percent:

$$\text{Normalized } dP/dt = Y_{max}\ (1\text{-EXP }[PESI_0 - PESI]/TC)$$

where Y_{max} *is the maximal or potentiated response,* $PESI_0$ *is the largest PESI that fails to produce a response and TC is the time constant of postextrasystolic potentiation.*

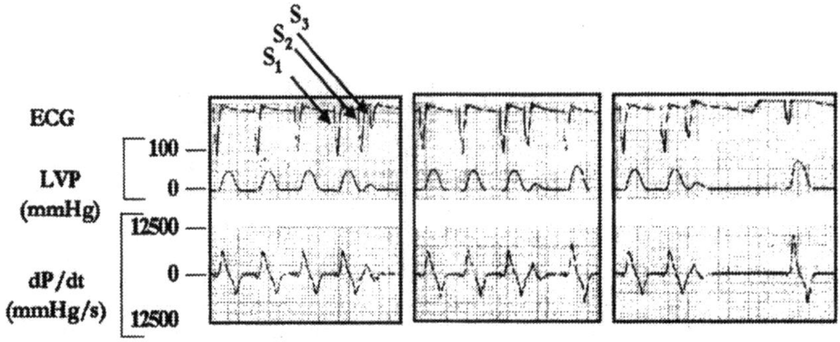

Figure 4. Double extrastimulus protocol for postextrasystolic potentiation. S_1 is the basic cycle length, S_2 is the fixed ESI and S_3 is the variable PESP.

Volume Infusion Protocol ([23]) LV pressure is increased by incremental steady-state infusion of Hespan at a rate of 10 ml/hr. Measurements are taken at 0.1 ml increments until ~1.4 ml are infused, and end diastolic pressure-dP/dt_{max} and end diastolic pressure-dP/dt_{max} relations are generated (Figure 5). Preload reserve is assessed by measurement of $\Delta dP/dt_{max}/\Delta LVEDP$.

Catecholamine Infusion Protocol ([23])
To assess the dynamic range of the system, beta-adrenergic responsiveness is determined using increasing doses of dobutamine, administered at a constant infusion of incremental doses given for a total of 3 minutes for each dose. Each dose is delivered via a femoral or jugular vein catheter at a rate of 0.1 µl/g BW with a microliter syringe pump. Responsiveness is estimated by the % change from baseline to maximal infusion, and the EC_{50}.

Baroreceptor Function Protocol ([24])
Systolic blood pressure and heart rate in the unpaced, instrumented animal are averaged over 5 minute measurements at baseline, after phenylephrine (0.3 µg/g i.p.) and phenylbiguanide (30 µg/g) injections into the right atrium.

Ryanodine Infusion ([20])
To study SR function, we administer an intravenous infusion of ryanodine at a rate of 0.5 ng/g•min for a total dose of 4 ng/g. This dose produces a serum concentration in the nanomolar range that binds specifically to high-affinity sites, locks the SR Ca^{2+} channel in a subconductance state, and depresses indices of LV isovolumic contraction and relaxation [25].

Effect of the Type of Preparation
A number of investigators have used cardiac catheterization techniques to assess the

physiological phenotype in engineered mice [20,21,23,26-30]. Many of these models and their hemodynamics are described in detail in other chapters of this monograph. Since both open and closed chest animals are employed in these studies, we compared the hemodynamics in both preparations from a group of wild type FVB/N mice [19]. The intrinsic (unpaced) heart rate, LV systolic and end-diastolic pressures and $\pm dP/dt_{max}$ were significantly greater, and at a matched heart rate of 420 bpm, the LV systolic pressure and $\pm dP/dt_{max}$ were significantly greater in the closed than open chest animals (Table 2).

Table 2. Hemodynamics at Matched (Paced) Heart Rates

	Open Chest (n=6)	Closed Chest (n=6)
Heart Rate (bpm)	410±12	420±0
LV Systolic Pressure (mmHg)	52±11	98±18*
LV End-Diastolic Pressure (mmHg)	3.3±2.1	5.5±3.0
LV (+) dP/dt_{max} (mmHg/s)	5777±2473	9240±2459*
LV (-) dP/dt_{max} (mmHg/s)	3753±1170	8375±2551*

Data are mean ± SD. * $p < 0.05$ vs. open chest (unpaired t-test).

Effect of Preload
LV systolic pressure and LV $+dP/dt_{max}$ both increased significantly with Hespan[R] infusion (10 ml/hr to ~ 1.4 ml), although end diastolic pressures greater than 15 mmHg could not be consistently achieved (Figure 5). Lorenz and Robbins have also demonstrated utilization of the Frank-Starling mechanism in these animals [23].

Effect of Heart Rate
Although incremental atrial pacing increased $+dP/dt_{max}$ significantly in both open and closed chest animals, the force-frequency curve in the latter was shifted upward and to the right, resulting in a delayed descending limb of the force-frequency relation (Figure 6). As a consequence, the HR_{crit} was significantly greater in the closed than open chest animals (432±42 vs. 318±42 bpm, $p<0.05$).

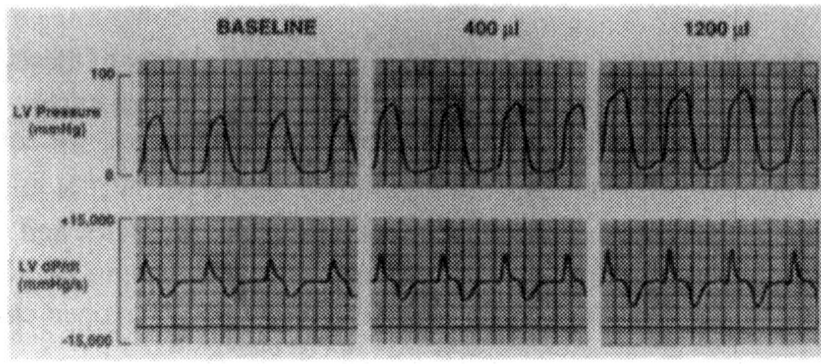

Figure 5. Hemodynamic records showing LV pressure and dP/dt at three levels of end-diastolic pressure (EDP) from a representative animal. Reproduced with permission from Hoit BD et al: Invasive hemodynamics and force-frequency relations in open versus closed chest mice. Am J Physiol 273:H2531 Figure 3A, 1997.

In both groups, incremental pacing produced significant increases in LV systolic pressure and decreases in LV end diastolic pressure, and as pacing progressed beyond the critical heart rate, both LV systolic and end diastolic pressures decreased. Palakodeti reported similar force-frequency relations in anesthetized mice, but were unable to demonstrate a descending limb in mice awakening from anesthesia; moreover, infusion of dobutamine increased both the average LV dP/dt_{max} and the HR_{crit} [8].
These studies indicate that in the normal mouse left ventricle, 1) there is a biphasic force-frequency relation in both the open and closed chest, anesthetized state, 2) hemodynamics and force-frequency relations are depressed in the open-chest state, and 3) steady-state alterations in left ventricular end-diastolic pressure demonstrate preload dependence of $+dP/dt_{max}$.

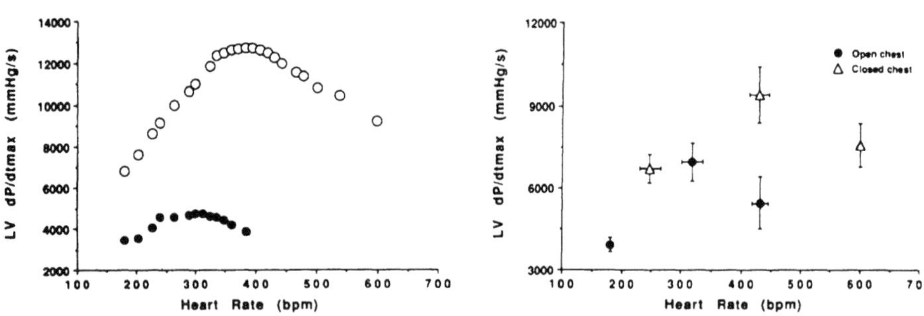

a. b.

Figure 6. Force-frequency relations in a single dog (a) and averaged data (b) in the open chest (n=6, dark symbols) and closed chest (n=6, light symbols) preparations. The relation in the closed chest is shifted upward and rightward, delaying the onset of descending limb. The critical heart rate is significant greater in the closed chest animals. Data are mean ± SE. Reproduced with permission from Hoit BD et al: Invasive hemodynamics and force-frequency relations in open versus closed chest mice. Am J Physiol 273:H2530, Figures 1a and 1b, 1997.

Effect of Mouse Strain

The influence of the strain used to generate a genetically altered mouse is a variable that may complicate phenotypic analysis. Therefore, we compared hemodynamics in two strains (FVB/N and SvJ129) of wild type mice. Comparisons were made at matched atrially paced heart rates. Critical heart rates were higher and $\pm dP/dt_{max}$ were significantly greater in the FVB/N than SvJ129 mice; in contrast, LV systolic pressure was similar in both groups of animals [9].

Table 3. Effect of Strain on Hemodynamics

	SvJ129 (n=6)	**FVB/N (n=8)**
LV Systolic Pressure (mmHg)	80±12	92±8
LV (+) dP/dt_{max} (mmHg/s)	5085±560	9180±330*
LV (-) dP/dt_{max} (mmHg/s)	4280±425	7920±250*
HR_{crit} (bpm)	360±18	440±25*
Tau (ms)	8.2±0.8	7.9±0.6

LV pressure and $\pm dP/dt_{max}$ at paced HR of 300 bpm; HR_{crit}=critical heart rate; mean±se; $p<0.05$ vs SvJ129.

Effect of Phospholamban

Under resting conditions, increases in heart rate are associated with an increase in Ca^{2+} availability to the contractile proteins as a result of: 1) an increase in transarcolemmal Ca^{2+} influx resulting from the increased number of action potentials per minute, leading to a rise in sarcoplasmic reticulum (SR) calcium storage with augmentation of subsequent Ca^{2+} transients, 2) reduced efflux of Ca^{2+} via the Na^+/Ca^{2+} exchanger in diastole, and 3) a delay in the Na+/K+ ATPase pump causing a decrease in the Na^+ gradient and increased intracellular Ca^{2+}.

We recently showed that the critical heart rate is dependent on the relative abundance of the sarcoplasmic reticulum protein, phospholamban [9]. Transgenic mice overexpressing phospholamban (PLBOE), gene targetted mice without phospholamban (PLBKO), and isogenic controls were studied at paced heart rates (300 bpm). As predicted, $+dP/dt_{max}$ was greatest in PLBKO (14570±950 mmHg/s), least in PLBOE (7210±836 mmHg/s) and intermediate in controls, and Tau was greatest in PLBOE (10.0±0.5 ms), least in PLBKO (3.9±0.6 ms) and intermediate in controls. The force-frequency relation was shifted upward and to the right in the PLBKO mice, whereas the relation was shifted downward and to the left in the PLBOE, conpared with their respective isogenic controls. Thus, the critical heart rate for $+dP/dt_{max}$ was significantly higher in the PLBKO mice than isogenic controls (427±20 bpm vs 360±18), whereas the critical heart rate for $+dP/dt_{max}$ was significantly lower in PLBOE than isogenic controls (340±30 vs 440±25 bpm). Thus, this study indicates that the level of phospholamban attenuates the force-frequency relation and indices of basal contraction and relaxation and further emphasizes the importance of using isogenic controls in transgenic experiments.

Mechanical Restitution and Postextrasystolic Potentiation

The time constants of restitution in PLBKO, PLBOE, and their respective wild type controls are shown below in Table 4.

Table 4. Parameters of Mechanical and Postextrasystolic Restitution in Mice with Variable Amounts of Phospholamaban

	$Y_{max\ MR}$ (%)	TC_{MR} (ms)	Y_{maxPP}(%)	TC_{PP}(ms)
FVB/N	109±9	54.1±9.2	108±2	75.2±5.5
PLBOE	129±7	80.3±7.6*	139±3*	107.8±7.0*
SvJ129	100±6	47.7±7.6		
PLBKO	103±12	6.3±1.2*		

Y_{maxMR}=response of the mechanical restitution curve; TC_{MR}= time constant of mechanical restitution; Y_{maxPP}= maximal or potentiated response; TC_{PP}= time constant of postextrasystolic potentiation; mean±SE; * $p<0.05$ vs isogenic control

Thus: 1) the rate of mechanical restitution is inversely related to the level of phospholamban, confirming the critical role of the SR calcium load on mechanical activity in the intact animal; and 2) overexpression of phospholamban slows the kinetics of PESP restitution, but enhances the *magnitude* of PESP. The data also demonstrate considerable species variability in the force-frequency parameters [9], emphasizing the need to use isogenic controls for these studies.

Limitations and Future Directions

A limitation of the majority of invasive studies is that they are performed in the anesthetized animal. General anesthetics induce vasodilation and hypotension and when given in sufficient doses, cause myocardial depression, affect cardiac control mechanisms (e.g. the Frank Starling mechanism and the force-frequency relation) and influence the mechanisms involved in the baroreceptor response and reflex control. Although studies in "conscious" mice awakening from anesthesia confirm the cardiodepressant effects of general anesthetics, these data should be interpreted with a healthy degree of caution; unless the experimental animals are acclimatized by a period of training, the associated stress leads to high circulating catecholamines and may alter the chronotropic, dromotropic and inotropic state of the preparation. Telemetry should prove useful in this regard (see Chapter 13). Finally, it is assumed that the trauma of surgery and anesthesia do not alter the cardiovascular response to a specific experimental perturbation.

In addition to continued miniaturization of catheters, newer developments in the field include the combination of pressure measurements with imaging modalities in order to derive indices of contractility that are relatively independent of load. For example, pressure-volume loops, end systolic elastance estimates, and end-diastolic volume-dP/dt relations can be made with the conductance catheter and miniaturized sonomicrometers adhered to the LV epicardium [31,32]. (See Chapter 11). These techniques are highly invasive (open chest, ventilated, ventricular stab), technically demanding, and necessitate proper placement of either the catheter in the LV cavity or sonomicrometers on the epicardium and endocardium.

Despite these limitations, a variety of techniques used to evaluate cardiovascular function in larger animals can be applied to evaluate the phenotypic expression and complex physiological traits in the intact transgenic or gene-targetted mouse. The studies reviewed herein indicate that LV performance measurements vary depending on the particular state in which they are acquired (i.e. type of preparation, LV filling pressure, heart rate), and emphasize the need for appropriate controls. Insofar as all animal models have difficulties and limitations, understanding the influence of the preparation will minimize interpretative errors and facilitate recognition of true physiologic or pathologic differences.

REFERENCES

1. Hibberd M, Jewell B. Calcium- and length-dependent force production in rat ventricular muscle. *J Physiol.* 1982;329:527-540.
2. Quinones M, Gaasch W, Alexander J. Influence of acute changes in preload, afterload, contractile state and heart rate on ejection and isovollumic indices of myocardial contractility in man. *Circulation.* 1976;53:293-302.
3. Little W. The left ventricular dP/dt_{max}-end-diastolic volume relation in closed chest dogs. *Circ Res.* 1985;56:808-815.
4. Freeman G, Colston J. Evaluation of left ventricular mechanical restitution in closed-chest dogs based on single-beat elastance. *Circ Res.* 1990;67:1437-1445.
5. Wier WG, Yue DT. Intracellular calcium transients underlyng the short-term force-interval relationship in ferret ventricular myocardium. *J Physiol (Lond).* 1986;376:507-530.
6. Prabhu SD, Freeman GL. Effect of tachycardiac heart failure on the restitution of left ventricular function in closed-chest dogs. *Circulation.* 1995;91:176-185.
7. Johnson EA. The cardiovascular system I. In: Berne RM, ed. *Handbook of Physiology.* Bethesda, MD: American Physiology Society; 1979:475-496.
8. Palakodeti V, Oh S, Oh BH, Mao L, Hongo M, Peterson KL, J. Ross J. Force-frequency effect is a powerful determinant of myocardial contractility in the mouse. *Am J Physiol.* 1997;273:H1283-H1290.

9. Kadambi VJ, Ball N, Kranias EG, Walsh RA, Hoit BD. Modulation of the force frequency relation by phospholamban in genetically engineered mice. *Am J Physiol*. 1999;276:H2245-H2250.
10. Yue DT, Burkhoff D, Franz M, Hunter W, Sagawa K. Postextrasystolic potentiation of the isolated canine left ventricle: relationship to mechanical restitution. *Circ Res*. 1985;56:340-350.
11. Morad M, Goldman Y. Excitation-contraction coupling in heart muscle: Membrane control of development of tension. *Prog Biophys Mol Biol*. 1973;27:257-313.
12. Seed WA, Noble MIM, Walker JM, Miller GAH, Pidgeon J, Redwood D, Wanless R, Franz MR, Schoettler M, Schaeffer J. Relationships between beat-to-beat interval and the strength of contraction in the healthy and diseased human heart. *Circulation*. 1984;70:799-805.
13. Elzinga G, Lab MJ, Noble MIM, Papadoyannis DE, Pidgeon J, Seed WA, Wohlfart B. The action-potential duration and contractile response of the intact heart related to the preceding interval and the preceding beat in the dog and cat. *J Physiol*. 1981;314:481-500.
14. Prabhu SD, Freeman GL. Postextrasystolic mechanical restitution in closed-chest dogs. *Circulation*. 1995;92:2652-2659.
15. Hardman SM, Noble MI, Seed WA. Postextrasystolic potentiation and its contribution to the beat-to-beat variation of the pulse during atrial fibrillation. *Circulation*. 1992;86:1223-32.
16. Lenihan DJ, Gerson MC, Hoit BD, Walsh RA. Mechanisms, diagnosis, and treatment of diastolic heart failure. *Am Heart J*. 1995;130:153-166.
17. Lenihan DJ, Gerson MC, Dorn GWI, Hoit BD, Walsh RA. Effects of changes in atrioventricular gradient and contractility on left ventricular filling in human diastolic cardiac dysfunction. *Am Heart J*. 1996;132:1179-88.
18. Weiss JL, Frederickson JW, Weisfeldt ML. Hemodynamic determinants of the time course of fall in canine left ventricular pressure. *J Clin Invest*. 1976;58.
19. Hoit BD, Ball N, Walsh RA. Invasive hemodynamics and force-frequency relationships in open- versus closed-chest mice. *Am J Physiol*. 1997;273:H2528-33.
20. Hoit BD, Kadambi VJ, Tramuta DA, Ball N, Walsh RA, Kranias EG. Influence of sarcoplasmic reticulum calcium loading on mechanical and relaxation restitution. *Am J Physiol*. 2000;47:H958-H963.
21. Hoit BD, Tramuta DA, Kadambi VJ, Dash R, Ball N, Kranias EG, Walsh RA. Influence of transgenic overexpression of phospholamban on postextrasystolic potentiation. *J Mol Cell Cardiol*. 1999;31:2007-2015.
22. Prabhu SD, Freeman GL. Kinetics of restitution of left ventricular relaxation. *Circ Res*. 1992;70:29-38.
23. Lorenz J, Robbins J. Measurement of intraventricular pressure and cardiac performance in the intact closed-chest anesthetized mouse. *Am J Physiol*. 1997;272:H1137-H1146.
24. Paton J, Butcher J. Cardiorespiratory reflexes in mice. *J Auton Nerv Syst*. 1998;68:115-124.

25. Rousseau E, Smith JS, Meisser G. Ryanodine modifies conductance and gating behavior of single Ca2+ release channel. *Am J Physiol.* 1987;253:C364-8.
26. Milano CA, Allen LF, Rockman HA, Dolber PC, McMinn TR, Chien KR, Johnson TD, Bond RA, Lefkowitz RJ. Enhanced myocardial function in transgenic mice overexpressing the beta 2-adrenergic receptor. *Science.* 1994;264:582-6.
27. Schmidt A, Kadambi V, Ball N, Sato Y, Walsh R, Kranias E, Hoit B. Cardiac specific overexpression of calsequestrin results in left ventricular hypertrophy, depressed force frequency relation and pulsus alternans in vivo. *J Mol Cell Cardiol.* 2000;37:1735-1744.
28. Hunter JJ, Tanaka N, Rockman HA, Ross J, Jr., Chien KR. Ventricular expression of a MLC-2v-ras fusion gene induces cardiac hypertrophy and selective diastolic dysfunction in transgenic mice. *J Biol Chem.* 1995;270:23173-8.
29. Sato Y, Ferguson D, Sako H, Dorn G, Kadambi V, Yatani A, Hoit B, Walsh R, Kranias E. Cardiac-specific overexpression of mouse cardiac calsequestrin is associated with depressed cardiovascular function and hypertrophy in transgenic mice. *J Biol Chem.* 1998;273:28470-28477.
30. Murphy A, Kogler H, Georgakopoulos D, McDonough J, Kass D, Van Eyk J, Marban E. Transgenic mouse model of stunned myocardium. *Science.* 2000;287:488-491.
31. Miyamoto M, Kim C, Guerrero J, Rosenzweig A, Gwathmey J, Hajjar R. Ventricular pressure and dimension measurements in mice. *Lab Anim Sci.* 1999;49:305-307.
32. Feldman M, Erikson J, Mao Y, Korcarz C, Lang R, Freeman G. Validation of a mouse conductance system to determine LV volume: comparison to echocardiography and crystals. *Am J Physiol.* 2000;279:H1698-H1707.

14
PRESSURE-VOLUME RELATIONS

Dimitrios Georgakopoulos, Ph.D. and David A. Kass, M.D.
Johns Hopkins Medical Institutions, Baltimore, Maryland 21287, USA

INTRODUCTION

The physiologic characterization of an intact ventricular chamber ideally employs analyses yielding integrated functional assessments as well as separate quantification of the major factors (i.e. load, contractility, diastolic function) that determine net performance. More than a century ago, Otto Frank initiated studies to assess ventricular function by relations between chamber pressure and volume, an approach already familiar to 19th century engineers. In one diagram, both the active and passive filling properties of the pump, its mechano-energetics, and its interaction with vascular loading systems could all be quantified. The pioneering work of Suga, Sagawa and colleagues in the 1970's, performed in isolated yet physiologically ejecting canine ventricles (1), established the pressure-volume relationship as a powerful means to achieve the major goals of function analysis. Beyond the utility of the end-systolic pressure-volume relationship as a load-independent index of contractility, pressure-volume analysis proved insightful and valuable to both laboratory research and clinical studies (2,3). Recently, this method has been extended by means of miniaturized catheter technology to the mouse (4). The capacity to genetically alter mice combined with in-depth hemodynamic characterization by *in-vivo* pressure-volume analysis presents unparalleled potential for interfacing molecular science with cardiac systems physiology. This chapter reviews the instrumentation and methods involved in *in vivo* pressure-volume relation measurement in mice. Cardiovascular physiology of normal mice are compared with published reference human data, and examples from transgenic models are shown to highlight the utility of the method.

PRESSURE-VOLUME MEASUREMENT by CONDUCTANCE CATHETER

Determination of pressure-volume relations requires three principal elements: intraventricular chamber pressure, simultaneous cavity volume, and the capacity to vary the loading conditions of the heart to assess these parameters over some span of arterial pressures and/or filling volumes. Miniaturization of micromanometers was among the early technical developments that helped extend organ physiologic methods to mice, and current catheter-based manometers have a cross sectional diameter of < 0.5 mm.

Loading maneuvers used in larger animals and humans have been mechanical - i.e. transient obstruction of the vena cava to vary preload or manipulation of the aorta to alter afterload, or pharmacological - i.e. use of constrictor or dilator agents. Both are relatively easy to apply to the mouse. This leaves the determination of instantaneous chamber volume as the remaining factor, and this has historically been the most difficult measurement. Methods have included echocardiographic or MRI image analysis, implanted sonomicrometer dimension-volumes, and the conductance catheter. Image based analysis, while non-invasive, are limited by resolution at rapid heart rates, and the requirement of extensive post-processing of multiple cycles. The crystal method involves implantation of sensors within the LV wall. The conductance method has several advantages. It provides an extremely rapid response so digitization at 2 kHz yields a very smooth signal, it is easy to use with little manipulation of the LV wall itself, and it facilitates recording rapid transient responses. The latter is important, as the mouse heart rate is typically about 550 min^{-1}, so that the 10-20 beats involved in a typical pressure-volume relation assessment occurs within 1-2 secs of loading change.

The conductance catheter method was introduced in the early 1980's by Baan and colleagues (5), who employed a multi-electrode catheter within the LV cavity, placed along the longitudinal axis. A high frequency (20 kHz) constant current (20-30 μA) electric field was generated between base and apex electrodes (Figure 1), and the time-varying voltage was measured between pairs of intervening sense electrodes. This voltage is directly proportional to changes in the resistivity of the conducting material spanning the electrodes (i.e. Ohm's law) - namely the blood pool in the heart, and the surrounding structures (muscle wall, potentially RV blood and more distant factors).

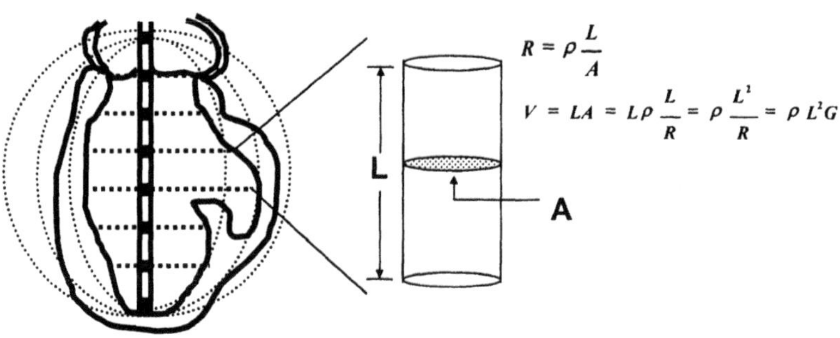

Figure 1. Schematic of cross-section of the left ventricle showing placement of conductance catheter along the ventricular long axis and associated electrical field lines between the outermost electrodes at the base and apex. Equipotential lines at intervening pairs of sensing electrodes perpendicular to the catheter divide the cavity into imaginary cylindrical segments. One segment is expanded to show the relationship between the volume of the segment and the electrical resistance across it. R= resistance, G= conductance, L= length, A= cross-sectional area, V= volume, ρ= resistivity of material (blood).

Since resistivity of blood is lower than in the surrounding tissue, altering blood volume in the LV cavity results in a reciprocal change in resistance. Thus cavity volume becomes directly proportional to conductance (G, in Fig. 1). G combines both the conductivity of the blood pool, and that of surrounding structures (i.e. $G_{tot} = G_{blood} +$

G_p), the latter (G_p) being often referred to as the parallel conductance(5). In larger animals, several sense segments are typically employed as shown in the figure, (3-5 for rabbit - to - human sized ventricles), and signals from each segment are added to generate total cavity volume. In the mouse, the chamber is small, and practical limitations of catheter manufacturing as well as requirements to provide an adequate signal/noise ratio have limited the number of segments. Currently, most systems employ a single segment of 4.5 mm length, with a second pair of outer stimulation electrodes.

The conductance signal is itself uncalibrated, combining both the offset term G_p, and a slope term (or gain) related to current field non-linearity. Both require determination to generate absolute volume. G_p can be determined by obtaining an absolute reference volume (i.e. EDV via echocardiography or MRI), or by using indicator dilution methods as described by Baan et al (3), and Langford et al (5,6). These techniques analyze the volume signal during an induced transient change in blood conductivity, assuming stable ventricular contraction (i.e. no physiologic effect from the change in conductivity). Typically, a small bolus of hypertonic saline is rapid injected to raise blood conductivity (or alternatively cold glucose to lower conductivity (5)), and the transient response in signal recorded. Details of the method and its implementation are provided below.

In addition to G_p term, the signal includes a slope factor. As shown in the equation in Fig. 1, part of this relates to the blood resistivity, and part to the segment length. In addition, deviation of the real current field from the one assumed (i.e. one with perfectly parallel equi-potential planes perpendicular to the catheter axis) results in an additional gain factor. As shown in Fig 1, the real current field is curvilinear, and this leads to some slight non-linearity in the signal. The highest concordance between real and assumed fields lies close to the catheter itself, and as the mouse heart has a small short axis diameter, this helps reliability in this species. Nonetheless, absolute

$$\text{GAIN} = \frac{\text{reference stroke volume}}{\text{conductance stroke volume}}$$

calibration also requires a comparison to an alternative flow measure: e.g. flow probe, thermodilution, Doppler echo. Such data are then used to assess the signal *gain* to be applied to the volume signal as follows:

GENERAL METHODOLOGY

We have adapted anesthesia protocols to mice to maintain heart rate and blood pressure in a physiologic range, even in animals with an open thoracotomy. Anesthesia is initiated with methoxyflurane (a 1:4 mixture of isoflurane and light paraffin oil may be used as a substitute) followed by an intraperitoneal injection of urethane (800 mg/kg), etomidate (20 mg/kg) and morphine (1-2 mg/kg). After placing the animal on a heating pad, the trachea is cannulated with a smoothed 19G needle and ventilation is initiated with 100% O_2 at a tidal volume of 200μL and a rate of 140 breaths/min. A custom designed, closed circuit ventilator employing solenoid valves (7) provides a constant flow at peak airway pressures of 5-10 cmH_2O.

The external jugular vein is exposed and directly cannulated using a 30G needle. Volume expansion is achieved by infusion of 150μL, 12.5% human albumin. Following stabilization, a heat cautery is used to make a lateral incision on either side of the xyphoid exposing the LV apex. An apical stab using a 26G needle allows placement of a 1.4F pressure-volume catheter (Millar, SPR-719)(Figure 2) along the long axis, such that the proximal electrode is just inside the endocardium. The catheter occupies a total volume of 0.76μL and approximately 20% of the cross-sectional area of the aortic valve at the widest part. The catheter is connected to a custom designed signal conditioner which provides a constant current of 30μA at a frequency of 20kHz. Manual compression of the inferior vena cava and thoracic aorta provides transient changes in preload and afterload respectively.

Figure 2. Pressure-volume catheter used in the mouse. The catheter is 1.4F (0.4 mm diameter), and contains four electrodes (E1-E4), and a single pressure sensor. Electrodes E1 and E4 are positioned at the base and apex of the heart respectively. Current is injected between electrodes E1 and E4 and the potential difference is measured across electrodes E2 and E3.

VOLUME SIGNAL CALIBRATION

While echocardiography and MRI may be used to match ventricular volume to the conductance volume, measuring absolute volume is limited by the small size and rapid heart rate of the murine heart and an assumption of ventricular geometry. To estimate the parallel conductance in the mouse, the hypertonic saline dilution method is used in our lab. A 10μL bolus of hypertonic saline (30%) is rapidly injected by hand into the external jugular vein. The saline transient is used to derive the parallel conductance (Figure 3). The assumption is that the saline has only altered the conductivity of the blood pool and not surrounding structures, and that the physiology of the heart is unaltered during this change. This is shown in Panel 3A by the dP/dt curve, which has a constant peak value despite an apparent rise in preload volume (Panel 3B). The latter does not reflect volume but rather an increased volume *signal* due to the rise in blood conductance. As previous shown by Baan et al (5), and Lankford et al (6), the time varying component of the transient rise in catheter signal can be dissected from the constant offset (i.e. G_p) by several regression methods (Panels 3C and 3D). In both methods, a value of apparent conductance at zero-blood conductivity (a situation that would be physically impossible to achieve) is estimated.

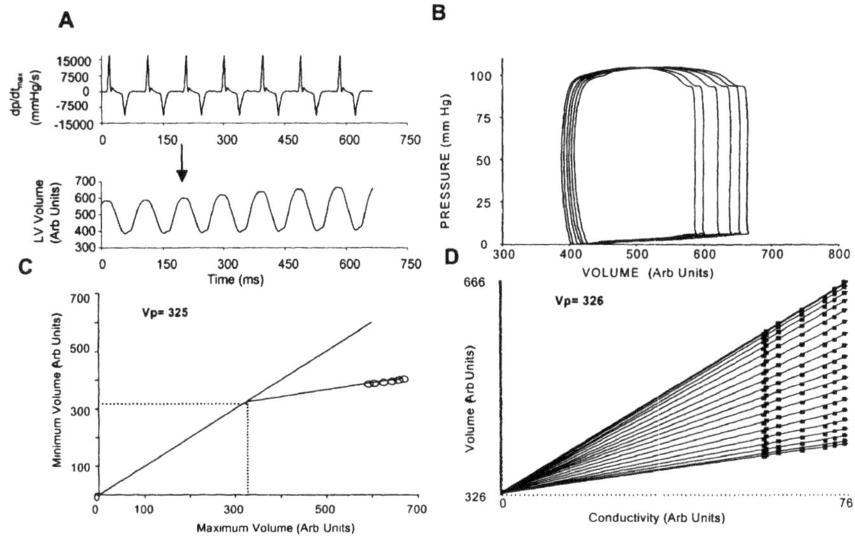

Figure 3. Determination of parallel conductance in the mouse. **A.** Time series tracings of dp/dt_{max} (top) and LV volume (bottom). Arrow indicates the onset of hypertonic saline wash-in with subsequent change in the position and stroke volume of the conductance signal due to the increased blood conductivity. Lack of change in dp/dt_{max} indicates no alteration in contractility or LV volume. **B.** Pressure-volume loops during the saline wash-in. **C.** Regression of maximum and minimum conductance from the transient shown in panel A is used to determine the volume value at which both signals would be identical (i.e. zero apparent stroke volume). This is an estimate of the parallel conductance (Vp), 325 in this example. **D.** Same data analyzed by the method of Lankford et. al. (6) whereby the volume signal from the multiple cycles is divided into equal segments during ejection, and each individually regressed back to the zero-conductance intercept. The value of Vp is in perfect agreement with the method in C, indicating the constancy of the parallel conductance throughout the cardiac cycle. See Lankford et al (6) for full descriptions of both methods.

Recent analysis of the source of the parallel conductance in the mouse (8) has shown that it arises primarily from the the LV myocardium and not additional extrinsic structures in the chest. Interventions that altered blood volume of the right ventricle had no effect on the conductance signal (8). Even more extreme, surrounding the heart with saline produced no change in the conductance signal (Figure 4). This unique property present in the mouse likely results from the relatively thick wall of the LV and close proximity of the sensing and stimulating electrodes, thereby tending to localize the electric field (8). This results in a signal that is extremely stable and allows for estimate of the parallel conductance to be made without the use of hypertonic saline (8).

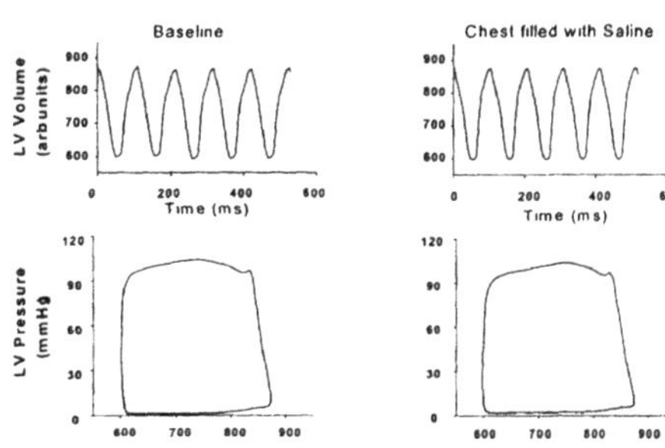

Figure 4. Effect of surrounding the mouse heart *in situ* with warmed physiologic saline. Baseline volume vs time and pressure-volume data are shown (left) and after the saline maneuver (right). Despite this dramatic change in external conductance to the heart, there was negligible change in the LV volume signal or pressure-volume loop.

After subtraction of the offset term as described above, the volume signal must be corrected to reflect accurate stroke volume. An independent measure of cardiac output is used for correction of stroke volume by the catheter. With the catheter in place, the mouse is turned on its side and an incision is made between the ribs at level T3-T4. The thoracic aorta is dissected free from the spinal column, and a perivascular, ultrasound

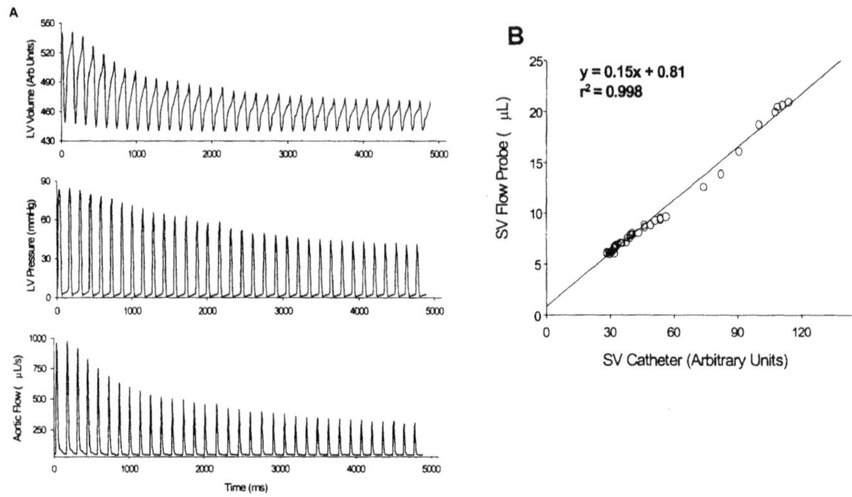

Figure 5. A. Time series tracings of LV Volume, LV pressure and thoracic aortic flow during inferior vena caval occlusion. **B.** Comparison of beat to beat stroke volume determined by flow probe and conductance catheter. The two variables are highly correlated and the gain in this example is 0.15.

flow probe (1RB, Transonic) is placed around it. With the catheter and flow probe in place, the inferior vena cava is occluded yielding multiple cycles with different stroke volumes (Figure 5A). Stroke volumes are calculated by beat by beat integration of the flow signal and compared with simultaneous stroke volumes from the uncalibrated conductance signal. Figure 5B shows that the two estimates of stroke volume are highly and linearly correlated over a broad range. Pooled correlation coefficients from all animals yielded r^2=0.983, a mean intercept of -1.23 µL (p<0.0001), with a mean gain of 0.13±0.015 (coefficient of variation of 11.5%). Application of this gain yields relative volume in microliters. It is important to note that the particular gain (and offset from the parallel conductance) obtained here are specific to the settings of *our* instrumentation. Recalibration for both gain and offset is necessary each time instrumentation settings are changed. The small, but significant offset obtained from the regression of the stroke volumes implies non-linearity of the underlying catheter volume vs true volume relation, consistent with electric field theory (9). However, it has recently been shown that in the mouse, the relationship is well approximated by linear relation over the operating volumes of the heart (8). Overall, the regressions obtained from comparing stroke volumes are highly linear over a broad range and likely result from the relatively small dimensions of the murine heart where the electric field is relatively linear. Thus, the application of the conductance catheter in the mouse appears to have been optimized such that the signals obtained are highly stable, free of artifact and linearly correlated with actual ventricular volume; making this approach an extremely powerful tool to assess hemodynamics in this small animal.

SYSTOLIC FUNCTION

Figures 6A and B display typical pressure-volume relations in the mouse derived by varying the loading conditions on the heart. Manual compression of the inferior vena cava results in preload decline (6A) while compression of the thoracic aorta results in afterload increase and systolic pressures in excess of 200 mmHg (6B). These relations can be used to derive the end-systolic pressure-volume relationship (ESPVR), which is often curvilinear in the mouse. The local slope of this relation (Ees), or end-systolic elastance, reflects chamber end-systolic *stiffness,* and has been used as an index of contractility. ESPVR non-linearity observed in mice, (and larger mammals at higher contractilities) can be fit using a logarithmic function (18). Restriction of analysis to the physiologic range of end-systolic pressures can assist in deriving a linear estimate of Ees. While quantitative assessment of Ees is useful, it is more important to examine the actual position and shape of the relation, as this itself conveys a great deal of information about systolic cardiac performance (2). In addition to the ESPVR,

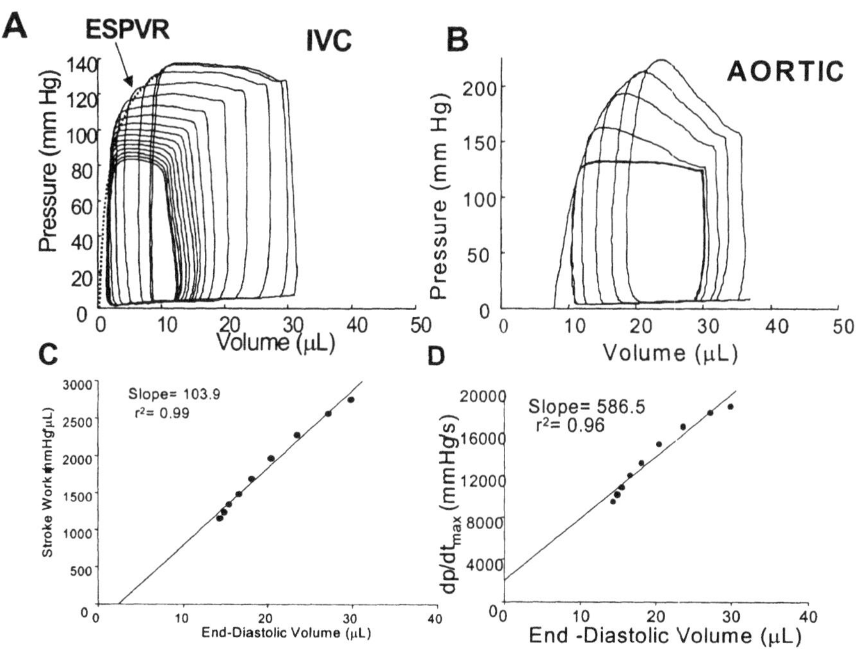

Figure 6. A. Pressure-volume relations derived from preload reduction and afterload increase used to derive the ESPVR. From the multiple beats in panel A, other indices of systolic function can be derived (C) stroke work versus end-diastolic volume and (D) dp/dt_{max} versus end-diastolic volume.

pressure-volume data measured during preload reduction can be used to derive other valuable indices of systolic function. The stroke work versus end-diastolic volume relation (Figure 6C) or preload recruitable stroke work (PRSW) is another useful parameter (19). While generally reflecting systolic function, this index integrates systolic and diastolic function, and changes in diastolic function alone can also influence its slope. Its advantages are that it is typically quite linear, and the slope has force units (e.g. mmHg) making it chamber size independent, unlike end-systolic elastance which has units of mmHg/μL. The PRSW is also more statistically robust than the ESPVR and thus can be useful in quantifying changes in function. Another functional index is the relationship between dp/dt_{max} and end-diastolic volume (Figure 6D) (20). The relationship can reflect somewhat different properties of the heart, since dP/dt_{max} is an isovolumetric parameter (rather than an ejection phase parameter). This index tends to be more sensitive to changes in systolic function than either the ESPVR or the PRSW. Importantly, all of these parameters can be assessed from the same data set, providing a comprehensive view of systolic performance.

The most widely used assessment of systolic function in mice is typically dp/dt_{max} alone. While preload dependence of this parameter is well known and established in larger mammals, the dependence appears to be exacerbated in the mouse. Figure 7 shows the effect of preload reduction on dp/dt_{max}. The change in end-diastolic volume and end-diastolic pressure is also shown. While the end-diastolic volume tracks the change in $dpdt_{max}$, the pressure falls by less than 2 mmHg and falls within the noise range of the catheter. This effect is the result of the very flat EDPVR in the mouse. Thus, indexing changes in dp/dt_{max} by changes in end-diastolic pressure to account for preload effects can be very difficult and it is clearly advantageous to be able to measure ventricular

Figure 7. The effect of preload change on dp/dt_{max} in the mouse. A chnange in LV volume alone will have dramatic changes in dp/dt_{max} regardless of changes in contractile function. However, without measurement of volume, it is difficult to account for this dependence since end-diastolic pressure is almost unchanged.

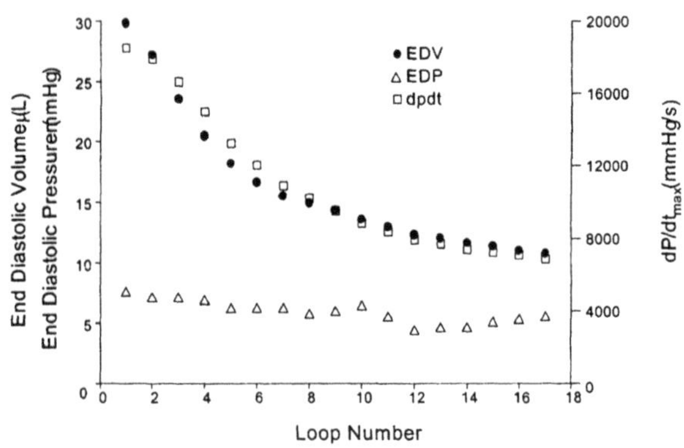

volume. If pressure is all that is available, then normalization of dP/dt_{max} for preload can be accomplished by dividing by the instantaneous pressure at which dp/dt_{max} occurs ($dP/dt_{max/IP}$).

STRESS-STRAIN RELATIONS

The ability to acquire pressure-volume data allows the application of mathematical models to these variables to derive relations between stress and strain. The advantage of stress-strain relations is that they are chamber size independent and they provide information about *myocardial* properties rather than chamber properties. A mathe-

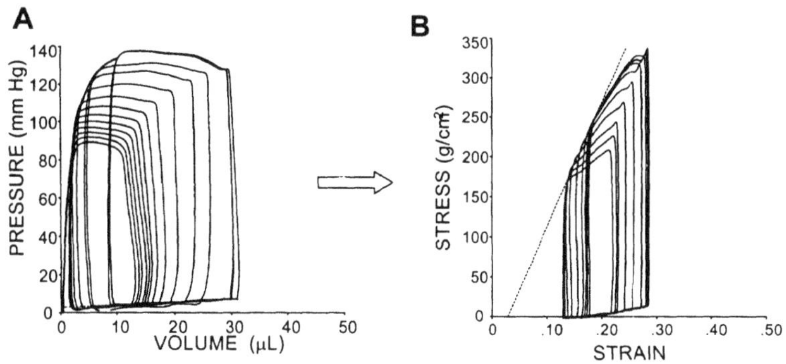

Figure 8. Conversion of chamber pressure-volume relations (A) to myocyte stress-strain relations (B). Note that peak stress occurs at the onset of ejection and falls during ejection.

matical model has been derived relating LV pressure and the ratio of cavity volume to wall volume to muscle fiber stress and strain during systole (21). This model is based on experimental data which show that muscle fiber stress and strain are homogeneously distributed and that the ventricle is rotationally symmetrical. The shape of the ventricle is unimportant. The model was applied to the mouse pressure-volume data to derive systolic stress-strain relations (Figure 8).Since fiber stress at zero volume is unknown, the strain calculated is a relative strain i.e. only the width of the stress-strain is calibrated, not the absolute position. The end systolic stress-strain relation (ESSSR) is highly linear and its slope reflects myocyte stiffness. Values for peak stress and ESSSR obtained with this method were similar to those recently derived in the mouse using sonomicrometer crystals placed on the LV endocardium (22) to derive an elipsoidal-based stress estimate.

DIASTOLIC FUNCTION

The method of using pressure-volume relations and altered loading conditions to the heart has also proven very valuable to the study diastolic chamber function. Figure 9A shows diastolic pressure-volume curve (EDPVR) from wild-type mice. The diastolic data points during late filling , before atrial systole, are used and are fit by a mono-exponential function, $P=\alpha(e\beta^{V}-1) + P_o$ where P is pressure, V is volume, P_o is the pressure at zero volume, α is a geometric coefficient and β reflects chamber stiffness. Like the slope of the ESPVR, β is chamber size dependent (Table 2). In panel 9B, the EDPVR is derived in a mouse with increased diastolic stiffness - but also with increased right ventricular pressures, and pulmonary hypertension. Note the clear difference between the steady state loop and that obtained during preload reduction. Diastolic pressure-volume data for the steady state cycle are elevated, primarily due to the influence of the pericardium and the elevated right heart pressures which impact the LV through the sharing of a common septum. By minimizing external forces, the end-diastolic PV relation points represents the closest measure of passive chamber properties in the intact heart. In addition to the information obtained from the EDPVR,

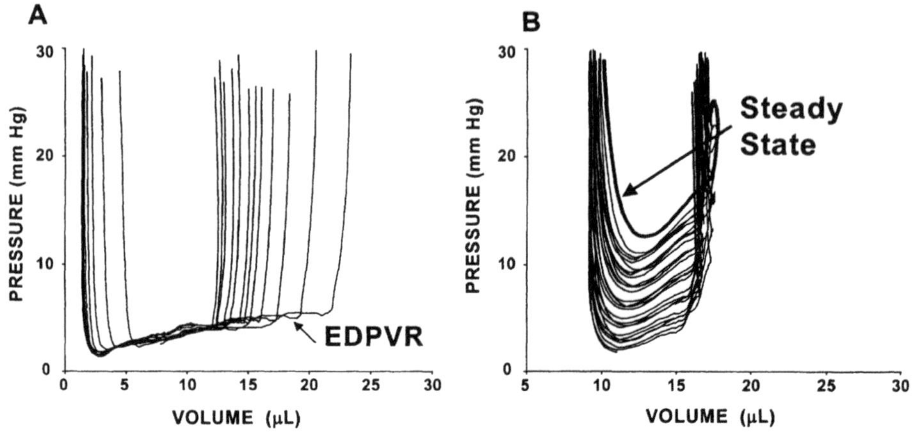

Figure 9. A. Pressure-volume loops and preload reduction used to generate the end-diastolic pressure-volume relationship (EDPVR) in wild-type mice. **B**. In a model of increased diastolic stiffness, there is an initial downward displacement of the EDPVR, and a disparity between steady state diastolic pressure. This is due to external constraining forces from the right heart and pericardium which are removed with preload reduction. This latter portion describes the true passive stiffness of the ventricle and is not as severe as initially seen at steady state.

the measurement of pressure and volume allows one to derive tau, the time constant of isovolumic relaxation of LV pressure. Various formulations of tau have been developed (see Table 2). In this table, Tau (Ln) and tau (N) represent mono-exponential decay time constants derived from the pressure curve, assuming either a zero or non-zero pressure asymptote, respectively. Tau Tau (L) employs a logistic-growth function to model the fall in pressure and has been shown to be a more robust estimate, minimally influenced by loading conditions, shifts in the signal or inotropic state (16). This formulation is commonly used in our lab in diseased ventricles. Further insights into diastolic function can be obtained from the volume signal. Differentiation (dV/dt) can used to estimate early (E wave) and late (A wave) filling properties of the ventricle (Figure 10).

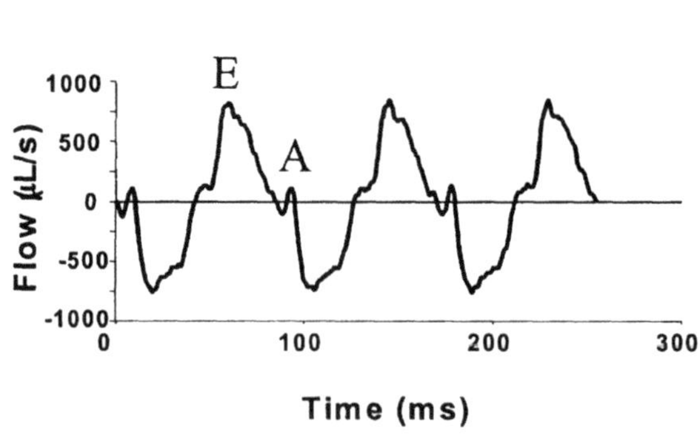

Figure 10. Filling patterns of the mouse ventricle obtained by differentiation of the volume signal. At physiologic heart rates, a small or negligible A wave is usually observed.

Unlike Doppler signals, dV/dt is not affected by the area of the mitral valve since flow and not flow velocity is measured. At physiologic heart rates (>550 beats/min.), a very small A wave is observed in the mouse. In studies where ventricular volume is not measured (i.e. only pressure is recorded), use of relaxation time constants is preferable to $dp/dt_{min,}$ which while commonly reported, is itself quite load dependent. Also, it has been shown in larger mammals that there is no correlation between dp/dt_{min} and tau (23).

HEART-ARTERIAL INTERACTION

The output of the ventricle is not only determined by its systolic and diastolic properties, but also by its interaction with the arterial system. Using the pressure-volume approach and the elastance concept, Sunagawa et al (24) proposed a framework for analysis of coupling of the ventricle and arterial system. Arterial system properties can be represented by Ea, effective arterial elastance which embodies the mean resistance and pulsatile features (as well as heart rate interaction) of the arterial load (25). This is approximated by P_{es}/SV where SV is stroke volume. The Ea/Ees ratio has been used as an index of ventricular-arterial coupling, much like impedance matching of a electronic circuit, and this coupling has been shown to be an important determinant of pump performance and efficiency (26). To examine this relation in the mouse, we transiently occluded the aorta (Figure 6B) to vary Ea more than two-fold while maintaining Ees constant, and assessed the effect of varying coupling on stroke work (Figure 11).

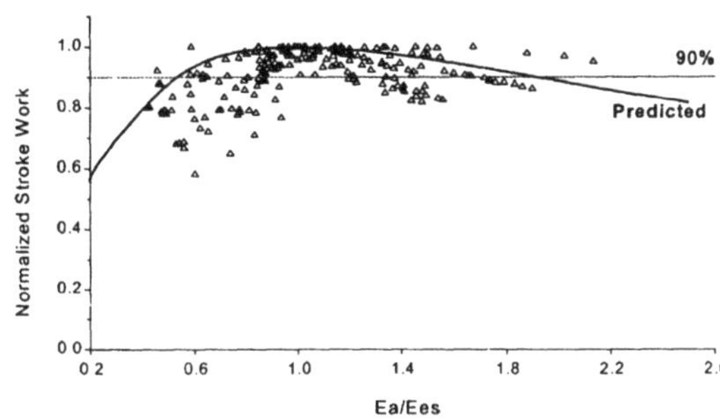

Figure 11. Assessment of ventricular-vascular coupling in the mouse by Ea/Ees during aortic occlusion. Normalized stroke work peaks at an Ea/Ees value of 1, and the relation observed in the mouse is well similar to that predicted by theoretical modeling.

The relation obtained in the mice is similar to that predicted by theory and observed in conscious dogs (27). As in the dog, the actual data underestimate the predicted value because of the assumption of a linear ESPVR (28). At steady state, the Ea/Ees value is 0.5 (Table 1), which is still approximately 90% of the maximal stroke work.

PRESSURE-VOLUME HEMODYNAMICS in NORMAL MICE

The following tables summarize hempodynamic and pressure-volume relation parameters derived from various wildtype mouse strains (e.g. FVB/N, C57Bl/6, 129SV/EV, Black Swiss, C3H/He). These data are meant to provide a reference for individuals interested in applying this approach, particularly as reported data in control animals remains relatively scant. For comparison, the corresponding value obtained from the literature in humans is also provided. Values for heart rate, systolic pressure and cardiac output obtained with our protocol are very similar to those described in conscious mice (10,11). Interestingly, many parameters between mouse and human are quite similar when either normalized for heart rate (time dependent ratio is between 7-8.5), or for stroke volume (internal chamber size). One noteworthy difference is the ratio of LV end-diastolic volume to wall volume. The mouse (0.31±0.04) appears to be lower than normal human, and more in the range observed in patients with hypertrophied hearts (0.5±0.1) (12). This may contribute to relatively higher ejection fractions in mice versus larger mammals.

Table 1. Hemodynamic parameters in the mouse derived from pressure-volume relations

PRESSURE-VOLUME RELATIONS AND CARDIAC DISEASE

Parameter	Mouse	Human	Mouse/Human
Heart Rate, min^{-1}	597±58	71	8.4
Systole, % of Cardiac Cycle	36.4±3.5	35	1
Diastole, % of Cardiac Cycle	63.6±3.5	65	1
End-Diastolic Pressure, mmHg	5.73±2.1	12.5	2.0
End-Systolic Pressure, mmHg	117.5±10.1	121	1
End-Diastolic Volume, µL	29.3±3.1	108.1×10^{3}	1/3700
End-Systolic Volume, µL	6.1±4.1	43.9×10^{3}	1/7200
Stroke Volume, µL	23.2±2.6	64.9×10^{3}	1/2800
Cardiac Output, mL/min.	13.8±2.5	4.6×10^{3}	1/333
Arterial Elastance (Ea), mmHg/µL	4.8±0.91	2.3×10^{-3}	2100
Body Mass, g	25.3±4.6	70×10^{3}	1/2800
Heart Mass, mg	118.4±18.3	325×10^{3} [14]	1/2800
LV Mass, mg	95.4±9.6	130×10^{3} [14]	1/1400
Heart Mass / Body Mass, mg /g	4.3±0.42	4.6	1
End-Diastolic Volume / LV Wall Volume	0.31±0.04	0.8-1 [12]	0.31

In the classification cardiac dysfunction, abnormalities are usually divided into dilated cardiomyopathy, hypertrophic cardiomyopathy and restrictive cardiomyopathy. Figure 12 displays an example from each type of disease, using examples drawn from genetically engineered animals. The dilated cardiomyopathy example was obtained in a mouse with a homozygous mutation in myosin binding protein C. It shows typical chamber dilation, depression of the ESPVR slope, and reduced ejection fraction. Interestingly, dP/dt_{max} was unaltered from controls in this model, showing that not all aspects of cardiac systolic failure will be reflected in isovolumic contraction phase

Table 2. Indices of systolic and diastolic function derived from pressure-volume relations

Parameter	Mouse	Human	Mouse/ Human
Diastole			
Diastolic Compliance (β^{-1}), μL/mmHg	11.1±5.4	5.1×10^{3}	NA
Peak Filling Rate/End-Diastolic Volume, s^{-1}	31.5±6.8	3.6	8.75
Time to Peak Filling Rate, ms	25.1±4.4	200[15]	1/8
Pressure Half Time, ms	3.1±0.73	26[16]	1/8.4
Tau (Ln), ms	3.9±0.96	38[16]	1/9.7
Tau (G), ms	6.53±1.2	45[16]	1/7
Tau (L), ms	3.01±0.76	21[16]	1/7
Tau (N), ms	6.37±0.18	46[16]	1/7.2
Peak Early/Peak Late filling, E/A ratio	no late filling	1.5-15[a]	NA
dP/dt_{min}, mmHg/s	-15432±2189	-1900	8.1
Systole			
Ejection Fraction, %	80±6.8	61	1.3
Stroke Work (SW), mmHg×μL	2732±289.7	9.6×10^{6}	1/3500
dP/dt_{max}, mmHg/s	16894±2132	2100	8
dP/dt_{max}/Pressure at dP/dt_{max}, s^{-1}	251.6±39.5	35	7.2
$Power_{MAX}$/End-Diastolic Volume, mmHg/s×100	44.5±9.9	6.3	7.1
End-Systolic Elastance (Ees), mmHg/μL	10.1±4.3	2.1×10^{-3}	1/4800
Ea/Ees	0.48±.06	0.5-1[17]	0.5-1.0
Stroke Work/End-Diastolic Vol. relation, mmHg	100.5±29.1	82	1.2
dP/dt_{max}- End Diastolic Vol. relation, mmHg/s/mL	610±178.2	20×10^{-3}	30×10^{3}

parameters. (28).Hypertrophic cardiomyopathy is most often associated with a leftward shift and steep slope of the ESPVR and reduced chamber volumes. The steepness of the ESPVR is likely not due to improved myocardial function but instead to small chamber volumes, hypertrophy and fibrosis (29). Indeed this model, involving a heterozygous mutation of myosin heavy chain (R403Q), displayed reduced cardiac output (29). Our analysis such far has found that abnormalities of the EDPVR are often less dramatic, even with a prolonged time constant of relaxation in such models. Restrictive cardiomyopathy is associated with a rightward shifted ESPVR and a very steep EDPVR. In this particular transgenic model, over-expression of MKK-p38 pathway resulted in profound stiffening of the ventricle. End-diastolic pressure was in excess of 30 mmHg, yet relaxation time constant were normal. Generation of the EDPVR by preload reduction displayed the pattern shown on an expanded pressure scale in Fig 9B, suggesting an increased role of external forces contributing to the steady state EDPVR.

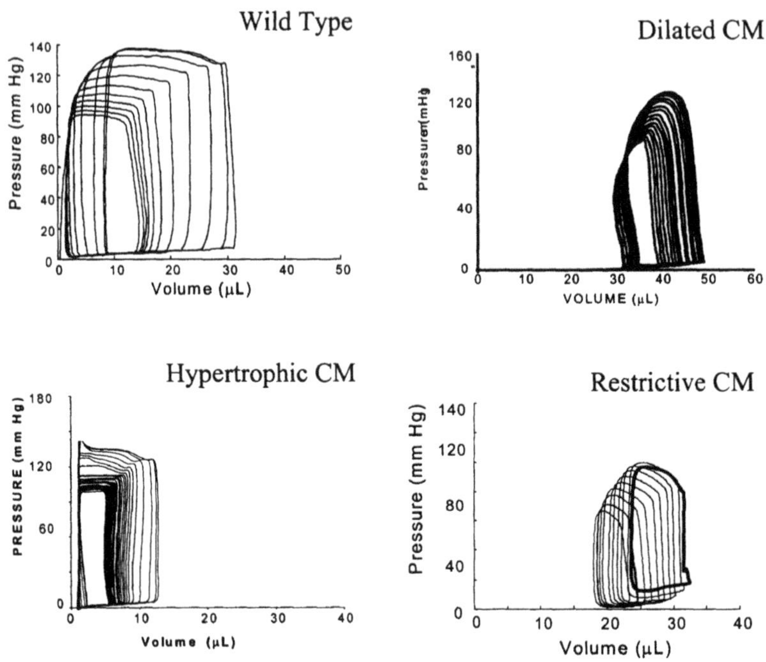

Figure 12. Pressure-volume relations for example normal and diseased ventricles in mice.

LIMITATIONS AND FUTURE DIRECTIONS

While the data provided here are rather physiologic, particularly when contrasted to pressures and flows measured in conscious mice, the methodology employed remains invasive and provides data at only one point in time. Furthermore, the effects of anesthetics, particularly in the mouse, are very complex and can affect vascular and myocardial properties. Our lab is currently developing experimental procedures aimed at chronically instrumenting a mouse with a non-tethered, pressure-volume catheter. While this method may not be employed in every study, it should provide the ability to follow animals over a period of time and to validate results in the conscious mouse. Despite these limitations, the assessment of cardiovascular function in the mouse has now reached a level which is near-comparable to larger mammals, including humans. With the rapid development of molecular and genetic techniques which allow for very specific targeting of pathways and molecules in the mouse, pressure-volume relations can be used to translate the effects of these precise changes to the complex function of the intact heart. It is hoped that this interaction between molecular genetics and systems physiology will lead to a fundamental understanding of the function of the heart in normal and disease states.

REFERENCES

1. Sagawa K, Maughan WL, Suga H, Sunagawa K: *Cardiac Contraction and the Pressure-Volume Relationship*, New York, Oxford University Press, 1988.
2. Kass DA, Maughan WL: From 'Emax' to pressure-volume relations: a broader view. *Circulation* 1988; 77:1203-1212.
3. Little WC. Assessment of Cardiac Function. In Braunwald's Heart Disease: A Textbook of Cardiovascular Medicine. E. Braunwald, editor. Saunders, Philadelphia, PA. pp421-445, 1997.
4. Georgakopoulos D, Mitzner WA, Chen CH, Byrne BJ, Millar HD, Hare JM, Kass DA: In vivo murine left ventricular pressure-volume relations by miniaturized conductance micromanometry. *Am.J.Physiol.* 1998;274:H1416-H1422
5. Baan J, van der Velde ET, de Bruin HG, Smeenk GJ, Koops J, van Dijk AD, Temmerman D, Senden J, Buis B: Continuous measurement of left ventricular volume in animals and humans by conductance catheter. *Circulation* 1984;70:812-823
6. Lankford EB, Kass DA, Maughan WL, Shoukas AA: Does volume catheter parallel conductance vary during a cardiac cycle? *Am.J.Physiol.* 1990;258:H1933-H1942
7. Ewart S, Levitt R, Mitzner W: Respiratory system mechanics in mice measured by end-inflation occlusion. *J.Appl.Physiol.* 1995;79:560-566
8. Georgakopoulos D, Kass DA: Estimation of parallel conductance by dual frequency conductance catheter in Mice. *Am.J.Physiol.* 2000;279: H443-450.
9. Steendijk P, van der Velde ET, Baan J: Dependence of anisotropic myocardial electrical resistivity on cardiac phase and excitation frequency. *Basic.Res.Cardiol.* 1994;89:411-426
10. Janssen BJ, Leenders PJ, Smits JF: Short-term and long-term blood pressure and heart rate variability in the mouse. *Am.J.Physiol.Regul.Integr.Comp.Physiol.* 2000;278:R215-R225
11. Mills PA, Huetteman DA, Brockway BP, Zwiers LM, Gelsema AJ, Schwartz RS, Kramer K: A new method for measurement of blood pressure, heart rate, and activity in the mouse by radiotelemetry. *J.Appl.Physiol.* 2000;88:1537-1544
12. Mirsky I, Cohn PF, Levine JA, Gorlin R, Herman MV, Kreulen TH, Sonnenblick EH: Assessment of left ventricular stiffness in primary myocardial disease and coronary artery disease. *Circulation* 1974;50:128-136
13. Liu CP, Ting CT, Lawrence W, Maughan WL, Chang MS, Kass DA: Diminished contractile response to increased heart rate in intact human left ventricular hypertrophy. Systolic versus diastolic determinants. *Circulation* 1993;88:1893-1906
14. Little WC: Assessment of Cardiac Function., in Braunwald E (ed): *Braunwald's Heart Disease: A textbook of Cardiovascular Medicine*. ed4. Philadelphia, PA, Saunders, 1997, pp 421-445
15. Little WC, Downes TR: Clinical evaluation of left ventricular diastolic performance. *Prog.Cardiovasc.Dis.* 1990;32:273-290

16. Senzaki H, Fetics B, Chen CH, Kass DA: Comparison of ventricular pressure relaxation assessments in human heart failure: quantitative influence on load and drug sensitivity analysis. *J.Am.Coll.Cardiol.* 1999;34:1529-1536
17. Asanoi H, Kameyama T, Ishizaka S: Ventriculo-Arterial Matching of Failing Hearts., in LeWinter M, Suga H, Watkins MW (eds): *Cardiac Energetics: From Emax to Pressure-Volume Area*. Boston, Kluwer, 1994, pp 157-169
18. Mirsky I, Tajimi T, Peterson KL: The development of the entire end-systolic pressure-volume and ejection fraction-afterload relations: a new concept of systolic myocardial stiffness. *Circulation* 1987;76:343-356
19. Glower DD, Spratt JA, Snow ND, Kabas JS, Davis JW, Olsen CO, Tyson GS, Sabiston DCJ, Rankin JS: Linearity of the Frank-Starling relationship in the intact heart: the concept of preload recruitable stroke work. *Circulation* 1985;71:994-1009
20. Little WC: The left ventricular dP/dtmax-end-diastolic volume relation in closed-chest dogs. *Circ.Res.* 1985;56:808-815
21. Arts T, Bovendeerd PH, Prinzen FW, Reneman RS: Relation between left ventricular cavity pressure and volume and systolic fiber stress and strain in the wall. *Biophys.J.* 1991;59:93-102
22. Esposito G, Santana LF, Dilly K, Santos Cruz J, Mao L, Lederer WJ, Rockman HA: Cellular and functional defects in a mouse model of heart failure. *Am.J.Physiol.Heart Circ.Physiol.* 2001;279:H3101-H3112
23. Leite-Moreira AF, Gillebert TC: Nonuniform Course of Left Ventricular Pressure Fall and Its Regulation by Load And Contractile State. *Circulation* 1994;90:2481-2491
24. Sunagawa K, Maughan WL, Burkhoff D, Sagawa K: Left ventricular interaction with arterial load studied in isolated canine ventricle. *Am.J.Physiol.* 1983;245:H773-H780
25. Kelly RP, Ting CT, Yang TM, Liu CP, Maughan WL, Chang MS, Kass DA: Effective arterial elastance as index of arterial vascular load in humans. *Circulation* 1992;86:513-521
26. Burkhoff D, Sagawa K: Ventricular efficiency predicted by an analytical model. *Am.J.Physiol.* 1986;250:R1021-R1027
27. Little WC, Cheng CP: Effect of exercise on left ventricular-arterial coupling assessed in the pressure-volume plane. *Am.J.Physiol.* 1993;264:H1629-H1633
28. McConnell BK, Jones KA, Fatkin D, Arroyo LH, Lee RT, Aristizabal O, Turnbull DH, Georgakopoulos D, Kass D, Bond M, Niimura H, Schoen FJ, Conner D, Fischman DA, Seidman CE, Seidman JG: Dilated cardiomyopathy in homozygous myosin-binding protein-C mutant mice [In Process Citation]. *J.Clin.Invest.* 1999;104:1771
29. Georgakopoulos D, Christe ME, Giewat M, Seidman CM, Seidman JG, Kass DA: The pathogenesis of familial hypertrophic cardiomyopathy: early and evolving effects from an alpha-cardiac myosin heavy chain missense mutation *Nat.Med.* 1999;5:327-330

15
ASSESSMENT OF LEFT VENTRICULAR DIASTOLIC FUNCTION IN MICE

Michael R. Zile, M.D., Anne Hauet, M.D., Elizabeth Hapke, BS, Janet Simsic, M.D.

Medical University of South Carolina, Gazes Cardiac Research Institute, Ralph H. Johnson Department of Veterans Affairs Medical Center, Charleston, South Carolina.

INTRODUCTION

Diastolic dysfunction occurs when the ventricle is unable to fill to an adequate volume at low pressure. When diastolic dysfunction increases pulmonary venous pressures, symptoms of diastolic congestive heart failure develop. This can occur at rest, in the most extreme case or only with exertion, in the most common case. Diastolic dysfunction results from abnormalities in active relaxation and/or passive stiffness. These abnormalities in diastolic function can occur during the development of hypertension, valvular stenosis, ischemia, and advanced age. (1)

The basic mechanisms by which these diseases alter diastolic function can be divided into two categories: cellular mechanisms, consisting of those structures and processes intrinsic to the cardiac muscle cell itself; and extracellular mechanisms, consisting of those structures and processes within the extracellular matrix (ECM) which surrounds the cardiomyocyte. (2) These cellular and ECM mechanisms are dependent on a specific set of gene products which may change during disease or which may be changed by transgenic technology.

The purpose of this chapter is to describe the *in vivo* and *in vitro* methods which can be used to characterize the changes in diastolic function produced by the genetic alterations which develop in murine models of cardiovascular disease or which result from genetic engineering in transgenic mice.

BASIC MECHANISMS

Active Relaxation

Diastole is the process by which the myocardium loses its ability to generate force and shorten and then returns to resting force and length. Diastole occurs in a series of energy consuming steps beginning with release of calcium from troponin C, detachment of the actin-myosin cross-bridge, phosphorylation of phospholamban, sarcoplasmic reticulum calcium ATPase induced calcium sequestration in the sarcoplasmic reticulum, sodium/calcium exchanger induced extrusion of calcium from the cytoplasm, the resultant slowing of cross-bridge cycling rate and extension of the sarcomere to rest length. Adequate energy supplies and the mechanisms to regenerate them must be present for this process to occur at a sufficient rate and extent. These intracellular energy consuming processes are discussed in detail in Chapters 2, 5 and 6.

Passive Stiffness

In addition to active processes, passive viscoelastic processes contribute to returning the myocardium to resting force and length. These passive viscoelastic properties are dependent both on intracellular and extracellular structures. During contraction, energy is gained by intracellular structures as they undergo shape change. For example, cytoskeletal proteins such as titin and microtubules are deformed by actin-myosin cross-bridge cycling and sarcomere contraction. (3-5) During diastole, these proteins act like viscoelastic springs, expend this stored potential energy and provide a recoiling force to restore the myocardium to its resting length. In addition, it is likely that proteins within the extracellular matrix contribute to the establishment of resting force and length. For example, fibrillar extracellular matrix proteins, such as collagen, uncoil during diastole, act to limit cell lengthening during diastole, and protect the myocardium from being stretched beyond resting length. (6-8)

The methods described in this chapter provide the means to assess the resultant effects of changes in these cellular or ECM mechanisms on diastolic function and cardiovascular phenotype.

MEASUREMENT OF FUNCTION

In this chapter, diastole will be defined as that time period from aortic valve closure (AVC) to mitral valve closure (MVC). This definition provides a practical platform from which to develop methods to quantitate the active and passive processes of diastole, force decline, filling and stiffness.

Active Relaxation

The rate and extent to which the cellular and extracellular processes described above occur determines the rate and extent of active relaxation. At the chamber level, this process results in LV pressure decline between AVC and mitral valve opening (MVO), a time during which volume is constant (isovolumic relaxation). Then, when LV pressure falls below left atrial pressure, the mitral valve opens, chamber filling occurs until MVC. During this filling process there are variable changes in LV pressure. Auxotonic relaxation (filling of variable pressure) describes the entire filling period. Auxotonic relaxation is affected both by active relaxation and passive stiffness.

Using *in vitro* preparations of an isolated cardiac muscle or cardiomyocyte, this process of active relaxation results in isometric force decline followed by isotonic lengthening.

Passive Stiffness

The rate and extent of auxotonic filling also depends on LV chamber stiffness and on the material properties of the myocardium. Chamber stiffness can be assessed by examining the pressure and volume relationship during diastole. Myocardial stiffness can be assessed by examining the stress and strain relationship during diastole.

Using *in vitro* preparations of an isolated cardiac muscle or cardiomyocyte, material viscoelastic stiffness can be assessed using uniaxial variable rate stretches and examining the relationship between stress, strain, and strain rate.

METHODS OF MEASUREMENT

In vivo Quantitation of LV Chamber Diastolic Function

1) **Isovolumic Relaxation:** Quantitating isovolumic relaxation requires direct measurement of LV pressure. In the mouse, this is done using a 1.4 French high fidelity micromanometer tipped catheter placed into the carotid artery by direct arterial dissection and advanced into the ascending aorta and then into the LV cavity. (Also see Chapter 10) Care must be taken to place the catheter in the mid portion of left ventricle in order to avoid apical entrapment. There are a number of challenges performing catheterization in a mouse. First, the catheter must be carefully calibrated. This is especially important in measuring diastolic pressures where small changes in pressure carry large implications. Because this 1.4 fr catheter can not accommodate a fluid filled lumen, LV pressures can not be simultaneously calibrated *in vivo*. Calibration should be done by placing the catheter in warm saline and calibrating it against an internal electronic reference and against an external mercury column at 0 and 200 mmHg both prior to and following each catheterization. Second, hemodynamic and respiratory compromise can occur in a mouse if it not kept warm at 37° C. Therefore, the mouse must be kept normothermic using heat lamps or heated surgical tables and the rectal temperature should be monitored. Third, the mouse should receive supplemental oxygen. Fourth, anesthesia should be kept at a minimum acceptable level to maintain a heart rate > 350 bpm and systolic blood pressure >100 mmHg. Fifth, blood loss during surgical dissection should be carefully controlled. One milliliter of blood loss in a mouse is significant considering the total blood volume of a mouse. Any blood loss should be replaced with saline. Finally, data should be taken from cardiac cycles which occur at or near end expiration when LV pressure changes are physiologic.

Invasively derived LV pressures are used to describe the rate of isovolumic relaxation by calculating the peak instantaneous rate of LV pressure decline, peak (-) dP/dt and the time constant of isovolumic LV pressure decline, tau (τ). ′9-13) (Also see Chapter 10)

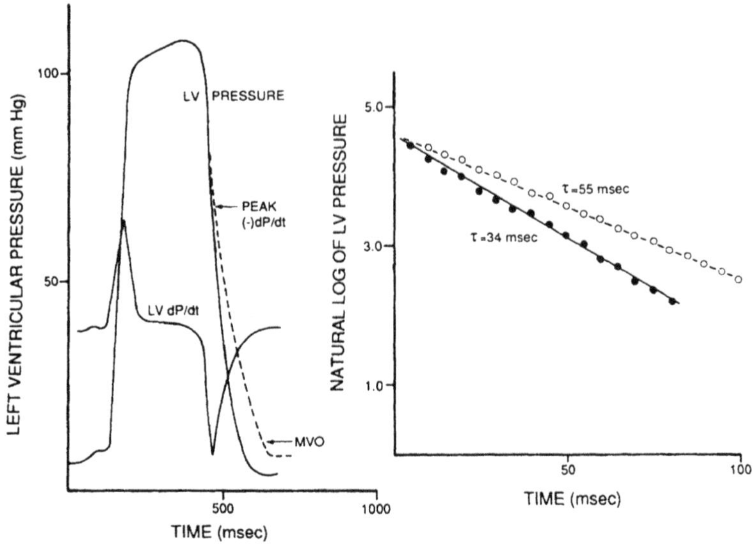

Figure 1: Schematic drawing of LV pressure and dP/dt showing methods used to quantitate isovolumic relaxation. Normal relaxation shown as solid lines, slowed relaxation as dashed lines.

Peak (-)dP/dt is calculated as the first derivative of LV pressure with respect to time. It reflects a single instantaneous event that occurs very early during LV pressure decline. This value not only reflects changes in isovolumic relaxation rate but also is affected by changes in mechanical factors such as preload, afterload, and heart rate.

The relaxation time constant, tau, reflects the time course of LV pressure decline from peak (-)dP/dt to MVO. Tau is generally quantitated using a method which assumes that pressure declines in monoexponential fashion. Because LV pressure is fit to a monoexponential decay, the plot of the natural log of LV pressure versus time over this period forms a linear relation. Tau is equal to the inverse slope of this linear relation. This relation can be expressed mathematically as $P = P_0 e^{-t/\tau}$, where P is LV pressure, e is the base of the natural logarithm, t is the time in milliseconds after peak (-)dP/dt, and P_0 is the pressure at the time of peak (-)dP/dt. (14)Stated in more conceptual terms, tau is the time that it takes for LV pressure to fall by 1/e from its initial pressure, ie to fall by approximately 2/3 of its initial value. When isovolumic pressure decline is slowed, tau is prolonged and the numeric value of tau increases (dashed lines in Figure 1). Because the heart rate in mice is so rapid, LV pressures must be recorded and analyzed at a very rapid sampling rate (at least 1 kHz). Because the exact time of AVC and MVO can not be easily determined in mouse preparations, tau is calculated using LV pressures from peak (-) dp/dt to an LV pressure value in early diastole equivalent to that obtained at end diastole. This method of calculating tau makes three assumptions: LV pressure decline is monoexponential, LV cavity pressure is equal to transmural pressure, and LV pressure will fall asymptotically to zero. (15) Assuming a monoexponential pressure decline is probably the only practical analytic approach in a mouse because isovolumic relaxation is so short. However, to be certain this assumption is appropriate, LV pressure decline should be inspected to be certain that the correlation coefficient of the natural log of pressure vs time relationship is • 0.98.

Alternative methods of calculating tau in which the zero pressure asymptote is not assumed to be zero include the derivative method (16) and the use of a three-constant nonlinear regression analysis (17). Comparison of these 3 methods have not demonstrated significant differences in interpretation of the effects of pathologic state or transgenic construct on measurements of active relaxation. (17) Tau, like peak (-)dP/dt is influenced by loading condition (18). Therefore, tau can be considered an index of "intrinsic" relaxation rate only when loading conditions (and other modulators) are held constant or at least specified. One practical way to overcome this limitation is to examine indices of relaxation over a range of loads as shown in Figure 2. Afterload can be altered acutely using mechanical or pharmacologic methods. Abnormal relaxation (dashed lines) is indicated by the shift in the position of the relaxation rate versus afterload relationship where tau is prolonged at any equivalent systolic stress. (19) Therefore, these indices of active relaxation must be measured at equivalent loading conditions or at least interpreted in light of differing loading conditions.

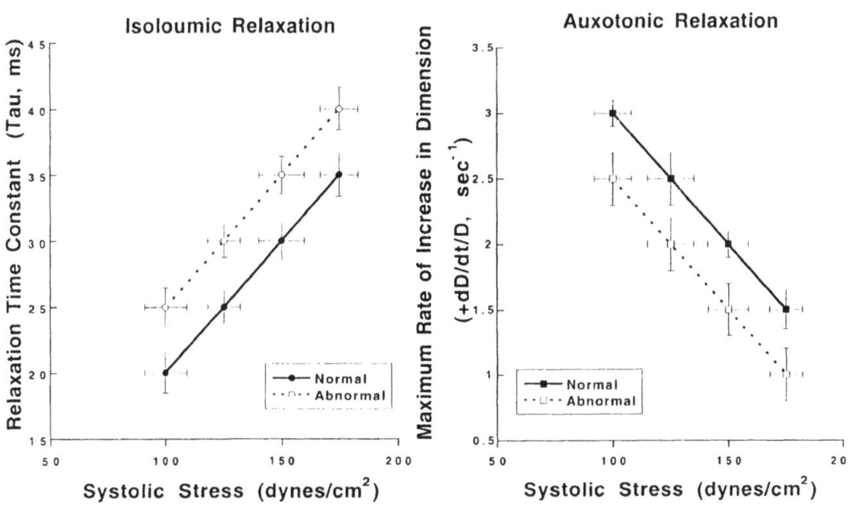

Figure 2: Example of relationship between indices of relaxation rate and afterload when relaxation is normal (solid lines) and abnormal (dashed lines).

2) **Auxotonic Relaxation:** The auxotonic LV filling phases of diastole can be characterized using techniques to estimate LV volume. Volume and volume transients can be measured using echo-doppler, conductance, MRI, or sonomicrometry techniques. (20-24) Each of these techniques have specific advantages and disadvantages in their application to mouse physiology. For example, sonomicrometry requires surgical implantation and thus far is limited to acute open chest anesthetized preparations in mice. Conductance catheter placement can be done using the same methods as the micromanometer catheter. Volumes derived from this technique must be normalized for parallel conductance and must be calibrated against a second independent technique such as aortic flow probe based measures of cardiac output. Wall thickness and therefore wall stress and myocardial stiffness can not be calculated using conductance techniques alone. MRI while noninvasive, will not allow simultaneous measurement of pressure. Each of these techniques are described in detail for mice application in chapters 9,10,11, and 25. This chapter will focus on echo-doppler methods because it is noninvasive, avoids some of the limitations listed above, and is readily available in many murine research laboratories.

Two dimensional (2-D) echocardiography, 2-D directed M-mode echocardiography, ultrasonic contrast, echo-doppler techniques and their specific application to mice are discussed in detail in chapter 9. Application of these techniques to the assessment of diastolic function in mice involves measuring indices of volume or velocity which reflect LV filling rate. M-mode echocardiograms are digitized to provide LV dimension vs time (dD/dt) or LV wall thickness vs time (dH/dt) data. Peak (+)dD/dt (dimension increasing) and peak (-) dH/dt (wall thinning) can be used as indices of early auxotonic relaxation rate. In addition, LV cavity area from 2-D images can be digitized on a frame by frame basis with sufficient sample frequency. Cavity area vs time (dA/dt) can also be used. If carried throughout diastolic filling, both early and late (atrial contraction induced LV filling) filling rates can be quantified.

Transmitral LV inflow velocity during diastole can be assessed using Doppler echocardiography in mice. (Also see Chapter 9) This technique provides the means to assess temporal distribution of LV filling, provides indices of auxotonic filling rates, and provides indirect indices of LV chamber compliance. Early diastolic filling velocity (E wave), late diastolic filling velocity consequent to atrial contraction (A wave), the E/A ratio, E wave deceleration time, and A wave duration can be used to assess auxotonic relaxation. However, like all of the isovolumic and auxotonic indices described above, interpretation of changes in the indices must be done in light of simultaneous changes in afterload (Figure 2). In addition, changes in afterload and filling load (load present during filling) must be assessed when examining indices of filling. For example, transmitral flow velocity is particularly effected by the atrial-ventricular pressure gradient (filling load)(25,26). The precise pattern of early and late diastolic transmitral flow velocities will depend on factors governing instantaneous atrial and LV pressures before and after mitral valve opening. It is thus not surprising that interventions or pathologic conditions that increase left atrial pressure increase early transmitral flow velocities while interventions which reduce LAP reduce early filling velocities.(25-28) To correctly interpret changes in transmitral flow velocities, concomitant changes in filling load must be considered. Three additional indices which may be less sensitive to load are rapidly being tested and applied to murine physiology. These include: pulmonary venous flow rates, transmitral propagation velocity, and tissue Doppler velocity. Their utility to murine studies is being examined.

3) **Chamber Stiffness:** Chamber stiffness can be quantified by examining the relationship between diastolic LV pressure and volume. LV diastolic pressure can be changed either by a volume dependent change in instantaneous operating stiffness or by a volume independent change in the overall chamber stiffness(figure 3). The operating stiffness at any point along a given pressure-volume curve is equal to the slope of a tangent drawn to the curve at that point. Operating stiffness changes throughout filling, stiffness (dP/dV) is less at smaller volumes (point A) and greater at larger volumes (point B).
Obtaining sufficient data to characterize the entire LV diastolic pressure vs volume curve may be difficult using currently available methods in mice. At a minimum, an index of end diastolic operating stiffness may be determined by examining the ratio of end diastolic pressure (EDP) vs end diastolic volume (EDV). For example, if EDP is increased with a normal EDV, end diastolic operating stiffness (and likely overall chamber stiffness) is increased. If EDP is normal and EDV is increased end diastolic operating stiffness (and likely overall chamber stiffness) is decreased. If EDP is increased and EDV is also increased, stiffness may be normal, increased or decreased. In this later condition, measuring the entire pressure-volume relationship becomes essential in determining stiffness.

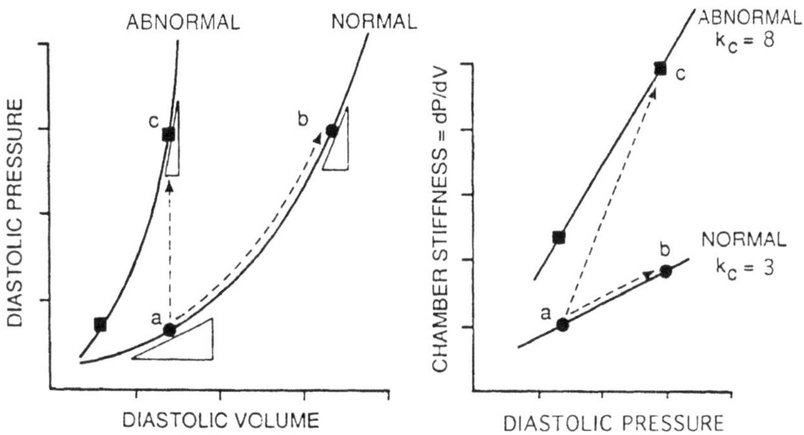

Figure 3: Schematic drawing of the LV diastolic pressure vs volume relationship showing methods used to quantitate chamber stiffness.

The shape and position of the entire pressure-volume relationship can be used to calculate an overall chamber stiffness constant. (Figure 3) Because the diastolic pressure-volume relation is curvilinear and generally exponential, the relation between dP/dV and pressure is linear. The slope of this relation (K_c), is called the modulus of chamber stiffness (or stiffness constant) and can be used as a single numeric value to quantitate chamber stiffness. K_c can be represented mathematically using a monoexponential equation: $P = Ae^{K_c V}$,where P is LV pressure, V is LV volume, A is a constant and K_C is the chamber stiffness constant. When overall chamber stiffness is increased, the pressure-volume curve shifts to the left, the slope of the dP/dt versus pressure relation becomes steeper, and K_C is increased. (6,29,30)

Diastolic pressure vs volume (and stress vs strain, described below) relationships can be derived from a single cardiac cycle or from multiple cardiac cycles. Using a single cardiac cycle, pressure vs volume coordinates during a single diastole from MVO to MVC must be obtained. Using multiple cardiac cycles, end diastolic values are obtained from cardiac cycles in which end diastolic volume or pressure are altered using pharmacologic or mechanical means. The difficulty applying these principles to murine physiology is that ambient heart rate is so rapid, that the number of points obtainable during a single cardiac diastole may be limited. In addition, the only currently available method of volume determination to accomplish this task with sufficient sample frequency may be the conductance method. Therefore, the optimal method to calculate K_C is to use multiple end diastolic pressure vs volume data during a volume infusion with saline or during a caval occlusion.

To accurately compare indices of chamber stiffness in mice with pathologic states or mice with different genetic backgrounds, these indices must be derived from data which share common ranges of pressure. Specifically, the range of pressure over which stiffness is calculated must be similar in each mouse.

4) **Myocardial Stiffness:** The muscle of the left ventricle behaves as an elastic material, developing a resisting force as it is stretched by ventricular filling. The greater the change in muscle length (strain), the greater the increase in the force (wall stress, σ) that resists this deformation and the stiffer the myocardium becomes. Myocardial stiffness can be quantitated by examining the relation between LV wall stress and strain during diastole.

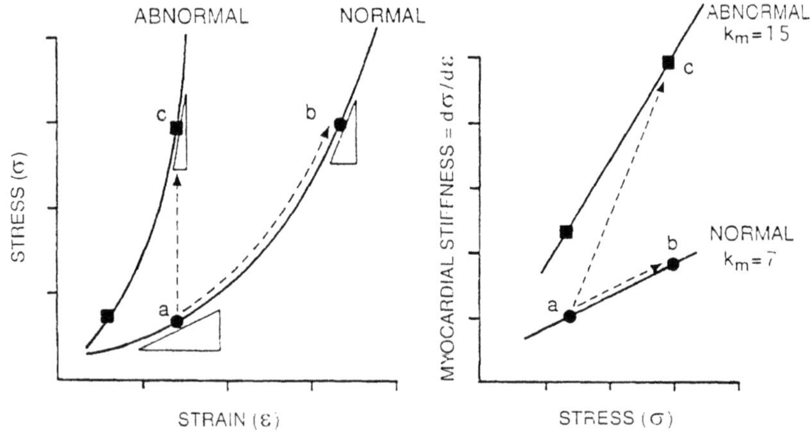

Figure 4: Schematic drawing of the myocardial diastolic stress vs strain relationship showing methods used to quantitate myocardial stiffness.

Strain is the deformation (lengthening) of the muscle produced by the application of a force (increase in wall stress). Strain (ε) is generally expressed as the percent change in length (L) from the unstressed length. Using Lagrangian strain, the unstressed length (L_0) is equal to the equilibrium length (i.e., the length of the muscle when subjected to no external forces). At any given strain throughout filling, myocardial stiffness is equal to the slope ($d\sigma/d\varepsilon$) of a tangent drawn to the stress-strain curve at that strain. Because the stress-strain relation is curvilinear and exponential, the relation between $d\sigma/d\varepsilon$ and stress is linear. The slope of this relation, K_m, is the modulus of myocardial stiffness (or stiffness constant). When myocardial stiffness is increased, the stress-strain relation shifts to the left, the slope of the $d\sigma/d\varepsilon$ versus stress relationship (Figure 4) becomes steeper and K_m increases. K_m can be represented mathematically using a monoexponential equation: $\sigma = A\ (e^{K_m \varepsilon})$, where σ = LV wall stress, ε = strain, A is a constant and K_m is the elastic stiffness constants. (6,29,30) Comparison of myocardial stiffness between mice using these methods is only appropriate when the calculations are made over common levels of LV wall stress.

In Vitro Quantitation of Diastolic Function

1) **Active Relaxation:** The process of relaxation can be measured *in vitro* in isolated muscle tissue or isolated cardiomyocytes. Isometric force decline and isotonic lengthening can be measured in an isolated mouse muscle. In contrast, only isotonic lengthening rate can be assessed in an isolated mouse cardiomyocyte. (31-34) These methods are discussed in detail in chapter 5. However, it should be noted that these *in vitro* indices of active relaxation like the *in vivo* indices, must be interpreted in light of simultaneous changes in load. For example, cardiomyocyte isotonic lengthening rate may be decreased by two possible mechanisms: 1) a decrease in restoring force or 2) abnormal myocardial inactivation (Figure 5). Restoring forces are created during contraction when potential energy is produced by deformation in cellular structures. This potential energy is released during diastole and acts to restore the cardiomyocyte to rest or equilibrium length. Changes in myocardial inactivation alter lengthening rate independent of concomitant changes in restoring force. Studies have shown that examining the relaxation velocity vs shortening extent relationship (Figure 5), provides an analytic method to distinguish between these two factors (7).

If a given pathological process causes the relaxation velocity to move from point A to point B along a common relationship between relaxation velocity and shortening extent, this indicates that these changes were caused by alterations in restoring forces that occurred concomitantly with alterations in systolic function. In contrast, if a pathophysiological process causes the relaxation velocity to move from point A to point C along a distinctly different relationship, this indicates that there had been changes in the rate of myocardial inactivation.. Two interventions (hypothermia and isoproterenol) known to produce primary alterations in myocardial inactivation, were examined in normal cardiomyocytes (7,35,36).(Figure 5). During hypothermia, the slope of the relaxation velocity vs shortening extent relationship decreased such that for any given shortening extent, relaxation velocity was less at 30° C than at 37° C (7,35,36). During isoproterenol infusion, the slope of the relaxation velocity vs. shortening extent relationship increased significantly (7,35,36). Thus, the slope of the relaxation velocity vs. shortening extent relationship is sensitive to and dependent on changes in the rate of myocardial inactivation. This analytic method can be applied to isolated murine cardiomyocyte and muscle preparations.

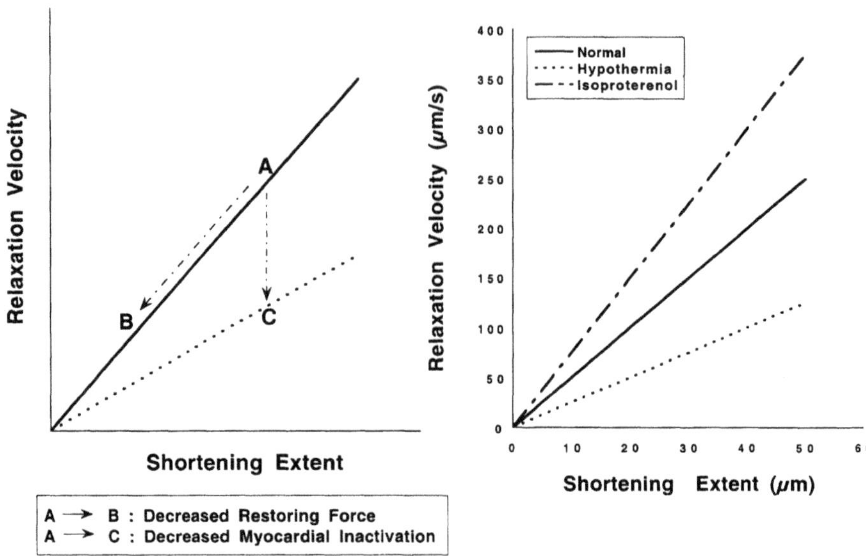

Figure 5: Example of relationship between indices of cellular relaxation rate and shortening extent when relaxation is normal (solid lines) and altered (dashed lines).

2) **Passive viscoelastic stiffness:** A number of mechanical and bioengineering models and experimental designs have been used to describe the constitutive material viscoelastic properties of myocardial tissue and cardiomyocytes (5,37). One such method, applicable to both isolated murine muscle and cardiomyocyte, consists of a two element model composed of a non-linear spring in parallel with a non-linear viscous damper. Conceptually, the passive elastic spring consists of structures or processes which resist deformation in a time independent manner, i.e., it resists deformation to the same extent irrespective of the rate of deformation. The viscous damper consists of structures or processes which resist stretch in a time dependent manner, i.e., it resists deformation to a greater extent when the rate of deformation is increased. While this kind of analysis is difficult to perform *in vivo*, especially in mice, it can be done *in vitro*. There are a number of important reasons to examine elastic stiffness and viscous damping, separately and independently. This kind of analysis may provide clues to which specific cellular or extracellular mechanism is altered by a particular pathologic state or transgenic construct and which specific mechanism leads to changes in diastolic function and cardiovascular phenotype.

To calculate elastic stiffness (β) and viscous damping (η) constants, measurements of muscle stress, strain, and strain rate are obtained using uniaxial variable rate stretches (Figure 6). These data can be fit by a constitutive equation for a non-linear viscoelastic composite biomaterial where: $\sigma(\varepsilon,\varepsilon') = (Ae^{\beta\varepsilon} + B) + (C - De^{\eta\varepsilon'})$, A,B,C, and D are curve fitting constants. The elastic stiffness constant is assessed using the slowest uniaxial stretch. Under these conditions, strain rate approximates zero, the second portion of the equation $(C - De^{\eta\varepsilon'})$ becomes negligible, and stress becomes a function of strain alone and β can be determined. When elastic stiffness is increased, the σ vs ε relationship shifts up and left and β increases (Figure 6). The viscous damping constant is assessed using all four uniaxial variable rate stretches. From these stretches, the relationship between stress and strain rate is defined at constant strain. At any selected, constant strain, the relationship between stress and strain rate is curvilinear. Under these experimental conditions, strain is constant, the first portion of the equation $(Ae^{\beta\varepsilon} + B)$ becomes constant, and stress becomes a function of strain rate alone. Therefore, the stress versus strain rate data can be fit to the equation, and η determined. When viscous damping is increased, the σ vs ε' relationship shifts up and η increases (figure 6).

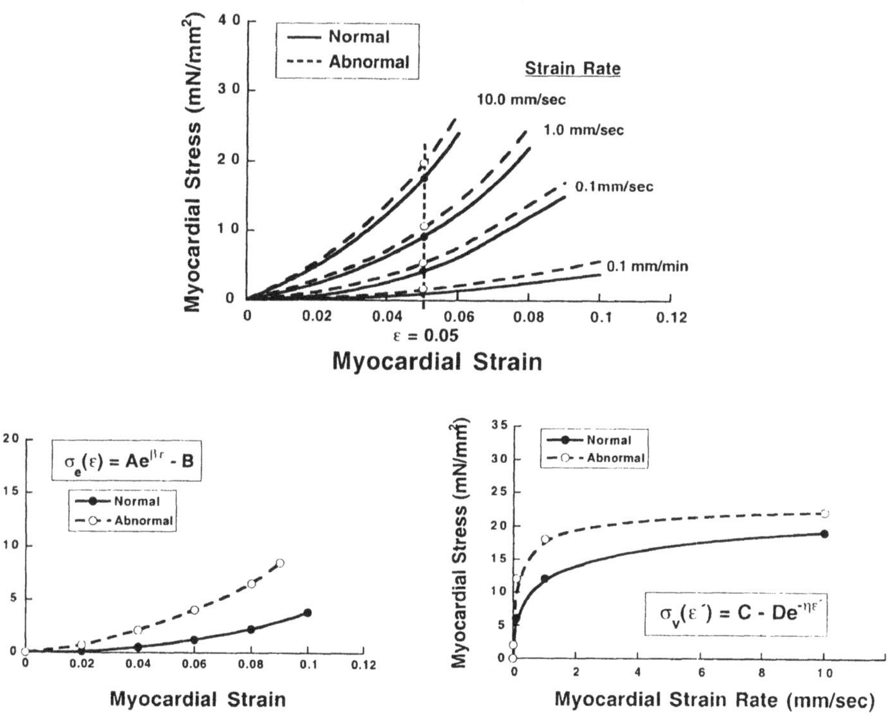

Figure 6: Schematic showing uniaxial variable rates stretches. Stress, strain and strain rate data obtained from these methods can be used to calculate elastic stiffness and viscous damping constants.

CONCLUSION

This chapter described the *in vivo* and *in vitro* methods which can be used to characterize the changes in diastolic function produced by the genetic alterations which develop in murine models of cardiovascular disease or which result from genetic engineering in transgenic mice.

REFERENCES

1. Zile MR, Simsic J: Diastolic Heart Failure: Diagnosis and Treatment. Excerpta Medica, *Clinical Cornerstone*, In press, 2000.
2. Zile MR: Diastolic Heart Failure: Diagnosis, Mechanisms, and Treatment. Cardiology Rounds as Presented in the Rounds of the Cardiovascular Division of Brigham and Women's Hospital, Boston, Massachusetts, 3: 1-7, November, 1999.
3. Bell SP, Nyland L, Tischler MD, McNabb M, Granzier H, LeWinter MM: Alterations in the determinants of diastolic suction during pacing tachycardia. Circ Res, 87(3):235-240, 2000 Aug 4.
4. Cazolla O, Freiburg A, Helmes M, Centner T, McNabb M, Wu Y, Trombitas K, Labeit S, Granzier H: Differential expression of cardiac titin isoforms and modulation of cellular stiffness. Circ Res, 86(1):59-67, 2000 Jan 7-21.
5. Zile MR, Richardson K, Cowles MK, Buckley JM, Koide M, Cowles B, Gharpuray V, Cooper G IV: Constitutive properties of adult mammalian cardiac muscle cells. Circulation, 98:567-579, 1998.
6. Kato S, Spinale FG, Tanaka R, Johnson WS, Cooper G, Zile MR: Inhibition of collagen cross-linking: Effects on extracellular matrix structure and left ventricular diastolic function. Am J Physiol, 269:J863-H868, 1995.
7. Zile M, Spinale FG: Diastolic Dysfunction And Tachycardia-Induced Heart Failure; International Symposium on the Physiology of Diastole in Health and Disease, in Grossman and Lorell (eds), Diastolic Relaxation of the Heart, Kluwer Academic Press, Boston, 1994, pp 243-263.
8. Zile MR: The Development of Diastolic Dysfunction, in Spinale and Bishop(eds), Pathophysiology of Tachycardia-Induced Heart Failure, Futura Publishing, 1996.
9. McConnell BK, Jones KA, Fatkin D, Arroyo LH, Lee RT, Aristizabal O, Turnbull DH, Georgakopoulos D, Kass D, Bond M, Niimura H, Schoen FJ, Conner D, Fishman DH, Seidman CE, Seidman JG: Dilated cardiomyopathy in homozygous myosin-binding protein-C mutant mice. J Clin Invest 104:1235-1244, 1999.
10. Schmidt A, Kadambi V, Ball N, Sato Y, Walsh R, Kranias E, Hiot B: Cardiac-specific overexpression of calsequestrin results in left ventricular hypertrophy, depressed force-frequency relation and pulsus aternans in vivo. J Mol Cell Cardiol 32:1735-1744, 2000.
11. Hoit BD, Ball N, Walsh RA: Invasive hemodynamics and force-frequency relationships in open versus closed-chest mice. Am J Physiol 273:H2528-H2533, 1997.
12. Kadambi VJ, Ball N, Kranias EG, Walsh RA, Hiot BD: Modulation of force-frequency relation by phospholamban in genetically engineered mice. Am J Physiol 276:H2245-H2250, 1999.
13. Hoit BD, Kadambi VJ, Tramuta DA, Ball N, Kranias EG, Walsh RA: Influence of sarcoplasmic reticulum calcium loading on mechanical and relaxation restitution. Am J Physiol 278 :H958-963, 2000.

14. Weiss JL, Fredericksen JW, Weisfeldt ML: Hemodynamic determinants of the time course of fall in canine left ventricular pressure. J Clin Invest 58:83-95, 1976.
15. Smith VE, Zile MR: Relaxation and Diastolic Properties of the Heart, in Fozzard et al (ed), The Heart and Cardiovascular System, New York, Raven Press, 1992, pp1353-1367.
16. Mirsky I: Assessment of diastolic function: Suggested methods and future considerations. Circulation 69:836-941, 1984.
17. Mirsky I and Pasipoularides A: Clinical assessment of diastolic function. Prog Cardiovasc Dis 32:291-318, 1990.
18. Zile MR, Nishimura RA, Gaasch WH: Hemodynamic Loads and Left Ventricular Diastolic Function: Factors Affecting the Indices of Isovolumetric and Auxotonic Relaxation, in Gaasch and LeWinter (eds), LV Diastolic Dysfunction and Heart Failure, Philadelphia, Lea and Febiger, 1994, pp 219-242.
19. Zile MR, Blaustein AS, Gaasch WH: The effect of acute alterations in left ventricular afterload and beta-adrenergic tone on indices of early diastolic filling rate. Circ Res 65:406-416, 1989.
20. Franco F, Thomas GD, Giroir B, Bryant D, Bullock MC, Chwialkowski MC, Victor R, Peshock RM: Magnetic resonance imaging and invasive evaluation of development of heart failure in transgenic mice with myocardial expression of tumor necrosis factor-α. Circulation 99:448-454, 1999.
21. Feldman MD, Erikson JM, Mao Y, Korcarz CE, Lang RM, Freeman GL: Validation of a mouse conductace system to determine LV volume: comparison to echocardiograpy and crystals. Am J Physiol (Heart Circ Physiol) 279:H1698-H1707, 2000.
22. Gao X, Dart AM, Dewar E, Jennings G, Du X: Serial echocardiographic assessment of left ventricular dimensions and function after myocardial infarction in mice. CV Res 45:330-338, 2000.
23. Hiot BD:Echocardiographic assessment of the mouse heart, in Hoit BD, Walsh RA (eds), Cardiovascular Physiology in the Genetically Engineered Mouse, Norwell, MA, Kluwer Academic Publishers, 1998:97-110.
24. Murphy AM, Kogler H, Georgakopoulos D, McDonough JL, Kass DA, Van Eyk JE, Marban E: Transgenic mouse model of stunned myocardium. Science 287:488-491, 2000.
25. Ishida Y, Meisner JS, Tsujioka K, Gallo JI, Yoran C, Frater RWN, Yellin EL: Left ventricular filling dynamics: Influence of left ventricular relaxation and left atrial pressure. Circulation 74:187-196, 1986.
26. Zile MR: Hemodynamic determinants of echocardiography derived indices of left ventricular filling. Echocardiography, 9:289-300, 1992.
27. Tsukamoto T, Kakeuchi M, Fukuzaki H: Effects of lower body negative pressure and volume loading on transmitral flow velocity pattern by pulsed Doppler echocardiography. Am Heart J 53:377-386, 1989.
28. Yellin EL, Hori M, Yoran C, Sonnenblick EH, Gabbay S, Frater RWM:

Ventricular relaxation in the filling and nonfilling intact canine heart. Am J Physiol H620-H629, 1986.
29. Zile MR, Tomita M, Ishihara K, Nakano K, Spinale F, Swindle M, Carabello B: Changes in diastolic function during the development and correction of chronic LV volume overload produced by mitral regurgitation. Circulation 87:1378-1388, 1993.
30. Tomita M, Spinale FG, Crawford FA, Zile MR: Changes in left ventricular volume, mass, and function during the development and regression of supraventricular tachycardia-induced cardiomyopathy: Disparity between recovery of systolic versus diastolic function. Circulation 83:635-644, 1991.
31. Zile MR, Conrad CH, Gaasch WH, Robinson KG, Bing OHL: Preload does not affect relaxation rate in normal, hypoxic or hypertrophic myocardium. Am J Physiol 258:H191-H197, 1990.
32. Zile MR, Gaasch WH, Weigner AW, Robinson KG, Bing OHL: Mechanical determinants of the rate of isotonic lengthening in rat left ventricular myocardium. Circ Res 60:815-823, 1987.
33. Brooks WW, Conrad CH: Differences between mouse and rat myocardial contractile responsiveness to calcium. Comp Bioch and Physiol Part A 124:139-147, 1999.
34. Li L, Desantiago J, Chu G, Kranias EG, Bers DM: Phosphorylation of phospholamban and troponin I in β-adrenergic-induced acceleration of cardiac relaxation. Am J Physiol 278:H769-H779, 2000.
35. Tsutsui H, Urabe Y, Mann D, Tagawa H, Carabello B, Cooper G, Zile MR: Effects of chronic mitral regurgitation on diastolic function in isolated cardiocytes. Circ Res 72:1110-1123, 1993.
36. Zile MR, Mukherjee R, Clayton C, Kato S, Spinale FG: Effects of chronic supraventricular pacing tachycardia on relaxation rate in isolated cardiac muscle cells. Am J Physiol 268: H2104-H2113, 1995.
37. Zile MR, Cowles MK, Buckley JM, Richardson K, Cowles B, Baicu CF, Cooper G IV, Gharpuray V: Gel stretch method: A new method to measure constitutive properties of cardiac muscle cells. Am J Physiol, 274: H2188 - H2202, 1998.

16

IN VIVO ELECTROPHYSIOLOGIC STUDIES IN MICE

Charles I. Berul, MD[1] and Paul J. Wang, MD[2]

Departments of Cardiology

[1]Children's Hospital, Boston, Harvard Medical School

[2]New England Medical Center, Tufts University School of Medicine

Boston, Massachusetts

INTRODUCTION

Molecular biological and genetic approaches to the evaluation of cardiac rhythm disturbances have led to new understanding regarding disease mechanisms and therapies. Genetic bases for electrophysiological disorders are increasingly being recognized. Single-gene defects that produce electrophysiological abnormalities are difficult to study in humans due to small pedigree sizes, complexities of the genome, and confounding technical, practical, and ethical considerations. Therefore, genetically-engineered animals are essential models of human cardiovascular diseases. Advances in breeding and husbandry, molecular biological techniques and the diversity of transgenic and genetically-manipulated strains make mice an important model to study the effects of specific mutations that result in cardiovascular and electrophysiological phenotypes. To this end, miniaturized electrophysiology studies (EPS) for use in genetically-engineered mice have been developed. Based on analogous clinical protocols used to evaluate cardiac conduction in humans, this method allows phenotypic assessment of the conduction characteristics of the murine heart in normal and transgenic animals. The mouse *in vivo* EP study is comprehensive and includes evaluation of conduction and refractoriness by responses to pacing, programmed electrical stimulation and assessment of pharmacologic agents. Monitoring of ambient ectopy during normal activities and graded exercise via implantable telemetry devices is also discussed. This chapter reviews the mouse electrophysiology model and several illustrative applications of murine models important to human electrophysiologic diseases and genetic syndromes associated with cardiac arrhythmias, including familial hypertrophic and dilated cardiomyopathies, inherited arrhythmia disorders such as long QT syndrome, and developmental cardiac conduction abnormalities.

MURINE *IN VIVO* ELECTROPHYSIOLOGY METHODS

Animal Preparation

A surface 12-lead electrocardiogram can be obtained by placing subcutaneous electrodes under the skin in each limb and across the precordium. For an open-chest epicardial approach, mice are anesthetized and intubation is achieved via direct laryngoscopy, visualizing the vocal cords with transillumination of the ventral neck. The mice are mechanically ventilated using a rodent respirator (Harvard Apparatus), administering a mixture of air and supplemental oxygen. A midline sternotomy is performed under an operating microscope and epicardial pacing/recording wires are placed on the right ventricle, left ventricle and right atrial surfaces (1). The wires can be externalized through the posterior neck, the lungs re-expanded with positive end-expiratory pressure, and the sternotomy incision sutured closed (Figure 1).

Alternatively, an endocardial closed-chest approach can be performed (2,3). Endocardial access is obtained via a right jugular vein cutdown approach and an octapolar electrode catheter (CIBer mouse EP catheter, NuMed, Inc., Hopkinton, NY) with precise interelectrode distances, allows for simultaneous atrial and ventricular endocardial recording and pacing (Figure 2). Electrocardiograms and intracardiac recordings are displayed continuously and simultaneously acquired to computer via an analog to digital converter and stored for analysis and measurement with on-line calipers. A warming light is used to maintain body between 34-37°C.

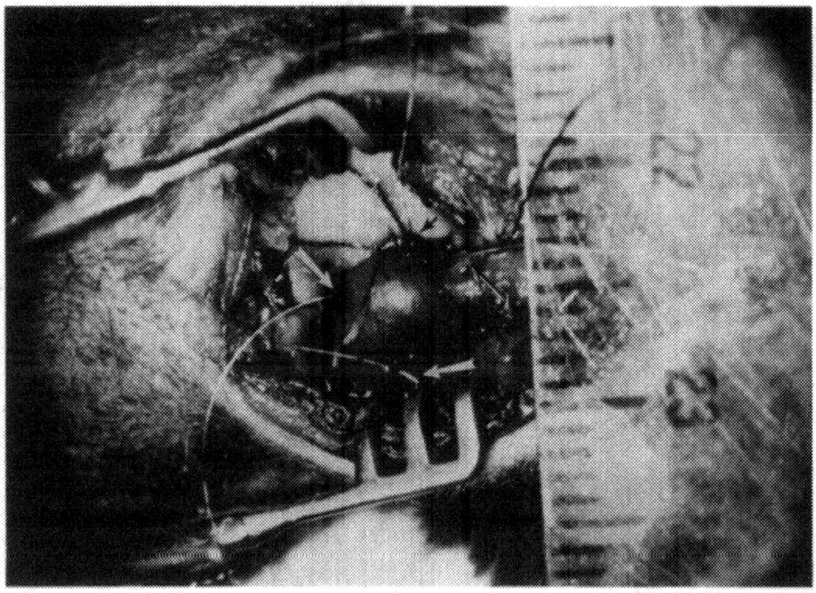

Figure 1: Photograph of epicardial lead placement. Anesthetized mouse undergoing acute operative procedure. Viewed through the operating microscope (24X), the chest is opened and 4 epicardial leads are positioned. The airway is maintained via a tracheotomy for mechanical ventilation. Two epicardial 0.003" pacing leads are sutured with 7-0 silk to the atrial and ventricular surfaces.

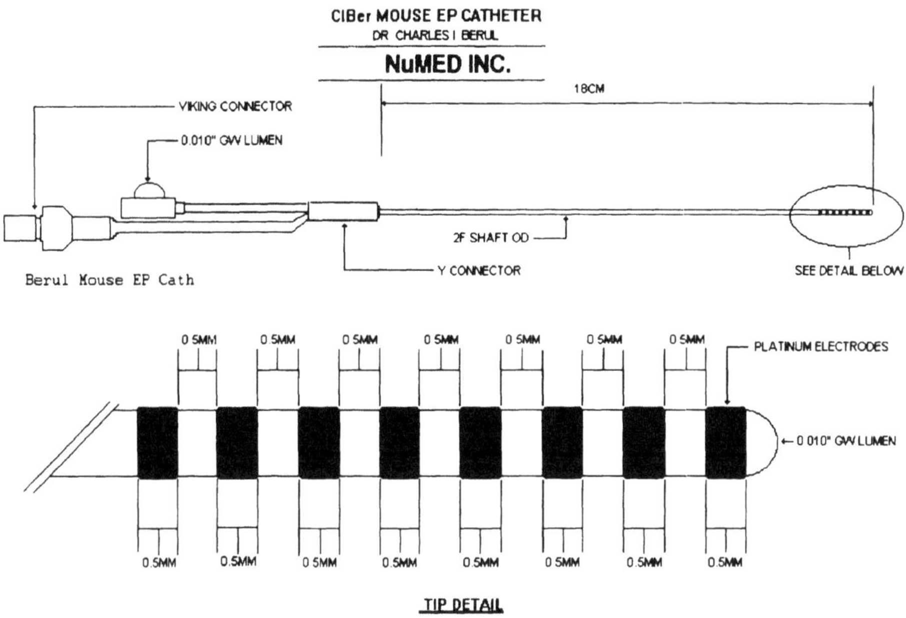

Figure 2: Endocardial Octapolar Electrophysiology Catheter. The catheter is 1.7 French, capable of transvenous pacing and intracardiac electrogram recording, and luminal for administration of medications and blood sampling. The catheter is typically inserted via a right jugular vein cutdown.

Electrophysiology Study Protocol

Pacing thresholds (in mAmps) are determined for each lead and bipolar stimulation performed for 1.0 msec pulse widths at twice the diastolic capture threshold. The electrophysiologic stimulator has been modified to allow pacing at coupling intervals as short as 10 msec for rapid stimulation rates necessary in mice. The cardiac rhythm is continuously recorded and all ECG frontal axes (P and QRS) and time intervals (PR, QRS, QT, JT, QTc, JTc, RR) calculated (4-7). The PR interval is marked from the beginning of the surface P wave to the beginning of the QRS complex. The JT interval is marked from the end of the QRS complex (J point) to the end of the T wave, defined as the point where it returns to the isoelectric baseline (Figure 3). The QT intervals may be rate-corrected using Bazett's formula (8) or an alternate formula more applicable to mouse heart rates (9).

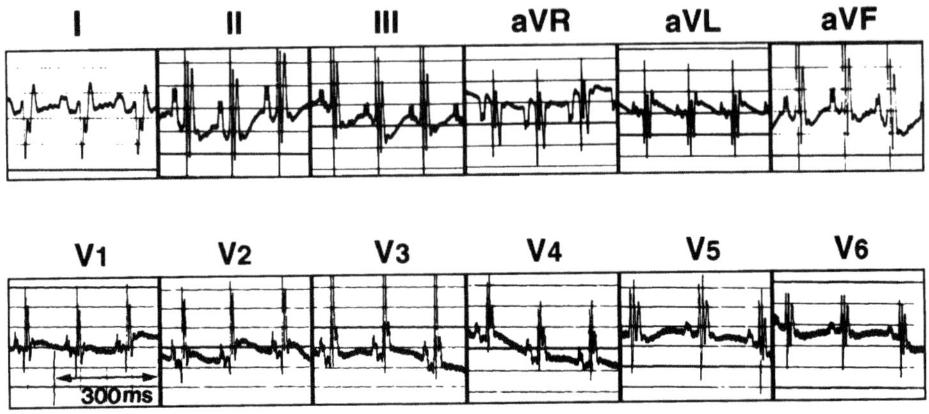

Figure 3: Murine 12-lead surface electrocardiogram. A surface 12-lead murine ECG recorded at 50 mm/sec paper speed, with a cycle length of 155 msec. The P and QRS axes are +30°. Vertical calibration lines are 75msec markers (ECG display amplification: 0.1 mV/cm).

Standard clinical electrophysiologic pacing protocols are used to determine all basic electrophysiologic parameters (1,10). Sinus node function is evaluated by indirectly measuring sinus node recovery time (SNRT) by pacing for 15-30 seconds at two cycle lengths and measuring the duration of the return cycle (Figure 4a). The maximum return CL from both pacing drives is used in calculations of SNRT, analogous to human studies (11). The corrected SNRT is calculated as the SNRT less the sinus cycle length (SCL) and SNRT/SCL percentages are determined by adjusting the maximum absolute SNRT and correcting for sinus cycle length. The atrioventricular (A-V) conduction properties are assessed by rapid atrial pacing at rates up to 1200 beats per minute (BPM). The minimum cycle length which allows 1:1 A-V conduction, the Wenckebach paced cycle length, and the maximum paced cycle length causing 2:1 A-V block (Figure 4b) are determined (12,13). Programmed atrial stimulation is performed to determine A-V and atrial effective refractory periods. Extra-stimulation techniques down to a minimum coupling interval of 20msec are performed to attempt to induce potential atrial arrhythmias (1,10). By recording a His bundle electrogram, it is possible to determine the effective and functional A-V nodal refractory periods (Figure 4c) and to identify infraHisian conduction abnormalities (14,15).

Sinus Node Recovery Time

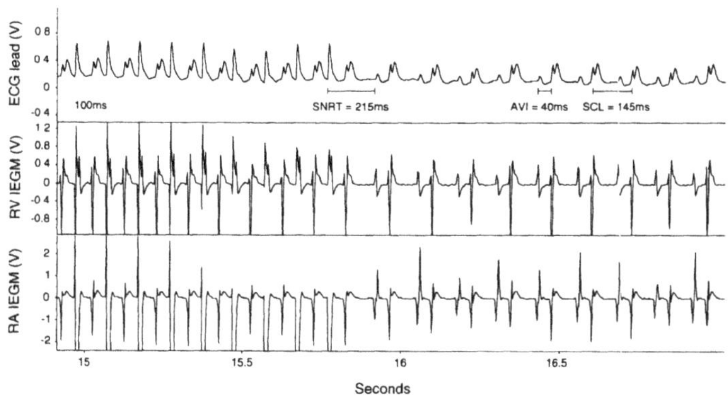

Figure 4a: Sinus node recovery time. The top tracings are surface ECG lead I, middle tracings are RV intracardiac electrogram (IEGM) and the lower tracings are atrial IEGM. Right atrial stimulation at 100 ms for 30 seconds is followed by the sinus node recovery time (SNRT), the point from the last paced atrial signal to the first sinus atrial response. The SNRT can be corrected by normalizing for sinus cycle length (SCL).

Endocardial Right Atrial Pacing

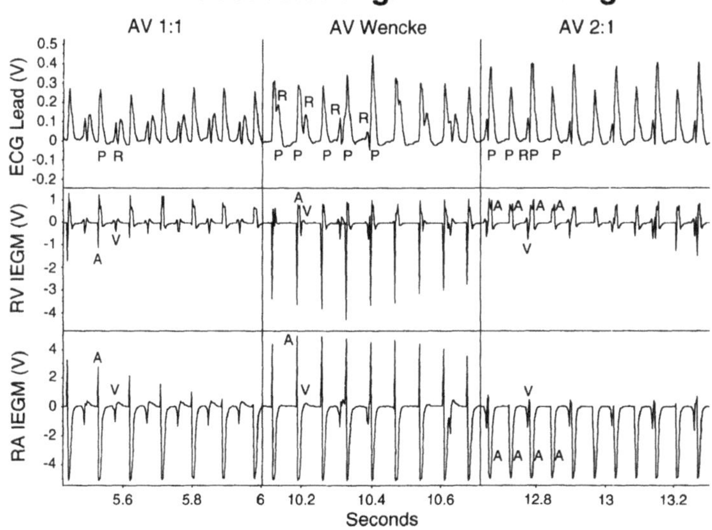

Figure 4b: Atrioventricular Conduction. The top tracings are surface ECG lead I, middle tracings are RV intracardiac electrogram (IEGM) and the lower tracings are atrial IEGM. The left side illustrates rapid atrial pacing at 90 ms, allowing 1:1 A-V conduction. The middle panel shows atrial pacing at 80 ms, demonstrating A-V nodal Wenckebach periodicity, and the right panel is atrial pacing at 70 ms with 2:1 A-V block.

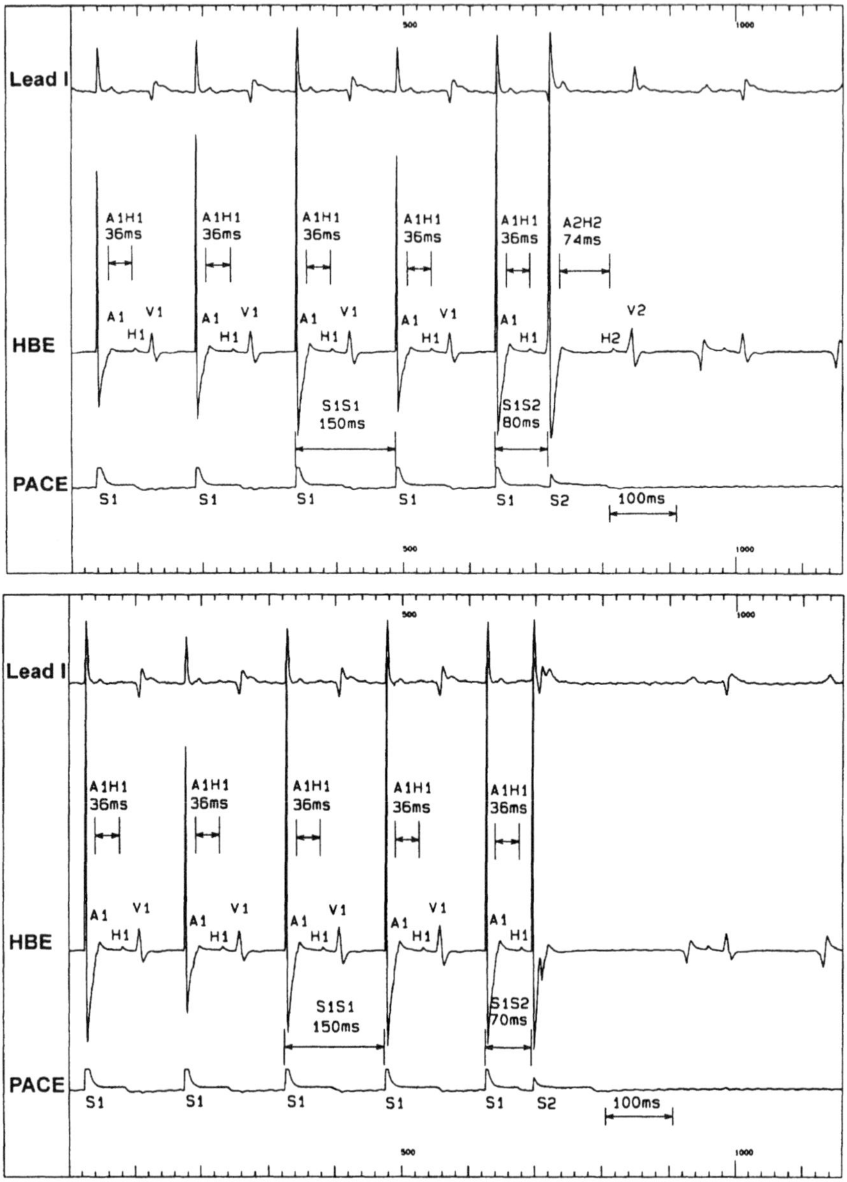

Figure 4c: Atrioventricular Nodal Refractoriness. The top tracings are surface ECG lead I, middle tracings are His Bundle electrogram (HBE) and the lower tracings are from the atrial pacing (PACE) site. Programmed atrial stimulation reveals stable A_1H_1 and H_1V_1 intervals and a decremental A_2H_2 interval ($S_1S_2 = 80$ ms, top panel) to the point of atrioventricular nodal effective refractory period (AVNERP) at $S_1S_2 = 70$ ms illustrated in the bottom panel.

Ventricular burst pacing is performed at rates of 250-1200 BPM to assess retrograde ventriculo-atrial conduction and ventricular effective refractory periods using programmed stimulation at several paced drive rates using single extrastimuli (Figure 4d). Double and triple extrastimulation techniques are then performed to attempt induction of ventricular arrhythmias, similar to the atrial pacing protocol (1-3). Dispersion of refractoriness can be calculated to assess for heterogeneity of regional repolarization (16-18).

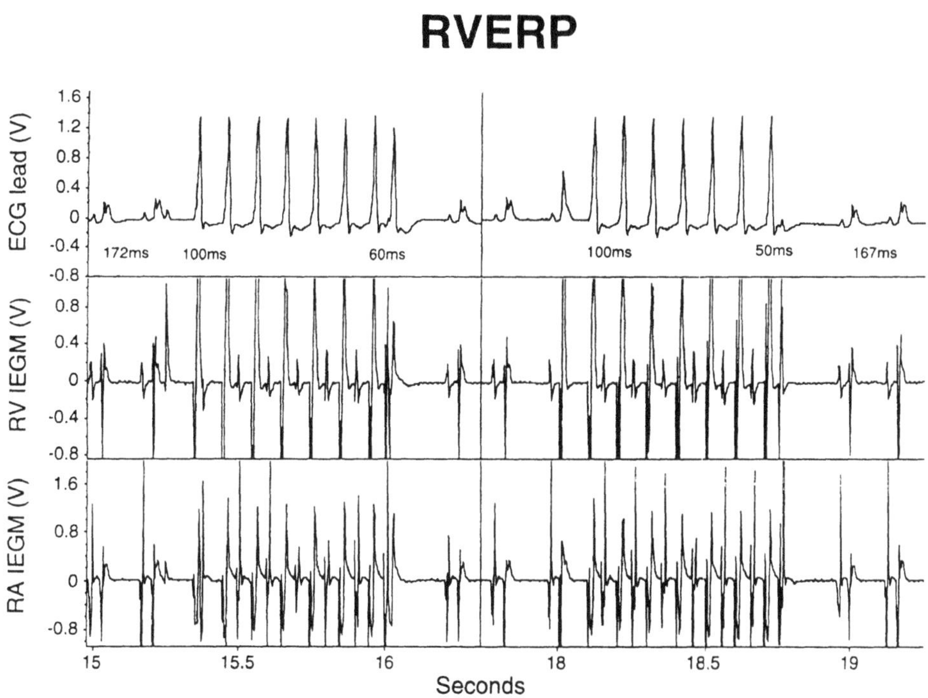

Figure 4d: Ventricular Pacing and Refractoriness. The top tracings are surface ECG lead I, middle tracings are RV intracardiac electrogram (IEGM) and the lower tracings are atrial IEGM. The left side illustrates programmed ventricular stimulation at a drive train of 100 ms followed by a single extrastimulus coupled at 60 ms. This rate permits 1:1 V-A conduction, except for V-A block with the extrastimulus. The right panel reveals the right ventricular effective refractory period (RVERP) at an extrastimulus coupling interval of 50 ms.

Transesophageal Stimulation

Using anesthesia, transesophageal stimulation may be performed in lieu of an intracardiac catheter to achieve atrial stimulation. This technique may be used to assess sinus recovery times and atrioventricular Wenckebach pacing cycle length. Atrial pacing may be used to assess intra-atrial conduction times and atrioventricular conduction times. In addition, atrial arrhythmias may be induced via this technique.

Monophasic Action Potentials

In order to correlate the intracardiac electrograms and surface ECG with alterations in ionic membrane currents, a monophasic action potential (MAP) catheter is positioned on the right and left ventricle for simultaneous recording (Figure 5). The MAP catheter is 4 French with a pair of orthogonally-positioned silver-silver chloride electrodes. The amplifier filters are opened completely, with a range of 1 to 5000 Hertz. The local activation times, action potential durations, and repolarization times can be measured, and potential afterdepolarizations can be identified from MAP catheter recordings (17).

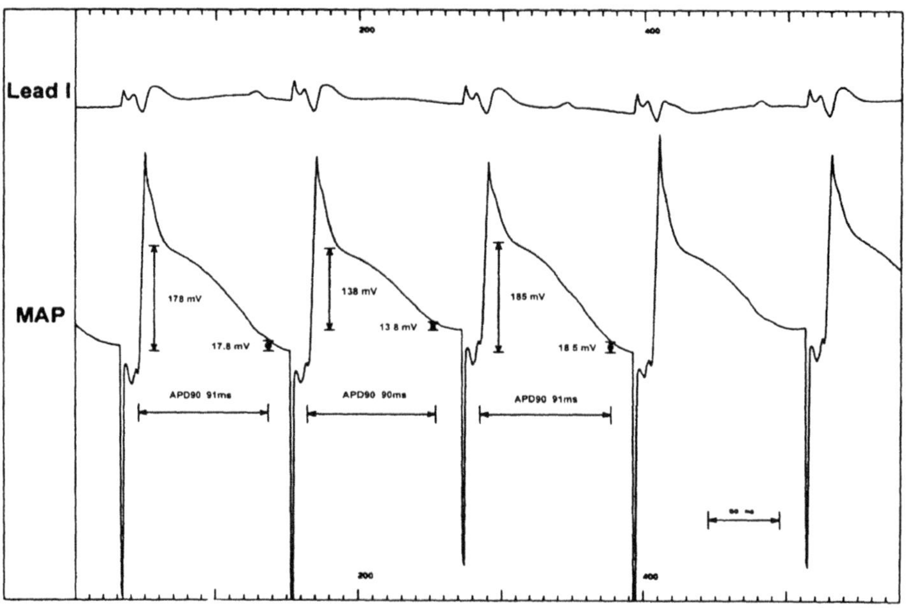

Figure 5: Monophasic action potential (MAP) recording. The top channel is surface ECG lead I, and the lower channel is an epicardial ventricular MAP signal, illustrating measurements at 90% repolarization of the action potential duration (APD_{90}). MAP amplification: 1 V/cm).

Ambient Ectopy and Exercise-Stress Induced Arrhythmias

The detection of spontaneous ectopy can be directly compared with the presence of inducible arrhythmias or other EPS findings by use of ambulatory ECG monitoring. Analogous to Holter monitoring in humans, a telemetric transmitter (DSI, Inc., St. Paul, MN) is implanted and a single-lead ECG signal is obtained by placing 2 leads from the transmitter, tunneled through the subcutaneous space, as a lead II frontal plane ECG vector (Figure 6). The ECG rhythm strip can be recorded continuously for up to 6 months at a time without battery depletion. Additionally, the effects of exercise-stress on arrhythmia induction can be evaluated using the implantable transmitter while subjecting the mouse to a standard graded exercise protocol.

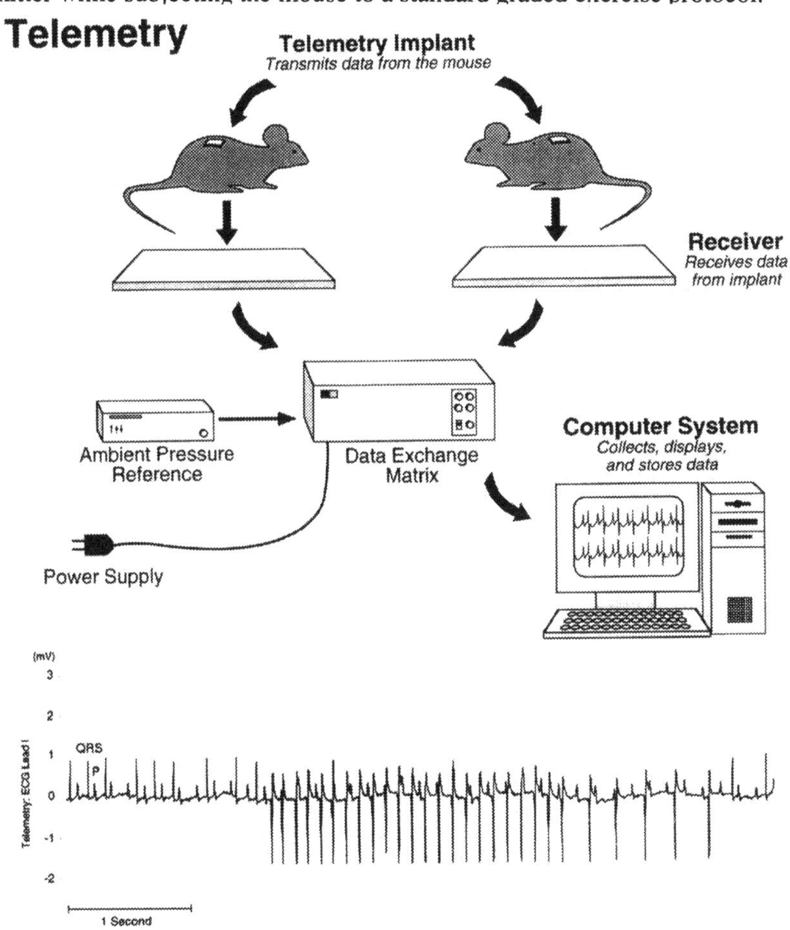

Figure 6: Murine ambulatory telemetered electrocardiogram. The diagram illustrates the transmitter implanted in the back. The mice are placed over individual receivers and the information is telemetered and stored on computer system. The lower panel demonstrates an example ECG obtained via an implantable transmitter in a freely-moving mouse. This ECG demonstrates an example of complete A-V block with a run of nonsustained ventricular tachycardia in a mutant mouse. The signal is being acquired overnight at a sampling rate of 1000/second.

Arrhythmia models can be produced in normal mice, using several techniques. Chronic pacing or administration of medications may promote arrhythmogenesis in wild-type mice (19). Creation of ischemia or myocardial infarction has been performed in the mouse, allowing the study of electrophysiologic sequelae (20). In summary, the *in vivo* mouse electrophysiology study allows a comprehensive evaluation of the murine cardiac conduction system. The complete study includes obtaining a 12-lead surface resting ECG, transvenous endocardial atrial and ventricular electrograms, epicardial electrograms, monophasic action potentials, ambulatory long-term ECG recordings at rest and during exertion, and the response to pacing, programmed stimulation and pharmacologic agents.

EXAMPLE APPLICATIONS OF MOUSE ELECTROPHYSIOLOGY STUDY

Familial Hypertrophic Cardiomyopathy

Familial hypertrophic cardiomyopathy (FHC) is a relatively common inheritable disease characterized clinically by arrhythmias, syncope, exercise intolerance, and sudden cardiac death. The clinical course of FHC can range from the asymptomatic carrier to premature sudden death in childhood. Risk factors for sudden death in FHC include younger age, syncope, arrhythmias, and a malignant family history (21-24). In FHC patients resuscitated from sudden cardiac arrest, EP testing has revealed varying degrees of sinus node dysfunction, electrogram fractionation, and inducible ventricular arrhythmias (25-27). However, clinical EP testing has had variable prognostic value in FHC, perhaps as it is not feasible to perform large-scale EP studies in genetically-similar human cohorts (28,29). Pathological features of familial hypertrophic cardiomyopathy in humans include ventricular hypertrophy, myocyte disarray, interstitial fibrosis, and electrophysiological abnormalities (29,30). The abnormal genes identified in FHC patients encode for mutations in sarcomeric proteins (30-32).

Analogous mutations in sarcomeric protein genes have been engineered in mice. For example, the cardiac myosin heavy chain gene mutation $Arg^{403}Gln$ has been genetically-engineered to create a murine model of FHC (33). The homozygous mutation (α-$MHC^{403/403}$) is uniformly fatal, while the heterozygous α-$MHC^{403/+}$ mice survive, but with a histological and hemodynamic FHC phenotype, including sudden death during exercise (33). The $Arg^{403}Gln$ mutation is associated with a relatively high risk of arrhythmia and mortality in humans (34). The mouse EP method has been used to demonstrate that mice with this mutation have distinct ECG and electrophysiologic abnormalities, including prolonged sinus node recovery, heterogeneous ventricular conduction, and inducible ventricular ectopy (35). Figure 7 demonstrates an example of inducible ventricular arrhythmias. These findings demonstrate specific EP abnormalities associated with the α-$MHC^{403/+}$ mutation and raise the possibility that prospective evaluation may help electrophysiologic risk-stratification in patients with FHC. In addition, other sarcomeric gene mutations leading to an FHC phenotype have also been characterized with electrophysiological phenotyping (36,37), allowing for potential gene-specific diagnostics, prognostication, and therapeutics.

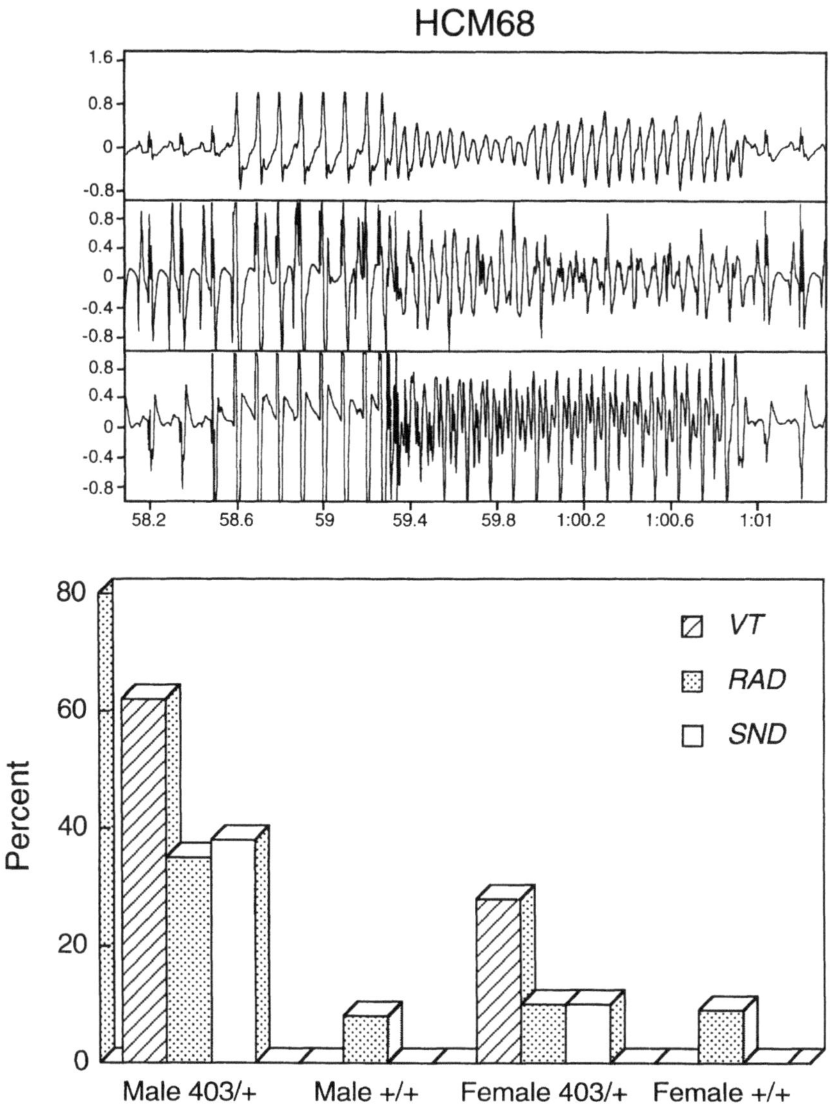

Figure 7: Ventricular tachycardia in heterozygous α-MHC$^{403/+}$ mice. A 3-channel recordings (ECG lead I, intracardiac right atrial (RA) and right ventricular (RV) electrograms) during ventricular tachycardia (VT) induction. The VT is reproducibly induced by ventricular programmed electrical stimulation. The histogram illustrates the frequency of electrical abnormalities such as VT, right axis deviation (RAD), and sinus node dysfunction (SND) identified in FHC mice.

Familial Dilated Cardiomyopathy

Congestive heart failure represents one of the most common clinical cardiovascular conditions. Among non-ischemic cardiomyopathies, idiopathic dilated cardiomyopathy remains the most common example. Idiopathic dilated cardiomyopathy is characterized by dilatation of the left ventricle and impaired left ventricular systolic function. The molecular mechanisms of this primary myocardial disease remain unknown. CREB, a 43-kD basic leucine zipper transcription factor, is responsible for regulating gene expression in response to a variety of extracellular signals. Transgenic mice with a dominant-negative mutant form of CREB, containing a Ser^{133}Ala substitution, develop dilated cardiomyopathy with many of the anatomical and pathophysiological features of human idiopathic dilated cardiomyopathy (38). Manifestations of heart failure including ventricular dilatation, impaired left ventricular systolic and diastolic function, hepatic congestion, peripheral edema, and premature death were present. The EP mouse model has been used to demonstrate that these CREB mutant mice develop abnormalities of A-V nodal conduction. As these mice age and heart failure worsens, the degree of A-V nodal conduction abnormalities becomes more severe and the HV interval and QRS duration become prolonged, consistent with abnormalities of intraventricular and His-Purkinje conduction. In addition, the EP mouse model demonstrated that ventricular arrhythmia inducibility also occurs only in CREB mutant mice with advanced heart failure. Ambulatory ECG at the time of death revealed high grade A-V block. These data suggest that CREB mutant mice display many of the important electrophysiologic properties of human heart failure due to idiopathic dilated cardiomyopathy (38).

Congenital Long QT Syndrome

As in the cardiomyopathies, the congenital long QT syndromes (LQTS) are due to multiple distinct genotypes leading to a similar phenotype (39). The LQTS phenotype is characterized by a combination of prolongation of the rate-corrected QT interval (QTc) on the ECG, stress-induced syncope, ventricular tachycardia (characteristically with a torsades de pointes pattern), and sudden cardiac death. Before the identification of specific mutations, congenital LQTS was differentiated into three groups: autosomal dominant inheritance with variable penetrance (Romano-Ward syndrome), autosomal recessive inheritance with associated congenital neural deafness (Jervell-Lange-Nielson syndrome) and sporadic mutations without clear inheritance patterns. The best characterized is Romano-Ward type, which accounts for the majority of identified congenital long QT syndrome cases. Presently, at least 6 different mutations responsible for LQTS have been elucidated, with genetic mechanisms leading to distinct mutations in ion channel function in each case (40,41). These mutations have been aptly named LQT1, LQT2, LQT3, etc. The two more common defects (LQT1 and LQT2) are due to potassium channel mutations, leading to prolonged repolarization times, and the third (LQT3) is caused by a sodium channel defect. Like the FHC mutations, there are likely to be many more mutations leading to the LQTS clinical phenotype that have yet to be recognized.

Murine models of LQTS have specific ion channel defects, in particular potassium channel subtypes (I_{Kr}, I_{Ks}, I_{to}, I_{K1}) and the sodium-potassium ATPase pump (42). Increased dispersion of action potential duration may account for the susceptibility to ventricular arrhythmias in many of these long QT mice (43). Mouse models of Jervell and Lange-Nielsen syndrome have also been developed. Such a model has been created by knocking out the isk gene, likely involved in the adaptation of QT to heart rate variation (44). These mouse models of LQTS are presently being studied using the *in vivo* EP methods described here and other phenotyping techniques in order to help elucidate the electrophysiological basis for the congenital LQTS and attempt to develop mutation-specific therapies for the wide spectrum of the LQTS phenotype (45).

Myotonic Dystrophy

Myotonic muscular dystrophy is the most common type of muscular dystrophy. The disease affects smooth, skeletal, and cardiac muscles and leads to varying degrees of myotonia, skeletal wasting, and abnormalities in myocardial and conduction system function. The cardiac symptoms relate to progressive atrio-ventricular block and bradycardia. Myotonic dystrophy is likely due to a putative mutation in a protein kinase (DMPK) gene (46). A murine model of myotonic dystrophy has been designed recently by targeted disruption of the myotonic dystrophy protein kinase gene (47). The DMPK-deficient mice ($DMPK^{-/-}$) display abnormalities in skeletal muscle with a progressive skeletal myopathy (47). In order to characterize cardiac conduction, *in vivo* electrophysiology (EP) studies were performed via epicardial and endocardial pacing/recording in DMPK mutant mice. Complete 12-lead ECGs and EP data were collected, including sinus node recovery times, atrioventricular (A-V) and V-A conduction properties, and atrial, A-V, and ventricular effective refractory periods. On ECG, a prolonged PR interval was seen compared with wild-type mice, analogous to 1° A-V block in humans. Some DMPK-deficient mice had 2° and 3° A-V block during atrial pacing (Figure 8). There were no inducible tachyarrhythmias in the DMPK-deficient mice, indicating distinct electrophysiological abnormalities of atrioventricular conduction (48). These changes appear to be age-related, as they are not yet evident in immature mutant mice (49). Further studies have demonstrated using His bundle electrogram recordings that the conduction abnormalities are localized to both the A-V node and the infraHisian regions (50).

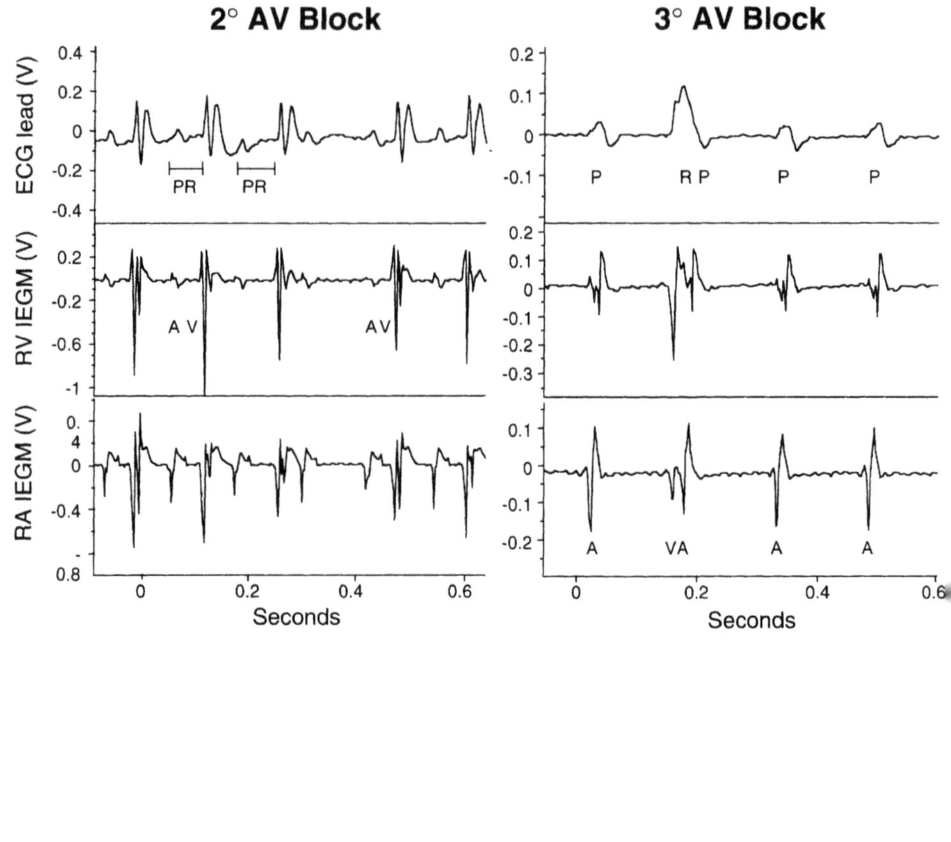

Figure 8: Three-channel recording during atrioventricular block in DMPK$^{-/-}$ mouse. Displayed are surface ECG lead I, right atrial (RA) and left ventricular (LV) electrograms during spontaneous complete heart block in an epicardial open-chest preparation from a DMPK$^{-/-}$ mutant mouse. The left panels demonstrate Wenckebach type 2° A-V block, with progressive PR prolongation. The right panels illustrate 3° A-V block, with an atrial cycle length of 135 msec and a ventricular cycle length during complete heart block is 220 msec.

Connexin Defect Disorders

Gap junction proteins are critical to cardiac electrical signal conduction from cell to cell. There are multiple connexin proteins, which are responsible for intermyocyte communication of electrical impulses (51). Connexin subtypes are distributed in anatomically distinct regions of the heart (atrium versus ventricle versus specialized conduction tissue). Disorders of the cardiac gap junction channel proteins may lead to distinct forms of cardiac conduction abnormalities dependent on the spatial concentration of the individual connexin protein subtypes (52). Targeted mutagenesis of connexin43, the most prevalent gap junction protein, has been performed, with homozygous mice surviving fetal life, but dying at birth from congenital heart disease affecting the right ventricular outflow tract (53). However, heterozygote mutant mice survive, without obvious structural heart anomalies. Some studies have suggested that defects in connexin43 may be implicated in heterotaxy syndrome, a form of complex congenital heart and visceral abnormalities, while other authors have not found that association (54,55). Mice with connexin43 mutations have been shown to have QRS prolongation on ECG, corresponding with delayed intraventricular conduction time (56), although this finding has not been consistently observed (57), likely dependent on the preparation and experimental conditions (58). In sinus rhythm or with pacing, there were no other abnormalities on surface ECG or during EP testing.

In contrast to homozygous connexin43 mutants, homozygous connexin 40 mutant mice ($Cx40^{-/-}$) survive into adulthood. We and others have observed significant abnormalities of A-V nodal function and His-Purkinje conduction in $Cx40^{-/-}$ mice (Figure 9). Interestingly, these changes become more pronounced with aging of the mice (59,60). Sinus heart rate regulation may additionally be adversely affected in $Cx40^{-/-}$ mice (61). Transesophageal and epicardial pacing have also been used to assess conduction in Connexin 40 deficient mice (62,63). Atrial burst pacing has been shown to induce atrial tachyarrhythmias and atrial conduction velocity is reduced in these mice (62,63). The effects of connexin mutations on potential arrhythmias and sudden cardiac death, and their clinical relevance remain to be determined.

Cardiac Development Murine Models and Electrophysiology

The developing heart and its conduction system can be studied using molecular biological and electrophysiological techniques. Recently, the molecular precursors of the murine embryologic cardiac electrical system have been identified. Mouse models of abnormal formation have been designed using targeted mutagenesis and transgenic techniques. These models allow the electrophysiological analysis in vivo of the maturing conduction system, and well as the developing responses in the immature heart. Examples of these models include mutations in transcription factor genes, such as Nkx2.5/Csx, TBX-5, dHand, and Irx-4, some of which lead to distinct electrophysiologic phenotypes (64-67).

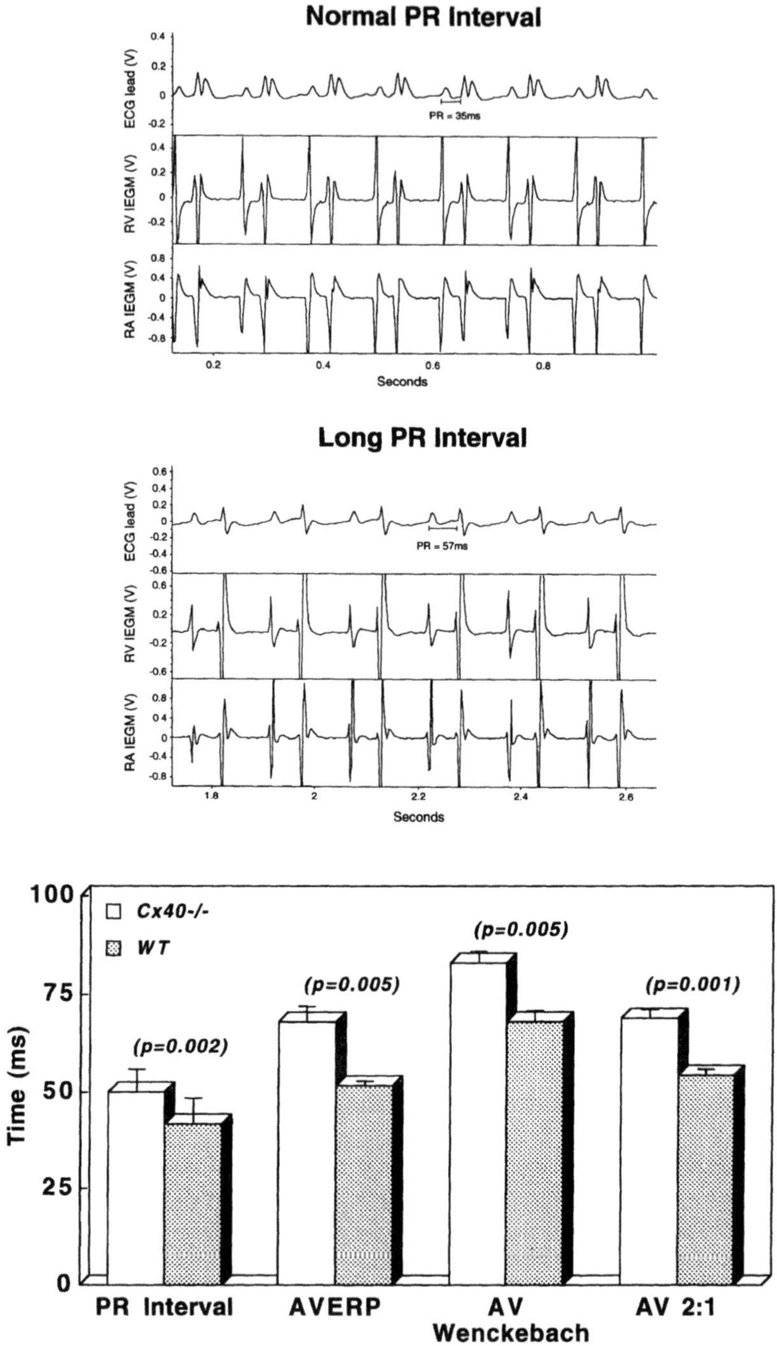

Figure 9: Abnormalities of atrioventricular conduction in connexin 40 deficient mice. Homozygous knockout $Cx40^{-/-}$ mice had prolonged PR intervals on ECG and longer A-V conduction properties during *in vivo* EP studies.

SUMMARY

Genetically-manipulated animals have become efficient models for the study of human cardiac electrophysiological diseases. These examples highlighted above illustrate experiences with *in vivo* murine electrophysiology. Electrophysiological properties have been evaluated in many species by a variety of methods, from patch-clamping of an isolated myocyte for assessment of ion channel function, to transmembrane potential measurement in myocardial tissue strips, to Langendorff model whole-heart preparations and intact whole-animal studies. The *in vivo* mouse EP model can be used to obtain EP parameters by pacing and recording directly from the epicardium and endocardium of intact mice simultaneously with monophasic action potentials, and surface ECG recordings. Potential therapies may be appraised with murine EP studies, including assessment of drugs, pacing, and novel gene replacement approaches. Specific disease models can be studied to further understand the relationship between gene defects and their electrophysiologic consequence in vivo. Mice harboring mutations in genes encoding proteins that are highly expressed in myocytes or that lead to cardiovascular abnormalities may prove to have interesting and informative electrophysiologic phenotypes.

REFERENCES

1. Berul CI, Aronovitz M, Wang PJ, Mendelsohn ME. *In vivo* cardiac electrophysiology studies in the mouse. *Circulation* 1996;94:2641-2648.
2. Gehrmann J, Berul CI. Cardiac electrophysiology in genetically-engineered mice. *J Cardiovasc Electrophysiol* 2000;11:354-368.
3. Berul CI, Mendelsohn ME. Molecular Biology and Genetics of Cardiac Disease Associated with Sudden Death. In: Estes NAM, Salem DN, Wang PJ, editors. *Sudden Cardiac Death in the Athlete*. New York: Futura; 1998, 465-482.
4. Davignon A, Rautaharju P, Boisselle E, Soumis F, Megelas M, Choquette A. Normal ECG standards for infants and children. *Pediatr Cardiol* 1980;2:123-131.
5. Berul CI, Sweeten TL, Dubin AM, Shah MJ, Vetter VL. Use of the rate-corrected JT interval for prediction of repolarization abnormalities in children. *Am J Cardiol* 1994;74:1254-1257.
6. Zipes DP. 1992. Specific arrhythmias: Diagnosis and treatment. In Heart disease: A textbook of cardiovascular medicine. E. Braunwald, editor. WB Saunders, Philadelphia. 667-725.
7. Hiss RG, Lamb L. Electrocardiographic findings in 122,043 individuals. *Circulation* 1962;25:947-8.
8. Bazett HC. 1920. An analysis of the time relationships of the heart. *Heart* 7:353-370.
9. Mitchell GF, Jeron A, Koren G. Measurement of heart rate and Q-T interval in the conscious mouse. *Am J Physiol* 1998;274:H747-751
10. Gillette PC, Buckles DS, Harold M, Garson A. Intracardiac electrophysiology studies. In Pediatric Arrhythmias: Electrophysiology and Pacing. Gillette PC and Garson A, eds. WB Saunders, Philadelphia. 216-248.
11. Mandel W, Hayakawa H, Danzig R, Marcus HS. Evaluation of sino-atrial node function in man by overdrive suppression. *Circulation* 1971;44:59-66.
12. Denes P, Wu D, Dhingra R, Pietras RJ, Rosen KM. The effects of cycle length on cardiac refractory periods in man. *Circulation* 1974;49:32-37.
13. Bisset JK, Kane JJ, DeSoyza N, Murphy M. Electrophysiological significance of rapid atrial pacing as a test of atrioventricular conduction. *Cardiovasc Res* 1975;9:593-600.

14. Maguire CT, Bevilacqua LM, Wakimoto H, Gehrmann J, Berul CI. Maturational atrioventricular nodal physiology in the mouse. *J Cardiovasc Electrophysiol* 2000;11:557-564.
15. VanderBrink BA, Link MS, Aronovitz MJ, Saba S, Sloan SB, Homoud MK, Estes NAM, Wang PJ. Assessment of atrioventricular nodal physiology in the mouse. *J Intervent Cardiovasc Electrophysiol* 1999;3:207-212.
16. Han J, Moe GK. Non-uniform recovery of excitability in ventricular muscle. *Circ Res* 1964;14:44-60.
17. Zabel M, Portnoy S, Franz M. Electrocardiographic indices of dispersion of ventricular repolarization: An isolated heart validation study. *J Am Coll Cardiol* 1995;25:746-752.
18. Bevilacqua LM, Maguire CT, Seidman CE, Seidman JG, Berul CI. QT dispersion in α-myosin heavy-chain familial hypertrophic cardiomyopathy mice. *Pediatr Res* 1999;45:643-647.
19. Wakimoto H, Maguire CT, Kovoor P, Gehrmann J, Berul CI. Induction of atrial tachycardia and fibrillation in the mouse heart. *Cardiovasc Res* 2001;50:463-473.
20. Gehrmann J, Frantz S, Maguire CT, Vargas M, Ducharme A, Wakimoto H, Lee RT, Berul CI. Electrophysiological characterization of murine myocardial ischemia and infarction. *Basic Res Cardiol* 2001;96:237-250.
21. Fananapazir L, Chang AC, Epstein SE, McAreavy D. Prognostic determinants in hypertrophic cardiomyopathy: prospective evaluation of a therapeutic strategy based on clinical, Holter, hemodynamic, and electrophysiological findings. *Circulation* 1992;86:730-740.
22. Epstein ND, Cohn GM, Cyran F, Fananapazir L. Differences in clinical expression of hypertrophic cardiomyopathy associated with two distinct mutations in α-myosin heavy chain gene. *Circulation* 1992;86:345-352.
23. Buja G, Miorelli M, Turrini P, Melacini P, Nava A. Comparison of QT dispersion in hypertrophic cardiomyopathy between patients with and without ventricular arrhythmias and sudden death. *Am J Cardiol* 1993;72:973-6.
24. Wigle ED, Rakowski H, Kimball BP, Williams WG. Hypertrophic cardiomyopathy: Clinical spectrum and treatment. *Circulation* 1995;92:1680-1692.
25. Fananapazir L, Epstein SE. Hemodynamic and electrophysiologic evaluation of patients with hypertrophic cardiomyopathy surviving cardiac arrest. *Am J Cardiol* 1991;67:280-287.
26. Maron BJ, Fananapazir L. Sudden death in hypertrophic cardiomyopathy. *Circulation* 1992;85:157-163.
27. Watson RM, Schwartz JL, Maron BJ, Tucker E, Rosing DR, Josephson ME. Inducible polymorphic ventricular tachycardia and ventricular fibrillation in a subgroup of patients with hypertrophic cardiomyopathy at risk for sudden death. *J Am Coll Cardiol* 1987;10:761-774.
28. Cannon CR, Reeder GS, Bailey KR, Melton III LJ, Gersh BJ. Natural history of hypertrophic cardiomyopathy: *Circulation* 1995;92:2488-2495.
29. Seidman CE, Seidman JG. Gene mutations that cause familial hypertrophic cardiomyopathy. In Molecular Cardiovascular Medicine. E. Haber, editor. 1995. Scientific American Press, New York. 193-210.
30. Watkins H, McKenna W, Thierfelder L, Suk HJ, Anan R, O'Donoghue A, Spirito P, Matsumori A, Moravec CS, Seidman JG, Seidman CE. Mutations in the genes for cardiac troponin T and α-tropomyosin in hypertrophic cardiomyopathy. *N Engl J Med* 1995;332:1058-1064.
31. Thierfelder L, Watkins H, MacRae C, Lamas R, McKenna W, Vosberg HP, Seidman JG, Seidman CE. Mutations in α-tropomyosin and in cardiac troponin T cause hypertrophic cardiomyopathy: A disease of the sarcomere. *Cell* 1994;77:701-712.
32. Watkins H, Anan R, Coviello DA, Spirito P, Seidman JG, Seidman CE. A *de novo* mutation in α-tropomyosin that causes hypertrophic cardiomyopathy. *Circulation* 1995;91:2302-2305.
33. Geisterfer-Lowrance A, Christe M, Conner D, Ingwall J, Schoen F, Seidman CE, Seidman JG. A mouse model of familial hypertrophic cardiomyopathy. *Science* 1996;272:731-734.
34. Watkins H, Rosenzweig A, Hwang D, Levi T, McKenna W, Seidman CE, Seidman JG. Characteristics and prognostic implications of myosin missense mutations in familial hypertrophic cardiomyopathy. *N Engl J Med* 1992;326:1108-1114.
35. Berul CI, Christe ME, Aronovitz MJ, Seidman CE, Seidman JG, Mendelsohn ME. Electrophysiological abnormalities and arrhythmias in α-MHC mutant familial hypertrophic cardiomyopathy mice. *J Clin Invest* 1997;99:274-280.

36. Berul CI, Fatkin D, McConnell BK, Maguire CT, Gehrmann J, Rakhit A, Jones K, Seidman JG, Seidman CE. Electrophysiological characterization of familial hypertrophic cardiomyopathy mice with mutant myosin binding protein C gene. *Circulation* 1999;100:I207.
37. Bevilacqua LM, Berul CI. Familial Hypertrophic Cardiomyopathy Genetics. In: Berul CI, Towbin JA, eds. *Molecular Genetics of Cardiac Electrophysiology.* Boston: Kluwer Academic Publishers; 2000, 185-200.
38. Zhu W, Saba S, Lepore J, Link MS, Homoud MK, Estes NAM, Wang PJ, Leiden J. Cardiac electrophysiologic abnormalities in a mouse model of idiopathic dilated cardiomyopathy. *Circulation* 2000 Abstract.
39. Roden DM, Lazzara R, Rosen M, Schwartz PJ, Towbin JA, Vincent GM, and SADS Foundation Task Force on LQTS. Multiple mechanisms in the long-QT syndrome: Current knowledge, gaps, and future directions. *Circulation* 1996;94:1996-2012.
40. Towbin JA, Hua L, Taggart T, Lehmann MH, Schwartz PJ, Satler CA, Ayyagari R, Robinson JL, Moss A, Hejtmancik JF. Evidence of genetic heterogeneity in Romano-Ward long QT syndrome. *Circulation* 1994;90:2635-2644.
41. Vincent GM, Timothy K. Congenital Long QT Syndromes. In: Berul CI, Towbin JA, eds. *Molecular Genetics of Cardiac Electrophysiology.* Boston: Kluwer Academic Publishers; 2000, 119-136.
42. O'Brien SE, Apkon MA, Berul CI, Patel HT, Saupe K, Spindler M, Ingwall JS, Zahler RX. Phenotypic features of long QT syndrome in transgenic mice expressing human α3 Na,K-ATPase isoform in hearts. *Am J Physiol* 2000;279:H2133-2142.
43. Jeron A, Mitchell GF, Zhou J, Murata M, London B, Buckett P, Wiviott SD, Koren G. 2000. *Am J Physiol* 278:H1891-1898.
44. Drici MD, Arrighi I, Chouabe C, Mann JR, Lazdunski M, Romey G, Barhanin J. 1998. *Circ Res* 83:95-102.
45. Tanaka T, Nagai R, Tomoike H, Takata S, Yano K, Yabuta K, Haneda N, Nakano O, Shibata A, Sawayama T, Kasai H, Yazaki Y, Nakamura Y. Four novel *KVLQT1* and four novel *HERG* mutations in familial long-QT syndrome. *Circulation* 1997;95:565-567.
46. Housman DE, Shaw DJ. Expansion of an unstable DNA region and phenotypic variation in myotonic dystrophy. *Nature* 1992;355:545-546.
47. Reddy S, Smith DBJ, Rich MM, Leferovich JM, Reilly P, Davis BM, Tran K, Rayburn H, Bronson R, Cros D, Balice-Gordon RJ, Housman D. Mice lacking the myotonic dystrophy protein kinase develop a late-onset progressive myopathy. *Nature Genet* 1996;13:325-335.
48. Berul CI, Maguire CT, Aronovitz MJ, Greenwood J, Miller C, Housman D, Mendelsohn ME, Reddy S. DMPK dosage alterations result in atrioventricular conduction abnormalities in a mouse myotonic dystrophy model. *J Clin Invest* 1999;103:R1-R7.
49. Berul CI, Maguire CT, Gehrmann J, Reddy S. Progressive atrioventricular conduction block in a mouse myotonic dystrophy model. *J Intervent Cardiac Electrophysiol* 2000;4:351-358.
50. Saba S, VanderBrink BA, Luciano B, Aronovitz MJ, Berul CI, Reddy S, Housman D, Mendelsohn ME, Estes NA 3rd, Wang PJ. Localization of the sites of conduction abnormalities in a mouse model of myotonic dystrophy. *J Cardiovasc Electrophysiol* 1999;10:1221-1223.
51. Saffitz JE, Davis LM, Darrow BJ, Kanter HL, Laing JG, Beyer EC. The molecular basis of anisotrophy: Role of gap junctions. *J Cardiovasc Electrophysiol* 1995;6:498-510.
52. Davis LM, Kanter HL, Beyer EC, Saffitz JE. Distinct gap junction protein phenotypes in cardiac tissues with disparate conduction properties. *J Am Coll Cardiol* 1994;24:1124-1132.
53. Reaume AG, deSousa PA, Kulkarni S, Langille BL, Zhu D, Davies TC, Juneja SC, Kidder GM, Rossant J. Cardiac malformation in neonatal mice lacking connexin43. *Science* 1995;267:1831-1834.
54. Britz-Cunningham SH, Shah MM, Zuppan CW, Fletcher WH. Mutations of the connexin43 gap junction gene in patients with heart malformations and defects of laterality. *N Engl J Med* 1995;332:1323-1329.
55. Gebbia M, Towbin JA, Casey B. Failure to detect connexin43 mutations in 38 cases of sporadic and familial heterotaxy. *Circulation* 1996;94:1909-1912.

56. Thomas SA, Schuessler RB, Berul CI, Beardslee MA, Beyer EC, Mendelsohn ME, Saffitz JE. Disparate effects of deficient connexin43 expression on atrial and ventricular conduction: evidence for chamber-specific molecular determinants of conduction. *Circulation* 1998;97:686-691.
57. Morley GE, Vaidya D, Samie FH, et al: A characterization of conduction in the ventricles of normal and heterozygous Cx43 knockout mice using optical mapping. *J Cardiovasc Electrophysiol* 1999;10:1361-1375
58. Berul CI. A gap in understanding the connection between connexins and conduction. *J Cardiovasc Electrophysiol* 1999;10:1376-1379.
59. Bevilacqua LM, Simon AM, Maguire CT, Gehrmann J, Wakimoto H, Paul DL, Berul CI. A targeted disruption in connexin40 leads to distinct atrioventricular conduction defects. *J Intervent Card Electrophysiol* 2000;4:459-467.
60. Zhu W, Saba S, Sellitto C, Paul DL, VanderBrink B, Link MS, Homoud MK, Estes NAM, Wang PJ.. Age-related changes in A-V conduction in connexin-40 deficient mice. *J Am Coll Cardiol* 2000, Abstract.
61. Gehrmann J, Maguire CT, Hammer PE, Bevilacqua LM, Simon AM, Triedman JK, Paul D, Berul CI. A targeted disruption of connexin 40 leads to abnormal heart rate regulation in mice. *Proceed Eur Soc Cardiol* 1999;21:129-133
62. Hagendorff A, Schumacher B, Kirchhoff S, Luderitz B, Willecke K. Conduction disturbances and increased atrial vulnerability in Connexin40-deficient mice analyzed by transesophageal stimulation. *Circulation* 1999;99:1508-1515.
63. Verheule S, Van Batenburg CA, Coenjaerts FE, Kirchhoff S, Willecke K, Jongsma HJ. Cardiac conduction abnormalities in mice lacking the gap junction protein connexin40. *J Cardiovasc Electrophysiol* 1999;10:1380-1389.
64. Berul CI, Tanaka M, Maguire CT, Gehrmann J, Bevilacqua LM, Izumo S. Arrhythmia vulnerability in a murine model of transcription factor Nkx2.5-deficiency associated with familial atrial septal defect. *Circulation* 1999;100:I208 Abstract.
65. Wakimoto H, Kasahara H, Maguire CT, Vargas MM, Lawitts J, Izumo S, Berul CI. Developmental cardiac conduction phenotype in Csx/Nkx2.5 transgene mice. *Circulation* 2000;102:II-210 Abstract.
66. Bruneau BG, Bao ZZ, Fatkin D, Xavier-Neto J, Georgakopoulos D, Maguire CT, Berul CI, Kass DA, Kuroski-deBold ML, deBold AJ, Conner DA, Rosenthal N, Cepko CL, Seidman JG, Seidman CE. Cardiomyopathy in Irx4-deficient mice is preceded by abnormal ventricular gene expression. *Molec Cell Biology* 2001;21:1730-1736.
67. Biben C, Weber R, Kesteven S, Stanley E, McDonald L, Elliott DA, Barnett L, Koentgen F, Robb L, Feneley M, Harvey RP. Cardiac septal and valvular dysmorphogenesis in mice heterozygous for mutations in the homeobox gene Nkx2-5. *Circ Res* 2000;87:888-895.

17

CARDIOVASCULAR PHYSIOLOGY IN MICE: CONSCIOUS MEASUREMENTS AND EFFECTS OF ANESTHESIA

Stephen F. Vatner, M.D., Gen Takagi, M.D.
Kuniya Asai, M.D., *Richard P. Shannon, M.D.

Cardiovascular Research Institute
University of Medicine & Dentistry of New Jersey
New Jersey Medical School
Newark, New Jersey
**Department of Medicine*
Allegheny General Hospital
Pittsburgh, Pennsylvania

Transgenic murine models have been employed with increasing frequency over the past several years for the elucidation of mechanisms of normal and abnormal cardiovascular function. While these models have been used extensively for biochemical and cellular physiological studies, their use in studies of *in vivo* cardiovascular physiology has until recently been limited by technical considerations. Recent advances in surgical and catheterization techniques, imaging modalities and other technologies have allowed for detailed *in vivo* cardiovascular studies in mice. These developments recall the technical advances made several decades ago that allowed for detailed study of cardiovascular physiology in large animals. These investigations in large animal models demonstrated not only the importance of *in vivo* studies for understanding complex physiological relationships, but also the profound effects of anesthesia on measures of cardiovascular function [1-9]. Ideally, studies of transgenic murine cardiovascular physiology would similarly be carried out in conscious, chronically instrumented animals. Limitations in current technology, however, have confined conscious studies to only a few techniques. This chapter reviews the currently available methods for the measurement of cardiovascular parameters in conscious mice, including ranges for normal values, and discusses the effects of anesthesia on cardiovascular function in the mouse which make the need for further progress on conscious measures so compelling.

SURGICAL TECHNIQUES

Anesthesia and Surgical Preparation

A variety of anesthetic regimens have been used for the instrumentation or surgical manipulation of mice for the purpose of later conscious cardiovascular study. Anesthetic agents should ideally provide an adequate anesthetic plane while allowing for the animal to breathe spontaneously (if the surgery does not involve thoracotomy), as well as allowing for recovery of the animal within a reasonable time frame.

Surgical preparation should in general follow the norms for veterinary surgery. Meticulous attention to sterile technique is essential to ensure recovery of the animal and a return to a healthy baseline state prior to beginning study. Perioperative prophylactic antibiotic therapy (cephalothin 30 mg subcutaneously for 3-7 days[10]) is recommended. Small rodents are susceptible to hypothermia during surgical instrumentation and efforts should be made to maintain normal body temperature. This can be accomplished by measuring rectal temperature continuously, and keeping temperature constant with heating lamps or with commercially available isothermic pads.

Specific surgical techniques are described in the individual sections on physiological measurements.

ARTERIAL PRESSURE

Non-Invasive Measurement

In humans and large animals, non-invasive measurement of arterial blood pressure can be performed using auscultation or plethysmography during institution of and release from arterial occlusion. The small size of mice precludes the use of auscultation, but both plethysmographically- and photoelectrically-determined non-invasive measures of arterial blood pressure have been reported in mice. While plethysmography during arterial occlusion with a tail cuff has been used in rodents for several decades [11], recent advances in equipment and automation have allowed for wider applicability, including use in conscious mice and more consistent results. Recently several investigators have reported the use of a non-invasive computerized tail cuff system for measuring arterial pressures in conscious mice [12,13]. The system employed records the systolic arterial blood pressure as the externally applied pressure at which blood flow to the tail is eliminated. Blood flow is detected by placing the mouse in a restraint with the tail between a light source and photo resistor. The photo resistor yields an oscillating waveform readout of blood flow. A tail cuff attached to a manometer is inflated and the pressure at which blood flow is eliminated is recorded. Adjustments are made in the threshold amplitude (as a percentage of baseline) and cycle number used to define elimination of blood flow, allowing for consistent measurements even in the presence of modest levels of interference.

Using the tail cuff method, Krege et al. [12] reported control systolic blood pressure values of 123±2 mmHg and control heart rates (measured by waveform frequency prior to cuff inflation) of 700±10 min^{-1}. Systolic pressure following 3 months of enalapril therapy had fallen to 104±6 mmHg. There was excellent reproducibility of tail cuff-derived systolic pressure and heart rate and the authors reported a tight correlation between systolic pressure measured by tail cuff and mean arterial pressure measured by carotid artery catheter. Other investigators [13-15], using a similar technique, have reported slightly lower pressures in control mice and the ability to detect significant changes in arterial pressure in mice genetically engineered or surgically manipulated to produce alterations in determinants of blood pressure. Importantly, there is considerable variability in different murine strains, which may contribute independently to differences in resting blood pressure measurements in normal mice. Table 1 summarizes results from several studies utilizing either the tail cuff technique or direct measurement of arterial pressure in conscious mice.

Non-invasive measurement of systolic pressure and heart rate in conscious mice by tail cuff appears to be an accurate and reproducible technique. Mice may require warming to allow for adequate vasodilation of tail vasculature. Care must be taken to avoid overwarming to prevent excessive vasodilation and relative hypotension, and to properly train the mice to rest quietly in the restraint during warming and cuff inflation so as not to unduly elevate pressure and heart rate. Indeed, even with careful training, some [12], though not all [13,15], investigators have reported consistently higher pressures and heart rates as measured by tail cuff versus indwelling catheter. This, however, may have reflected residual anesthetic effects in the catheterized mice or a baroreceptor-mediated effect due to carotid catheter placement. Given that it is non-invasive and reproducible, the tail cuff technique for measuring blood pressure in conscious mice is likely to find its greatest utility in long term studies of cardiovascular responses to genetic manipulation or chronic interventions.

Invasive Measurement

The majority of reports of conscious blood pressure measurements in mice have involved surgical placement of arterial catheters, variable periods of post-operative recovery, and measurements conducted at baseline and during acute interventions. Catheters have been placed in the carotid [12,15-21] and femoral [10,14,22-26] arteries. Catheters can be purchased from vendors or constructed by flame stretching PE50 or PE10 tubing. Surgical placement is generally accomplished with the aid of a dissecting microscope. The mouse is anesthetized, the area over the carotid or femoral artery is shaved, and the mouse is placed supine on a warmed surface. The artery is isolated and ligated distally and a microsurgical clip is placed proximally. One or more sutures are loosely placed proximally. An arteriotomy is performed with microsurgical scissors or a 25 gauge or smaller needle and the catheter is threaded into the vessel lumen. The sutures are tightened and the clip is carefully released, checking for adequate hemostasis. The catheter is tunneled subcutaneously and the incision is closed in layers. The catheter is flushed with a dilute heparin solution, heat sealed, and stored on the dorsum in a retaining device (small plastic cap or pouch) sutured to the animal. Intravenous catheters can be placed at the same time (usually in the internal jugular vein) for later delivery of fluid or drugs. Prophylactic antibiotics are recommended following surgery. At the time of recording, catheters are freed, flushed and connected to modified strain gauge transducers or micromanometers. Mice are lightly restrained and allowed to accommodate to the restraining chamber.

Numerous studies have reported directly-measured conscious arterial pressure in wild type and transgenic mice [10,12-26]. The results of a number of these are summarized in Table 1. Normal values for systolic blood pressure measured by arterial catheter in wild type mice appear to range from 120 to135 mm Hg, with mean arterial pressures ranging from 95 to 120 mm Hg. These contrasts with the lower values reported in most studies of anesthetized mice. While inter-strain differences make comparison difficult, it is evident from published reports and unpublished observations that inadequate post-operative recovery time yields lower blood pressures and either lower or higher heart rates as compared to recordings made after more prolonged periods of recovery (see below). This seems to be true even when the animal appears to have recovered fully from anesthesia as judged by a return to normal motor activity. With regard to catheters for measurement of carotid and femoral arterial pressures and heart rate, there is some evidence that placement of a carotid catheter may yield abnormally low heart rates due to carotid sinus-mediated reflex slowing. This is one limitation of studies in mice where catheters are almost the same size as the arterial lumen.

A number of strains of mice with specific genetic manipulations have been reported to have alterations in resting arterial pressure. Hypertension has been documented in mice overexpressing the α subunit of the stimulatory G protein [10] or expressing human renin and angiotensinogen genes [21,22], D79Nα_{2A}-adrenoceptor [19], in mice with targeted disruption of the bradykinin B_2- receptor gene [15], in mice deficient in insulin-like growth factor 1 [23], and in mice with eNOS knockout [20]. Hypotension has been seen in mice overexpressing atrial natriuretic factor [24] and in mice with targeted disruption of the type 1A angiotensin II receptor gene [13] or the angiotensin converting enzyme gene [18]. Several transgenic murine strains have been found to have altered responses to acute administration of vasoactive agents in the absence of any abnormality in resting arterial pressure [16-18]. These observations illustrate the importance of techniques for the measurement of blood pressure in conscious mice to the full understanding of the potentially complex cardiovascular effects of specific genetic manipulations.

HEART RATE

Noninvasive Measures

As mentioned, the non-invasive tail cuff method of determining blood pressure can also be used for measurement of heart rate (Table 1). Heart rate is derived from the pulsatile waveform generated by the photosensor. This method appears to yield heart rates slightly greater than invasive methods, likely due to agitation of the mice caused by heating or cuff inflation.

Invasive measures

The simplest technique for recording conscious heart rate in mice is via surgically placed subcutaneous electrodes. The animal is anesthetized, the lateral thorax and interscapular regions shaved and sterilely prepared. Three 10 cm lengths of 0.0014" stainless steel wire are stripped of insulation over the distal 10 mm and inserted subcutaneously via small skin incisions, one each on either side of the thorax and the third tunneled subcutaneously along the back. The thoracic leads are tunneled subcutaneously to the interscapular region where all leads are externalized and stored in a small plastic cap sutured to the animal's back. The animal is allowed to recover for at least 48 hrs. At the time of study the animal is placed in a loose restraint, the leads freed from the cap and connected to an ECG monitor with or without an online tachometer. A sample electrocardiographic tracing is shown in Figure 1.

Conscious heart rate can also be measured from arterial pressure tracings taken from indwelling arterial catheters (Table 1). The frequency responses of the strain gauges or micromanometers used to transduce pressure are generally adequate for heart rate determinations over the range seen in conscious mice. The inconsistency of signal quality makes this method less well suited to the analyses required for heart rate variability studies (see below).

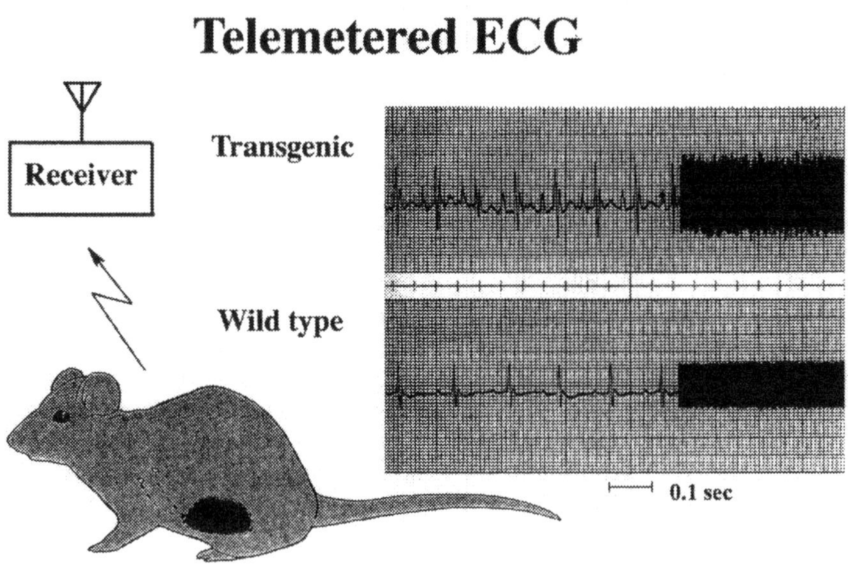

Figure 1. ECG recording using a telemetric system from a free-ranging transgenic mouse (top) and a wild-type control mouse (bottom). The transgenic implant is shown in the abdominal cavity of the mouse. The recorded heart rates were higher in the transgenic mice.

Conscious heart rate is measured best with the use of implanted telemetry devices. Commercially available units are surgically implanted in the peritoneal cavity with paired wire electrodes placed subcutaneously over the thorax. Following recovery, conscious heart rate can be recorded on a multi-channel tape recorder [10,27,28]. Reported heart rates in conscious wild type mice have ranged from 425 to 700 min^{-1} (Table 1). In transgenic mice overexpressing Gsα, conscious heart rates approaching 740 min^{-1} have been reported [10]. The effect of the various methods of measurement on the values for heart rate obtained have been examined to a limited extent. Krege et al. [12] noted a 10-15% greater heart rate as determined by the non-invasive tail cuff technique as compared to the intra-arterial catheter technique. Uechi et al. [10] found slight decreases in heart rate measured using a telemetry system in unrestrained mice as compared to values measured using subcutaneously placed electrodes in lightly restrained mice (transgenic mice overexpressing Gsα, 696±13 from 736±15 min^{-1}; wild-type controls 568±28 from 580±18 min^{-1}). This difference was significant in the transgenic mice overexpressing Gsα but not in wild type controls. These data suggest that light restraint, when the mice are properly trained and acclimatized, has only a modest effect on resting heart rate. Recently, implantable pressure telemetry devices are available. Using this methodology, arterial pressure and heart rate can be monitored for longer times (up to 5 months) in conscious mice without any restraint or infection. In addition, this technique eliminates risk from intravascular coagulation and thromboemboli [29].

RESPIRATORY RATE

Conscious respiratory rate in the mouse can be measured using surgically implanted sonomicrometry crystals to record thoracic dimension changes during the respiratory cycle. The mouse is anesthetized and prepared as for implantation of subcutaneous electrodes. Piezoelectric ultrasonic dimension crystals are placed on either side of the thoracic cage and connected to an ultrasonic transit-time dimension gauge. Crystal position is adjusted to optimize the signal and sutured in place. Following signal calibration the crystal wires are tunneled subcutaneously, externalized and stored. After recovery, the wires are reattached to the gauge and recalibrated using impulses of known duration from a crystal-controlled pulse generator. A sample waveform is shown in Figure 2. The respiratory rates of lightly restrained conscious mice have been reported to range from 160 to 270 min^{-1} [10,30].

MEASUREMENT OF AUTONOMIC FUNCTION

There has been a great deal of interest of late in measurement of autonomic influences on cardiovascular function. Because autonomic tone is markedly altered by anesthesia, it is imperative that these studies be conducted in the fully conscious state. Autonomic function in mice has been assessed by evaluating heart rate and blood pressure responses to autonomic blocking agents, measuring baroreflex-mediated responses of heart rate to changes in blood pressure, and by examining heart rate variability data.

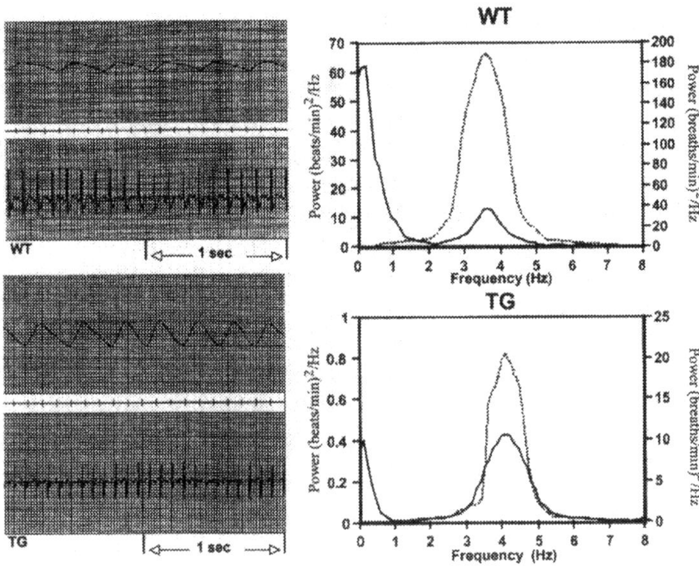

Figure 2. ECG and respiration traces from wild-type (WT) control mice (top left) and transgenic (TG) mice (bottom left). The power spectra for respiratory frequency (dotted line) are compared with heart rate spectra (solid line) in a WT control mouse (top right) and a TG mouse (bottom right). The power spectra of respiration were coupled with the high frequency component of the ECG power spectra.

Responses to Autonomic Blocking Agents

Heart rate and blood pressure responses to blockade of components of the autonomic nervous system can be used to shed light on the relative import of sympathetic and parasympathetic influences on resting and stimulated hemodynamics. A significant fall in resting heart rate following administration of the β-adrenergic antagonist propranolol indicates a relatively high degree of resting sympathetic tone. Conversely, a significant rise in resting heart rate following administration of the muscarinic antagonist atropine indicates a relatively high level of resting parasympathetic tone. Ishii et al. [27] demonstrated relatively greater sympathetic versus parasympathetic influences on conscious resting heart rate in wild type mice. Intrinsic heart rate (heart rate after administration of combined beta- and muscarinic blockade) was approximately 15% lower than intact heart rate. Uechi et al. [10] found that transgenic mice overexpressing Gsα had apparently higher levels of sympathetic and lower levels of parasympathetic tone at baseline as compared to wild type controls. Intrinsic heart rate was not different between groups, in contrast to the markedly greater heart rate found in transgenic mice in the absence of autonomic blockade. Ganglionic blockade with hexamethonium resulted in a similar equalization of heart rate between the groups, underscoring the importance of autonomic influences in the expression of the heart rate phenotype.

Baroreflex-Mediated Responses

Baroreflex function has been evaluated by measuring simultaneous arterial pressure and heart rate responses to acute pharmacologic manipulation of the arterial pressure. Arterial pressure is increased by infusing the α-adrenoreceptor agonist phenylephrine and decreased by infusing the direct vasodilator nitroprusside. Plotting the heart rate versus the mean arterial pressure generally yields a sigmoidal curve (Figure 3). Merrill et al. [22] demonstrated a rightward shift in this curve in mice expressing both the human renin and angiotensinogen genes as compared to wild type or spontaneously hypertensive controls. This was thought to result from resetting of the baroreflex curve to a higher pressure level without changing the sensitivity or gain of the reflex. Uechi et al. [10] demonstrated a flattened sigmoidal response curve in mice overexpressing Gsα as compared to wild type controls, indicating both a resetting of the reflex and depression of sensitivity and/or gain. Thus, straightforward plots of pressure and heart rate during vasoactive drug infusion can yield significant insights into the effects of specific genetic manipulations on important baroreceptor-mediated responses.

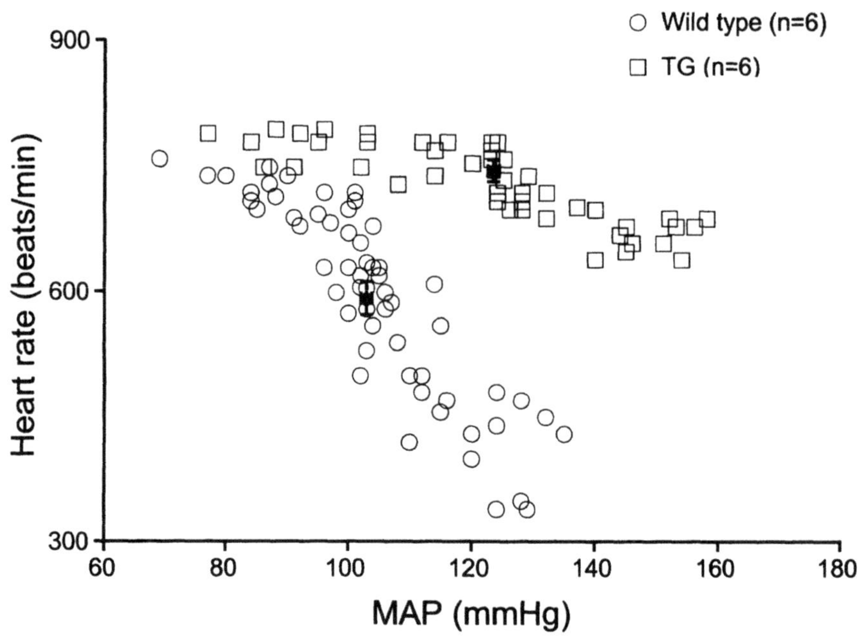

Figure 3. The relationship between pharmacologically induced hypertension and hypotension and reflex changes in heart rate in conscious transgenic (TG) mice and wild-type controls. MAP indicates mean arterial pressure. Note the depressed arterial baroreflex regulation of heart rate both in response to acute hypertension and acute hypotension. Baseline values are noted by the solid symbols with standard errors. Note the higher baseline heart rate and arterial pressure in the TG mice.

Heart Rate Variability
Measures of heart rate variability can be made using data generated from an implanted telemetry transmitter or from implanted electrodes. The transmitter is implanted in the peritoneum; electrodes and thoracic crystals are placed subcutaneously as described above. Mice are kept in thermostatically controlled cages and exposed to 12 hour light-dark cycles to maintain normal circadian rhythm. ECG and respiratory signals are recorded and filtered, and tachograms constructed. Autoregressive spectral analysis is used to analyze power spectra and define low and high frequency bands. These bands (and their ratio) provide important insights into the relative degree of sympathetic and parasympathetic influences on the heart.

Relatively few studies of heart rate variability in mice have been reported [10,27,28]. Mansier et al. [28] assessed heart rate variability in mice overexpressing atrial β_1-adrenergic receptors and their wild type littermates. They noted decreased heart rate variability in the transgenic mice, marked by decreases in the low and high frequency spectra and in their ratio. Uechi et al. [10] found similar decreases in heart rate variability in transgenic mice overexpressing Gsα also marked by decreases in both the low frequency and high frequency spectra and their ratio (Figure 2). While these observations suggest relatively greater parasympathetic influences on heart rate variability in these transgenic strains, an effect of altered baroreflex function or respiratory influences cannot be excluded. More likely, amplification of sympathetic input to the heart, either at or distal to the β-receptor, may produce relative saturation and apparently diminished sympathetic contribution to variability of heart rate. This possibility is supported by the observation that administration of the β-agonist dobutamine to wild type mice resulted in heart rates and power spectra indistinguishable from mice with overexpression of Gsα.

Each of the above studies found the low frequency peak in wild type mice to occur at 0.1-1.75 Hz and the high frequency peak to occur at 1.75-5.0 Hz. The ratio of low to high frequency in these studies has ranged from 2-7, a range potentially contributed to by strain differences. Given the apparent complexity of the cardiovascular effects of many genetic manipulations, use of these techniques for the elucidation of alterations in autonomic function is likely to add importantly to our assessment of the cardiovascular phenotype of transgenic murine models.

CARDIAC OUTPUT

Cardiac output can be measured in mice using a number of techniques. Measurements in anesthetized mice using transthoracic echocardiographically determined aortic dimensions and Doppler-derived aortic velocities have been reported, but this technique is not feasible in the conscious state [31]. The radioisotope-dilution technique was first employed in anesthetized mice several decades ago [32]. Barbee et al. [33] more recently reported measurements of cardiac output in conscious mice using radioactive microspheres. This technique, well established in large animal studies [34,35] and previously employed in rats [36] and anesthetized mice [37,38], involves central administration of a known quantity of radiolabeled microspheres with simultaneous withdrawal of a measured quantity of peripheral arterial blood. Radioactivity is then measured using a gamma counter, and cardiac output is calculated by an established formula. Prior work had demonstrated a high rate of inadequate mixing of microspheres [33,38]. This was largely overcome in Barbee's study by the use of specially constructed catheters with tip diameters of <400μm, allowing catheter placement into the LV without significant outflow tract obstruction or aortic insufficiency in the majority of cases. Using this technique, the authors reported a mean cardiac output of 16±1 ml/min (cardiac index 591±49 ml/min/kg) in conscious mice studied 4 hours after recovery from Avertin anesthesia and surgical placement of LV (via the right carotid artery) and femoral artery catheters. Stroke volume was calculated to be 26±2 μl, and post-mortem tissue sampling and analysis were performed for the calculation of regional cardiac output distribution, which largely mirrored prior reports in rats. The authors pointed out that the technique could be repeated acutely at least once without any significant impact on peripheral resistance. Concerns remain, however, regarding hemodynamic effects of blood withdrawal (Barbee reported a 10% fall in MAP during withdrawal) and potential vessel occlusion by the microspheres [39].

Recently, Vogel [40] reported measurement of cardiac output in anesthetized and conscious mice using a measure of blood conductivity during intravenous bolus injection of 20 μl of a 5% glucose solution. This technique involves surgical placement of a femoral arterial platinum electrode and femoral venous catheter as well as placement of a rectal reference electrode. Blood conductivity following the bolus of glucose is converted into cardiac output using an established formula. The author included validation studies in rats where the cardiac output determinations by blood conductivity correlated well with those determined by an indicator dilution technique.

The blood conductivity technique yielded a mean cardiac index of 592±78 ml/min/kg in the conscious state (one hour recovery from halothane/N_2O anesthesia). The author also reported decreases in cardiac index with hemorrhage-induced hypovolemia, but no change with saline-induced hypervolemia. While this method has the significant benefit of allowing virtually any number of determinations in a single animal during the course of an experiment, its utility for measurements in fully conscious unrestrained animals remains to be assessed.

Also, cardiac output measured in conscious mice using an echocardiographic method was reported [41]. This technique requires holding the mice directly by hand. Under these conditions, it is difficult to avoid agitation and altered sympathetic and parasympathetic tone. An ultrasonic flow probe was devised to measure the blood flow and cardiac output instantaneously [42]. However, this requires invasive implantation and a tether system.

RESPIRATORY GAS ANALYSIS

Determination of oxygen consumption (VO_2), carbon dioxide production (VCO_2) and the respiratory exchange ratio in conscious mice can be made using commercially available metabolic chambers. These chambers employ an open-circuit volumetric method of gas analysis. When coupled with a motorized treadmill, these chambers allow for the measurement of gas exchange at rest and during graded exercise. Desai et al. [43] reported conscious resting and exercise hemodynamics and gas exchange data in a number of strains of mice. Conscious resting heart rate and blood pressure values were comparable to previously reported values. VO_2 and VCO_2 at rest ranged from 48-58 ml O_2 /kg/min and 36-41 ml CO_2 /kg/min, respectively. During incremental treadmill exercise testing, VO_2 and VCO_2 increased until the mice reached exhaustion, and ranged at peak exercise from 69-106 ml O_2 /kg/min and 58-87 ml CO_2 /kg/min, respectively. These values are several fold what is seen in humans, reflecting the higher metabolic rates of small rodents. There were linear relationships between workload (expressed as a combination of treadmill speed and degree of incline) and both VO_2 and VCO_2. Given the important role of peak VO_2 in determining prognosis in human heart failure, this technique in mice should add importantly to the cardiovascular physiologic assessment of transgenic mice with phenotypic features of cardiomyopathy.

LEFT VENTRICULAR FUNCTION

Most techniques for the measurement of LV function in fully conscious mice still have limitations. Several investigators have reported ventricular function data from murine isolated working heart preparations [44,45]. Others have reported echocardiographic measures of ejection phase indices [46-49] and high fidelity pressure transducer-derived measures of isovolumic indices of LV contractility [49-53] in anesthetized mice. There are echocardiographic studies utilizing avertin anesthesia [54,55], which suggests less adverse effects of the anesthetic on sympathetic function. Recently, several studies have reported echocardiographic assessment of LV function in conscious mice. In most cases, heart rates are higher than normal, potentially due to excitement [41]. With training, however, it is possible to obtain echo measurements in conscious mice at a relatively normal heart rate. (Figure 4)

LV dP/dt measurements have been made in both open [50-52] and closed chested preparations [49,52,53]. Recently, relatively load-insensitive measures of LV function have been reported in closed chested mice [56,57]. Selected reported values for isovolumic measures of LV contractility are summarized in Table 2, highlighting the high degree of variability between studies. While these measures so far have not been reported in conscious mice, recently LV dP/dt values in mice recovering from anesthesia [58,59] and in mice anesthetized with a chloralose/urethane [60] were found to be nearly four- and three-fold greater, respectively, than those obtained in mice fully anesthetized with more commonly used agents. This suggests that the reported values for some measures of ventricular contractility in anesthetized mice may underestimate those in conscious mice by several fold, calling into question the utility of the reported ventricular responses to inotropic stimuli in these studies.

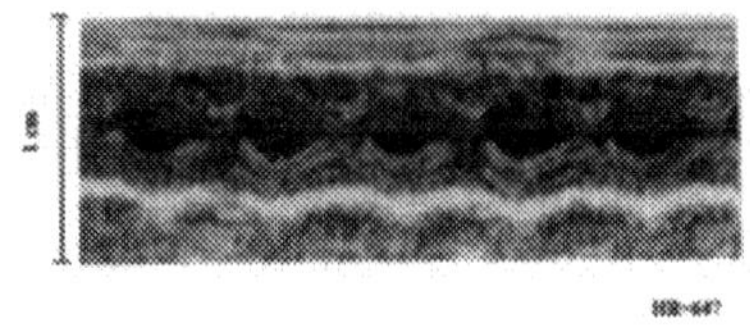

Figure 4. Representative M-mode echocardiogram from a conscious mouse studied in our laboratory. Note that with sufficient training heart rate can be relatively normal for a mouse (647 beats/min).

Effects of Anesthesia

A variety of techniques have been developed to accurately measure cardiac structure and function in anesthetized mice. It has long been recognized from large animal studies, however, that anesthesia has profound effects on cardiovascular functional parameters. While relatively little has been published addressing this issue in mice, there are a number of observations in the literature suggesting that anesthetic effects are at least as important in studies of murine cardiovascular physiology as they are in large animals.

It is the rule that measurements of arterial blood pressure made in anesthetized mice yield lower values compared to those performed in fully conscious mice. Little data exist, however, comparing blood pressure and heart rate in the same mice in the fully conscious state and following anesthesia. Lembo et al. [23] described significantly lower blood pressures in open-chest mice anesthetized with ketamine and xylazine versus conscious mice. They found an approximately 30% decline in mean arterial pressure following anesthesia, thoracotomy and carotid catheter placement versus the blood pressure measured via the femoral artery in the fully conscious state. The contribution of thoracotomy and acute carotid arterial cannulation to the fall in blood pressure is unclear, however, and the degree and time course of the onset of anesthesia-related hypotension in general is not well established. We have noted marked relative hypotension and bradycardia (approximately 30 and 50% declines from baseline mean arterial pressure and heart rate, respectively) within the 5 minutes following intraperitoneal administration of a similar anesthetic regimen (Figure 5). These parameters only began to return toward baseline by 45 minutes following the administration of the anesthetic. It is difficult to compare values from different studies owing to a host of potentially confounding factors, including differences in the strains of mice studied, anesthetic regimens, and the time course of study. Nevertheless it should be noted that studies in which measurements were made within 4-6 hours of anesthesia tend to report lower arterial pressures as compared to most studies performed after longer periods of recovery from anesthesia (Table 1). The relationship of heart rate to recovery time is less clear, with short duration of recovery appearing to yield either lower or higher than expected heart rates. The exact duration of recovery from anesthesia needed to insure a return to baseline hemodynamics is uncertain, but unpublished observations in our laboratory suggest that a minimum of 48 hours following ketamine/xylazine anesthesia is optimal.

Anesthesia is known to have profound effects on autonomic function and consequently would be expected to affect measures of heart rate variability. Uechi et al. [10] have confirmed this in mice, finding marked reductions in heart rate and both the low and high frequency spectra in anesthetized versus fully conscious mice (Figure 5). This was observed in both wild type mice and mice overexpressing Gsα, in whom these parameters were already depressed in the conscious state as compared to controls.

Barbee et al. [33] examined the effect of anesthesia on cardiac output and distribution of blood flow in mice. While heart rate and blood pressure, as expected, fell significantly following anesthesia, there was an increase in stroke volume that preserved cardiac output. Changes in blood flow distribution with anesthesia included shunting of flow away from skeletal muscle and into the splanchnic bed. Vogel [40], in contrast, found a modest (~10%) but significant fall in cardiac output in anesthetized mice compared to mice allowed to recover from anesthesia for one hour.

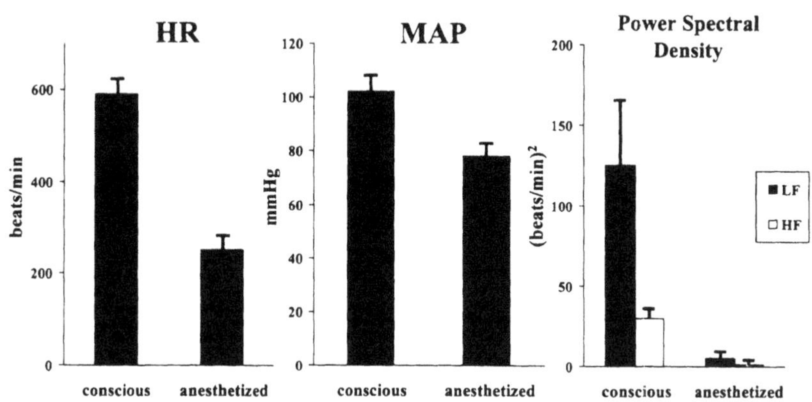

Figure 5. Effects of anesthesia on heart rate (HR), mean arterial pressure (MAP), and power spectral density in wild-type control mice. Anesthesia markedly reduced all three parameters when compared to conscious mice. LV indicates low frequency, HF, high frequency.

Desai et al. [43] compared VO_2 and VCO_2 in mice recovering from inhaled methoxyflurane anesthesia to mice in the fully conscious state. These parameters, along with heart rate and blood pressure, tended to be depressed shortly after cessation of anesthesia. The authors concluded that at least 12 hours were required for complete recovery from this anesthetic regimen.

It is evident from the above that cardiovascular function deviates significantly from baseline not only when the animal is fully anesthetized but also, importantly, during recovery from anesthesia. It is also evident that the exact recovery period required for a return to normal conscious hemodynamics is not well understood, and is likely to depend on a number of factors, including murine strain, anesthetic agent or agents, duration of anesthesia and procedure performed during anesthesia. Given the burgeoning importance of transgenic murine models in cardiovascular research, these uncertainties underscore the need for continued advances in the acquisition and analysis of conscious hemodynamics in mice.

FUTURE DIRECTIONS

Significant advances have permitted the acquisition of detailed cardiovascular physiologic data in conscious mice. There remain, however, several parameters for which we lack the ability to make consistent measures in the fully conscious state. We thus far do not have a reliable way to measure ventricular pressures or dimensions in conscious mice. Consequently studies of ventricular function in mice continue to be weakened by the confounding influences of anesthesia or recovery from anesthesia. It is likely that further refinement in equipment and technique will allow for surgical placement of intraventricular catheters (either solid state of fluid-filled) for the measurement of ventricular pressures in conscious mice. Sonomicrometry crystals can be placed epicardially and may allow for measurement of cardiac dimensions in the conscious state. These would, however, be limited by their location at the epicardial rather than the endocardial surface. It is conceivable that a recently described conductance catheter used for the simultaneous measurement of left ventricular pressure and volume in mice [60] could be used in conscious animals if it could be surgically implanted in the LV or reliably advanced via an indwelling arterial catheter. This device would have the important added advantage of providing relatively load-independent measures of ventricular function. These invasive techniques seem to hold more promise for measurement of LV function in the conscious state than do the non-invasive techniques that have been successfully employed in anesthetized preparations.

Less readily foreseeable are techniques for analyzing vascular function in conscious mice. Given recent descriptions of murine models of myocardial infarction [61] and myocardial ischemia and reperfusion [62], it would be of great utility to develop techniques to assess coronary flow in the mouse, and one would hope eventually to perform this in the conscious state. Other potentially important techniques, such as electrophysiologic studies [63] and pulmonary arterial catheterization [64], have been reported in anesthetized mice and conscious rats, respectively, and may eventually be feasible in conscious murine preparations. Continued expansion of the methods available for measuring physiologic parameters in conscious mice will allow for the full realization of the insights into cardiovascular function promised by transgenic techniques.

REFERENCES

1. Barlow G, H. KD. Hemodynamic alterations after 30 minutes of pentobarbital sodium anesthesia in dogs. *Am J Physiol.* 1964;207:754-761.
2. Vatner SF, Smith NT. Effects of halothane on left ventricular function and distribution of regional blood flow in dogs and primates. *Circ Res.* 1974;34:155-67.
3. Manders WT, Vatner SF. Effects of sodium pentobarbital anesthesia on left ventricular function and distribution of cardiac output in dogs, with particular reference to the mechanism for tachycardia. *Circ Res.* 1976;39:512-7.
4. Lang RM, Marcus RH, Neumann A, Janzen D, Hansen D, Fujii AM, Borow KM. A time-course study of the effects of pentobarbital, fentanyl, and morphine chloralose on myocardial mechanics. *J Appl Physiol.* 1992;73:143-50.
5. Vatner SF, McRitchie RJ, Braunwald E. Effects of dobutamine on left ventricular performance, coronary dynamics, and distribution of cardiac output in conscious dogs. *J Clin Invest.* 1974;53:1265-73.
6. Pagani M, Mirsky I, Baig H, Manders WT, Kerkhof P, Vatner SF. Effects of age on aortic pressure-diameter and elastic stiffness-stress relationships in unanesthetized sheep. *Circ Res.* 1979;44:420-9.
7. Zimpfer M, Vatner SF. Effects of acute increases in left ventricular preload on indices of myocardial function in conscious, unrestrained and intact, tranquilized baboons. *J Clin Invest.* 1981;67:430-8.
8. Young MA, Knight DR, Vatner SF. Parasympathetic coronary vasoconstriction induced by nicotine in conscious calves. *Circ Res.* 1988;62:891-5.
9. Morita H, Nishida Y, Motochigawa H, Uemura N, Hosomi H, Vatner SF. Opiate receptor-mediated decrease in renal nerve activity during hypotensive hemorrhage in conscious rabbits. *Circ Res.* 1988;63:165-72.
10. Uechi M, Asai K, Osaka M, Smith A, Sato N, Wagner TE, Ishikawa Y, Hayakawa H, Vatner DE, Shannon RP, Homcy CJ, Vatner SF. Depressed heart rate variability and arterial baroreflex in conscious transgenic mice with overexpression of cardiac Gsalpha. *Circ Res.* 1998;82:416-23.
11. Byrom FB, Wilson C. A plethysmographic method for measuring systomic blood pressure in the intact rat. *J Physiol.* 1938;93:301-304.
12. Krege JH, Hodgin JB, Hagaman JR, Smithies O. A noninvasive computerized tail-cuff system for measuring blood pressure in mice. *Hypertension.* 1995;25:1111-5.
13. Ito M, Oliverio MI, Mannon PJ, Best CF, Maeda N, Smithies O, Coffman TM. Regulation of blood pressure by the type 1A angiotensin II receptor gene. *Proc Natl Acad Sci U S A.* 1995;92:3521-5.
14. Johns C, Gavras I, Handy DE, Salomao A, Gavras H. Models of experimental hypertension in mice. *Hypertension.* 1996;28:1064-9.

15. Madeddu P, Varoni MV, Palomba D, Emanueli C, Demontis MP, Glorioso N, Dessi-Fulgheri P, Sarzani R, Anania V. Cardiovascular phenotype of a mouse strain with disruption of bradykinin B2-receptor gene. *Circulation*. 1997;96:3570-8.

16. Link RE, Desai K, Hein L, Stevens ME, Chruscinski A, Bernstein D, Barsh GS, Kobilka BK. Cardiovascular regulation in mice lacking alpha2-adrenergic receptor subtypes b and c. *Science*. 1996;273:803-5.

17. Hein L, Barsh GS, Pratt RE, Dzau VJ, Kobilka BK. Behavioural and cardiovascular effects of disrupting the angiotensin II type-2 receptor in mice. *Nature*. 1995;377:744-7.

18. Tian B, Meng QC, Chen YF, Krege JH, Smithies O, Oparil S. Blood pressures and cardiovascular homeostasis in mice having reduced or absent angiotensin-converting enzyme gene function. *Hypertension*. 1997;30:128-33.

19. Zhu QM, Lesnick JD, Jasper JR, MacLennan SJ, Dillon MP, Eglen RM, Blue DR, Jr. Cardiovascular effects of rilmenidine, moxonidine and clonidine in conscious wild-type and D79N alpha2A-adrenoceptor transgenic mice. *Br J Pharmacol*. 1999;126:1522-30.

20. Stauss HM, Godecke A, Mrowka R, Schrader J, Persson PB. Enhanced blood pressure variability in eNOS knockout mice. *Hypertension*. 1999;33:1359-63.

21. Davisson RL, Yang G, Beltz TG, Cassell MD, Johnson AK, Sigmund CD. The brain renin-angiotensin system contributes to the hypertension in mice containing both the human renin and human angiotensinogen transgenes. *Circ Res*. 1998;83:1047-58.

22. Merrill DC, Thompson MW, Carney CL, Granwehr BP, Schlager G, Robillard JE, Sigmund CD. Chronic hypertension and altered baroreflex responses in transgenic mice containing the human renin and human angiotensinogen genes. *J Clin Invest*. 1996;97:1047-55.

23. Lembo G, Rockman HA, Hunter JJ, Steinmetz H, Koch WJ, Ma L, Prinz MP, Ross J, Jr., Chien KR, Powell-Braxton L. Elevated blood pressure and enhanced myocardial contractility in mice with severe IGF-1 deficiency. *J Clin Invest*. 1996;98:2648-55.

24. Barbee RW, Perry BD, Re RN, Murgo JP, Field LJ. Hemodynamics in transgenic mice with overexpression of atrial natriuretic factor. *Circ Res*. 1994;74:747-51.

25. Just A, Faulhaber J, Ehmke H. Autonomic cardiovascular control in conscious mice. *Am J Physiol Regul Integr Comp Physiol*. 2000;279:R2214-21.

26. Mattson DL, Krauski KR. Chronic sodium balance and blood pressure response to captopril in conscious mice. *Hypertension*. 1998;32:923-8.

27. Ishii K, Kuwahara M, Tsubone H, Sugano S. Autonomic nervous function in mice and voles (Microtus arvalis): investigation by power spectral analysis of heart rate variability. *Lab Anim*. 1996;30:359-64.

28. Mansier P, Medigue C, Charlotte N, Vermeiren C, Coraboeuf E, Deroubai E, Ratner E, Chevalier B, Clairambault J, Carre F, Dahkli T, Bertin B, Briand P, Strosberg D, Swynghedauw B. Decreased heart

rate variability in transgenic mice overexpressing atrial beta 1-adrenoceptors. *Am J Physiol.* 1996;271:H1465-72.

29. Mills PA, Huetteman DA, Brockway BP, Zwiers LM, Gelsema AJ, Schwartz RS, Kramer K. A new method for measurement of blood pressure, heart rate, and activity in the mouse by radiotelemetry. *J Appl Physiol.* 2000;88:1537-44.
30. Onodera M, Kuwaki T, Kumada M, Masuda Y. Determination of ventilatory volume in mice by whole body plethysmography. *Jpn J Physiol.* 1997;47:317-26.
31. Hartley CJ, Michael LH, Entman ML. Noninvasive measurement of ascending aortic blood velocity in mice. *Am J Physiol.* 1995;268:H499-505.
32. Broulik PD, Kochakian CD, Dubovsky J. Influence of castration and testosterone propionate on cardiac output, renal blood flow, and blood volume in mice. *Proc Soc Exp Biol Med.* 1973;144:671-3.
33. Barbee RW, Perry BD, Re RN, Murgo JP. Microsphere and dilution techniques for the determination of blood flows and volumes in conscious mice. *Am J Physiol.* 1992;263:R728-33.
34. Domenech RJ, Hoffman JI, Noble MI, Saunders KB, Henson JR, Subijanto S. Total and regional coronary blood flow measured by radioactive microspheres in conscious and anesthetized dogs. *Circ Res.* 1969;25:581-96.
35. Hoffbrand BI, Forsyth RP. Validity studies of the radioactive microsphere method for the study of the distribution of cardiac output, orn blood flow, and resistance in the conscious rhesus monkey. *Cardiovasc Res.* 1969;3:426-32.
36. Ishise S, Pegram BL, Yamamoto J, Kitamura Y, Frohlich ED. Reference sample microsphere method: cardiac output and blood flows in conscious rat. *Am J Physiol.* 1980;239:H443-H449.
37. Quintana A, Raczka E, Bonaccorsi A. Cardiac output distribution measured with radioactive microspheres in the mouse. *Pharmacol Res Commun.* 1979;11:245-52.
38. Sarin SK, Sabba C, Groszmann RJ. Splanchnic and systemic hemodynamics in mice using a radioactive microsphere technique. *Am J Physiol.* 1990;258:G365-9.
39. Kobayashi N, Kobayashi K, Kouno K, Horinaka S, Yagi S. Effects of intra-atrial injection of colored microspheres on systemic hemodynamics and regional blood flow in rats. *Am J Physiol.* 1994;266:H1910-7.
40. Vogel J. Measurement of cardiac output in small laboratory animals using recordings of blood conductivity. *Am J Physiol.* 1997;273:H2520-7.
41. Yang XP, Liu YH, Rhaleb NE, Kurihara N, Kim HE, Carretero OA. Echocardiographic assessment of cardiac function in conscious and anesthetized mice. *Am J Physiol.* 1999;277:H1967-74.
42. Krivitski NM, Starostin D, Smith TL. Extracorporeal recording of mouse hemodynamic parameters by ultrasound velocity dilution. *Asaio J.* 1999;45:32-6.

43. Desai KH, Sato R, Schauble E, Barsh GS, Kobilka BK, Bernstein D. Cardiovascular indexes in the mouse at rest and with exercise: new tools to study models of cardiac disease. *Am J Physiol.* 1997;272:H1053-61.
44. Grupp IL, Subramaniam A, Hewett TE, Robbins J, Grupp G. Comparison of normal, hypodynamic, and hyperdynamic mouse hearts using isolated work-performing heart preparations. *Am J Physiol.* 1993;265:H1401-10.
45. Brooks WW, Apstein CS. Effect of treppe on isovolumic function in the isolated blood-perfused mouse heart. *J Mol Cell Cardiol.* 1996;28:1817-22.
46. Gardin JM, Siri FM, Kitsis RN, Edwards JG, Leinwand LA. Echocardiographic assessment of left ventricular mass and systolic function in mice. *Circ Res.* 1995;76:907-14.
47. Hoit BD, Khoury SF, Kranias EG, Ball N, Walsh RA. In vivo echocardiographic detection of enhanced left ventricular function in gene-targeted mice with phospholamban deficiency. *Circ Res.* 1995;77:632-7.
48. Tanaka N, Dalton N, Mao L, Rockman HA, Peterson KL, Gottshall KR, Hunter JJ, Chien KR, Ross J, Jr. Transthoracic echocardiography in models of cardiac disease in the mouse. *Circulation.* 1996;94:1109-17.
49. Iwase M, Bishop SP, Uechi M, Vatner DE, Shannon RP, Kudej RK, Wight DC, Wagner TE, Ishikawa Y, Homcy CJ, Vatner SF. Adverse effects of chronic endogenous sympathetic drive induced by cardiac GS alpha overexpression. *Circ Res.* 1996;78:517-24.
50. Milano CA, Allen LF, Rockman HA, Dolber PC, McMinn TR, Chien KR, Johnson TD, Bond RA, Lefkowitz RJ. Enhanced myocardial function in transgenic mice overexpressing the beta 2-adrenergic receptor. *Science.* 1994;264:582-6.
51. Hunter JJ, Tanaka N, Rockman HA, Ross J, Jr., Chien KR. Ventricular expression of a MLC-2v-ras fusion gene induces cardiac hypertrophy and selective diastolic dysfunction in transgenic mice. *J Biol Chem.* 1995;270:23173-8.
52. Hoit BD, Ball N, Walsh RA. Invasive hemodynamics and force-frequency relationships in open- versus closed-chest mice. *Am J Physiol.* 1997;273:H2528-33.
53. Lorenz JN, Kranias EG. Regulatory effects of phospholamban on cardiac function in intact mice. *Am J Physiol.* 1997;273:H2826-31.
54. Huang GY, Linask KK. Doppler echocardiographic analysis of effects of tribromoethanol anesthesia on cardiac function in the mouse embryo: a comparison with pentobarbital. *Lab Anim Sci.* 1998;48:206-9.
55. Kiatchoosakun S, Kirkpatrick D, Hoit BD. Effects of tribromoethanol anesthesia on echocardiographic assessment of left ventricular function in mice. *Comparative Medicine.* 2001;51:27-30.
56. Hoit BD, Khan ZU, Pawloski-Dahm CM, Walsh RA. In vivo determination of left ventricular wall stress-shortening relationship in normal mice. *Am J Physiol.* 1997;272:H1047-52.

57. Fentzke RC, Korcarz CE, Shroff SG, Lin H, Sandelski J, Leiden JM, Lang RM. Evaluation of ventricular and arterial hemodynamics in anesthetized closed-chest mice. *J Am Soc Echocardiogr.* 1997;10:915-25.
58. Palakodeti V, Oh S, Oh BH, Mao L, Hongo M, Peterson KL, Ross J, Jr. Force-frequency effect is a powerful determinant of myocardial contractility in the mouse. *Am J Physiol.* 1997;273:H1283-90.
59. Rockman HA, Choi DJ, Rahman NU, Akhter SA, Lefkowitz RJ, Koch WJ. Receptor-specific in vivo desensitization by the G protein-coupled receptor kinase-5 in transgenic mice. *Proc Natl Acad Sci U S A.* 1996;93:9954-9.
60. Georgakopoulos D, Mitzner WA, Chen CH, Byrne BJ, Millar HD, Hare JM, Kass DA. In vivo murine left ventricular pressure-volume relations by miniaturized conductance micromanometry. *Am J Physiol.* 1998;274:H1416-22.
61. Manning WJ, Wei JY, Katz SE, Douglas PS, Gwathmey JK. Echocardiographically detected myocardial infarction in the mouse. *Lab Anim Sci.* 1993;43:583-5.
62. Michael LH, Entman ML, Hartley CJ, Youker KA, Zhu J, Hall SR, Hawkins HK, Berens K, Ballantyne CM. Myocardial ischemia and reperfusion: a murine model. *Am J Physiol.* 1995;269:H2147-54.
63. Berul CI, Aronovitz MJ, Wang PJ, Mendelsohn ME. In vivo cardiac electrophysiology studies in the mouse. *Circulation.* 1996;94:2641-8.
64. Hess P, Clozel M, Clozel JP. Telemetry monitoring of pulmonary arterial pressure in freely moving rats. *J Appl Physiol.* 1996;81:1027-32.

18
EXERCISE AND OXYGEN CONSUMPTION IN THE MOUSE

Kavin H. Desai, M.D.
Daniel Bernstein, M.D.

Stanford University, Stanford, CA 94305 USA

INTRODUCTION

The mouse remains the most commonly used species for gene overexpression studies and continues to be the only species in which targeted genetic alterations are possible. The thorough phenotypic evaluation of genetically manipulated murine models of human cardiovascular disease requires the ability to study all components of murine cardiovascular physiology. The recent development of techniques allowing the evaluation of awake, unrestrained mice has advanced the study of cardiovascular dynamics during stressed states such as exercise. These studies provide valuable information about how individual organ systems respond under maximally stimulated conditions, with the possibility of uncovering subtle phenotypes not observed in at rest.

CARDIOVASCULAR PHYSIOLOGY DURING EXERCISE

Exercise provides one of the most severe physiologic stresses to the cardiovascular system. Although this stress results in coordinated responses of a multitude of organ systems, the cardiovascular system is most directly affected and is primarily responsible for the acute changes that allow an organism to perform exercise.

Determinants of Oxygen Consumption

The ultimate goal of cardiovascular adaptation to acute exercise is to deliver an adequate supply of oxygen and metabolic substrate to, and efficiently remove carbon dioxide and waste products from, working muscle. The success of the cardiovascular system in achieving this goal is reflected in the balance between oxygen delivery and oxygen demand. Oxygen consumption occurs at the tissue level (QO_2) but is typically measured through ventilation (VO_2). At equilibrium, these two values are identical. The *Fick principle* describes the relationships between tissue metabolism and cardiovascular performance:

$$VO_2 = CO \times (CaO_2\text{-}CvO_2) = HR \times SV \times (CaO_2\text{-}CvO_2)^{1}$$

---

[1] VO_2 = oxygen consumption; CO = cardiac output; CaO_2 = arterial oxygen content; CvO_2 = mixed venous oxygen content; HR = heart rate; SV = stroke volume.

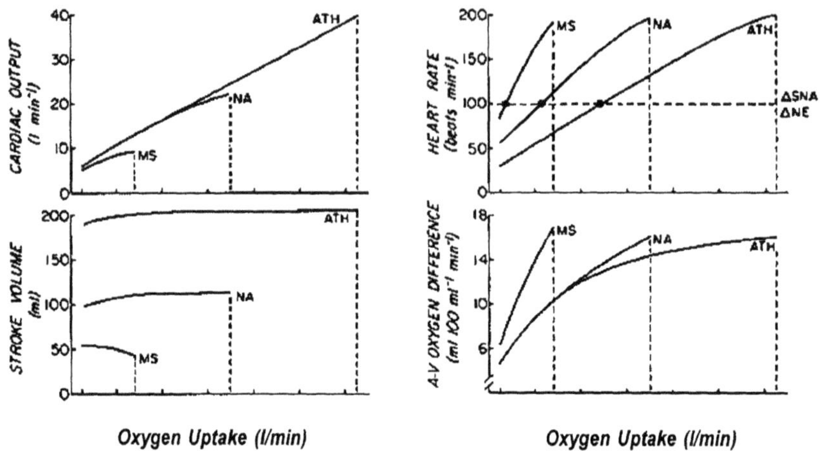

Figure 1. Graphs depicting the effects of conditioning on determinants of exercise capacity: cardiac output, stroke volume, heart rate, and arteriovenous oxygen difference. MS = patient with cardiac disease; NA = sedentary individual; ATH = athlete. Adapted from Rowell (1).

As exercise intensity increases, muscle energy usage and thereby tissue oxygen needs increases. To meet the increased oxygen demand, the cardiovascular system plays the primary role in maintaining oxygen delivery by increasing cardiac output (CO) and by increasing the extraction of oxygen from the blood, as quantified by the arterial-venous oxygen difference (CaO_2-CvO_2).

Ventricular Stroke Volume (SV). The most immediate cardiovascular response to exercise is an increase in cardiac output, the product of left ventricular stroke volume and heart rate. In humans, increases in stroke volume are responsible for minor increases in cardiac output primarily during low levels of exercise (see Figure 1). Stroke volume is regulated by ventricular filling (preload), the contractile state of the myocardium, and arterial vascular resistance (afterload).

Ventricular filling (preload). Ventricular filling is increased immediately at the onset of exercise by a combination of venoconstriction, caused by a sudden increase in sympathetic tone, and by the pumping action of contracting muscles (1-3). Since the pulmonary circulation normally has a relatively low resistance, increased venous return to the right ventricle translates to increased left ventricular filling. Echocardiographic studies have also suggested that enhanced ventricular relaxation (increased compliance) plays a role in improving stroke volume during exercise (2, 4, 5).

Ventricular contractility. Intrinsic ventricular contractility is thought to increase with exercise, although it has been difficult to measure directly. Using radionuclide angiography, intrinsic ventricular contractility was shown to increase independent of preload and afterload during exercise (1, 6) due to sympathetic stimulation. Extrinsic factors affecting contractility include the length-tension (Frank-Starling) relationship and the force-frequency relationship. The former is clinically evidenced as an increase in end-diastolic ventricular volume. The latter dictates that increasing the frequency of cardiac contractions improves myocardial contractility during exercise. There is a point beyond which, however, increasing heart rate will result in a decrease in contractility due to a reduced diastolic filling interval (7).

Ventricular afterload. Ventricular afterload is almost entirely dictated by systemic vascular resistance which decreases markedly during exercise. This is primarily due to a major redistribution of blood flow to skin and working muscle (8). Vasoconstriction occurs at the arteriolar level due to the release of circulating hormones such as epinephrine, vasopressin, and angiotensin (9). This is balanced by the more powerful local effects of vasodilators such as potassium, lactate, bradykinins, hypoxia, and hypercarbia in metabolically active areas such as skin and working muscle (10). The collective effect is a redistribution of blood flow away from metabolically less active areas such as the kidney, gastrointestinal tract, and liver and an increase in blood flow to working muscle by more than 10-fold(11).

It has been well established that during exercise systolic blood pressure increases significantly and diastolic blood pressure decreases mildly, resulting in a slight increase in mean blood pressure (12). This suggests that the large increase in cardiac output must be closely matched by a decrease in overall systemic vascular resistance during exercise in order to maintain a relatively stable mean blood pressure (ventricular-vascular coupling). Ventricular afterload is therefore finely adjusted to optimize cardiac output while still maintaining adequate perfusion pressure.

Heart Rate (HR). The mammalian heart rate is significantly influenced by the balance of parasympathetic and sympathetic autonomic output (Figure 2). Since exercise induces sympathetic and inhibits parasympathetic activity, the result is an overall stimulatory effect on heart rate. In humans performing low intensity exercise, the major cause of tachycardia (up to heart rates of about 100 bpm) is withdrawal of parasympathetic vagal tone. With increasing intensities of exercise a more gradual but proportional increase in heart rate ensues due to increasing sympathetic stimulation (13).

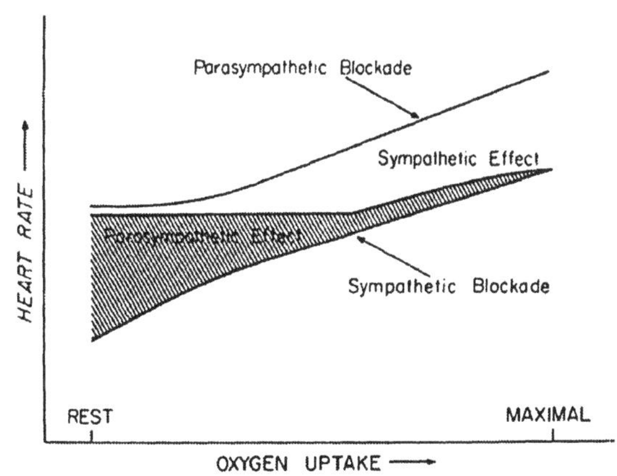

Figure 2. Representation of parasympathetic vs. sympathetic contribution to the heart rate response during exercise in humans. Adapted from Rowell (1).

Arterial-Venous Oxygen Difference (CaO_2-CvO_2). Increasing tissue oxygen extraction (reflected by the *arterial-venous oxygen difference*) is another means by which tissue oxygenation can be maintained during exercise. Arterial oxygen content is near maximal in normal individuals at rest and changes very little with increased exercise intensity. The mixed venous oxygen content, on the other hand, is normally high at rest, representing only a 25-30% extraction of available oxygen by the tissues. With high intensities of exercise, the mixed venous oxygen content can decrease significantly, representing an 80-85% oxygen extraction at maximum (Figure 1). This is primarily accomplished by the redistribution of blood flow to working muscle, where the QO_2 is high. The increase in blood flow to active tissues must be accommodated by opening previously constricted capillary beds so that near normal capillary transit times are maintained, allowing adequate time for extraction of oxygen at the cellular level. To further improve local oxygen extraction, the hemoglobin-oxygen dissociation curve is typically shifted rightward due to the influence of local factors such as acidemia, hypoxemia, and hyperthermia, allowing oxygen to be unloaded more easily from hemoglobin. These mechanisms optimize the balance between blood flow and oxygen extraction for a given tissue resulting in an increased overall arterial-venous oxygen difference during exercise.

Exercise Capacity

Exercise capacity can be defined as the maximal amount of work achievable by an organism and is described by such parameters as *maximal workload* achieved during exercise or *maximal duration* of exercise. Alternatively, exercise capacity can also be defined by assessing physiological performance, often described using metabolic concepts such as *aerobic exercise capacity.* Although the two definitions of exercise capacity are intricately related, it is possible for one parameter to be normal while the other is not. For example, highly motivated individuals may be able to achieve a normal maximal workload yet, due to underlying cardiovascular or pulmonary disease, their aerobic exercise capacities may be limited. Measures of aerobic exercise capacity, perhaps less influenced by motivational factors, may offer a more reliable assessment of the true physiologic response to exercise. Aerobic exercise capacity is most commonly determined by measuring the *maximal oxygen consumption (VO_{2max})* or the *anaerobic threshold (AT).*

Maximal Oxygen Consumption (VO_{2max}). The concept of *maximal oxygen consumption, or VO_{2max}* must be distinguished from *oxygen consumption at peak exercise, or maximum VO_2.* The subtle, yet very significant difference between these concepts is that the latter merely is the value of oxygen consumption at the time immediately before exercise ceases and the former represents a true metabolic limit to the consumption of oxygen. One commonly accepted indication that VO_{2max} has been reached is that the VO_2 fails to increase despite further increases in work intensity during incremental exercise (14-16). Exercise physiologists continue to disagree, however, about the validity of VO_{2max} measurements in representing oxidative capacity because of the difficulty and variability in determining when a subject can exercise no longer. In both humans and animals, motivational factors can significantly affect the maximum VO_2 measured during an exercise study. The VO_{2max}, however, is less subject to such influence if properly determined.

Anaerobic metabolism. The contribution of anaerobic metabolism to overall energy production during exercise can be estimated by calculation of the *respiratory exchange ratio* (RER), the ratio of VO_2 to VCO_2. Typical human RERs range from 0.8 at rest to over 1.0 during peak exercise. An RER of 1.0 represents the point at which carbon dioxide production exceeds oxygen consumption, and traditionally represents when anaerobic metabolism begins to significantly contribute to overall energy production due to an imbalance of oxygen demand and oxygen delivery (17). This point, also known as the *anaerobic threshold (AT)*, is commonly used as a means of evaluating exercise performance. It is particularly useful when VO_{2max} cannot be determined because of the subject's inability to complete an exercise protocol. Several methods of estimating anaerobic threshold have been proposed (17). A simple non-invasive method is to determine the point at which the RER exceeds 1.0 using gas analysis during incremental exercise. A more sophisticated non-invasive method requires breath-by-breath respiratory analysis to determine the point at which the minute ventilation to carbon dioxide production ratio (VE/VCO_2) begins to increase. The validity of these methods in determining AT, however, has been questioned because they do not always correspond with a rise in serum lactate (18, 19). Repetitive measurements of serum lactate levels during exercise and determination of the *lactate threshold*, is felt by some to be the most reliable measure of exercise performance (20). However, its invasive nature makes it less feasible to perform, particularly in smaller mammals such as mice.

The Effects of Chronic Exercise Conditioning

Conditioning induces physiologic adaptations within an organism to optimize exercise performance. The most obvious manifestation of this is an increase in exercise capacity. With respect to the cardiovascular system, these alterations result in more efficient delivery of oxygen to working muscle at a given work intensity. In humans, exercise conditioning produces a significant increase in cardiac output, primarily as a result of an increase in stroke volume (Figure 1). The maximal achievable heart rate and oxygen extraction, however, remain relatively constant between sedentary individuals and conditioned athletes (1).

The primary mechanism by which stroke volume increases after conditioning is believed to be through improved ventricular filling and a larger ventricular end-diastolic volume (21). Alterations in cardiac contractility are of such a small degree that they are almost immeasurable (22). Changes in afterload are also difficult to measure, but it is known that systemic blood pressures remain constant before and after training. As such, ventricular afterload must adjust to accommodate the drastic increase in blood flow that occurs with conditioning (8).

Although the overall arteriovenous oxygen difference does not increase with conditioning, skeletal muscle metabolism is altered such that local oxygen extraction and utilization are greatly improved. These changes usually result in improvements in such parameters as endurance and maximal power output, but have little impact on cardiovascular performance during exercise.

EXERCISE PHYSIOLOGY IN THE MOUSE

The murine response to exercise has generated recent interest primarily because of the need to evaluate the cardiovascular performance of genetically altered mice in response to stress and ultimately to apply these findings to human disease processes. To date, parameters such as heart rate, blood pressure, oxygen consumption, and carbon dioxide production (and thereby the respiratory exchange ratio) have been measured during treadmill exercise in mice (Figure 3). Indirect evaluation of cardiovascular performance also includes such measurements as maximal intensity of exercise attained, duration of exercise, and total distance run. Finally, post-exercise studies also provide a means of evaluating the cardiovascular response to exercise, particularly after conditioning.

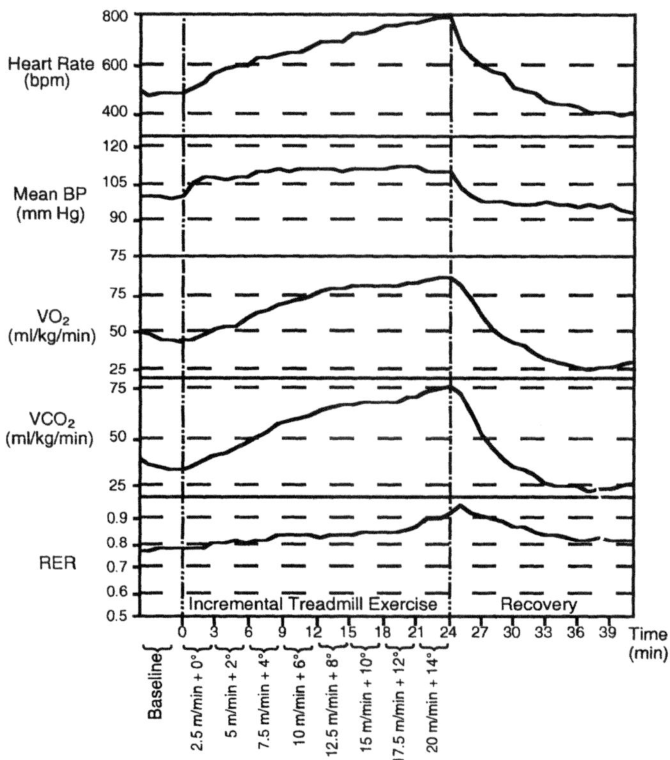

Figure 3. Cardiovascular parameters measured in a mouse during incremental treadmill exercise. Note that this mouse did not reach VO_{2max} but was close to its anaerobic threshold. Adapted from Desai, et al. (23).

The Murine Cardiovascular Response To Exercise

Murine heart rate. A discussion of the heart rate response to exercise must begin with an understanding of heart rate regulation in the mouse as well as an appreciation of the difficulties of measuring true "resting" heart rate in this species.

Resting Heart Rate. The resting heart rates of mice are significantly higher than those of larger mammals. However, there is disagreement over what constitutes "resting heart rate", with heart rates reported over as wide a range as 212-690 beats/min (Table 1), reflecting differences in recording techniques, the use of anesthesia, period of recovery from anesthesia before measurement, the use of restraint, as well as the activity state of the animal during recording. (24-31). Studies in awake mice without restraint and which observe an adequate period of recovery from anesthesia agree, however, that the true resting heart rate of mice is between 450-550 bpm (23, 32-35).

Study	Heart	Mean BP	Condition
Blizard and Welty, 1971	527-557	-	48° after Avertin
Christensen, et al., 1990	585 ± 15	112 ± 4	24° after mebumal, restrained
Sarin, et al., 1990	541 ± 15	97 ± 5	During pentobarbital
Barbee, et al. 1992	622 ± 20	96 ± 3	4° after Avertin, restrained
Kramer, et al., 1993	450-500	-	48° after pentobarbital or fentanyl/midazolam
Barbee, et al. 1994	652 ± 25	92 ± 3	4° after Avertin, restrained
Gardin, et al., 1995	390-690	-	During Avertin
Hoit, et al., 1995	253 ± 28	-	During ketamine/acepromaxine/xylazine
Merril, et. al., 1996	455 ± 95	111 ± 4	24-48° after Avertin, restrained
Iwase, et al. 1996	212-246	-	During ketamine/acepromazine/xylazine
Rohrer, et. al., 1996	520 ± 24	101 ± 10	24° after methoxyflurane
Lorenz, et. al. 1997	508 ± 17	77 ± 4	During thiobutabarbital and ketamine
Desai, et al., 1997	503 ± 52	106 ± 2	24° after methoxyflurane

Table 1. Resting heart rate and mean blood pressure values under various levels of consciousness and restraint reported in the literature. Values are represented as either mean ± SE or as a range.

Heart Rate Control. Separate studies by Desai and Kramer have shown that significant tachycardia, to rates as high as 800 bpm, can occur even with routine activities such as grooming and handling (23, 33). These relatively rapid alterations in heart rate suggest significant autonomic control. The very high resting rate in the mouse and the relatively small tachycardic response to the vagolytic drug atropine suggest a relatively low resting vagal tone (23, 35).

However, during the early phase of incremental exercise in mice, there is a proportionally more rapid increase in heart rate compared with subsequent exercise, from a resting rate of about 500 bpm up to about 600 bpm (Figure 3). Human studies have shown that this initial rapid increase in heart rate can be attributed to vagal inhibition and that the subsequent increase is due to sympathetic stimulation (1). Similarly, in mice, deletion of the gene for the murine ß1 adrenergic receptor appears to abolish the tachycardic response to exercise above heart rates of 600 bpm, suggesting that vagal tone accounts for the remaining one-third of the normal heart rate response to exercise (36).

Heart Rate During Exercise. The heart rate response to exercise in mice has only recently been directly studied. Kaplan, et al. used electrocardiographic recordings to show that a maximal heart rate of approximately 650 bpm was achieved in mice during forced swimming (37). This was achieved after only 3 minutes of exercise and was followed by a plateau, indicating that forced swimming is a form of constant workload exercise resulting in submaximal levels of exercise intensity. Desai, et. al. used indwelling fluid-filled catheters to measure heart rate and blood pressure during graded treadmill exercise (23) and demonstrated linear increases in heart rate during an incremental treadmill protocol. Peak heart rates observed with maximal exercise in some mice reached as high as 840 bpm. This is the highest in vivo heart rate observed in the mouse, irrespective of whether pharmacologic or exercise stimulation was used. Recent studies have begun to explore the cellular mechanisms accounting for these extremely high rates. Electrophysiologic studies have found very rapid recovery times and refractory periods in the atria, indicating that murine myocytes may have an intrinsic ability to function at much higher rates than those of larger mammals (38, 39).

Murine blood pressure. Murine blood pressure measurements were initially performed using indirect sphygmomanometry via a tail cuff. Although this methodology yields reproducible measurements, there has been some concern about their reliability because of the need for heating, restraint, or anesthesia. Schlager found systematically lower tail-cuff blood pressure measurements compared to direct arterial cannulation (40). Newer, more sophisticated computerized tail-cuff sphygmomanometers do correlate more closely with direct measurements (41). However, there are still potential problems with this methodology as indicated by significantly higher than normal heart rates in these studies, suggesting that blood pressure measurements may have also been affected due to the stress of restraint.

Direct measurement of arterial blood pressure in the mouse using an indwelling arterial catheter was first reported in 1990 (26, 42). However, because these studies were performed under anesthesia, the reported pressures are low compared with follow-up studies performed on awake mice by the same group (43, 44). Mean blood pressures in awake and unrestrained mice have subsequently been measured in the range of 100-120 mmHg (23, 28, 45). There are considerable differences in resting murine blood pressures depending on strain background (23). This suggests a genetic basis for blood pressure variability, which has been extensively studied in mice (46-48). The existence of strain-related differences in blood pressure underscores the need to perform control studies utilizing mice from the same strain background when studying any genetically altered model of cardiovascular disease.

The murine blood pressure response to a variety of vasoactive drugs is qualitatively similar to that in humans. Vasodilators such as isoproterenol, sodium nitroprusside, hydralazine, and nicotinamide cause dose-dependent hypotension and vasoconstrictors such as phenylephrine and angiotensin cause dose-dependent hypertension in mice (23, 43-45). The baroreflex is also intact in mice, with the notable difference being the higher heart rate set-point compared to larger mammals (45). It would seem reasonable, then, that the murine response to exercise should also be similar to the response of larger mammals. Indeed, when mice are exercised using an incremental treadmill protocol, they exhibit a qualitatively similar mean blood pressure response to that seen in larger animals (23) with a fairly sharp rise at the onset of exercise, followed by a plateau during exercise, and then a fall to resting levels with cessation of exercise (Figure 3). This response suggests that the increase in stroke volume at the onset of exercise is followed by vascular compensation through a redistribution of blood flow and hence a decrease in peripheral vascular resistance. Although this blood pressure response is somewhat different than peripheral blood pressure measurements in exercising humans, it is identical to that in humans when mean blood pressures are measured centrally (23, 49).

Murine Oxygen Consumption. Oxygen consumption was first measured in mice in the early 1930's by Davis, et al. who adapted an apparatus designed for dogs (50). Although the technical aspects of the measurement of murine oxygen consumption have changed dramatically through the years, the results obtained today are surprisingly similar to those obtained in these very early studies.

Study	VO2 (ml/kg/min) Resting	Peak	Exercise
Davis and van Dyke, 1966	20 - 27	-	-
Pennycuik, 1967	42	-	-
Pasquis, et al., 1969	-	132 ± 5	Treadmill
Lechner, 1978	30 - 45	155 -182	Treadmill
Seeherman, et al., 1981	-	261 ± 2	Treadmill
Chappel, 1983	60	176 ± 23	Treadmill
Fernando, et al., 1993	-	120 (submaximal)	Treadmill
MacNeil and Hoffman-Goetz, 1993	-	140 ± 3	Treadmill
Dohm, et al., 1994	-	212 - 247	Treadmill
Yoshimura, et al., 1996	46 ± 14	137 ± 14	Swimming
Desai, et. al., 1997	50	90 (submaximal)	Treadmill

Table 2. Murine resting and peak oxygen consumption values reported in the literature using treadmill and swimming exercise. Values are reported as mean, mean ± SE, or a range.

Resting oxygen consumption. As with any resting physiologic parameter, the measurement of resting VO_2 is ideally performed without anesthesia, restraint, or other manipulation. In their detailed review of the metabolic performance of various mammals, Jones and Lindstedt found that resting VO_2 generally increases as the body surface area-to-mass ratio increases (51). Consistent with this, studies in mice have found significantly higher weight-corrected VO_2 values than for larger mammals (23, 50, 52). The agreement amongst multiple studies on resting VO_2 values also attest to the reliability of these measurements (Table 2). Similar to the variation seen in murine heart rate measurements, routine activities, such as grooming, feeding, or handling, can also increase VO_2 by up to 50% (50).

VO_{2max}. Pasquis and coworkers described the first method of measuring VO_{2max} using the currently accepted standard for determination of VO_{2max} by gas analysis alone: an inability to increase VO_2 in response to increasing workloads (14). They found average VO_{2max} values in mice of 132.5 ml/kg/min. Their study, however, used a very short period of exercise, usually less than 2 minutes, and the inability to establish a steady state equilibrium between tissue oxygen utilization (QO_2) and ventilatory oxygen consumption (VO_2) may have biased their measurements. Seeherman and coworkers described a method of VO_{2max} determination in mice which requires measurement of steady-state VO_2 levels at each level of exercise (15). Mice are forced to run at increasing exercise workloads until VO_2 does not increase with further increases in workload. They determined VO_{2max} values for pygmy mice of 261 ml/kg/min, a value that was over 4 times higher than at rest. Because they also measured serum lactate levels, they could assure that above the VO_{2max}, exercise was limited by increases in lactate. A more recent study by Dohm and colleagues reported VO_{2max} values determined using a continuous incremental treadmill protocol in both wild and laboratory mice (53). They also found that VO_2 values plateau when maximal oxygen consumption was reached. Although they found that laboratory mice (197 ml/kg/min) have lower VO_{2max} values than wild mice (294 ml/kg/min), both values reasonably agree with those reported by Seeherman. VO_{2max} values for mice measured during swimming exercise have been performed but have never been validated with serum lactate measurements or by determining the exercise level at which VO_2 does not increase despite increases in workload. As such, the discrepancy between values of VO_{2max} reported during treadmill exercise and the significantly lower value (140 ml/kg/min) reported by Yoshimura and colleagues during swimming, may be due to the difficulty in forcing mice to perform maximal exercise while swimming (52, 54). In summary, reported VO_{2max} values appear to have a fair amount of variability with values ranging from 132-294 ml/kg/min (Table 2). Part of this may be explained by differences in the strains of mice studied and also whether they were wild or laboratory bred mice. Beyond these differences, we can only speculate that the different methods for determining VO_{2max} are not well enough standardized to be reproducible.

Murine RER and Anaerobic Threshold (AT). Several studies have shown that the murine resting RER matches that of humans at approximately 0.8 and reflects the predominant utilization of fat as an energy source (23, 55, 56). With increasing workloads of exercise, the RER of mice has been shown to increase in a fashion similar to that seen in larger animals (23). In mice, as in humans, when the RER increases above 1.0, there is evidence that anaerobic metabolism begins to play a major role in energy production in mice as well. Seeherman, et al. found significant increases in serum lactate at levels of exercise beyond 50% VO_{2max} (15). We have also shown that serum lactate, measured at 5 minute intervals during an incremental treadmill exercise protocol, increased beginning at approximately 50% of peak VO_2 values (Figure 4), suggesting that anaerobic metabolism increases in mice well before peak exercise workloads are reached, and that lactate measurements may be used to estimate the anaerobic threshold. Indeed, lactate levels have recently been used to demonstrate conditioning in mice (65).

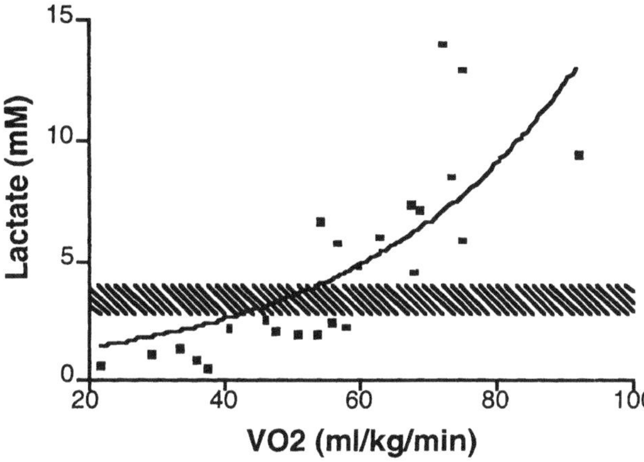

Figure 4. Lactate levels in mice undergoing an incremental treadmill exercise protocol. Note lactate begins to increase exponentially at a value of VO_2 which is approximately 50% of peak VO_2 values in untrained mice. Hatched area represents 2x basal $[La^-]$ to 4 mM $[La^-]$ (Desai, et al., unpublished data, 1997).

Murine Exercise Performance

Mice of different strain backgrounds and those bearing genetic manipulations have been shown by many researchers to have differential capacities to exercise. These studies, however, have often used non-standardized exercise protocols, and few have measured VO_{2max}. Dohm measured VO_{2max} to show differences in exercise capacities of wild and laboratory mice (53) and Chappell compared VO_{2max} values obtained with exercise to those obtained by cold exposure (55). Several studies have compared VO_{2max} values in mice to those of other species (14-16, 51) and some have used VO_{2max} as a maximal exercise standard with which to gauge submaximal exercise during swimming (52, 54). Beyond these studies, VO_{2max} has seldom been used as a measure of exercise capacity in mice primarily because of the difficulty in proving that the VO_2 value obtained is truly the maximal aerobic capacity, and not a submaximal value due to motivational differences.

Physical limits of endurance and exercise intensity have been measured during both swimming exercise and treadmill testing. Until recently, with almost any swimming protocol, the only quantifiable parameter used to assess exercise capacity was duration of exercise before exhaustion (54, 57-59). Some researchers have attached weights to swimming mice in order to alter their workloads, but even these studies relied on duration of swimming as the method of comparison (53). Recently, Kaplan, et. al. have attached electrocardiographic leads to swimming mice and calculated heart rates (37). They were able to quantitate the elevation in heart rate during swimming, and demonstrate that heart rate during exercise decreases by 20% after just 4 weeks of conditioning (Figure 5).

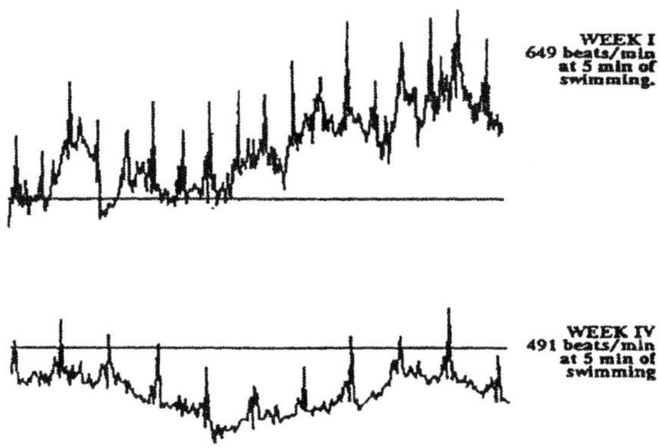

Figure 5. Representative electrocardiograms during swimming exercise. Note decreased heart rate after 4 wk. of training. Adapted from Kaplan, et al.(37).

Exercise performance has also been evaluated in mice using a treadmill apparatus. Duration of exercise using a given treadmill protocol has been used, in much the same way as with swimming, as a quantitative measure of exercise capacity (60-65) The intensity of exercise achieved during an incremental exercise protocol can also be readily quantified using a treadmill, either as the maximal workload achieved (23, 65) or by estimating actual work performed (66). Recently, metabolic parameters reflecting exercise performance have also begun to be able to be evaluated using treadmill exercise (23) (67). The ability to measure serum lactate during exercise using an indwelling catheter will make evaluation of the anaerobic threshold possible (65). Other measurements such as serum pyruvate levels and oxygen saturations (leading to calculation of arteriovenous oxygen difference) can now be performed in exercising mice. These and future advances in device miniaturization will further enhance our ability to accurately assess exercise performance in mice.

Post-Exercise Cardiovascular Evaluation – The Effects of Conditioning

The cardiovascular and metabolic responses to chronic exercise in the mouse have only recently begun to be evaluated. An early study of the cardiac effects of training on the murine heart showed that chronic exercise causes a decrease in overall body weight, an increase in heart weight and, therefore, a significant increase in heart-to-body weight ratio (68). Studies on skeletal muscle lactate metabolism have found that lactate transport is significantly increased after acute exercise (69). Chronic exercise was found to decrease blood lactate levels and increase the oxidative metabolism of lactate (65, 70) indicating that, similar to larger mammals, mice have adaptive mechanisms to reduce serum lactate, allowing a higher exercise capacity. In the first cardiac evaluation of mice during chronic swimming exercise, Kaplan, et al. found increased heart weight-to-body weight ratios, decreased heart rates during exercise, and increased skeletal muscle glycolytic capacity in mice that were trained for just 4 weeks (37). Ruwitch, et al, recently evaluated the cardiovascular and metabolic effects of conditioning in mice using a treadmill protocol (65). After a conditioning period of 12 weeks, mice demonstrated an increase in exercise capacity and maximal VO_2 (Figure 6) as well as a lower serum lactate level at equivalent submaximal exercise workloads. These results suggest

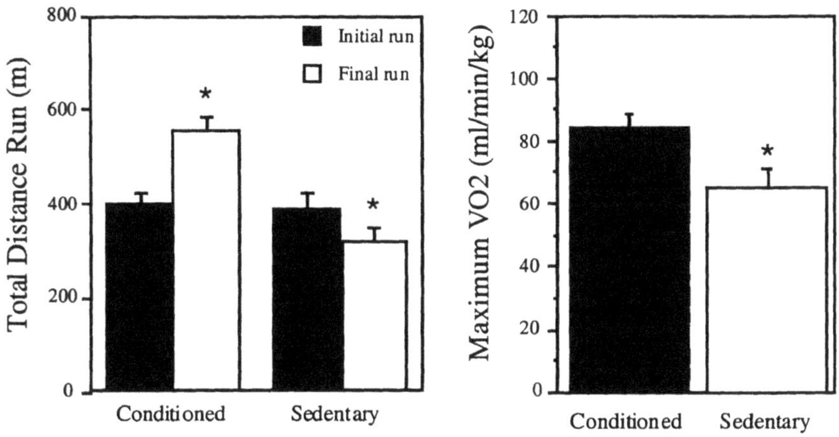

Figure 6. Increase in exercise capacity and maximum VO2 in mice after exercise conditioning (65).

that mice can be conditioned using protocols similar to humans and larger mammals.

PROTOCOLS FOR EXERCISING MICE

To date, studies of cardiovascular responses during exercise have primarily utilized two methods: swimming and treadmill running. Using these techniques, it has been possible to apply predetermined levels of work intensity and duration to mice and accurately and reproducibly evaluate cardiovascular and metabolic changes.

Swimming Exercise

Forced swimming has been used as a means of inducing physiologic stress in mice by exploiting the mouse's excellent capacity for swimming. Although, physiologic stress can be reproducibly induced with swimming exercise, assessment of cardiovascular physiology remains difficult with this protocol due to the technical limitations of performing sensitive physiologic measurements in an aquatic environment. Recently, however, several cardiovascular and metabolic parameters have been evaluated in mice during swimming exercise. Yoshimura et al. (52) adapted an apparatus first used in rats (71) to measure oxygen consumption in mice during swimming exercise. Although measurements of VO_2 in this study were only used confirm that a particular level of exercise had been reached, these researchers were able to demonstrate increases in VO_2 during swimming exercise. Their apparatus consisted of a circular Plexiglas water tank measuring 80 cm in diameter and 90 cm in height, filled to the 60 cm mark with 32-34°C water (Figure 7). Oxygen consumption was measured using a closed circuit technique by sealing the tank above the water level to maintain a constant volume of gas. Room air was pumped at a controlled rate and the gas above the water was constantly mixed by a blower. Expelled gas was collected in anesthesia bags for analysis by a commercially available gas analyzer. VO_{2max} was evaluated by placing a weight equivalent to 5% of the mouse's body weight around the neck and forcing it to swim for 3-4 minutes. Steady state VO_2 values were measured without the weight over 2 h which yielded values equivalent to approximately 60% of the measured VO_{2max}.

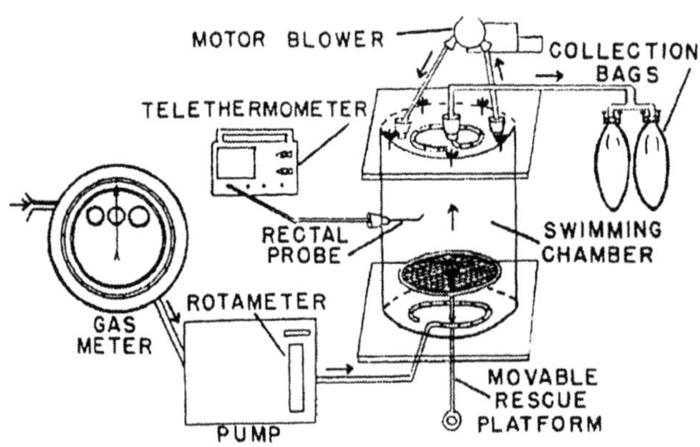

Figure 7. Swimming apparatus used to measure VO_2 in rodents. Adapted from Baker and Horvath (71).

Kaplan et al. (37) used a smaller tank with a surface area one-twentieth of Yoshimura's and a water depth of 15 cm. Using surgical steel sutures implanted onto the backs of the mice they were able to obtain electrocardiographic recordings to establish heart rates during exercise (Figure 5). Interestingly, they note that mice are far better swimmers than rats because they rarely dive, however they need to be forced to swim by the presence of other mice in the tank, otherwise they just float. It is not clear, however, how the presence of other mice alters the imposed stress of forced swimming.

Advantages/Limitations of Forced Swimming Protocols. One advantage to forced swimming exercise is that a significantly larger volume of comparative data exists for rats compared to treadmill data. Furthermore, it is much easier to induce mice to swim than it is to induce them to run. As has been encountered in treadmill studies with other species, at the highest levels of workload mice will choose the stimulus (shock, forced air, prodding) over treadmill running, whereas avoidance of drowning appears to be a stronger stimulus to continue exercising. Another advantage to swimming exercise is that the apparatus is relatively simple and inexpensive and requires no sophisticated timing motors or stimulus systems. The difference in cost between swimming and treadmill exercise systems becomes minimal, however, once more sophisticated evaluations such as electrocardiography and oxygen consumption measurements are required.

Major limitations exist to performing cardiovascular measurements in swimming mice. Even though measurements of VO_2 are possible using sealed water tanks, maintaining a stable gas phase, and stable temperature and humidity levels is difficult. These factors can influence gas analysis and may account for some of the differences found in measured VO_2 values using swimming versus treadmill protocols (Table 2). Although measurement of other cardiovascular parameters such as blood pressure are theoretically possible, the aquatic environment significantly complicates invasive procedures requiring implantation of foreign objects such as catheters or electrocardiographic leads, and may significantly increase mortality and morbidity.

Another limiting is the inability to precisely regulate exercise intensity during forced swimming. Although Yoshimura et al. were able to demonstrate that their exercise protocol resulted in a workload that was approximately 60% of maximal, the level of workload was static and could not be changed during exercise (52). One method by which exercise workloads can be altered during swimming is by the addition of extra weight to the animal. However, this precludes incremental workload protocols. The only other method by which to regulate swimming exercise is by altering its duration. However, the duration of forced swimming exercise has not been shown to alter VO_2 over durations of up to 8 h (52). Despite these limitations, forced swimming exercise continues to be a useful technique to induce non-pharmacological cardiovascular stress in mice, primarily because of the low cost and the relative ease of inducing mice to swim (37).

Treadmill Exercise

Treadmill exercise has been the gold standard for inducing controlled physiologic stress to the cardiovascular system. With the adaptation of the metabolic treadmill apparatus to mice coupled with recent advances in performing non-anesthetized physiologic measurements, there has been a resurgence of interest in using treadmill exercise to study cardiovascular performance in mice. Although the first reported use of the treadmill apparatus to induce stress in mice was in the 1960's, it was more than 30 years later that Seeherman and coworkers reported the first successful continuous measurement of VO_2 and VCO_2 in mice during treadmill exercise (15). Most subsequent studies using treadmill exercise since then have used a fundamentally similar system.

Using mice chronically instrumented with carotid arterial and jugular venous catheters and a commercially available treadmill (Columbus Instruments, Columbus, OH), Desai, et. al., measured heart rate, blood pressure, VO_2, and VCO_2 during incremental exercise (Figure 8) (23). This apparatus employs a moving belt encased in a Plexiglas enclosure, assuring that the mouse remains on the belt and allowing for oxygen consumption measurements. The speed of the belt is electronically controlled from 0 to 99 m/min., although the functional maximum for mice is at approximately 50 m/min. The slope of the treadmill can also be increased from 0°-45°. Most treadmills employ stimulus devices to assure that the animal does not remain at the rear of the moving belt: either a metallic grid which delivers a mild electric shock or forced air or cold water jets.

Figure 8. Mouse treadmill apparatus. This model allows metabolic measurements to be made in 4 mice simultaneously.

Metabolic measurements. For metabolic measurements, mice are placed into an Oxymax metabolic chamber integrated into the treadmill apparatus (Columbus Instruments, Columbus, OH) for measurements of VO_2 and VCO_2 using an open circuit volumetric method of gas analysis (Figure 9). Respiratory equivalent ratio (RER) is calculated on-line continuously as VCO_2/VO_2. During exercise, metabolic measurements are made in an isolated, but not airtight, treadmill chamber which has a rate of inflow gas high enough so that no entrained room air is inadvertently measured in the outflowing gas. Gas samples are automatically dried, measured and averaged every minute. A low dead-space circuit allows quick gas equilibration with a $t_{1/2}$ of 30 seconds. The open circuit gas measurement system is, however, inherently difficult to validate in a murine model since it is difficult to collect gas directly from the animal. The instrument has been validated using the "Oxygen Recovery Test" which employs combustion of a hydrocarbon gas of a known mass in the test chamber. The VO_2 value measured is then compared with the stoichiometric quantity of oxygen required to combust the hydrocarbon gas. For this system, oxygen recovery is greater than 97%.

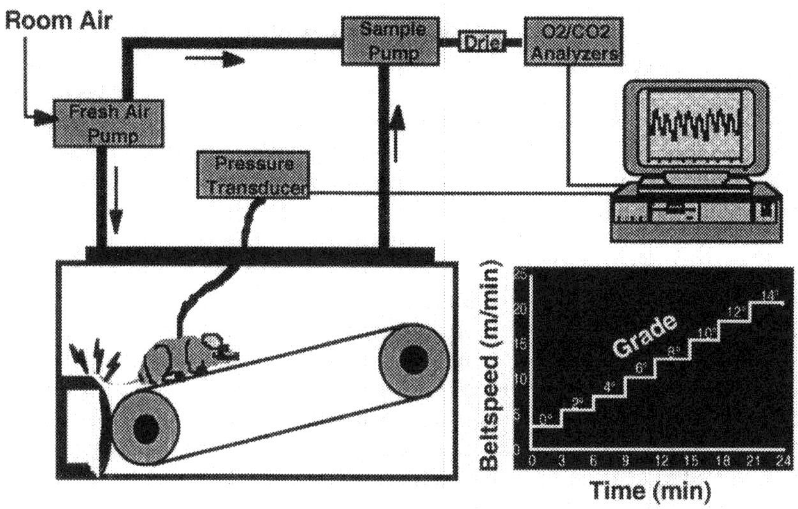

Figure 9. Diagram of metabolic treadmill system using an incremental exercise protocol during which heart rate, blood pressure, VO2 and VCO2 can be measured.

Incremental Workload Protocol. The incremental exercise protocol begins with a 45-60 min acclimation period followed by 2.5 m/min increases in running speed and 2° increases in slope every 3 minutes until exhaustion. At the highest exercise workloads using this technique, VO_2 does not always plateau, thus VO_{2max} is not always achieved. However lactate levels increase significantly at this level of exercise, suggesting that these peak VO_2 values are indicative of an increase in anaerobic metabolism (Figure 4). Modifications of this protocol, using higher increments of workload earlier in the protocol or beginning exercise immediately at high workloads, allow for the elucidation of true VO_{2max} values (14, 16, 55).

Constant Workload Protocol. An incremental protocol, such as discussed above, employs gradual increases in exercise intensity so that organ systems can be maximally stressed in a relatively short period of time. Constant workload exercise, on the other hand, occurs in two distinct phases. The first phase is a dynamic period during which the cardiovascular and pulmonary systems can be evaluated as they change to meet increased metabolic needs induced by exercise. This is followed by a steady state phase which allows evaluation of tissue metabolism, since an equilibrium between all systems has been reached. A constant workload protocol is performed by setting the predetermined workload (belt speed and slope) either immediately at the onset of exercise or gradually over a few minutes. The former allows evaluation of dynamic changes that occur at the onset of exercise whereas the latter induces less stress initially, allowing the subject to reach a steady-state more quickly due to a smaller VO_2 overshoot (oxygen debt).

Constant workload protocols can be performed at any level of exercise, depending upon the parameter of interest. Seeherman et al. used constant exercise of increasing workloads to reconstruct a workload vs. VO_2 graph and evaluate VO_{2max} (15). In this way, they were assured steady state levels of VO_2 at a particular workload, without concern of measuring transitional values of VO_2. Alternatively, a near maximal workload can be used to compare VO_2 between two groups of mice with the potential for determining which group has a greater aerobic exercise capacity (55). Another potential use for constant workload protocols is the evaluation of the early dynamic phase of cardiorespiratory adjustment to exercise. This has been utilized, for example, in humans with chronic obstructive pulmonary disease, to show significant differences in how rapidly the plateau in VO_2 is achieved (72, 73).

Advantages/Limitations of Treadmill Exercise Protocols. The major advantage of treadmill exercise over swimming is the ability to precisely regulate work intensity as well as duration. It allows 1) the application of uniform exercise workloads to all experimental groups, 2) the systematic evaluation of a physiologic response over the entire range of exercise intensities, and 3) the assessment of aerobic exercise capacity. Another advantage is that a non-aqueous environment allows easier maintenance of catheters, electrophysiologic leads and other hardware with less concern regarding infection or electrical interference.

One major disadvantage of treadmill exercise is the difficulty in motivating the animal to exercise beyond a certain point. A mouse will eventually tolerate negative stimuli over the alternative discomfort of continuing to exercise. Thus, it is not always clear whether the true maximal exercise capacity has been reached without the addition of more complicated testing such as serum lactate analysis. The equipment required for treadmill exercise is also more complex and expensive than that required for swimming. Overall, however, treadmill exercise affords significantly more control over the intensity of exercise induced and the ability to evaluate a larger array of physiologic parameters than is possible with swimming.

EXERCISE TESTING IN GENETICALLY ALTERED MURINE MODELS

ß-Cardiac Myosin Heavy Chain Mutant Mice

One of the earliest studies evaluating the cardiovascular response to exercise in genetically altered mice used forced swimming (74). Transgenic mice that were heterozygous for a mutant ß-myosin heavy chain gene exhibited gross and microscopic findings similar to those seen in human hypertrophic cardiomyopathy. Humans with the disease are known to experience ventricular arrhythmias and sudden death with exercise. To evaluate exercise capacity and the cardiac response to chronic exercise, mice were forced to swim daily for 90 min. Two of the transgenic mice did not tolerate this exercise protocol and died or became ill, with post-mortem findings in one mouse of gross cardiac enlargement and intramural thrombosis. Although ex vivo evaluations found abnormalities in contractility and relaxation in hearts from these mice, cardiovascular and metabolic parameters were not directly assessed during exercise.

Adenine Nucleotide Translocator 1 (Ant1) Deficient Mice

The first report of metabolic evaluation during exercise in a genetically altered model was by Graham et al. in mice deficient for the mitochondrial membrane protein, adenine nucleotide translocator 1 (67). Deficiency of Ant1 in mice results in muscle cytopathy including ragged-red fibers, hypertrophic cardiomyopathy, and mitochondrial proliferation, similar to that seen in humans. Using an incremental treadmill exercise protocol, knockout mice were found to have an almost 50% reduction in exercise duration and a severe reduction in VO_2 at peak exercise, resulting in a near doubling of their RER (Figure 10). Interestingly, VO_2 in their wildtype mice did not increase with exercise, a finding that differs significantly from other studies. This may have been due to a very short acclimation period with insufficient time for a true resting VO_2 value to be achieved. Indeed, the resting VO_2 in wildtype mice reported in this study was higher than reported by other groups. Moreover, RER values were always greater than 1 (evidence of predominantly anaerobic metabolism) even at rest, in contrast to previous studies reporting resting RER values of 0.8. This study demonstrates how incremental exercise can be used to more comprehensively define a phenotype, especially when not all characteristics of a phenotype are evident at rest. However, the data also caution the need for full acclimation to the exercise apparatus and the need to carefully define the "resting" state when performing exercise studies.

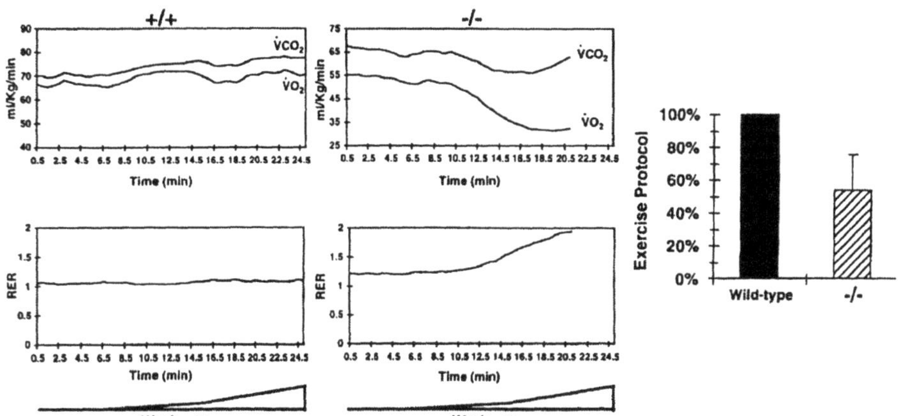

Figure 10. Metabolic analysis of Ant1 knockout (-/-) mice during graded treadmill exercise. RER increases significantly beginning at 10.5 min. A decrease in VO_2 also occurs after this point. Ant1(-/-) mice had an exercise capacity of approximately 50% that of wildtype mice (+/+). Adapted from Graham et al.(67).

ß-Adrenergic Receptor Deficient Mice

The first genetically altered murine model to undergo full cardiovascular evaluation during treadmill exercise was the ß1-adrenergic receptor deficient (ß1KO) mouse. Both in vivo and in vitro studies had revealed the absence of any chronotropic or inotropic response to ß-agonists in these mice (32). Despite this lack of cardiac sympathetic responsiveness, when exercise studies were performed, ß1KO mice were able to achieve similar workloads and reached similar total exercise capacities as their wildtype littermates (Figure 11) (36). At rest, ß1KO mice have normal heart rates, blood pressures, VO_2, VCO_2 and RER. Incremental exercise induced increases in VO_2, VCO_2 and RER identical to those seen in wildtypes, however, there was a marked attenuation of the exercise-induced tachycardia, with heart rate responses in the ß1KO only 20% above resting values. In order for these mice to maintain normal VO_2 during exercise, some compensation for their relative bradycardia must be present, mediated either through an increase in left ventricular stroke volume, by an enhancement of oxygen extraction, or both. In this study oxygen extraction was measured directly during exercise, and shown to be equivalent in ß1KO and wildtype mice, suggesting that the primary method of compensation must be through an increase in stroke volume. Evaluation of cardiovascular and metabolic parameters in these mice during exercise was a critical step to elucidate the mechanism by which they were able to achieve normal exercise capacities despite the absence of any response to ß-adrenergic agonist stimulation.

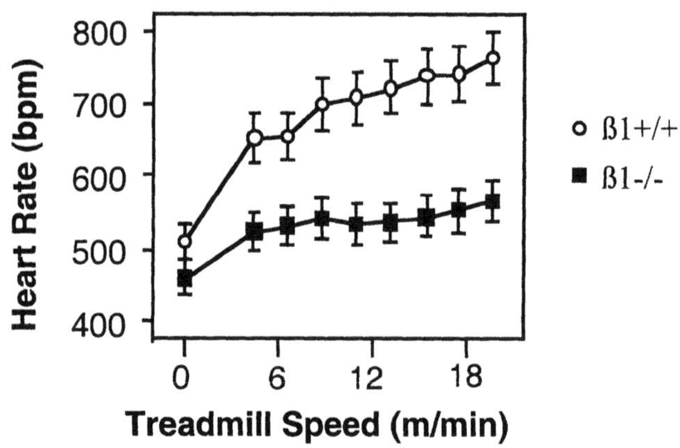

Figure 11. The cardiovascular response of ß1AR deficient mice to graded treadmill exercise. Despite normal exercise capacity and VO2, the tachycardic response is markedly blunted. Adapted from Rohrer, et al. (36).

Exercise studies were also performed in mice with targeted deletions of the ß2-receptor (ß2KO) (75) and in mice bred for deletions of both ß1 and ß2-receptors (ß1/ß2KO) (76). Interestingly, ß2KO mice have a greater exercise capacity than wildtypes. At comparable workloads they also had a lower RER, suggesting that they may use a greater ratio of fat to carbohydrate than wildtype mice during exercise. In this regard, ß2KO mice have a lower total body fat content than wildtypes (75). ß1/ß2KO mice, similar to ß1KO mice, were able to exercise to the same total workload as wildtype despite a blunted tachycardic response. In contrast to ß1KO mice, the double knockouts showed decreased VO2 and VCO2 at all workloads compared to wildtypes (76).

Phospholamban Deficient Mice

The gene for phospholamban, a sarcoplasmic reticulum protein that regulates cardiac contractility and relaxation, has been ablated in the mouse. Working heart preparations from these mice show increased basal contractility and relaxation as well as a lack of response to ß-adrenergic stimulation (77), suggesting that they are already performing at a maximal level of contractility. Treadmill exercise studies in these mice demonstrate normal exercise capacities compared to wildtype littermates (Figure 12) (78). Since their peak VO_2 and peak HR values are normal, and since these mice are unable to increase cardiac contractility via adrenergic mechanisms, these data suggest that compensation may be occurring in the form of improved stroke volumes via enhanced ventricular filling or reduced ventricular afterload during exercise. In vitro data suggest that ventricular filling is indeed improved in these mice (25) and ventricular afterload may also be reduced, since blood pressure drops in these mice at peak exercise. Exercise studies have thus shown that phospholamban deficient mice are able to compensate for a proven deficiency in cardiac adrenergic reserve while under physiologic stress.

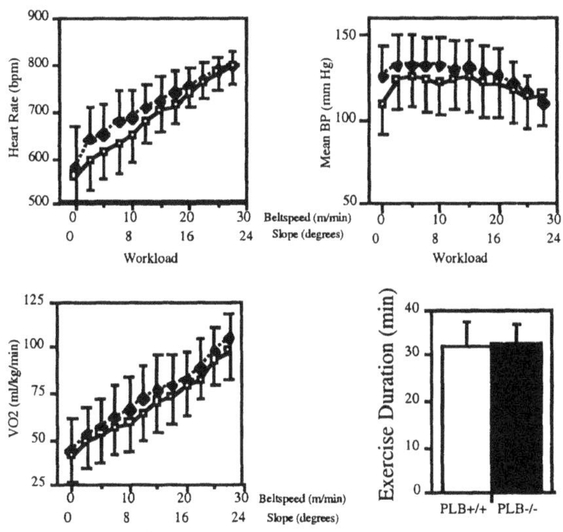

Figure 12. Phospholamban deficient (black, PLB-/-) vs. wildtype (white, PLB+/+) mice during a graded treadmill protocol. Heart rate (A), blood pressure (B), VO_2 (D), VCO_2 (E), RER (F), and exercise duration (C) were all normal. The only differences were in the resting blood pressures. Values are means ± SE for 7 mice in each group. Adapted from Desai, et al. (78).

Mutant Opioid Receptor Overexpressing Mice

Treadmill exercise has also been used to quantify decreased cardiac performance in genetically engineered mice. Mice overexpressing a genetically altered opioid receptor targeted to the myocardium (OR1) and resulting in over-stimulation of the inhibitory G protein, $G_{\alpha i}$, are prone to massive edema with echocardiographic evidence of poor ventricular function (79). Exercise evaluation was utilized to detect early abnormalities in cardiac function before the onset of overt heart failure. OR1 mice were found to have exercise capacities that were lower than wildtype, as determined by duration of exercise, workload achieved, and maximum VO_2 value attained. There was also a high level of correlation between echocardiographically determined ventricular function (LV fractional shortening) vs. their cardiac performance during exercise ($maxVO_2$) (Figure 13).

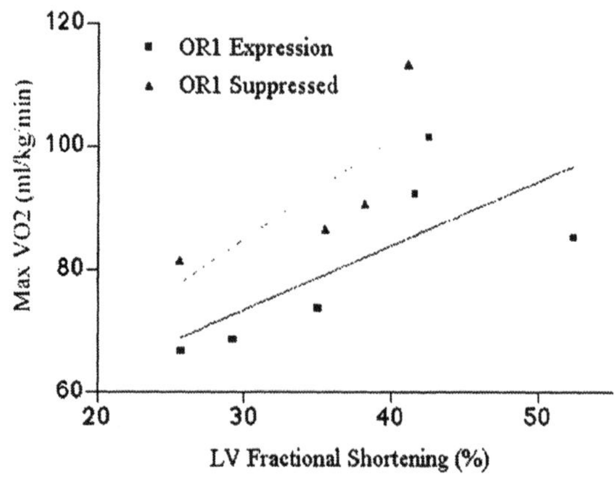

Figure 13. The relationship of LV fractional shortening to exercise capacity (Max VO_2) in opioid receptor overexpressing (OR1) mice is fairly linear. Expression of the receptor causes a downward shift in the regression line to a less favorable slope. (Desai, et al., unpublished data, 1997).

FUTURE DIRECTIONS

The advent of implantable telemetry systems has made possible continuous ECG measurements, even in the presence of the vigorous muscular activity induced by exercise (37). The recent miniaturization of transducer-tipped left ventricular pressure catheters (29) may soon make possible measurement of left ventricular contractile indices such as dP/dt during exercise.

Measurement of cardiac output remains difficult in exercising mice, although this may soon be achieved using the labeled microsphere technique (27). Due to the requisite placement of carotid and femoral arterial reference catheters, exercise may be difficult to perform using this technique. However, microspheres have been successfully used to measure regional distribution of blood flow during exercise in the mouse, as relative organ flows can be measured without the need for a lower body reference catheter (80). Another potential method for measuring cardiac output during exercise utilizes a cuffed Doppler flowprobe. Doppler flowprobes have been placed around various arteries in rats to measure regional blood flows (81) and around the aorta to measure cardiac output (82). We have successfully used Doppler flowprobes to measure cardiac output in the anesthetized mouse, however these techniques are still in the early stages of being adapted to exercising mice. If the recent strides in miniaturizing physiologic monitoring technology continue, this technique will also take its place amongst those other, once unimaginable, techniques that have become commonplace in the well-equipped mouse physiology lab.

REFERENCES

1. Rowell LB. Human circulation: regulation during physical stress. New York: Oxford University Press; 1986.
2. Sheriff DD, Zhou XP, Scher AM, Rowell LB. Dependence of cardiac filling pressure on cardiac output during rest and dynamic exercise in dogs. Am J Physiol 1993:H316-22.
3. Sheriff DD, Rowell LB, Scher AM. Is rapid rise in vascular conductance at onset of dynamic exercise due to muscle pump? Am J Physiol 1993:H1227-34.
4. Levine BD. Regulation of central blood volume and cardiac filling in endurance athletes: the Frank-Starling mechanism as a determinant of orthostatic tolerance. Med Sci Sports Exerc 1993;25(6):727-32.
5. Niess A, Roecker K, Mayer F, Heitkamp HC, Dickhuth HH. Effect of combined parasympathetic and sympathetic blockade on left ventricular relaxation at rest and during exercise in trained and untrained men. Int J Sports Med 1996:S180-3.
6. Poliner LR, Dehmer GJ, Lewis SE, Parkey RW, Blomqvist CG, Willerson JT. Left ventricular performance in normal subjects: a comparison of the responses to exercise in the upright and supine positions. Circulation 1980;62(3):528-34.
7. Katz AM. Physiology of the heart. 2 ed. New York: Raven Press; 1992.
8. Clausen JP. Effect of physical training on cardiovascular adjustments to exercise in man. Physiol Rev 1977;57(4):779-815.
9. Stebbins CL, Symons JD. Vasopressin contributes to the cardiovascular response to dynamic exercise. Am J Physiol 1993:H1701-7.
10. Rowell LB. Human cardiovascular adjustments to exercise and thermal stress. Physiol Rev 1974;54:75-159.
11. Bergman HL. Enzymatic and circulatory adjustments to physical training in middle aged men. Eur J Clin Invest 1973;3:414.
12. Wasserman K, Hansen JE, Sue DY, Whipp BJ. Principles of exercise testing and interpretation. Philadelphia: Lea & Febiger; 1987.
13. Robinson BF, Epstein SE, Beiser GD, Braunwald E. Control of heart rate by the autonimic nervous system: studies in man on the inter-relation between baroreceptor mechanisms and exercise. Circ Res 1966;19:400-11.
14. Pasquis P, Lacaisse A, Dejours P. Maximal oxygen uptake in four species of small mammals. Respir Physiol 1984;9:298-309.
15. Seeherman C, Taylor RC, Maloiy GMO, Armstrong RB. Design of the mammalina respiratory system. II. Measuring maximum aerobic capacity. Resp Physiol 1981;44:11-23.
16. Lechner AJ. The scaling of maximal oxygen consumption and pulmonary dimensions in small mammals. Respir Physiol 1978;34:29-44.
17. Wasserman K. The anaerobic threshold measurement to evaluate exercise performance. Am Rev Respir Dis 1984;129(suppl):S35-40.
18. Simon J, Young JL, Blood DK, Segal KR, Case RB, Gutin B. Plasma lactate and ventilation thresholds in trained and untrained cyclists. J Appl Physiol 1986;60:777-81.

19. Green HJ, Hughson RL, Orr GW, Ranney DA. Anaerobic threshold, blood lactate, and muscle metabolites in progressive exercise. J Appl Physiol 1983;54:1032-8.
20. Myers J, Ashley E. Dangerous curves. A perspective on exercise, lactate, and the anaerobic threshold. Chest 1997;111(3):787-95.
21. Rerych SK, Scholz PM, Sabiston DC, Jones RH. Effects of exercise training on left ventricular function in normal subjects: a longitudinal study by radionuclide angiography. Am J Cardiol 1980;45:244-52.
22. Blomqvist CG, Saltin B. Cardiovascular adaptations to physical training. Ann Rev Physiol 1983;45:169-89.
23. Desai KH, Sato R, Schauble E, Barsh GS, Kobilka BK, Bernstein D. Cardiovascular indexes in the mouse at rest and with exercise: new tools to study models of cardiac disease. Am J Physiol 1997:H1053-61.
24. Iwase M, Bishop SP, Uechi M, Vatner DE, Shannon RP, Kudej RK, et al. Adverse effects of chronic endogenous sympathetic drive induced by cardiac Gsa overexpression. Circ Res 1996;78:517-24.
25. Hoit BD, Khoury SF, Kranias EG, Ball N, Walsh RA. In vivo echocardiographic detection of enhanced left ventricular function in gene-targeted mice with phospholamban deficiency. Circ Res 1995;77(3):632-7.
26. Christensen KL, Horsman MR, Overgaard J. Invasive blood pressure measurements in restrained but non-anaesthetized mice. In Vivo 1990;4(1):13-6.
27. Barbee RW, Perry BD, Re RN, Murgo JP. Microsphere and dilution techniques for the determination of blood flows and volumes in conscious mice. Am J Physiol 1992;263:R728-33.
28. Barbee RW, Perry BD, Re RN, Murgo JP, Field LJ. Hemodynamics in transgenic mice with overexpression of atrial natriuretic factor. Circ Res 1994;74:747-51.
29. Lorenz JN, Robbins J. Measurement of intraventricular pressure and cardiac performance in the intact closed-chest anesthetized mouse. Am J Physiol 1997;272:H1137-H46.
30. Sarin SK, Sabba C, Groszmann RJ. Splanchnic and systemic hemodynamics in mice using a radioactive microsphere technique. Am J Physiol 1990;258:G365-9.
31. Gardin JM, Siri FM, Kitsis RN, Edwards JG, Leinwand LA. Echocardiographic assessment of left ventricular mass and systolic function in mice. Circ Res 1995;76(5):907-14.
32. Rohrer DK, Desai KH, Jasper JR, Stevens ME, Regula DP, Barsh GS, et al. Targeted disruption of the mouse ß1-adrenergic receptor gene: developmental and cardiovascular effects. Proc Natl Acad Sci 1996;93:7375-80.
33. Kramer K, van Acker S, Voss H, Grimbergen JA, van der Vijgh W, Bast A. Use of telemetry to record electrocardiogram and heart rate in freely moving mice. J Pharmacol Toxicol Methods 1993;30(4):209-15.
34. Blizard DA, Welty R. Cardiac activity in the mouse: strain differences. J Comp Phsyiol Psychol 1971;77(2):337-44.
35. Swynghedauw B, Jasson S, Chevalier B, Clairambault J, Hardouin S, Heymes C, et al. Heart rate and heart rate variability, a pharmacological target. Cardiovasc Drugs Ther 1997;10(6):677-85.
36. Rohrer DK, Schauble EH, Desai KH, Kobilka BK, Bernstein D. Alterations in dynamic heart rate control in the beta 1-adrenergic receptor knockout mouse. Am J Physiol 1998;274(4 Pt 2):H1184-93.
37. Kaplan ML, Cheslow Y, Vikstrom K, Malhotra A, Geenen DL, Nakouzi A, et al. Cardiac adaptations to chronic exercise in mice. Am J Physiol 1994:H1167-73.
38. Berul CI, Aronovitz MJ, Wang PJ, Mendelsohn ME. In vivo cardiac electrophysiology studies in the mouse. Circulation 1996;94(10):2641-8.
39. Berul CI, Christe ME, Aronovitz MJ, Seidman CE, Seidman JG, Mendelsohn ME. Electrophysiological abnormalities and arrhythmias in alpha MHC mutant familial hypertrophic cardiomyopathy mice. J Clin Invest 1997;99(4):570-6.
40. Schlager G, Weibust RS. Genetic control of blood pressure in mice. Genetics 1966;55:497-506.
41. Krege JH, Hodgin JB, Hagaman JR, Smithies O. A Noninvasive computerized tail-cuff system for measuring blood pressure in mice. Hypertension 1995;25:1111-5.
42. Steinhelper ME, Cochrane KL, Field LJ. Hypotension in transgenic mice expressing atrial natriuretic factor fusion genes. Hypertension 1990;16:301-7.
43. Horsman MR, Kristjansen PEG, Mizuno M, Christensen KL, Chaplin DJ, Quistorff B, et al. Biochemical and physiological changes induced by nicotinamide in a C3H mouse mammary carcinoma and CDF1 mice. J Radiation Oncology Biol Phys 1991;22:451-4.

44. Horsman MR, Christensen KL, Overgaard J. Relationship between the hydralazine-induced changes in murine tumor blood supply and mouse blood pressure. J Radiation Oncology Biol Phys 1991;22:455-8.
45. Merrill DC, Thompson MW, Carney CL, Granwehr BP, Schlager G, Robillard JE, et al. Chronic hypertension and altered baroreflex responses in transgenic mice containing the human renin and human angiotensinogen genes. J Clin Invest 1996;97:1047-55.
46. Piccini N, Knopf JL, Gross KW. A DNA polymorphism, consistent with gene duplication, correlates with high renin levels in the mouse submaxillary gland. Cell 1982;30:205-13.
47. Kim HS, Krege JH, Kluckman KD, Hagaman JR, Hodgin JB, Best CF, et al. Genetic control of blood pressure and the angiotensinogen locus. Proc Natl Acad Sci U S A 1995;92(7):2735-9.
48. Tian B, Meng QC, Chen YF, Krege JH, Smithies O, Oparil S. Blood pressures and cardiovascular homeostasis in mice having reduced or absent angiotensin-converting enzyme gene function. Hypertension 1997:128-33.
49. Hansen JE, Sue DY, Wasserman K. Predicted values for clinical exercise testing. Am Rev Respir Dis 1984;129(suppl.):S49-55.
50. Davis JE, van Dyke HB. The oxygen consumption of fasting white mice. J Biol Chem 1933;100:455-62.
51. Jones JH, Lindstedt SL. Limits to maximal performance. Annu Rev Physiol 1993;55(547):547-69.
52. Yoshimura A, Shimomura Y, Murakami T, Ichikawa M, Nakai N, Fujitsuka C, et al. Glycogen depletion of the intrafusal fibers in a mouse muscle spindle during prolonged swimming. Am J Physiol 1996:R398-408.
53. Dohm MR, Richardson CS, Garland TJ. Exercise physiology of wild and random-bred laboratory house mice and their reciprocal hybrids. Am J Physiol 1994:R1098-108.
54. Fushiki T, Matsumoto K, Inoue K, Kawada T, Sugimoto E. Swimming endurance capacity of mice is increased by chronic consumption of medium-chain triglycerides. J Nutr 1995;125(3):531-9.
55. Chappell MA. Maximum oxygen consumption during exercise and cold exposure in deer mice, Peromyscus maniculatus. Respir Physiol 1984;55(3):367-77.
56. MacNeil B, Hoffman GL. Chronic exercise enhances in vivo and in vitro cytotoxic mechanisms of natural immunity in mice. J Appl Physiol 1993;74(1):388-95.
57. Lu YH, Xin CL, Zhou YF, Liu XW, Chi JW, Chang X. [Effect of hedgehog hydnum on the delay of fatigue in mice]. Sheng Li Hsueh Pao 1996;48(1):98-101.
58. Leeuwenburgh C, Leichtweis S, Hollander J, Fiebig R, Gore M, Ji LL. Effect of acute exercise on glutathione deficient heart. Mol Cell Biochem 1996;156(1):17-24.
59. Shimura S, Hasegawa T, Saburo T, Suzuki T. Studies on the effect of octacosanol on motor endurance in mice. Nutr. Rep. Int. 1987;36(5):1029-38.
60. Hudecki MS, Pollina CM. Mdx mouse as therapeutic model system: development and implementation of phenotypic monitoring. In: Griggs R, Karpati G, editors. Myoblast Therapy. New York: Plenum Press; 1990. p. 251-65.
61. Hudecki MS, Pollina CM, Granchelli JA, Daly MK, Byrnes T, Wang JC, et al. Strength and endurance in the therapeutic evaluation of prednisolone-treated MDX mice. Res Commun Chem Pathol Pharmacol 1993;79(1):45-60.
62. Muntoni F, Mateddu A, Marchei F, Clerk A, Serra G. Muscular weakness in the mdx mouse. J Neurol Sci 1993;120(1):71-7.
63. Pachner AR, Itano A, Ricalton N, Choe S. Chronic murine experimental myasthenia gravis: strength testing and serology. Clin Immunol Immunopathol 1991;59(3):398-406.
64. McCall RD, Frierson DJ. Inheritance of hypoxic exercise tolerance in mice. Behav Genet 1997;27(3):181-90.
65. Ruwitch L, Schauble E, Desai K, Kobilka B, Bernstein D. Role of ß-Adrenergic Signaling in Exercise Conditioning in Mice. Pediatric Research 1999;45:30A.
66. Ingalls CP, Barnes WS, Smith SB. Interaction between clenbuterol and run training: effects on exercise performance and MLC isoform content. J Appl Physiol 1996;80(3):795-801.
67. Graham BH, Waymire KG, Cottrell B, Trounce IA, MacGregor GR, Wallace DC. A mouse model for mitochondrial myopathy and cardiomyopathy resulting from a deficiency in the heart/muscle isoform of the adenine nucleotide translocator. Nat Genet 1997;16(3):226-34.
68. Aigner R, Frick H. [The effect of training and anabolic agents on the weight of the heart and extremities]. Morphol Med 1981;1(1):43-52.
69. Bonen A, McCullagh KJ. Effects of exercise on lactate transport into mouse skeletal muscles. Can J Appl Physiol 1994;19(3):275-85.

70. Hatta H, Soma R. Effect of DCA administration or endurance training on lactate metabolism in mice during and after exercise. Ann Physiol Anthropol 1992;11(3):339-43.
71. Baker MA, Horvath SM. Influence of water temperature on oxygen uptake by swimming rats. J Appl Physiol 1964;19(6):1215-8.
72. Nery LE, Wasserman K, Andrews JD, Huntsman DJ, Hansen JE, Whipp BJ. Ventilatory and gas exchange kinetics during exercise in chronic airways obstruction. J Appl Physiol 1982;53(6):1594-602.
73. Weissman ML, Jones PW, Oren A, Lamarra N, Whipp BJ, Wasserman K. Cardiac output increase and gas exchange at start of exercise. J Appl Physiol 1982;52(1):236-44.
74. Geisterfer-Lowrance AAT, Christe M, Conner DA, Ingwall JS, Schoen FJ, Seidman CE, et al. A mouse model of familial hypertrophic cardiomyopathy. Science 1996;272(5262):731-4.
75. Chruscinski AJ, Rohrer DK, Schauble E, Desai KH, Bernstein D, Kobilka BK. Targeted disruption of the beta2 adrenergic receptor gene. J Biol Chem 1999;274(24):16694-700.
76. Rohrer DK, Bernstein D, Chruscinski A, Desai KH, Schauble E, Kobilka BK. The developmental and physiological consequences of disrupting genes encoding beta 1 and beta 2 adrenoceptors. Adv Pharmacol 1998;42:499-501.
77. Luo W, Grupp IL, Harrer J, Ponniah S, Grupp G, Duffy JJ, et al. Targeted ablation of the phospholamban gene is associated with markedly enhanced myocardial contractility and loss of beta-agonist stimulation. Circ Res 1994;75(3):401-9.
78. Desai KH, Schauble E, Luo W, Kranias E, Bernstein D. Phospholamban deficiency does not compromise exercise capacity. Am J Physiol 1999;276(4 Pt 2):H1172-7.
79. Redfern CH, Degtyarev MY, Kwa AT, Salomonis N, Cotte N, Nanevicz T, et al. Conditional expression of a Gi-coupled receptor causes ventricular conduction delay and a lethal cardiomyopathy. Proc Natl Acad Sci U S A 2000;97(9):4826-31.
80. Maxwell AJ, Niebauer J, Lin PS, Bernstein D. Hypercholesterolemia reduces aerobic capacity by altering NO synthase activity. Circulation 1996;94(8):I-347 (abstract).
81. Haywood JR, Shaffer RA, Fastenow C, Fink GD, Brody MJ. Regional blood flow measurement with pulsed Doppler flowmeter in conscious rat. Am J Physiol 1981;241(2):H273-8.
82. Gardiner SM, Compton AM, Bennett T, Hartley CJ. Can pulsed Doppler technique measure changes in aortic blood flow in conscious rats? Am J Physiol 1990;259:H448-56.

19
CLINICAL LABORATORY ANALYSIS OF THE GENETICALLY MANIPULATED MOUSE

William Lewis, M.D.
Emory University, Atlanta, Georgia 30322 USA

INTRODUCTION

The importance of analyzing routine chemistries in living genetically manipulated mice is increasingly clear. Analogous data bases of serum and/or plasma analytes exist for humans, are commonly used in screening human patients for the presence of health and disease, and serve as important clinical screening tools for the presence of cardiovascular diseases, their prognosis, and efficacy of therapy. These data are absent or incomplete for the mouse. In a correlative way, the availability of these data will amplify the ability scientists to screen for cardiovascular and other diseases in the genetically manipulated mouse.

A national database of such analytical data should be formulated. If it were available, it could establish the normal values for serum and/or plasma analytes in *healthy* inbred strains of mice with respect to age, sex, and diet. It will establish degrees of variation that are found between inbred strains, gender and age groups.

It is understood further that genetic manipulations may result in a variety of insults to the organism in secondary and correlative ways. These may be expected or unexpected. Accordingly, the effects of genetic manipulations that are *directed towards* targets in the murine cardiovascular system may have indirect effects on the murine kidney (for example). Logically, this may result in kidney dysfunction. Altered kidney function, although unexpected by the investigative team, would be detected in the screen serum and/or plasma analytes that is used to examine "general health" of the mouse.

In essence, biochemical screening of genetically manipulated mice offers a panoramic, "whole organism" view of the general health of the living, genetically manipulated mouse while maintaining a focus on cardiovascular diseases. The approach offers the advantages of broad screening for a variety of analytes (in a "murine general health" panel), for disease-related screening (in the "murine cardiac disease" panels) and to pool data and to retrieve pooled data (from the database).

The resultant data provide an important service to the scientific community. Since the test can be repeated in the same animal, it is feasible to obtain longitudinal data about a given mouse. With the advent of automated chemistry and the use of micro-volume analyses, serum and/or plasma chemistry values as indicators of cardiovascular diseases (including cardiomyopathy [CM], congestive heart failure [CHF], and atherosclerotic heart disease) may be available for screening and/or clinical monitoring of the genetically manipulated mouse. Using inbred, recombinant inbred (RI), transgenic (TG) and knockout (KO) mice, serum and/or plasma chemistry analyses offer relatively efficient screens for effects of genetic manipulations *while maintaining viability of the mouse*.

As mentioned, normal ranges for analytes of widely used murine strains in genetic experiments are not available. Accordingly, significant work needs to be performed to define "biochemical health" and variations within strains as a function of age and gender, diet and other important variables. Genetically manipulated TGs and KOs with representative cardiovascular diseases may be useful as "reference values" of serum and/or plasma analytical values for established models of important cardiovascular diseases.

Serum and/or plasma analytes from newly generated TGs and KOs may be analyzed and compared to the values obtained from inbred "normals" and from specific disease models. Ultimately, it may be possible to establish a database that is useful for the scientific community that receives input and that is accessible to investigators. This enhances understanding of important cardiovascular phenotypes and correlates the physiological and pathological data with various plasma and serum values.

The targeting of genes to create relevant murine TG and KO models of cardiovascular diseases has revolutionized biology and medicine. With the advent of these exciting developments, a need for rapid screening of these models has become clear. However, rapid screening for phenotypic changes in large numbers of TGs or KOs has been hampered. Among the factors that hamper rapid screening are the small size of the mouse, the labor-intensive and frequently terminal nature of detailed pathological or physiological studies, and importantly a relative lack of information about the normal physiological parameters in inbred and genetically manipulated mice. To that extent, clinical chemistry analysis offers rapid screening and non-lethal examination.

Genetically manipulated and transgenic mouse (TG) in cardiovascular research

The use of serum analytes in the evaluation of genetically manipulated mice that exemplify cardiovascular diseases follows the analogous approach to that used in screening humans for cardiovascular and other diseases in which markers of disease are available or usable in certain populations.

Over the past decade, the introduction and structural-functional evaluation of genetically manipulated mice afforded significant insights into the pathophysiologic and pathologic mechanisms of human disease. Manipulation of the murine genome by use of "gain of function" or "loss of function" strategies yielded insights into the mechanisms of atherosclerosis [1], hypertension [2], cardiac hypertrophy and congestive heart failure [3]. The strength of TGs comes from the ability to precisely overexpress or ablate a gene of interest and examine the phenotypic consequences during murine embryogenesis, development, maturation and senescence.

Classic experimental animal models of human disease impose some form of environmental stress to elicit an array of molecular and biochemical events. Genetic manipulation involves a single molecular perturbation with resultant physiological and pathological phenotype that may (or may not) recapitulate human disease. Despite these major scientific advances in understanding the genetically manipulated mouse, serum and/or plasma analytical values that indicate and/or monitor disease are rarely reported or available. However, such analyses are commonly and successfully used in diagnostic screening, in prognosis, and in monitoring effects of treatment of cardiovascular disease in humans. It seems reasonable that such analyses should be useful for screening disease in the genetically manipulated mouse.

The mouse is (and likely will be) a preferred mammalian system for genetic manipulation. Its small size, relatively low maintenance costs, rapid gestation, and large litter size are practical advantages. Extensive knowledge of the mouse genome exists. Over 10,000 genetic markers, including 6,000 microsatellite sequences have been described [4]. The mouse genome is essentially sequenced.

Many inbred mouse strains are available to facilitate the study of the influence of genetic background upon gene expression. Finally, germ line transmission has only been achieved with mouse embryonic stem cells [5]. Although detailed studies are elegant and technically challenging, they are impractical for screening large numbers of mice for evidence of genetic manipulation. It is for this reason that screening of serum and/or plasma analytes offers the scientist an effective window to explore the presence of cardiovascular disease in the genetically manipulated mouse. Using automated analysis and microvolume sampling, blood analysis is an efficient and practical way to screen for cardiovascular diseases in the genetically manipulated mouse.

Strain differences

Differences in normal cardiovascular performance were identified in different murine strains, by gender, and by age, and are becoming more apparent as we gather a better understanding of murine biology [6]. In early investigations, these subtle differences were frequently overlooked yet potentially confounding variables [7]. In one important study, mice responded with different +dp/dt and -dp/dt, important indices of cardiac contractility and relaxation. Other data indicate that variations in a multitude of biological phenomena (bone induction, cardiovascular function, development of vascular disease, testicular development) are dependent on the strain of mouse used experimentally. One study examined ectopic osteo-induction. That study indicated significant variability in analytes tested between various strains. The volume of newly formed tissue ranged from 1.3% in the A/J strain to 74.6% in the DBA/2J strain four weeks after implantation of rat bone gelatin [8]. Atherosclerosis susceptibility was tested in ten inbred strains of mice and revealed striking differences between strains [9]. After 5 weeks of atherogenic diet, plasma cholesterol ranged from 131 mg/dl to 328 mg/dl. After 14 weeks on the diet, frequency of lesion formation ranged from 0 to 1.8 lesions/mouse depending only on strain. Most recently, a 16-fold difference in susceptibility to interruption of reproductive development was found in commonly used strains of inbred mice [10]. To optimally define disease parameters, it appears crucial to search for differences in values of serum and/or plasma analytes that may exist in different murine strains.

Diagnostic and experimental pitfalls

False positive or false negative information regarding the existence and nature of a cardiovascular phenotype in genetically engineered mice may be derived from studies with inadequate attention to several variables. This suggests the importance of an integrated approach and speaks to the utility of analysis of a panel serum and/or plasma chemistry analytes in which broad-based information is obtained about the well being of the organism. By the same token, data obtained in screens of large numbers of mice offers information about the general well being of mice with presumed cardiovascular disease, but absent liver disease (for example).

The importance of examining the entire organism should not be underestimated when attempting to attribute very specific effects (like manipulation of a single gene driven by a specific promoter) on a complex living organism. The proposed screening tests offer the investigator an opportunity to target the cardiovascular system and its integrity through rapid non-lethal analyses while simultaneously examining the general well being of the mouse that is subjected to genetic manipulations.

The use of specific markers in the genetically manipulated mouse

These analytes would likely not fall under routine analysis, yet their important to the diagnosis of cardiovascular disease is implicit. As mentioned, the general health and cardiovascular health of the mouse has not been examined extensively using a panel of serum and/or plasma analytes. Cytokines play key roles in cardiac dysfunction, cardiomyopathy (CM), myocarditis, and CHF and other important cardiovascular diseases. It is reasonable to consider that serum and/or plasma cytokine determinations may be a logical screening point for detecting heart disease in the genetically manipulated mouse.

Evaluation of the presence of heart failure may be amenable to screening through plasma or serum analyses and immunoassay. Such approaches may define illness early (or even pre-clinically) in the genetically manipulated mouse. Circulating or locally acting cytokines including endothelin (ET-1), and tumor necrosis factor α (TNFα) are shown to be important in CHF and CM (reviewed in [11-13]). Left ventricle (LV) expression of atrial natriuretic factor (ANF) is a marker of LV hypertrophy (LVH) and cardiac dysfunction [14]. It follows that ANF serum and/or plasma levels may be effective markers for those conditions.

A cogent framework for understanding the effects of cytokines on the cardiomyocyte comes from Werdan and colleagues [15] who proposed that these substances have negative inotropic effects. Increased expression of both inducible nitric oxide syntheses (iNOS) mRNA and polypeptide was demonstrated *in vitro* in cardiac myocytes treated with TNFα, IL-1β, IFN-γ [16]. Myocyte death in culture paralleled increased nitric oxide (NO) synthesis. Myocyte death could occur as a result of NO produced by macrophages or within the myocyte.

Cytokines do not cause myocyte dysfunction or necrosis directly. However, evidence suggests that they alter myocyte function, possibly through the β_1-adrenoceptor-G protein-adenylyl cyclase axis [15, 17-19]. Interference with this axis has been studied with cardiac cells *in vitro*. Long-term treatment of cardiomyocytes with immune cell supernatants (containing IL-1 and TNFα) reduced contractility and cyclic adenosine monophosphate (cAMP) accumulation by inhibition of β–adrenergic responsiveness [20]. Myocardial depressant effects of TNFα have been studied extensively. Effects are established on cardiomyocytes in culture. TNFα infusion in dogs caused LV dysfunction [21]. Conversely, anti-TNFα antibodies reduced cardiac dysfunction during sepsis [22].

Plasma ET-1 concentrations are increased in patients with CHF [23-26] and suggest poor prognosis in patients with NYHA class III or IV CHF. ET receptors are located on several cell types in the myocardium, including myocytes, fibroblasts, and endothelial cells. *In vitro*, endothelin stimulates myocyte hypertrophy and expression of a fetal phenotype [27]. It also has important effects on the synthesis and degradation of the extracellular matrix [28].

The mechanism that underlies the progressive nature of cardiac dysfunction is termed *remodeling* [29]. Remodeling encompasses a complex of molecular and cellular events that lead to changes in the structure, function, and phenotype of the myocardium. These include hypertrophy and death of myocytes, regression to a molecular phenotype characterized by the expression of fetal genes and proteins, and alterations in the quantity and composition of the extracellular matrix. Since some of these alterations are associated with expression of fetal gene products (*e.g.*, ANF), it is possible that screening cohorts of the genetically manipulated mouse may be useful to detect this group.

ANF is a 28-amino-acid peptide that is normally synthesized in the atria and to a lesser extent in the ventricles and is released into the circulation during atrial distention [30]. In patients with CHF, plasma ANF concentrations rise as atrial pressure increase. Brain or B-type natriuretic peptide is a 32-amino-acid peptide that is synthesized primarily in the ventricles, and its release into the circulation is also increased in patients with CHF [31]. Because plasma concentrations of brain natriuretic peptide are increased in patients with early CHF, plasma brain natriuretic peptide also may be a sensitive diagnostic marker of CHF. Patients with CHF are resistant to the natriuretic effect of exogenously administered natriuretic peptides [32]. Increased secretion of atrial natriuretic peptide and brain natriuretic peptide in early CHF may attenuate or delay systemic and renal arterial vasoconstriction, venoconstriction with increased cardiac preload, and renal sodium retention. In animals with CHF, administration of antibodies to atrial natriuretic peptide or antagonists to atrial natriuretic peptide receptors decreases renal blood flow; increases right atrial pressure, plasma renin activity, and sodium retention; and worsens diastolic dysfunction by impairing myocardial relaxation [33]. The infusion of synthetic brain natriuretic peptide decreases pulmonary-capillary wedge pressure, diminishes systemic vascular resistance, and increases cardiac output in humans [34].

Increased sympathetic tone that occurs with ventricular dysfunction has several consequences. Increased myocardial contractility, tachycardia, arterial vasoconstriction and thus increased cardiac afterload, and venoconstriction with increased cardiac preload all result. β-adrenergic receptors in the heart either are down-regulated (β_1-adrenergic receptors) or have abnormalities in signal-transduction activity that effectively uncouple them from effector mechanisms (β_1- and β_2-adrenergic receptors; [34]. Increased local and circulating concentrations of norepinephrine (NE) may contribute to myocyte hypertrophy, either directly through stimulation of α_1- and β-adrenergic receptors or secondarily by activating renin-angiotensin-aldosterone [34]. NE's toxicity to myocardial cells is mediated through calcium overload, the induction of apoptosis, or both [12]. NE-induced death of myocytes can be prevented by β-adrenergic blockade or combined β- and α-adrenergic blockade [12]. Patients with plasma NE concentrations >800 pg per milliliter (4.7 nmol per liter) have a one-year survival <40%. Plasma catecholamine and NE levels may be effective as analytes to screen for CHF in mice.

Renin-angiotensin-aldosterone system activity is also increased in CHF. As with plasma NE and ANF, the degree of increase in plasma renin activity provides a prognostic index in these patients [35, 36]. If the assay were feasible in microvolume samples, this could serve as a useful screen of cardiovascular disease in mice.

Plasma analysis of general biochemical health: FVB/n mice age-matched (10-12 weeks of age), weight matched, both genders, using 100 μL of plasma

To examine the analysis of plasma chemistries in inbred mice, FVB/n mice were housed in AALAC approved barrier facilities. At weaning, mice were separated by gender and allowed to feed and drink *ad libitum*. At ≈12 weeks, animals were brought to the laboratory and subjected to light inhalation anesthesia. Retro-orbital bleeding was performed according to IACUC guidelines. Heparinized capillary tubes (2 tubes per mouse) were used to withdraw ≈120 μL/tube blood. Blood was rapidly transferred to Eppendorff tubes and subjected to centrifugation at 1500xg at 4°C. Plasma was separated rapidly, and 90 μL was transferred to rotors in a rotor based "point of care" blood analysis instrument (Hemagen Analyst, Waltham, MA). Other instruments of its kind are available. The following data were obtained from that procedure. *All animals survived.* Total bleeding time was 182 minutes for all animals.

Data are interesting and help point us to select some murine-specific tests and to refine some of the ranges for selected analytes. For example, creatinine levels (an important index of renal disease) were relatively low in all samples. This may relate to the relatively small muscle mass in the mouse (body weight ≈25g). Other analytical ranges (CK) also will require adjustment for these analyses. The following table presents the data obtained from this study (n=36 total). Overall, the reproducibility was fairly good. Coefficients of variation (CV) based on repetitive analyses ranged from under 1% to

6%.

Comparisons from literature values for serum chemistries of different inbred strains of mice

Table 1 is partially derived from data obtained in our laboratories and from the literature values available from the internet web site of Charles River Labs (CRL), one of the largest commercial breeders for animals used in biomedical experimentation. The strains used are our own FVB/n, ICR and CF1 (from CRL).

The ages of the mice in each strain were dramatically different. The FVB/n mice from Emory (right column) were younger than those from CRL. Of note, some values remain relatively constant amongst all groups, ages and both genders (e. g. BUN). Some demonstrate somewhat wider variation that occurs across strains (e.g., Alk Phos, GGT) or within strain between genders (e.g. GPT for FVB/n or Alk Phos for ICR and CF-1).

It is important to note that the sample size for FVB/n is relatively small. However, the data re-emphasize the fact that a relative paucity exists of normal serum and/or plasma chemistry values for mice. This strengthens our conviction about the importance of obtaining these data in the proposed experimental program. Based on initial studies, some channels may require refinement for use in screening the genetically manipulated inbred mice in these experiments. Adjustments to the rotor can be made to accommodate murine normal values.

"Point of experiment" analysis Vs central laboratory

The availability of data on serum and/or plasma analytes in the mouse is limited. Moreover, a paucity of such data are available in the literature. A major commercial source of inbred and outbred mice, Charles River (CRL), reported "base-line" normal data for some of their mouse strains in which three different age groups of mice were evaluated by gender. The data were reported on their website and we have included some of them (refer to Table 1, below).

Single or multiple analytes have been measured and reported in isolation, rarely does one screen a large number of routine serum and/or plasma analytes to assess the well being of the animal. If it is affirmed (as we and others suggest) that blood chemistry analyses are necessary, should there be a centralized laboratory to which investigators send samples or is a "point of experiment" instrument more effective and practical? In the latter case, the investigator determines an entire clinical chemistry panel. In the former, the investigator puts his experiment at risk of inherent problems of sending samples by courier ("down time", analytical variation owing to physical alterations in shipping such as temperature, breakage, loss, biochemical deterioration of sample, etc). from FVB/n in table)

Table 1: Murine values for selected serum and plasma chemistries: CRL values (≈20 weeks old; CRL strains) Vs FVB/n values (11-12 weeks old partial study of samples.

STRAIN	ICR_BR(Charles River)		CF-1-BR		FVB/n(Emory)	
AGE	19-21 WEEKS		19-21 WEEKS		11-12 WEEKS	
GENDER	F	M	F	M	F	M
SAMPLE SIZE	N=20	N=20	N=20	N-20	N=6	N=9
ANALYTE	Mean (range)	Mean (range)	Mean (range)	Mean (range)	(range)	(range)
ALP	162 (118-206)	166 (86-248)	192 (149-235)	145 (108-182)	131.2 (105-170)	95 (58-130)
GOT(AST) U/L	137 (87-187)	248 (24-472)	209 (67-351)	187 (86-308)	167.8 (114-199)	115.4 (63-186)
GPT(ALT) U/L	102 (56-159)	109 (28-190)	299 (96-482)	195 (63-307)	44.2 (31-80)	55.6 (27-241)
BUN MG/DL	28 (21-35)	33 (23-43)	27 (22-32)	22 (18-28)	24.8 (21.9-28.1)	23 (21.1-25.5)
GLU MG/DL	265 (207-320)	290 (185-270)	241 (185-297)	256 (178-334)	225.2 (213-242)	244 (214-273)
PHOS MG/DL	11.2 (7.4-15.0)	13.2 (7.8-18.6)	13.1 (9.7-16.5)	13.4 (10.9-15.9)	14.2 (12.9-16.5)	12.65* (11.3-14.9)
ALB G/DL	3.3 (2.9-3.7)	2.6 (1.8-3.4)	3.5 (3.3-3.7)	3.2 (2.9-3.5)	3.1 (2.9-3.2)	2.8 (2.4-3.1)
CHOL MG/DL	59 (41-77)	61 (31-91)	96 (80-110)	134 (117-151)	134.4 (124-146)	152 (126-183)
CREA MG/DL	0.6 (0.4-0.8)	0.5 (0.2-0.8)	0.6 (0.3-0.9)	0.5 (0.2-0.8)	<0.2	<0.2
TBIL MG/DL	0.3 (0-0.5)	0.2 (0.1-0.3)	0.2 (0.1-0.3)	0.2 (0.1-0.3)	<0.1	<0.1
TP G/DL	4.7 (4.2-5.2)	4.3 (3.0-5.5)	5.7 (5.3-6.1)	5.6 (5.2-6.0)	4.9 (4.8-5.1)	4.9 (4.7-5.2)
GLOB G/L	1.4 (1.2-1.6)	1.7 (1.2-2.2)	2.2 (1.8-2.6)	2.4 (2.1-2.7)	1.9 (1.6-2.0)	2.07 (1.6-2.8)
ALB/GLOB	1.3 (1.03-1.57)	1.6 (1.2-1.9)	1.59 (1.2-1.98)	1.32 (1.12-1.52)	1.7 (1.5-1.8)	1.4 (0.9-1.9)

ALB/GLOB=albumin/globulin
ALB=albumin
ALP=alkaline phosphatase
BUN=blood urea nitrogen
CHOL=cholesterol
GLOB=globulin
GLU=glucose
GOT(AST)=glutamic-oxaloacetic transaminase
GPT(ALT)=glutamic-pyruvic transaminase
PHOS=phosphate
TBIL=total bilirubin
TP=total protein

If a uniform system were available through a common use resource nationwide, a "point of experiment" system could be created in which each investigator obtains data immediately. The investigator then could quickly analyze and interpret data. These data from individual investigators could be fed into a centralized database or repository. Information in that repository would become available to all investigators. This vital system would provide each investigator with enormous databases, would allow rapid evaluation of the genetically manipulated mouse, and allow for unique and otherwise unavailable comparisons of amongst investigators throughout the scientific community.

Importantly, although it is understood that the focus of the chapter is on cardiovascular effects, data could become available from investigators both within and outside of this specialized area. Additionally, this approach offers the investigator the opportunity to define his control values in non-manipulated mice and to determine if any alterations have occurred as a result of extensive inbreeding (for example). Thus at any time, one has the ability to monitor the wild type mouse and compare the results to those archived in the database.

Logic supporting the "point of experiment" approach

If this is a sensible way to go, what parameters are necessary to develop a successful program using classical clinical chemistry measurements to monitor the health of the mouse? The following is a proposed list of requirements for success of the project:

(1) Point of Experiment Instrument

- a. low cost
- b. ease of use
- c. compatibility with the small volumes from mouse

(2) Clinical Chemistry Markers

- a. Available Now
    1. General Health Profile
    2. Cardiac Specific Profile
- b. Future Developments

(3) Establishment of Centralized Database

"Point of Experiment" Instrument

Today, many instruments are available from the clinical realm that are used for "point of care" analyses on micro-volume samples without the need for a centralized facility. Some include rotor based single specimen instrument, cleared by the FDA to be marketed in the US with a calibrated, disposable rotor. A plasma or serum volume of 90 μL or more is required to perform sixteen simultaneous assays. This volume of plasma is readily available from the mouse, even with sequential draws while maintaining the animal alive and well for a subsequent drawing several weeks later. A requirement is that the instrument is extremely easy to use and that tests are performed rapidly (in 10 minutes, for example). Cost of most rotor based analyses is approximately $9 per unit. Thus a single animal can be evaluated for under $10 expenditure.

General health of the genetically manipulated mouse

The value of low cost, point of experiment, easy to use instrumentation has been addressed. The following section addresses the rationale for having a general health rotor in the evaluation of the genetically manipulated mouse.

Although remarkable advances have been made and will continue in mouse genetic research, important concerns remain regarding the specificity of genetic manipulations of the cardiovascular system and other physiological systems. Do resultant changes yield true animal models of human disease? Perhaps more importantly, how accurately do the genetically manipulated mice display phenotypic characteristics of the disease one is trying to mimic? Are the appropriate non-targeted organ systems intact while the targeted organ(s) modified?

It should be expected that a therapeutic intervention (*e.g.* cholesterol lowering agents for use in the genetically manipulated mouse with certain forms of hypercholesterolemia) should result in some therapeutic effect and ultimately be manifested by the observed lowering of serum cholesterol (for example). Such a result would be expected in the physician's office where the patient being treated for hypercholesterolemia were monitored to define successful therapy. Another feature is the availability of data to assess the phenotypic characteristics of the mouse strain being studied and to evaluate the phenotypic changes that occur (or do not occur) as a result of a particular genetic manipulation. Many times an investigator and his collaborators are working in "relative isolation." The feedback from a well-qualified database will be extremely important.

Table 2: Panel of "general health" analytes for screening of the genetically manipulated mouse:

ALP	GGT	AST	ALT
AMY	BUN	GLU	PHOS
CA	ALB	CHOL	UA
CK	CREA	TBIL	TP

ALB=albumin
ALP=alkaline phosphatase
ALT=alanine aminotransferase
AMY=Amylase
AST=asparate aminotransferase
BUN=blood urea nitrogen
CA=calcium
CHOL=cholesterol
CK=creatine kinase
CREA=creatinine
GGT=gamma glutamyl transferase
GLU=glucose
PHOS=phosphate
TBIL=total bilirubin
TP=total protein
UA=uric acid

Table 2 describes the analytes evaluated in the mouse that indicate "general health" or that may indicate some form of illness of the cardiovascular or other systems (*e.g.*, liver or kidney).

Commercially available rotors are already developed to address the general health of domestic animals. Based on data presented below, little is necessary to modify the rotor for specific use in the mouse. Data from normal, 12 week old, male and female FVB/n mice fed routine lab chow in our institution suggest that most of the analyte determinations are quantifiable in FVB/n mice.

Blood sampling and sample handling

As mentioned previously, less than 100 µL of plasma or serum are necessary for the initial screening panel. Previous experience with obtaining blood from FVB/n and our own genetically manipulated mice (with FVB/n background) indicates that retro-orbital bleeding was rapid, safe and effective. Bleeding 200 µL blood into heparinized capillary tubes enables us to obtain approximately 100-120 µL of plasma. Of that total volume, ≈90 µL is useable for the analyte panel and approximately 10-30 µL plasma for other analyses. Bleeding a larger volume than 200 µL could pose a hazard to the mouse. Alternative blood collection methods include tail vein bleeding, which is more labor intensive. Sample is taken and subjected to centrifugation in an Eppendorff tube (1500*X*g; 4°C).

Effects of repetitive sampling in mice

One major advantage to the use of serum and/or plasma analysis as proposed is repetitive sampling as a function of age, disease development or in response to treatment of disease *while preserving the living organism.*

Cardiac-specific analytes

The following panel of analytes may serve as markers for cardiovascular disease to evaluate the genetically manipulated mouse and the controls (Table 3).

Table 3: Panel of "cardiac specific" analytes: ischemic heart disease

CK	CKMB	HDL cholesterol	CRP
LDH	Lactic Acid	ApoA1	ApoB
Troponin I	Troponin T		

CK=creatinine kinase
CKMB=creatinine kinase/MB isoform
HDL=high density lipoprotein
CRP=C-reactive protein
LDH=lactate dehydrogenase
ApoA1=apolipoprotein A1
ApoB=apolipoprotein B

All of the above assays fall into enzymatic colorimetric or immunoturbidometric assay formats and have been tested for feasibility (except for the troponins). An assay method may employ a Roche Cobas Mira or a Hitachi 917.

"Cardiac specific" analytes for CM and CHF

As mentioned above, some markers for CHF and CM are included in a screening program. It is increasingly clear that a panel of analytes from serum or plasma may be useful for the evaluation of CM or CHF in the genetically manipulated mouse.

Coupled with data for determination of "general health" of the genetically manipulated mouse (described above), this third group of analytes offers screening information for the condition of CHF like CM. The rationale for choosing the analytes listed below was described above. Their utility is established in evaluating CM and CHF in humans. Table 4 below is a selected series of analytes to be part of potential "cardiac specific" panel.

The remaining 20-30 µL of plasma from the original bleed is used in this panel of analyses. TNFα and catecholamines may be the most logical choices based on limitations of sample size, sample preparation requirements and detection limits. These are available in commercial kits (Pharmingen, Amersham and others). Details of the analyses for the two important and immediately feasible studies are included below.

Although their importance in the analysis of CHF and CM also is clear, ET-1 and ANF may pose some technical problems which are not resolved as of this writing.

Table 4: "Cardiac specific" analytes: CM and CHF

Analyte
Tumor necrosis factor alpha (TNFα)
Plasma catecholamines

Serum levels of TNFα in TGs and WTs

At present, TNFα is an important indicator of cardiovascular disease and for purpose of screening the genetically manipulated mouse, may serve as the most suitable analyte. This relates to the sensitivity assay resulting in the relatively small sample volume needed for analysis. TNFα levels in serum are determined at time of bleeding. Serum is collected via retro-orbital bleed (described above; ≤20 µL plasma used).

Cardiac levels of catecholamines in the genetically manipulated mouse

Elevated plasma catecholamines were an effective marker of CHF. The assay for catecholamines described here follows the basic principles of Passon and Peuler [37]. Commercially available assays of catecholamines use [^{3}H] radioenzymatic assay system for a reliable and precise quantitative determination of epinephrine (E), norepinephrine (NE) and dopamine, in plasma. The system utilizes the enzyme catechol-O-methyltransferase (COMT) to catalyze the transfer of a [^{3}H]-methyl group [^{3}H] S-adenosyl-L-methionine ([^{3}H]-SAM) to the OH in the 3-position of E, NE and dopamine. Separation of [^{3}H] products (metanephrine, normetanephrine, methoxytyramine) is by thin-layer chromatography. The band corresponding to [^{3}H]methoxytyramine is eluted and counted directly; bands containing [^{3}H]metanephrine and [^{3}H]normetanephrine are eluted, converted by oxidation to [^{3}H]vanillin, and counted. The assay system can alternatively measure 'total catecholamines'. E and NE are measured in the range of 2-5 pg and dopamine in the range 15-20pg. Ten to 20 µL plasma sample is used.

Additional assays and future developments

It is clear that if exsanguination is performed, adequate blood is available for studies that analyze ANF, ET-1 and plasma renin activity, all of which may be important analytes in defining CM and CHF and assays are available commercially for large sample volumes. An important step as we proceed into the grant period is to define the feasibility of ET-1 and ANF assays in the mouse using microvolumes of serum and/or plasma. If reverse phase is mandatory in ET-1 evaluations, this severely limits the feasibility of using such assays. More sensitive RIA and ELISA methods may satisfy the requirements of small sample size and limited detection levels in the mouse as they are developed and may make the laboratory analysis of the genetically manipulated mouse a common approach to screening and diagnosis.

REFERENCES:

1. Breslow JL. Mouse models of atherosclerosis. Science 1996; 272(5262):685-8.
2. Thompson MW, Merrill DC, Yang G, et al. Transgenic animals in the study of blood pressure regulation and hypertension. Am J Physiol 1995; 269(5 Pt 1):E793-803.
3. Chien KR. Cardiac muscle diseases in genetically engineered mice: evolution of molecular physiology. Am J Physiol 1995; 269(3 Pt 2):H755-66.
4. Paigen K. A miracle enough: the power of mice. Nat Med 1995; 1(3):215-20.
5. Mullins LJ, Mullins JJ. Transgenesis in the rat and larger mammals. J Clin Invest 1996; 97(7):1557-60.
6. Kiatchoosakun S, Kirkpatrick D, Olszens K, et al. Naturally Occurring Variation in Inbred Mouse Strains: A Mouse Model of the Athlete's Heart. Circulation 2000; 102(18):II-194.
7. Grupp IL, Subramaniam A, Hewett TE, et al. Comparison of normal, hypodynamic, and hyperdynamic mouse hearts using isolated work-performing heart preparations. Am J Physiol 1993; 265(4 Pt 2):H1401-10.
8. Marusic A, Katavic V, Grcevic D, Lukic IK. Genetic variability of new bone induction in mice. Bone 1999; 25(1):25-32.
9. Paigen B, Morrow A, Brandon C, et al. Variation in susceptibility to atherosclerosis among inbred strains of mice. Atherosclerosis 1985; 57(1):65-73.
10. Spearow JL, Doemeny P, Sera R, et al. Genetic variation in susceptibility to endocrine disruption by estrogen in mice [see comments]. Science 1999; 285(5431):1259-61.
11. Bristow MR. Tumor necrosis factor-alpha and cardiomyopathy [editorial; comment]. Circulation 1998; 97(14):1340-1.
12. Mann DL, Kent RL, Parsons B, Cooper Gt. Adrenergic effects on the biology of the adult mammalian cardiocyte. Circulation 1992; 85(2):790-804.
13. Wilbert-Lampen U, Seliger C, Zilker T, Arendt RM. Cocaine increases the endothelial release of immunoreactive endothelin and its concentrations in human plasma and urine: reversal by coincubation with sigma-receptor antagonists. Circulation 1998; 98(5):385-90.
14. Rosenzweig A, Seidman CE. Atrial natriuretic factor and related peptide

hormones. Annu Rev Biochem 1991; 60:229-55.

15. Werdan K, Muller U, Reithman C. Negative inotropic cascades in cardiomyocytes triggered by substances relevant to sepsis. *In* Schlag G, Redl H, eds. Pathophysiology of Shock, Sepsis, and Organ Failure. Berlin: Springer-Verlag, 1995. pp. 787-834.
16. Pinsky DJ, Cai B, Yang X, et al. The lethal effects of cytokine-induced nitric oxide on cardiac myocytes are blocked by nitric oxide synthase antagonism or transforming growth factor beta. J Clin Invest 1995; 95(2):677-85.
17. Levine B, Kalman J, Mayer L, et al. Elevated circulating levels of tumor necrosis factor in severe chronic heart failure. N Engl J Med 1990; 323(4):236-41.
18. Barry WH. Mechanisms of immune-mediated myocyte injury. Circulation 1994; 89(5):2421-32.
19. Lange LG, Schreiner GF. Immune mechanisms of cardiac disease. N Engl J Med 1994; 330(16):1129-35.
20. Gulick T, Chung MK, Pieper SJ, et al. Interleukin 1 and tumor necrosis factor inhibit cardiac myocyte beta- adrenergic responsiveness. Proc Natl Acad Sci U S A 1989; 86(17):6753-7.
21. Pagani FD, Baker LS, Hsi C, et al. Left ventricular systolic and diastolic dysfunction after infusion of tumor necrosis factor-alpha in conscious dogs. J Clin Invest 1992; 90(2):389-98.
22. Vincent JL, Bakker J, Marecaux G, et al. Administration of anti-TNF antibody improves left ventricular function in septic shock patients. Results of a pilot study. Chest 1992; 101(3):810-5.
23. Pacher R, Stanek B, Hulsmann M, et al. Prognostic impact of big endothelin-1 plasma concentrations compared with invasive hemodynamic evaluation in severe heart failure. J Am Coll Cardiol 1996; 27(3):633-41.
24. Cody RJ, Haas GJ, Binkley PF, et al. Plasma endothelin correlates with the extent of pulmonary hypertension in patients with chronic congestive heart failure [published erratum appears in Circulation 1993 Mar;87(3):1064]. Circulation 1992; 85(2):504-9.
25. Krum H, Goldsmith R, Wilshire-Clement M, et al. Role of endothelin in the exercise intolerance of chronic heart failure. Am J Cardiol 1995; 75(17):1282-3.
26. Wei CM, Lerman A, Rodeheffer RJ, et al. Endothelin in human congestive heart failure. Circulation 1994; 89(4):1580-6.
27. Shubeita HE, McDonough PM, Harris AN, et al. Endothelin induction of inositol phospholipid hydrolysis, sarcomere assembly, and cardiac gene expression in ventricular myocytes. A paracrine mechanism for myocardial cell hypertrophy. J Biol Chem 1990; 265(33):20555-62.
28. Weber KT. Extracellular matrix remodeling in heart failure. A role for de novo angiotensin II generation. Circulation 1997; 96:4065-82.
29. Colucci WS. Molecular and cellular mechanisms of myocardial failure. Am J Cardiol 1997; 80(11A):15L-25L.
30. Boland DG, Abraham WT. Natriuretic peptides in heart failure. Congestive Heart Fail 1998; 4:23-33.
31. Saito Y, Nakao K, Arai H, et al. Atrial natriuretic polypeptide (ANP) in human

ventricle. Increased gene expression of ANP in dilated cardiomyopathy. Biochem Biophys Res Commun 1987; 148(1):211-7.
32. Cody RJ, Atlas SA, Laragh JH, et al. Atrial natriuretic factor in normal subjects and heart failure patients. Plasma levels and renal, hormonal, and hemodynamic responses to peptide infusion. J Clin Invest 1986; 78(5):1362-74.
33. Drexler H, Hirth C, Stasch HP, et al. Vasodilatory action of endogenous atrial natriuretic factor in a rat model of chronic heart failure as determined by monoclonal ANF antibody. Circ Res 1990; 66(5):1371-80.
34. Abraham WT, Port JD, Bristow MR. Neurohormonal receptors in the failing heart. *In* Poole-Wilson PA, Colucci WS, Massie BM, et al., eds. Heart Failure. New York, NY: Churchill Livingstone, 1997. pp. 127-41.
35. Francis GS, Benedict C, Johnstone DE, et al. Comparison of neuroendocrine activation in patients with left ventricular dysfunction with and without congestive heart failure. A substudy of the Studies of Left Ventricular Dysfunction (SOLVD). Circulation 1990; 82(5):1724-9.
36. Levine TB, Francis GS, Goldsmith SR, et al. Activity of the sympathetic nervous system and renin-angiotensin system assessed by plasma hormone levels and their relation to hemodynamic abnormalities in congestive heart failure. Am J Cardiol 1982; 49(7):1659-66.
37. Passon PG, Peuler JD. A simplified radiometric assay for plasma norepinephrine and epinephrine. Anal Biochem 1973; 51(2):618-31.

20

SMOOTH MUSCLE STUDIES USING GENE-ALTERED MOUSE MODELS: A USERS' GUIDE

Roy L. Sutliff and Richard J. Paul
Department of Pathology
Emory University
Atlanta, GA 30322
and
Department of Molecular and Cellular Physiology
University of Cincinnati, College of Medicine,
Cincinnati, OH 45267-0576

INTRODUCTION

The coming of age of technology for development of specific gene-ablated or transgenic mice has opened new vistas toward understanding smooth muscle function. In conjunction with rapidly developing miniaturized techniques for physiological measurement of, for example, cardiovascular or respiratory function in mice [1], it is now possible to study the effects of gene modification at the smooth muscle cell, tissue and whole animal levels. Since publication of the first edition of this book, there has been a striking increase in the number of papers examining mouse smooth muscle contractility largely due to the proliferation of gene-altered mouse models. Over 60% of the articles (4500) categorized by the terms "mouse" and "smooth muscle" in the National Library of Medicine were published in the last decade and 40% in the past 5 years. As an organizational strategy, mouse smooth muscle studies may be divided into 3 categories

- Vascular (or smooth muscle) Biology
- Contractility
- Disease models.

These categories are not mutually exclusive, particularly overlapping in the area of disease models. Vascular Biology encompasses "signaling" meaning studies of the mechanisms of regulation of gene expression, and includes smooth muscle development and proliferation. On the disease side, models of Lipid metabolism and atherosclerosis are the most prominent, with those of inflammation and airway hyper responsiveness under this general umbrella. Contractility is common to all differentiated smooth muscles. "Signaling" in this context generally refers to those processes regulating contractile activity. Studies here include receptor-signaling pathways, role of kinase/phosphatase pathways, control of intracellular Ca^{2+} and function of contractile proteins and their isoforms. This function of smooth muscle is the end effector underlying many diseases of industrial society from asthma to hypertension. Not surprisingly, many mouse models are developed to elucidate mechanisms of disease processes.

Given the wide range and rapid development, it is impossible to be all encompassing in any review. Recently, Faraci and Sigmund [2] presented a detailed review of mouse smooth muscle studies from the perspective of vascular biology. Our laboratory has focused on regulatory mechanisms of smooth muscle contractility and its coordination with metabolism [3, 4]. Despite the rapid growth of the field, the number of studies on the level for the mouse is quite limited. Over the past several years, we have worked to refit our techniques for mechanics and energetics measurements in smooth muscle tissues of the mouse. Interestingly, because these studies are often first presented in Journals of wide circulation, such as *Science or Nature*, there are few detailed descriptions on preparing mouse smooth muscle tissues. *Our first goal* thus is to categorize methods for study of various mouse smooth muscle tissues. Using our experience and data from the literature, *our second goal* is to rigorously characterize the contractility of a variety of smooth muscle tissues and compare contractile parameters to other commonly used animal models. This aspect will become increasingly more important as quantitation of more subtle changes in contractility is required. *Our final goal* will be to briefly review the current literature on gene targeting to smooth muscle.

PREPARATION AND CHARACTERIZATION OF CONTRACTILITY OF MOUSE SMOOTH MUSCLE TISSUES

Aorta

This preparation is the most commonly used model for vascular tissue because it is relatively easy to dissect and its size is adequate for most commercial force measurement systems. In addition, it is the largest source of vascular mass and easily accessible thereby lending itself to biochemical measurements, such as immunoblot measurements of protein, an important consideration in studies of gene-altered mouse models.

Typically, mice age 10-14 weeks weighing approximately 25 g are studied. The length of the thoracic aorta is approximately 15 mm and weighs between 2-3 mg. The internal diameter is 600-800 μm with a wall thickness of approximately 80 μm. We use 100 μm wires for mounting a segment isometrically to a force transducer. Adjusting the tension on a 5-7 mm segment to a resting tension of 30 mN sets the muscle length in the range for optimal force development, generating 15-30 mN when stimulated.

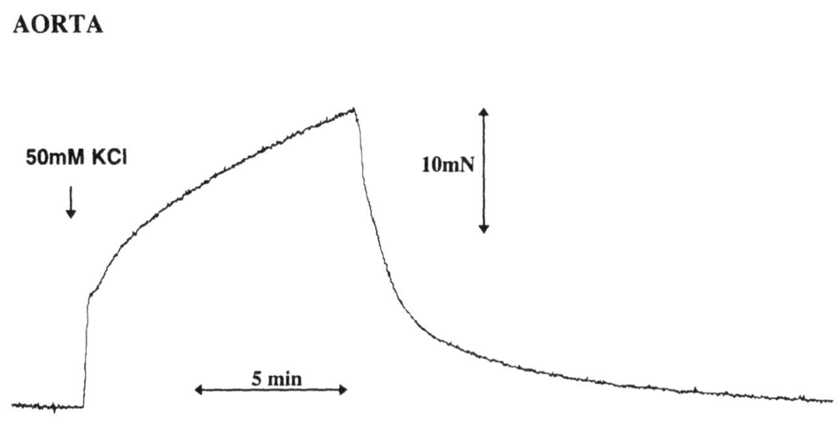

Fig. 1. Isometric contraction of mouse aorta in response to 50 mM KCl.

This yields a force/cross sectional area of approximately 20 to 40 mN/mm^2, a range consistent with rat aorta (Table 1). A typical recording of isometric force in response to KCl is shown in Fig. 1.

This contraction is typical of tonic vascular smooth muscle with half-times ($t_{1/2}$) for force development of approximately 60s for force development and for relaxation. Interestingly, we have found that adrenergic stimulation (norepinephrine or phenylephrine) is the only receptor-dependent agonist that elicits significant and maintained force. A detailed pharmacology of mouse aorta from a strain commonly used for producing knockouts (C57BL/6J) has recently been published [6]. Acetylcholine and A23187 produce an endothelium-dependent relaxation, which can be completely blocked by the NO synthase inhibitor, l-NNA [7]. One clearly would require an entire chapter to focus on endothelial-vascular interactions and the use of gene-altered mouse models; typical examples involving Ca^{2+} regulation [8-10]. Forskolin and SNP can completely relax mouse aorta, indicating that both cAMP and cGMP-mediated relaxation pathways are active [11]. In a number of reported mouse models, neonates are born, but die shortly thereafter, it is of interest that studies can be made on neonate aorta. Forces produced by a 3 mm neonatal aortic ring are in the 2-4 mN range (Shelly & Paul, unpublished observations).
Recent studies involving mouse aorta include [5, 6, 12-16]. Table 1 provides a guide to the average mechanical behavior of various mouse smooth muscle tissues and comparative values in the rat in terms of the contractile parameters of isometric force per cross sectional area, force development and relaxation half-times. It should be noted that the strain of mouse significantly affects the values of contractile parameters.

Table 1. Mouse & Rat Smooth muscle contractile parameters

	MOUSE			RAT		
	F_{max}	Force Development	Relax.	F_{max}	Force Development	Relax.
	(mN/mm^2) ‡	$t_{1/2}$ (s)	-$t_{1/2}$ (s)	(mN/mm^2) ‡	$t_{1/2}$ (s)	-$t_{1/2}$ (s)
AORTA	16.4[5] -51	60	66	31.8[17]	180	210
PULMONARY ARTERY	20.2	22	39	22[18]	200	50
TRACHEA	8 [19]	7.7	11.8	19[20]	30	---
PORTAL VEIN †	3.8[21]	0.65	0.49	5.4[22]	0.88	0.60
BLADDER	16	4.2	15.7	12.4[23]	7.6	7.2

† Mean value for integrated spontaneous mechanical activity

KCl stimulation except trachea (acetylcholine) and portal vein (none)

‡cross section areas calculated from tissue weight and dimensions.

Trachea

In a typical 25-30 g mouse, trachea of between 3-5 mm can be dissected which weigh 5.5 mg on average. Tracheal circumference averages 2.5 mm unloaded. Smooth muscle is limited to a band of about 1 mm and has an estimated cross-sectional area of 1.5 mm^2. Tracheal rings can be mounted on 0.3 mm wires for isometric force studies. A rapid contraction and relaxation permit multiple contraction/relaxation cycles in a relatively short period of time for a tonic smooth muscle (see Fig. 2). Thus, one can determine the optimal length for force generation in each study; this is not readily done with the aorta, which is typically much slower (Fig. 1). A passive isometric force of 4 mN places tracheal rings of the above dimensions at a length in the range where maximal force is generated. This is calculated to be equivalent to 3.7 mm circumference. Acetylcholine produces concentration-dependent increases in isometric force with an ED_{50} for stimulation between 1-3 μM. Maximum forces produced are between 5 and 15 mN, averaging 8 mN/mm^2 [24].

Isoproterenol produces a concentration-dependent relaxation in the mouse trachea. Substance P (10 nM) and ATP (10 μM) produce transient relaxations, which are significantly reduced by gently rubbing a silk thread through the lumen. By analogy to removal of vascular endothelium by similar protocols, it appears that these relaxations are epithelium-dependent [19]. Apparatus for perfusion of mouse trachea have been developed which permit measurement of constriction in terms of pressure or resistance differences. This is useful in that agonists can be introduced in a site-specific fashion [25]. Recent studies include [26, 27].

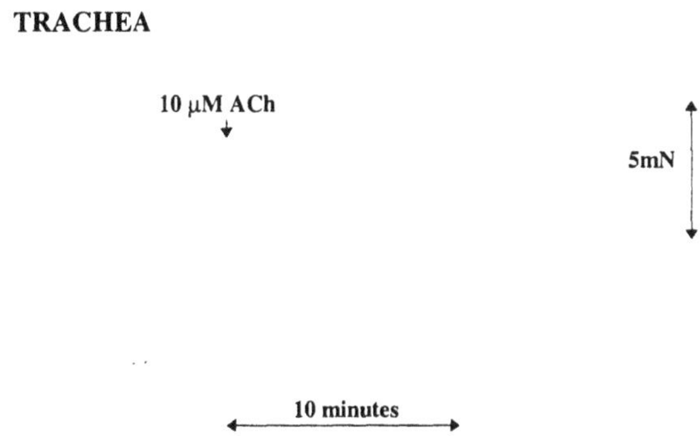

Fig. 2. Isometric myogram of a mouse tracheal ring.

Pulmonary artery

This preparation is often used in the study of pulmonary vasoconstriction in response to hypoxia. This vessel is approximately 3 mm long when measured between the heart and its first branching point. The cross-sectional area of this vessel is approximately 0.5 mm^2 making the force generated in response to phenylephrine 30 mN/mm^2. Under isometric conditions (Fig. 3), 15 mN is a preload which sets the muscle length in the range for optimal force development. The pulmonary artery contracts in response to phenylephrine and the thromboxane A_2 analog, U46619. This artery relaxes in response to acetylcholine mediated via activation of nitric oxide synthetase [28].

Fig. 3. Isometric myogram of mouse pulmonary artery in response to 50 mM KCl.

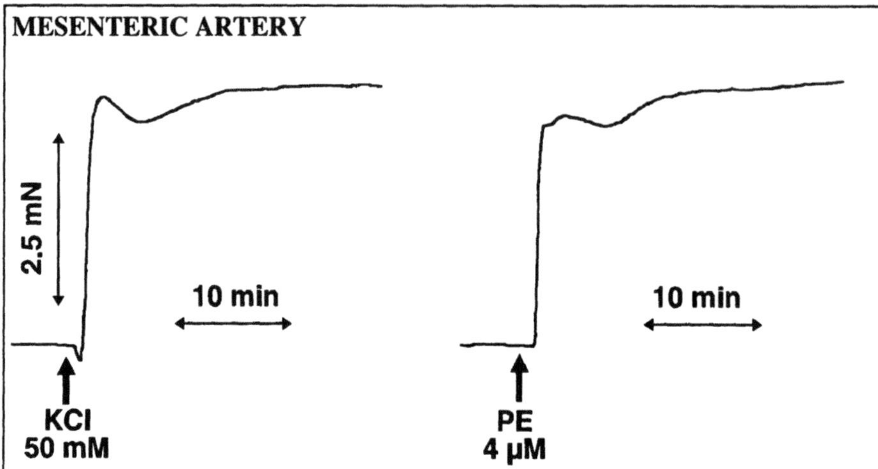

Fig. 4. Isometric myogram of a mouse mesenteric resistance vessel.

Mesenteric resistance vessels

These arteries comprise the vessels most associated with the determination of peripheral vascular resistance. This preparation is the preferred vessel for studies correlating *in vivo* hemodynamic effects and vascular tone. The resistance vessels are approximately 150 µm in diameter at the secondary level of the mesentery and are isometrically mounted using 16 µm tungsten wires. These vessels contract maximally at a resting tone of approximately 0.5-1.0 mN (Fig. 4).

Portal vein

This longitudinal smooth muscle is the archetypal phasic smooth muscle exhibiting spontaneous activity and is often used as a model for the resistance vessel. The length of this vessel between the hepatic bifurcation and the anterior mesenteric vein is approximately 3.5 mm with a weight of 0.3 mg. This vessel is mounted isometrically in a longitudinal orientation by securing either end of the vessel with 4-0 suture. Optimal tension is based on adjusting the vessel length to a point where maximum peak-peak oscillations of the spontaneous isometric contractions are observed at a tension of approximately 4-5 mN. A typical tracing from the portal vein and the data that can be determined from this spontaneously active vessel are shown in Fig. 5. The basal amplitude observed in these vessels is approximately 2.5-3.0 mN at a frequency of 3-3.5 Hertz. The force generated by this vessel as measured with the tension-time integral for one minute is equal to approximately 225 $mN*s/mm^2$ with an average force of 3.75 mN/mm^2. The force of the portal vein increases in a concentration-dependent manner in response to acetylcholine, endothelin and phenylephrine [29]. It is relaxed by isoproterenol, parathyroid hormone related peptides, and forskolin [8]. There does not appear to be any endothelium dependent components to relaxation in the portal vein [30]. Recent publications for mouse portal vein include [14, 31, 32].

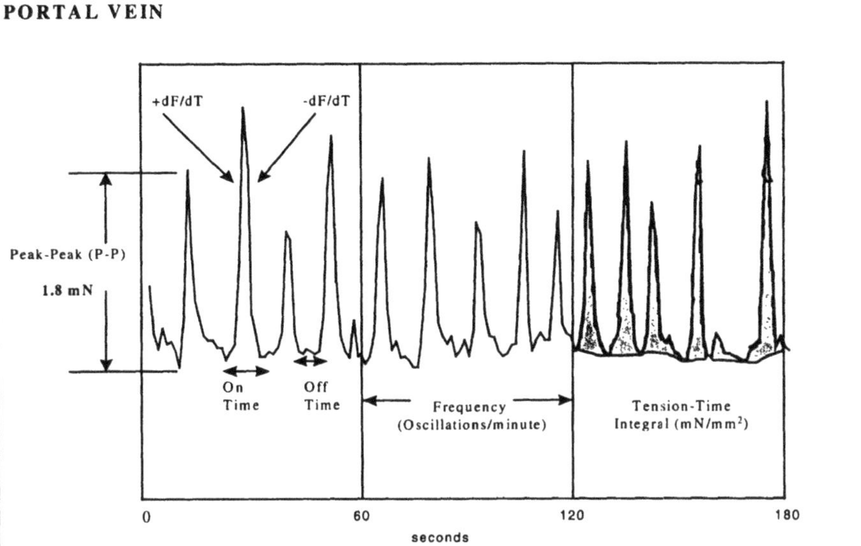

Fig. 5. Isometric myogram of mouse portal vein.

Bladder

This smooth muscle is of considerable interest because its mass facilitates biochemical studies and is very useful for studies in terminal neonates [33], and permeabilized preparations [34]. Bladder from a 25g mouse dissected free of mucosa weighs 20-30 mg and can generate an isometric force of 30 mN or 15

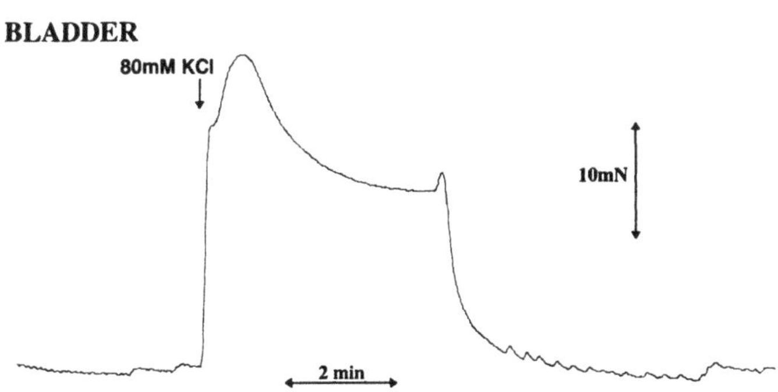

Fig. 6. Isometric myogram of mouse bladder smooth muscle. Record of a longitudinal strip of 3.5-4.5 mm wide weighing 3-4 mg.

mN/mm^2 along its longitudinal axis when depolarized with KCl. Muscarinic stimulation with acetylcholine or carbachol, elicits a maximal response that is approximately 2/3 of KCl. Both A- and G-kinase pathway activation are efficient mechanisms of relaxation. The latter has been convincingly shown to be modulated exclusively by cGMP-dependent protein kinase I using gene-targeting [35] This group and others [36, 37] have published methodology for cystometric measurements in the intact mouse to assess bladder capacity.

Carotid Artery

Contraction and relaxation of this vessel have been characterized by Faraci et al [38]. A maximum contraction of approximately 7 mN occurs in response to the thromboxane analog U-46619. This vessel is also the preferred vessel for studying vascular injury responses in genetically altered mice [29, 39-41].

Other mouse smooth muscles

Though not studied by our laboratory in detail, the following figures show typical isometric force records of ileum and uterine muscle. Contractility studies for the latter have been reported [42, 43]. Additional functional studies have examined contractility of the corpora cavernosa [44], fundus [45] and colon [46].

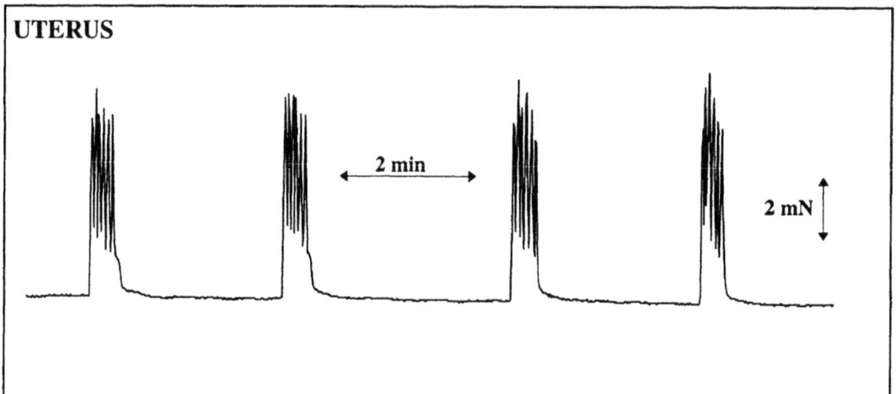

Fig. 8. Isometric myogram of mouse uterus.

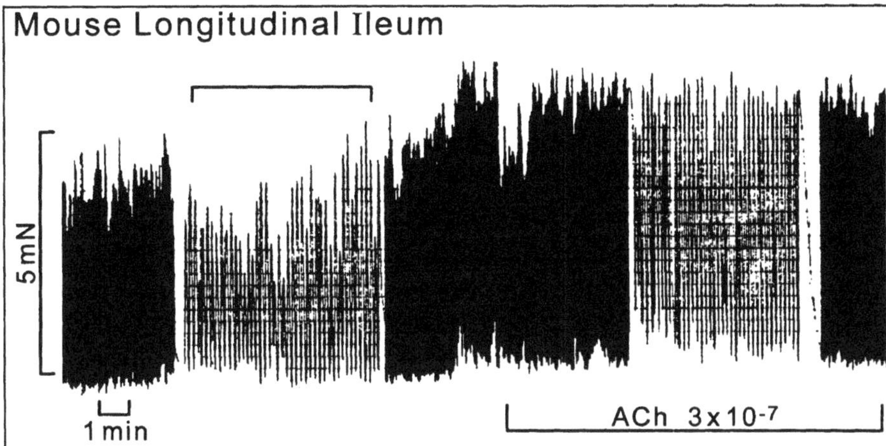

Fig. 7. Isometric myogram of mouse longitudinal ileum smooth muscle. Frequencies averaged 42 ± 7 in 11 preparations. Acetylcholine (ACh) increased forces but not frequency at intermediary concentrations and inhibited both at high levels.

SMOOTH MUSCLE SPECIFIC PROMOTERS

The initial smooth muscle studies using gene-altered mice were made with gene-targeted or gene-ablated (knockout) mice. Knockout mice have advantages due to the absolute nature of lacking a specific protein. A disadvantage is that the protein will be absent in all tissues, and the response of smooth muscle, for example, may be attributable to the alteration of function in another, for example the heart. The alternative is to add a transgene specific to smooth muscle and "overexpress" rather than eliminate the protein in question. The development of a specific and robust promoter for heart muscle, the cardiac α-myosin promoter led to an explosion of mouse models, expressing various transgenes for cardiovascular study [47, 48]. The smooth muscle field lagged in this aspect, but recently several promoters with smooth muscle specificity have been developed. In addition to specificity, the robustness or capacity to drive protein expression, is an important characteristic for a promoter. Many promoters have been shown to drive poor expression of protein, when suitable reporter genes, such as green fluorescent protein or β-galactosidase are used. Therefore some promoters may not produce suitable expression of a transgene. Smooth muscle specific proteins include SMP8 α-actin, smooth muscle myosin heavy chain, γ-actin, telokin, calponin, caldesmon, SM22, alpha- and beta-tropomyosins, and alpha1 integrin. These genes are transcriptionally regulated. Transcription of these genes (except for the α-smooth muscle actin gene) is upregulated in differentiated SMCs, but downregulated in dedifferentiated SMCs

[49]. To our knowledge, promoters derived from only the first three proteins have been used for expression of smooth muscle specific transgenes in mice.

The **α-actin promoter**, SMP8 of Strauch and colleagues [50], appears to be the most widely utilized. It is specific to smooth muscle, with the exception of appearing in neonatal heart for a short time. Its distribution has been extensively documented in terms of insulin-like growth factor (IGF-1) expression [51, 52]. Its hierarchy of expression is as follows: bladder > stomach > aorta = uterus > intestine, with 10-fold in the bladder and 5-fold in the aorta. It has also been used to express the IGF binding peptide [53], parathyroid hormone related peptide (PTHrP) and receptor [15, 31], and the β2-adrenergic receptor [24]. Our laboratory has used this promoter to express phospholamban, a protein associated with the sarcoplasmic reticulum Ca^{2+}-ATPase [54]. This protein is expressed in a similar-fold excess as that reported for the amplified receptor or hormones cited above, attesting to the robustness of this promoter.

Telokin is a 17-kDa protein with an amino acid sequence that is identical to the COOH terminus of the 130 kDa myosin light chain kinase (MLCK). Telokin mRNA is transcribed from a second promoter, located within an intron, in the 3' region of the MLCK gene. The telokin promoter has been used to express the SV40 large T antigen [55] and appears to be targeted primarily to visceral smooth muscle [56].

Smooth muscle myosin heavy chain promoter driving the lacZ reporter is widely expressed in smooth muscle tissues [57], but is reported to be largely in the vasculature with a preference for the arterial system, based on the expression of luciferase [58]. This promoter has been recently used to express the cre recombinase [59].

FINAL THOUGHTS

Even at this early stage, the use of gene-altered mouse models has produced many novel insights into smooth muscle function. We anticipate that research using mouse smooth muscle preparation will dramatically accelerate in the upcoming years. There are several major reasons for this prediction. First, is the ease of and database for production of gene-altered mouse models. Second, despite the likely development of gene-alteration techniques for other animals, such as guinea pigs and rats, routinely used for physiological studies, the cost and time involved for studies using these models is prohibitive. Promoters are rapidly being developed which enable specific smooth muscle tissues to be targeted, such as vascular smooth muscle. Finally, with the development of cre recombinase technology, specific smooth muscle tissue knockouts will be developed. We hope that this brief technical review is helpful to the development of this exciting area of smooth muscle research.

ACKOWLEDGMENTS

Supported in part by NIH HL54829 (RJP), HL61974 (Shull, RJP), HL58511 (Doetschman, RJP), DN54216 (Fagin, RJP) and NIH HL09781 (RLS), AHA 00324N (RLS). We also thank Peggy Sue Bowman, Craig Weber for technical support in conducting mouse smooth muscle tissue studies.

REFERENCES

1. Lorenz, J.N. and J. Robbins, *Measurement of intraventricular pressure and cardiac performance in the intact closed-chest anesthetized mouse.* Am J Physiol, 1997. 272(3 Pt 2): p. H1137-46.
2. Faraci, F.M. and C.D. Sigmund, *Vascular biology in genetically altered mice : smaller vessels, bigger insight.* Circ Res, 1999. 85(12): p. 1214-25.
3. Hardin, C.D., T.J. Allen, and R.J. Paul, eds. *Metabolism and energetics of vascular smooth muscle.* 4th ed. In Physiology and Pathophysiology of the Heart, ed. N. Sperelakis, *et al.* 2000, Academic Press: San Diego, CA. 571-598.
4. Paul, R.J. and P. de Lanerolle, *Regulation of Smooth Muscle Contractility*, in *Genetics of Asthma*, S. Liggett and D. Meyers, Editors. 1996, Marcel Dekker, Inc.: New York, NY. p. 91-117.
5. Lalli, J., J.M. Harrer, W. Luo, E.G. Kranias, and R.J. Paul, *Targeted ablation of the phospholamban gene is associated with a marked decrease in sensitivity in aortic smooth muscle.* Circ Res, 1997. 80(4): p. 506-13.
6. Russell, A. and S. Watts, *Vascular reactivity of isolated thoracic aorta of the C57BL/6J mouse.* J Pharmacol Exp Ther, 2000. 294(2): p. 598-604.
7. Furchgott, R.F. and J.V. Zawadzki, *The obligatory role of endothelial cells in the relaxation of arterial smooth muscle by acetylcholine.* Nature, 1980. 288(5789): p. 373-6.
8. Sutliff, R.L., C.S. Weber, J. Qian, M.L. Miller, T.L. Clemens, and R.J. Paul, *Vasorelaxant properties of parathyroid hormone-related protein in the mouse: evidence for endothelium involvement independent of nitric oxide formation.* Endocrinology, 1999. 140(5): p. 2077-83.
9. Sutliff, R.L., J.B. Hoying, V.J. Kadambi, E.G. Kranias, and R.J. Paul, *Phospholamban is present in endothelial cells and modulates endothelium-dependent relaxation. Evidence from phospholamban gene-ablated mice.* Circ Res, 1999. 84(3): p. 360-4.
10. Liu, L.H., R.J. Paul, R.L. Sutliff, M.L. Miller, J.N. Lorenz, R.Y. Pun, J.J. Duffy, T. Doetschman, Y. Kimura, D.H. MacLennan, J.B. Hoying, and G.E. Shull, *Defective endothelium-dependent relaxation of vascular smooth muscle and endothelial cell Ca2+ signaling in mice lacking sarco(endo)plasmic reticulum Ca2+-ATPase isoform 3.* J Biol Chem, 1997. 272(48): p. 30538-45.

11. Lalli, M.J., S. Shimizu, R.L. Sutliff, E.G. Kranias, and R.J. Paul, *[Ca2+]i homeostasis and cyclic nucleotide relaxation in aorta of phospholamban-deficient mice.* Am J Physiol, 1999. 277(3 Pt 2): p. H963-70.
12. Schildmeyer, L.A., R. Braun, G. Taffet, M. Debiasi, A.E. Burns, A. Bradley, and R.J. Schwartz, *Impaired vascular contractility and blood pressure homeostasis in the smooth muscle alpha-actin null mouse [In Process Citation].* Faseb J, 2000. 14(14): p. 2213-20.
13. Galvin, K.M., M.J. Donovan, C.A. Lynch, R.I. Meyer, R.J. Paul, J.N. Lorenz, V. Fairchild-Huntress, K.L. Dixon, J.H. Dunmore, M.A. Gimbrone, Jr., D. Falb, and D. Huszar, *A role for smad6 in development and homeostasis of the cardiovascular system.* Nat Genet, 2000. 24(2): p. 171-4.
14. Sjuve, R., A. Arner, Z. Li, B. Mies, D. Paulin, M. Schmittner, and J.V. Small, *Mechanical alterations in smooth muscle from mice lacking desmin.* J Muscle Res Cell Motil, 1998. 19(4): p. 415-29.
15. Qian, J., J.N. Lorenz, S. Maeda, R.L. Sutliff, C. Weber, T. Nakayama, M.C. Colbert, R.J. Paul, J.A. Fagin, and T.L. Clemens, *Reduced blood pressure and increased sensitivity of the vasculature to parathyroid hormone-related protein (PTHrP) in transgenic mice overexpressing the PTH/PTHrP receptor in vascular smooth muscle.* Endocrinology, 1999. 140(4): p. 1826-33.
16. Rubanyi, G.M., A.D. Freay, K. Kauser, D. Sukovich, G. Burton, D.B. Lubahn, J.F. Couse, S.W. Curtis, and K.S. Korach, *Vascular estrogen receptors and endothelium-derived nitric oxide production in the mouse aorta. Gender difference and effect of estrogen receptor gene disruption.* J Clin Invest, 1997. 99(10): p. 2429-37.
17. McMahon, E.G. and R.J. Paul, *Metabolic and mechanical properties of aortas from aldosterone-salt hypertensive rats.* Circ Res, 1984. 55(3): p. 349-57.
18. Zhao, Y., R.A. Rhoades, and C.S. Packer, *Hypoxia-induced pulmonary arterial contraction appears to be dependent on myosin light chain phosphorylation.* Am J Physiol, 1996. 271(5 Pt 1): p. L768-74.
19. Kao, J., C.N. Fortner, L.H. Liu, G.E. Shull, and R.J. Paul, *Ablation of the SERCA3 gene alters epithelium-dependent relaxation in mouse tracheal smooth muscle.* Am J Physiol, 1999. 277(2 Pt 1): p. L264-70.
20. Peiper, U., S.C. Knipp, B. Thies, and R. Henke, *Activation of protein kinase C accelerates contraction kinetics of airway smooth muscle.* Pflugers Arch, 1996. 432(3): p. R47-52.
21. Sutliff, R.L., E.G. Kranias, and R.J. Paul, *Phospholamban gene ablation is associated with alterations in portal vein contractility.* FASEB J, 1997. 11(: p. A25.
22. Hellstrand, P. and R.J. Paul, *Phosphagen content, breakdown during contraction, and O2 consumption in rat portal vein.* Am J Physiol, 1983. 244(3): p. C250-8.

23. Yu, H.I., A.J. Wein, and R.M. Levin, *Contractile responses and calcium mobilization induced by muscarinic agonists in the rat urinary bladder: effects of age.* Gen Pharmacol, 1997. 28(4): p. 623-8.
24. McGraw, D.W., S.L. Forbes, L.A. Kramer, D.P. Witte, C.N. Fortner, R.J. Paul, and S.B. Liggett, *Transgenic overexpression of beta(2)-adrenergic receptors in airway smooth muscle alters myocyte function and ablates bronchial hyperreactivity.* J Biol Chem, 1999. 274(45): p. 32241-7.
25. Fortner, C.N., J.N. Lorenz, and R.J. Paul, *Chloride channel function is linked to epithelium-dependent airway relaxation.* Am J Physiol, 2001. 280(2): p. L334-341.
26. Stengel, P.W., J. Gomeza, J. Wess, and M.L. Cohen, *M(2) and M(4) receptor knockout mice: muscarinic receptor function in cardiac and smooth muscle in vitro.* J Pharmacol Exp Ther, 2000. 292(3): p. 877-85.
27. Hamelmann, E., K. Takeda, J. Schwarze, A.T. Vella, C.G. Irvin, and E.W. Gelfand, *Development of eosinophilic airway inflammation and airway hyperresponsiveness requires interleukin-5 but not immunoglobulin E or B lymphocytes [see comments].* Am J Respir Cell Mol Biol, 1999. 21(4): p. 480-9.
28. Steudel, W., F. Ichinose, P.L. Huang, W.E. Hurford, R.C. Jones, J.A. Bevan, M.C. Fishman, and W.M. Zapol, *Pulmonary vasoconstriction and hypertension in mice with targeted disruption of the endothelial nitric oxide synthase (NOS 3) gene.* Circ Res, 1997. 81(1): p. 34-41.
29. Zhou, M., R.L. Sutliff, R.J. Paul, J.N. Lorenz, J.B. Hoying, C.C. Haudenschild, M. Yin, J.D. Coffin, L. Kong, E.G. Kranias, W. Luo, G.P. Boivin, J.J. Duffy, S.A. Pawlowski, and T. Doetschman, *Fibroblast growth factor 2 control of vascular tone.* Nat Med, 1998. 4(2): p. 201-7.
30. Feletou, M., U. Hoeffner, and P.M. Vanhoutte, *Endothelium-dependent relaxing factors do not affect the smooth muscle of portal-mesenteric veins.* Blood Vessels, 1989. 26(1): p. 21-32.
31. Maeda, S., R.L. Sutliff, J. Qian, J.N. Lorenz, J. Wang, H. Tang, T. Nakayama, C. Weber, D. Witte, A.R. Strauch, R.J. Paul, J.A. Fagin, and T.L. Clemens, *Targeted overexpression of parathyroid hormone-related protein (PTHrP) to vascular smooth muscle in transgenic mice lowers blood pressure and alters vascular contractility [see comments].* Endocrinology, 1999. 140(4): p. 1815-25.
32. Mancinelli, R., P. Tonali, R. Romani, A. Tringali, R. Vargiu, and G.B. Azzena, *Mechanical properties of smooth muscle portal vein in normal and dystrophin-deficient (mdx) mice.* Experimental Physiology, 1999. 84(5): p. 929-40.
33. Morano, I., G.X. Chai, L.G. Baltas, V. Lamounier-Zepter, G. Lutsch, M. Kott, H. Haase, and M. Bader, *Smooth-muscle contraction without smooth-muscle myosin.* Nat Cell Biol, 2000. 2(6): p. 371-5.
34. Matthew, J.D., A.S. Khromov, M.J. McDuffie, A.V. Somlyo, A.P. Somlyo, S. Taniguchi, and K. Takahashi, *Contractile properties and proteins of smooth muscles of a calponin knockout mouse.* J Physiol, 2000. 529(Pt 3): p. 811-824.

35. Persson, K., R.K. Pandita, A. Aszodi, M. Ahmad, A. Pfeifer, R. Fassler, and K.E. Andersson, *Functional characteristics of urinary tract smooth muscles in mice lacking cGMP protein kinase type I.* Am J Physiol Regul Integr Comp Physiol, 2000. 279(3): p. R1112-20.
36. Lemack, G.E., P.E. Zimmern, D. Vazquez, J.D. Connell, and V.K. Lin, *Altered response to partial bladder outlet obstruction in mice lacking inducible nitric oxide synthase.* Journal of Urology, 2000. 163(6): p. 1981-7.
37. Lemack, G.E., Z. Szabo, Z. Urban, C.D. Boyd, K. Csiszar, E.D. Vaughan, Jr., and D. Felsen, *Altered bladder function in transgenic mice expressing rat elastin.* Neurourology & Urodynamics, 1999. 18(1): p. 55-68.
38. Faraci, F.M., C.D. Sigmund, E.G. Shesely, N. Maeda, and D.D. Heistad, *Responses of carotid artery in mice deficient in expression of the gene for endothelial NO synthase.* American Journal of Physiology, 1998. 274(2 Pt 2): p. H564-70.
39. Lijnen, H.R., B. Van Hoef, F. Lupu, L. Moons, P. Carmeliet, and D. Collen, *Function of the plasminogen/plasmin and matrix metalloproteinase systems after vascular injury in mice with targeted inactivation of fibrinolytic system genes.* Arteriosclerosis, Thrombosis & Vascular Biology, 1998. 18(7): p. 1035-45.
40. Cheung, W.M., M.R. D'Andrea, P. Andrade-Gordon, and B.P. Damiano, *Altered vascular injury responses in mice deficient in protease- activated receptor-1.* Arterioscler Thromb Vasc Biol, 1999. 19(12): p. 3014-24.
41. Zaidi, S.H., X.M. You, S. Ciura, S. O'Blenes, M. Husain, and M. Rabinovitch, *Suppressed smooth muscle proliferation and inflammatory cell invasion after arterial injury in elafin-overexpressing mice.* Journal of Clinical Investigation, 2000. 105(12): p. 1687-95.
42. Naghashpour, M., M.I. Rosenblatt, I.M. Dickerson, and G.P. Dahl, *Inhibitory effect of calcitonin gene-related peptide on myometrial contractility is diminished at parturition.* Endocrinology, 1997. 138(10): p. 4207-14.
43. Matsui, M., D. Motomura, H. Karasawa, T. Fujikawa, J. Jiang, Y. Komiya, S. Takahashi, and M.M. Taketo, *Multiple functional defects in peripheral autonomic organs in mice lacking muscarinic acetylcholine receptor gene for the M3 subtype.* Proc Natl Acad Sci U S A, 2000. 97(17): p. 9579-84.
44. Hedlund, P., A. Aszodi, A. Pfeifer, P. Alm, F. Hofmann, M. Ahmad, R. Fassler, and K.E. Andersson, *Erectile dysfunction in cyclic GMP-dependent kinase I-deficient mice.* Proc Natl Acad Sci U S A, 2000. 97(5): p. 2349-54.
45. Ny, L., A. Pfeifer, A. Aszodi, M. Ahmad, P. Alm, P. Hedlund, R. Fassler, and K.E. Andersson, *Impaired relaxation of stomach smooth muscle in mice lacking cyclic GMP- dependent protein kinase I.* Br J Pharmacol, 2000. 129(2): p. 395-401.
46. Azzena, G.B. and R. Mancinelli, *Nitric oxide regenerates the normal colonic peristaltic activity in mdx dystrophic mouse.* Neurosci Lett, 1999. 261(1-2): p. 9-12.

47. James, J., A. Sanbe, K. Yager, L. Martin, R. Klevitsky, and J. Robbins, *Genetic manipulation of the rabbit heart via transgenesis.* Circulation, 2000. 101(14): p. 1715-21.
48. Palermo, J., J. Gulick, W. Ng, I.L. Grupp, G. Grupp, and J. Robbins, *Remodeling the mammalian heart using transgenesis.* Cell Mol Biol Res, 1995. 41(6): p. 501-9.
49. Sobue, K., K. Hayashi, and W. Nishida, *Expressional regulation of smooth muscle cell-specific genes in association with phenotypic modulation.* Mol Cell Biochem, 1999. 190(1-2): p. 105-18.
50. Min, B.H., D.N. Foster, and A.R. Strauch, *The 5'-flanking region of the mouse vascular smooth muscle alpha-actin gene contains evolutionarily conserved sequence motifs within a functional promoter.* J Biol Chem, 1990. 265(27): p. 16667-75.
51. Lund, P.K., *The alpha-smooth muscle actin promoter: a useful tool to analyse autocrine and paracrine roles of mesenchymal cells in normal and diseased bowel.* Gut, 1998. 42(3): p. 320-2.
52. Wang, J., W. Niu, Y. Nikiforov, S. Naito, S. Chernausek, D. Witte, D. LeRoith, A. Strauch, and J.A. Fagin, *Targeted overexpression of IGF-I evokes distinct patterns of organ remodeling in smooth muscle cell tissue beds of transgenic mice.* J Clin Invest, 1997. 100(6): p. 1425-39.
53. Wang, J., W. Niu, D.P. Witte, S.D. Chernausek, Y.E. Nikiforov, T.L. Clemens, B. Sharifi, A.R. Strauch, and J.A. Fagin, *Overexpression of insulin-like growth factor-binding protein-4 (IGFBP- 4) in smooth muscle cells of transgenic mice through a smooth muscle alpha-actin-IGFBP-4 fusion gene induces smooth muscle hypoplasia.* Endocrinology, 1998. 139(5): p. 2605-14.
54. Sutliff, R.L., V.J. Kadambi, C.S. Weber, E.G. Kranias, and R.J. Paul, *Smooth muscle-targeted overexpression of phospholamban (PLB) increases sensitivity to contractile agents.* Biophys. J., 1999. 76(: p. A12.
55. Herring, B.P., A.M. Hoggatt, A.F. Smith, and P.J. Gallagher, *Targeted expression of SV40 large T-antigen to visceral smooth muscle induces proliferation of contractile smooth muscle cells and results in megacolon.* J Biol Chem, 1999. 274(25): p. 17725-32.
56. Herring, B.P., G.E. Lyons, A.M. Hoggatt, and P.J. Gallagher, *Telokin expression is restricted to smooth muscle tissues during mouse development.* Am J Physiol Cell Physiol, 2001. 280(1): p. C12-C21.
57. Madsen, C.S., C.P. Regan, J.E. Hungerford, S.L. White, I. Manabe, and G.K. Owens, *Smooth muscle-specific expression of the smooth muscle myosin heavy chain gene in transgenic mice requires 5'-flanking and first intronic DNA sequence.* Circ Res, 1998. 82(8): p. 908-17.
58. Franz, W.M., O.J. Mueller, M. Fleischmann, P. Babij, N. Frey, M. Mueller, U. Besenfelder, A.F. Moorman, G. Brem, and H.A. Katus, *The 2.3 kb smooth muscle myosin heavy chain promoter directs gene expression into the vascular system of transgenic mice and rabbits.* Cardiovasc Res, 1999. 43(4): p. 1040-8.

59. Regan, C.P., I. Manabe, and G.K. Owens, *Development of a smooth muscle-targeted cre recombinase mouse reveals novel insights regarding smooth muscle myosin heavy chain promoter regulation.* Circ Res, 2000. 87(5): p. 363-9.

21

MODELS OF CARDIAC DISEASE IN THE MOUSE

CARDIAC HYPERTROPHY AND HEART FAILURE

Yasuhiro Ikeda M.D., Ph.D.* and John Ross Jr. M.D*[+]

**Division of Cardiology, Department of Medicine*

[+]Institute of Molecular Medicine

University of California San Diego

La Jolla, CA,

INTRODUCTION

Over the past decade the use of genetically engineered mouse models has led to considerable progress in understanding the pathophysiology of a variety of human diseases. Ability to manipulate the genome has allowed analysis of effects not only of gain or loss of the target gene function, but also actions of mutant alleles identified in human diseases. With regard to the cardiovascular system, although the mouse exhibits differences from humans in physiological variables and intracellular structure and function (e.g. isoforms of myosin heavy chain, relative contribution of sarcoplasmic reticulum (SR) to intracellular calcium regulation), a variety of approaches have been miniaturized for physiological studies in the mouse heart along with application to the mouse of methods for studying isolated cardiac muscle and cardiomyocytes. These approaches are listed in Table 19-1.

In this overview, we will focus in particular on genetic murine models of cardiac hypertrophy and heart failure which exhibit adaptations and maladaptations resembling those observed in human cardiovascular disease. The clinical syndrome of heart failure is the final common pathway through which various types of myocardial dysfunction converge (Figure 19-1), and it is usually characterized by cardiac pump failure, arrhythmia, neurohormonal maladaptations and subsequent death. Several reviews have been concerned with genetic mechanisms and stress pathways involved in heart failure (1), cytoskeletal abnormalities in cardiomyopathies (2), and signaling pathways for cardiac hypertrophy and failure (3). In addition, a number of genetically engineered models of cardiomyopathy have been reported recently, particularly in the mouse, which we will attempt to categorize according to the subcellular localization and potential functional importance of the gene products involved, primarily in models that have been characterized in vivo both morphologically and functionally. Discrepancies in phenotypic features between murine models and human diseases in hypertrophic and

Table 19-1. Miniaturized Technology Available to Assess Murine Cardiac Phenotype

- Echocardiography 2D mode, M-mode, doppler - non-invasive cardiac morphology and function
- Digital contrast ventricular microangiography - cardiac volume and function (regional and global)
- Magnetic resonance imaging – non-invasive cardiac volume and function (regional and global)
- Tail cuff, indwelling carotid catheters - aortic pressure manometry
- Transaortic high fidelity catheter tip micromanometers - cardiac pressures, contractility, velocity
- Conductance pressure catheters – open or closed chest cardiac pressure-volume relation
- Cardiac output measurement-thermodilution, microspheres
- Activity testing – treadmill, activity monitoring cages
- Electrocardiography – 12 lead under anesthesia, telemetry under conscious conditions
- Work performing and non-work performing isolated, perfused heart preparations
- Muscle contractility/relaxation – ventricular muscle strips
- Electrophysiological studies (open chest) – monophasic action potentials, programmed stimulation
- Single cell assays – cell shortening, relaxation, calcium transients, patch clamp studies

dilated cardiomyopathies will be considered, and potential therapeutic approaches associated with genetic modifications and viral gene transfer techniques will be briefly discussed.

Inherited Cardiomyopathies

Hypertrophic cardiomyopathy (HCM) is a well described primary disorder of the myocardium, the well known cardiac phenotype being characterized by increased cardiac mass with thickened ventricular walls and normal or reduced chamber volume associated with myocyte and myofibrillar disarray (Figure 19-1). Genetic studies of familial HCM have revealed many different mutations occurring in at least 8 genes responsible for disease including β-myosin heavy chain, ventricular myosins (1 and 2), cardiac troponin T, cardiac troponin I, α-tropomyosin, myosin binding protein C, and α-cardiac actin (Table 19-2) (4). The disease is associated with genetic mutations in over 50% of its patients (5).

Dilated cardiomyopathy (DCM) has a well defined phenotype characterized by impaired systolic and diastolic function, ventricular chamber enlargement and relative thinning of the ventricular walls, associated with myocyte hypertrophy (Figure 19-1). About 10 % of patients with HCM eventually develop left ventricular (LV) dilatation characteristic of DCM (6). In about 25% of patients with so-called idiopathic DCM (those not due to coronary heart disease) the cause is hereditary (7). At present, only 5 genes causing DCM in humans have been identified (8), including cardiac actin, dystrophin, desmin, δ-sarcoglycan, and lamin A/C, although genetic linkage analyses have identified more than 12 disease loci associated with DCM (9) (Table 19-3). So far, partial population analysis for some of the identified gene mutations has suggested that they are responsible for only a few percent of the total DCM population in humans (10). To help elucidate mechanisms responsible for the pathophysiological features of DCM and to establish potential treatment strategies, a number of animal models of DCM have been developed and investigated, and most mimic many phenotypic and clinical features of DCM in humans.

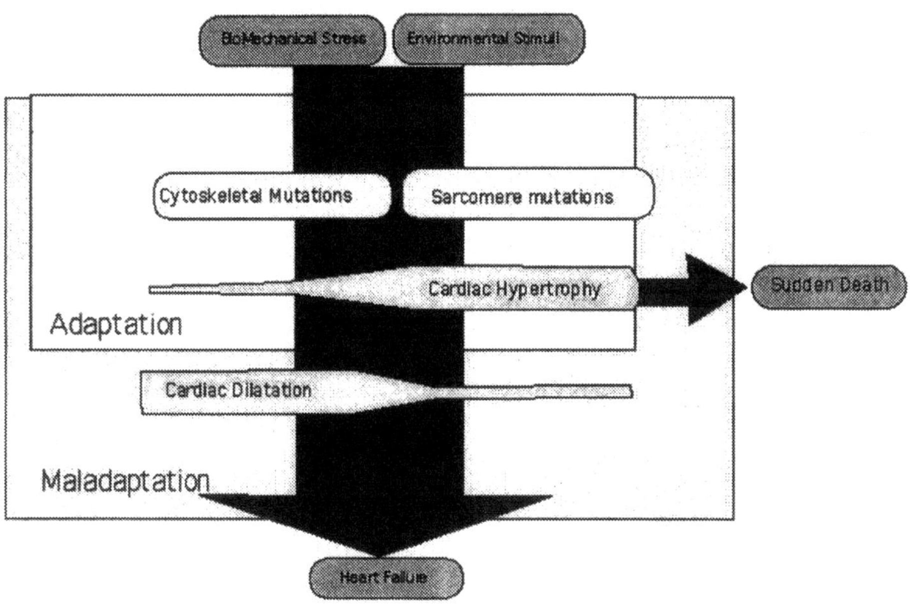

Figure 19-1. Overview of time course of adaptation and maladaptation in inherited cardiomyopathy. Mutations in sarcomeric gene usually are associated with a hypertrophic cardiomyopathy, whereas mutations in components of the extra-sarcomeric cytoskeleton can lead to a dilated cardiomyopathy.

Sarcomere mutations

The finding that gene mutations within the sarcomere are primarily involved in causing HCM in humans (Table 19-2) (5), and the discovery that mutations in the extra-sarcomeric cytoskeleton of the cell may be mainly causative of DCM (Table 19-3) have lead to the hypothesis that abnormalities in the force-generating apparatus

Table 19-2. Genes Responsible for Hypertrophic Cardiomyopathy and Related Murine Models

Gene Mutations	Clinical Characteristics	Population	Inherit-ance type	Mouse model	Inherit-ance type
β-Myosin heavy chain	Severe Hypertrophy, prognosis is depending on mutation siteArg403Gln <Val606Met	35%	AD	Mutation of α-myosin heavy chain causes hypertrophic cardiomyopathy (75)	AD
Ventricular myosin light chain 1 (essential chain)	Rare	<1%	AD	None	N/A
Ventricular myosin light chain 2 (regulatory chain)	Rare	<1%	AD	MLC2v null mutation cause embryonic dilated cardiomyopathy (76) and heterozygote (77) show normal cardiac function	N/A
Cardiac troponin T	Less hypertrophy than myosin mutation, poor prognosis, sudden death	~15 %	AD	Hypertrophic cardiomyopathy (78-80)	AD
Cardiac troponin I		?	AD	Mutant TG cause hypertrophic cardiomyopathy (81)	AD
α-Tropomyosin	Better prognosis, less hypertrophy	<5%	AD	Mutant TG mouse show hypertrophic cardiomyopathy with LV dysfunction (82,83)	AD
Myosin binding protein C	Late onset hypertrophic cardiomyopathy, relatively better prognosis	~15 %	AD	Homogenous mutation cause dilated cardiomyopathy (15), mutant TG cause hypertrophic cardiomyopathy (84)	AD
α-Cardiac actin	Not well described yet	?	AD	Null mutation cause embryonic DCM (13)	N/A

generally lead to the adaptive mechanism of cardiac hypertrophy, whereas abnormalities in the force transmission apparatus (primary cytoskeleton) result in progressive cardiac dilatation and heart failure (DCM). However, the underlying mechanisms may not be so readily classified, since mutations in a single gene can cause either DCM or HCM (11). For instance, mutations in the α-cardiac actin gene are reported to cause divergent forms of human cardiomyopathy, with a mutation in the myosin binding site leading to HCM (12) as in the α-cardiac actin knockout

Table 19-3. Genes Responsible for Dilated Cardiomyopathy and Related Murine Models

Gene Mutations	Clinical Characteristics	Populat ion	Inherit-ance type	Mouse model	Inherit-ance type
Dystrophin (19-21)	Dilated cardiomyopathy with or without DMD, Becker muscular dystrophy.	?	X-linked	Mouse lacking dystrophin and utrophin shows severe muscular dystrophy and cardiomyopathy, mouse lacking dystrophin only does not show cardiomyopathy due to the over-expression of utrophin in the heart (26)	X-linked
Tafazzin (G4.5) (85)	the infantile-onset DCM (Barth syndrome)	Very rare	X-linked	None	N/A
α-Cardiac actin (14)	Adult onset dilated cardiomyopathy	<1.2%, rare (10)	AD	Null mutation cause embryonic DCM (13)	N/A
Desmin (34,86)	Dilated Cardiomyopathy with or without Skeletal muscle involvement	<1.6%, rare (10)	AD	Cardiomyopathy with myopathy (33)	AR
LaminA/C (37)	Dilated cardiomyopathy with variable skeletal muscle involvement, related to Emery-Dreifuss Muscular Dystrophy	?	AD	None	N/A
δ-Sarcoglycan (25)	Familial and Sporadic Dilated Cardiomyopathy	?	AD	Null mutation of δ-sarcoglycan gene can cause progressive cardiomyopathy (30), similar to the hamster cardiomyopathy	AR

mouse model (13), whereas mutations in the immobilized end of the action filament are associated with human hereditary DCM (14). Moreover, mice with abnormalities of intra-sarcomeric proteins such as myosin binding protein-C can exhibit diverse phenotypes, depending on the design of transgenesis; thus, mice homozygous for mutant myosin binding protein C mice show features typical of DCM (15), whereas other mice with mutant myosin binding protein-C (which has been linked to human familial HCM) express typical features of HCM. Another intra-sarcomeric protein, tropomodulin, when over-expressed in mice (16) produces a DCM phenotype, without a hypertrophic response. Based on such evidence, it appears that phenotypes generated in murine cardiomyopathies may depend on multiple factors including genetic background, positional effects and environmental factors, such that the phenotype does not necessarily match that in humans. Nevertheless, the view that in humans mutations associated with DCM primarily involve defective force transmission, whereas mutations in HCM primarily involve impaired force generation remains an intriguing concept (14).

Cytoskeleton Mutations

Dystrophin and associated membrane glycoproteins of the muscle cytoskeleton link intracellular actin fibers to laminin α_2 of the extracellular matrix (17). This complex, the dystrophin-glycoprotein complex (DGC) is comprised of at least 10 subunits, and deficiencies of various components of the DGC are responsible for several hereditary forms of human muscular dystrophy, some which are associated with DCM. For example, among patients with Duchenne's progressive muscular dystrophy, which is caused by dystrophin deficiency, a high percentage eventually develop DCM (18), and DCM also is frequent in X-linked dystrophin deficiency (19-21) In humans, in contrast to animals, mutations in the sarcoglycan components of the DGC which are responsible for limb girdle muscular dystrophies usually cause only mild cardiac involvement (22-24). However, recently, a single mutation of the δ-sarcoglycan gene was reported to be associated with familial and sporadic cases of late-onset DCM, without symptoms of skeletal muscle dystrophy. In those cases, familial disease was transmitted in an autosomal dominant manner; such transmission was not observed in the mouse, or in patients with limb girdle muscular dystrophy, and the DCM of the cardiomyopathic hamster is transmitted in an autosomal recessive manner. These observations suggest the possibility that the mutant δ-sarcoglycan in human DCM may be acting in a dominant negative manner, but the mechanism for the DCM remains to be evaluated (25).

In mice with disruption of the dystrophin gene (*mdx* mice) there is only a mild skeletal muscular dystrophy without cardiomyopathy, whereas mice lacking both dystrophin and utrophin develop a fatal cardiomyopathy along with severe skeletal muscle dystrophy (26), suggesting that utrophin may have a compensatory role in mice. Also, mice lacking both dystrophin and MyoD (a skeletal muscle-specific transcription factor) develop a severe DCM and skeletal myopathy (27). Mice lacking β- (28), γ- (29), and δ-sarcoglycan (30) (but not α-sarcoglycan) show progressive dilated cardiomyopathy and heart failure with mild muscular dystrophy, as in the naturally occurring Syrian hamster cardiomyopathy. The mechanism of progression of dilated cardiomyopathy in β and δ-sarcoglycan deficient mice has been attributed to lack of sarcoglycan proteins in vascular smooth muscle, which leads to coronary microvascular abnormalities and inducible myocardial ischemia (30). γ-Sarcoglycan also is located in vascular smooth muscle, and null mutant mice exhibit cardiac myocyte loss through apoptosis (29) with subsequent heart failure, suggesting a mechanism different from that in other types of DGC-related DCM and muscular dystrophies. In the hamster DCM, increased myocardial cell membrane permeability and subsequent cardiac myocyte necrosis in cardiac muscle correlates with progression of left ventricular dysfunction, suggesting that loss of δ-sarcoglycan may lead to rupture of susceptible sarcolemmal membranes in cardiac muscle (31), although the vascular mechanism observed in δ-sarcoglycan deficient mice noted above also may cause vascular spasm described in the hamster DCM (32).

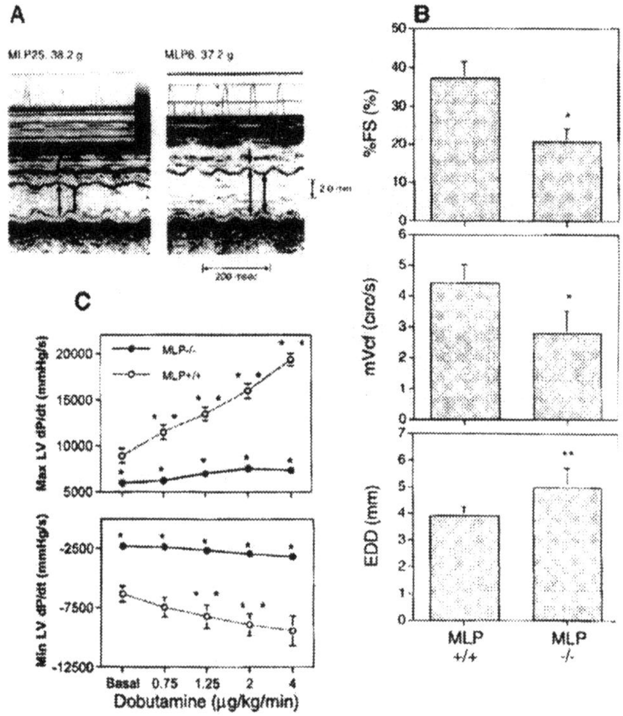

Figure 19-2. The cardiac phenotype of MLP-deficient mice exhibits features of dilated cardiomyopathy with heart failure. (A) Transthoracic M-mode echocardiographic tracings in a wild type MLP(+/+) (left panel); and MLP(-/-) mouse(right panel) The upper tracing in each panel is the electrocardiogram. The double-sided vertical arrows in each panel indicate the internal dimensions of the left ventricular chamber at end-diastole (larger arrow) and the end-systole (smaller arrow). The upper small vertical arrows in each panel mark the anterior edge of the anterior wall, showing wall thinning in the MLP(-/-) mouse. (B) Summary of echocardiographic findings in MLP(+/+) and MLP(-/-) mice. %FS: percent fractional shortening of LV wall.; mVcf: rate of wall shortening; EDD:end-diastolic diameter; *: $p<0.001$, **$p<0.002$. (C) Hemodynamic measurements under basal conditions, and in response to graded doses of the β-adrenergic agonist dobutamine. maxdP/dt and mindP/dt demonstrate marked β-adrenergic desensitization. Reproduced by permission from Arber et al (35).

Deficiency of desmin, which forms the intracellular intermediate filaments in myocytes, has been shown to be associated with abnormalities of cardiac, skeletal and smooth muscle in mice; the heart shows severe loss of myocardial architecture, degeneration and calcification (33). Missense mutations in muscle desmin also recently have been reported to cause familial forms of DCM(34), although the mechanisms leading to DCM remain to be identified.

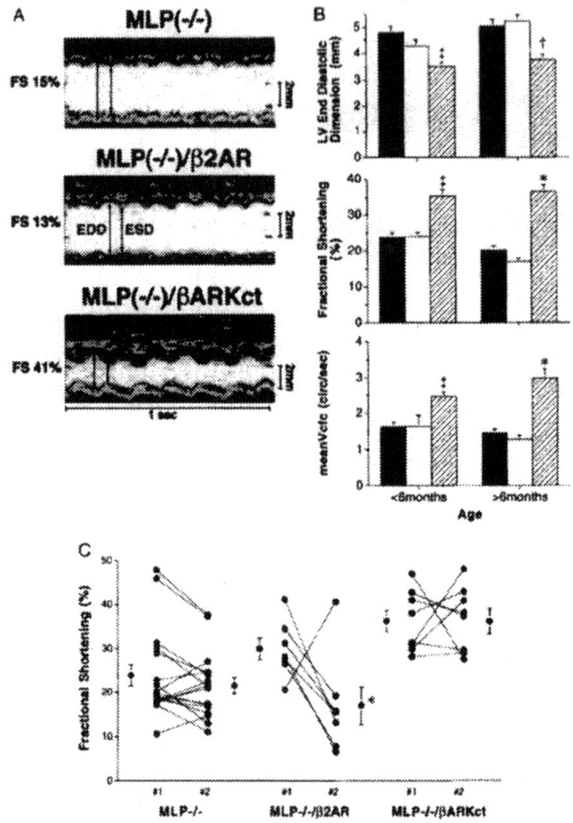

Figure 19-3. Summary of improved cardiac function in βARKct/MLP knockout(KO) mice. (A) Transthoracic M-mode echocardiographic tracings in a MLPKO (Top), MLPKO/β_2AR (Middle), and a MLPKO/βARKct mouse (Bottom). Left ventricular dimensions are indicated by the double-sided arrows. EDD, end diastolic dimension; ESD, end-systolic dimension. Both the MLPKO and MLP MLPKO/β_2AR mice have chamber dilatation with reduced wall motion indicating depressed cardiac function, whereas chamber size and cardiac function are normal in the MLPKO/βARKct mouse. (B) Echocardiographic findings in mice under 6 months of age: MLPKO (solid bar, MLPKO/β_2AR (open bar), and MLPKO/βARKct (hatched bar); and more than 6 months of age: MLPKO (solid bar), MLPKO/β_2AR (open bar), and MLPKO/βARKct (hatched bar). *, $P < 0.005$; †, $P < 0.01$; ‡, $P < 0.05$ MLPKO/βARKct vs. MLP MLPKO and MLPKO/β_2AR mice. (C) Change in cardiac function over a 2.5- to 3-month period in the three groups of gene-targeted mice. #1, early study (mean age, 3.9 months); #2, later study (mean age, 6.4 months). Reproduced by permission from Rockman et al.(40)

The muscle LIM protein (MLP) knockout mouse is one of the first engineered murine models of DCM and has many phenotypic features of human DCM with heart failure (35) (Figure 19-2). A mutation of MLP in human cardiomyopathy has not been identified, although MLP expression is reduced in the myocardium of the patients with idiopathic DCM and ischemic cardiomyopathy. MLP protein is a component of the cytoskeleton which colocalizes with vinculin at the z-lines and costameres. The mechanisms by which DCM is produced by MLP deficiency in the mouse are not yet fully clarified, although a defect in Ca^{2+} cycling have been identified (see below).

Although not yet studied in animal models, the nuclear localizing protein lamin A/C has been reported to cause Emery-Dreifuss muscular dystrophy in humans (36) and mutations in the rod-domain of the lamin A/C gene to cause DCM (37). The related nuclear membrane protein emerin also was reported to cause Emery-Dreifuss muscular dystrophy; the molecular function of lamin A/C and emerin are not well understood.

Potential Therapeutic Approaches in Genetic DCM

Strategies for the pharmacological management of patients with congestive heart failure have improved significantly in recent years (38), although the mortality

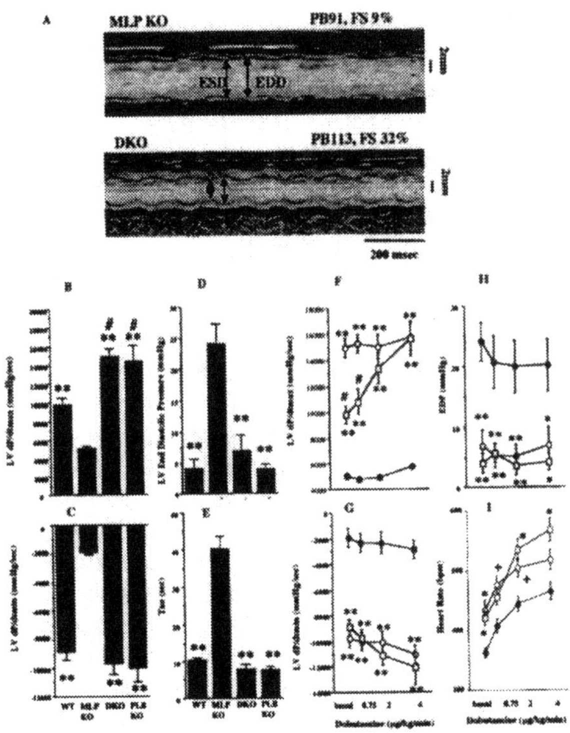

Figure 19-4. Prevention of DCM in MLP/phospholamban(PLB) double knockout(DKO) Mice. (A) Transthoracic M-mode echocardiographic tracings in a

MLPKO mouse (top) and a DKO mouse (bottom). Left ventricular (LV) dimensions are indicated by the double-sided arrows. EDD, end diastolic dimension; ESD, end systolic dimension. MLPKO mouse has chamber dilatation with reduced wall motion, indicating depressed cardiac function and increased wall stress, whereas chamber size and cardiac function are normal in the DKO mouse. (B–I) Maximal in vivo cardiac contractility and rescue of diastolic dysfunction in DKO mice. Cardiac catheterization was performed at basal condition in intact, anesthetized mice of wild type (WT, MLPKO, DKO, and PLBKO. (B) maxLV dP/dt, (C) minLVdP/dt, (D) LV end diastolic pressure, (E) Tau. (F–I) β-adrenergic responsiveness by graded infusion of dobutamine. Wild-type (open squares), MLPKO (filled circles), and DKO (open circles) mice. (F) LV dP/dtmax. (G) LV dP/dtmin, (H) LV end-diastolic pressure, (I) heart rate. Data were expressed as mean ± SEM. MLPKO vs. DKO or wild type, or PLBKO; +,$p<0.05$; *, $p<0.01$; **, $p<0.001$, WT versus DKO or PLBKO; #, $p< 0.01$. Reproduced by permission from Minamisawa, Hoshijima et al.(41)

associated with heart failure remains high. Novel therapeutic approaches have been sought using genetic engineering or viral gene transfer techniques in murine models of heart failure. In the MLP deficient mouse model, the pathophysiological manifestations and time course are sufficiently predictable to make this model useful for studying heart failure rescue or treatment strategies, such as with administered growth hormone (39), or by genetic strategies. When mice overexpressing the GRK inhibitor β-adrenergic receptor kinase were mated with MLP DCM mice, the development of DCM in the double transgenic offspring was prevented (Figure 19-3) (40). Recently reported by Minamisawa, Hoshijima et al. (41) that correction of a calcium cycling defect in the MLP model of DCM by crossing them with mice having phospholamban deficiency was associated with complete prevention of the adult DCM phenotype (Figure 19-4). It is notable that not only hemodynamic abnormalities but also subcellular structure and interstitial fibrosis noted in MLP deficient mice were normalized by phospholamban ablation, indicating a crucial role of intracellular calcium cycling regulation in the failing heart. However, crossing of tropomodulin overexpressing mice with PLB knockout mice failed to prevent the development of DCM (42).

Biomechanical stress: cardiac hypertrophy to heart failure

Surgically induced transverse aortic constriction (TAC) has been used to produce short-term pressure overload and induce cardiac hypertrophy (43). In normal mice, when TAC produces a significant pressure gradient significant cardiac hypertrophy occurs by 1 week after operation, and this approach can be useful for determining how biomechanical stress is tolerated in genetically engineered mouse models. For example, several cytokine receptors (including IL-6 and cardiotrophin-1) share GP130 as a signal transducing receptor component which activates the JAK-STAT and Ras/MAPk signaling pathways; these pathways are considered important for the physiological response to biomechanical stimuli. Overexpression of an active form of the GP130 protein in mice caused cardiac hypertrophy (44), while its disruption resulted in embryonic lethality due to skeletal, neural and metabolic defects (45). Hirota et al. then produced mice with a cardiac restricted conditional knockout of

GP130 using Cre-lox technology (46); these mice exhibited normal cardiac function at baseline but after TAC, rapid development of DCM and heart failure occurred, associated with extensive myocardial apoptosis. Thus, Gp130 appears to mediate a crucial survival pathway for protecting myocytes from biomechanical stress-related apoptosis, providing a potential therapeutic area target. In this regard, mice overexpressing tumor necrosis factor α have been shown to exhibit DCM, further suggesting the importance of cytokine regulation in the progression of heart failure (47).

Failure of adaptation to biomechanical stress also was reported in transgenic mice overexpressing Gαq, discussed in a subsequent section. When Gαq transgenic mice underwent TAC, LV dilation and dysfunction developed, associated with increased mortality.

Ischemic cardiomyopathy due to coronary artery disease has become the most common form of DCM in the United States and a mouse model of myocardial infarction with cardiac dilation and heart failure has been developed (48). Such models prove useful in understanding the effects of gene deletions, as in the decorin knockout mouse, as well as for investigating potential forms of therapy (49)

Abnormal Intracellular calcium regulation

Although mutations of the intracellular calcium regulatory proteins have not yet been described in human DCM, mice lacking cardiac ryanodine receptor (RyR2: the calcium release channel) have been reported to cause embryonic death due to heart development defect (50). Also, mice lacking FKBP-12, which is known to bind and regulate the skeletal muscle ryanodine receptor, have been reported to exhibit severe DCM, along with ventricular septal defects (51). It has been claimed that FKBP12.6 is a cardiac specific isoform, which selectively binds to the cardiac ryanodine receptor (RyR2), but no compensatory expression of FKBP12.6 was observed in FKBP-12 deficient mice, leaving open the question regarding the inter-relation between these FKBP isoforms and ryanodine receptors. In this regard, hyperphosphorylation of cardiac ryanodine receptor (RyR2) with dissociation of FKBP12.6 was reported to cause dysregulation of the calcium release channel in heart failure, defining important features of calcium releasing abnormality in the failing heart (52). Mice lacking a novel SR junctional protein, junctophilin-2, recently were found to exhibit embryonic lethality due to DCM with dissociation of coupling between the plasma membrane and sarcoplasmic reticulum, suggesting an important role for cell organization in intracellular calcium regulation (53). Overexpression in mice of calsequestrin, a high affinity Ca^{2+} binding protein located in the junctional SR complex which regulates calcium-induced calcium release, produces cardiac hypertrophy with subsequent development of DCM and premature death associated with marked suppression of calcium-induced calcium release (54). Thus, each intracellular calcium regulatory protein has a pivotal role for E-C coupling and myocyte contraction, which along with the evidence for rescue of phenotype in MLP DCM by PLB ablation (55), emphasizing the importance of intracellular calcium regulation in the failing heart.

Expression of a constitutively active form of the Ca^{2+}/calmodulin-dependent protein phosphatase calcineurin induces cardiac hypertrophy and subsequent DCM through the activation of NFAT3 which associates with GATA4 in the nucleus, (56), and these effects could be prevented by administration of calcineurin inhibitors such as cyclosporin A and FK506 (56,57). Whether this mechanism is operative in pressure overload hypertrophy has been controversial. Also, overexpression of Ca^{2+}/calmodulin-dependent protein kinase IV was shown to induce dilated cardiomyopathy with a marked hypertrophic response through activation of the MEF2 transcriptional factor by a posttranslational mechanism, suggesting two important parallel pathways that can induce maladaptive cardiac hypertrophy (58).

Abnormal G-protein coupled receptor pathways

Gαq: As mentioned earlier, overexpression of Gαq (which is involved in α_{1}-adrenergic, angiotensin II and endothelin-1 receptor signaling) causes cardiac hypertrophy in transgenic mice which, depending on the level of overexpression and whether or not aortic banding is imposed, can lead to cardiac dilation and dysfunction (59); Gαq overexpression also can lead to postpartum cardiomyopathy, associated with extensive myocardial apoptosis (60).

Gsα: Overexpression of Gsα in transgenic mice results in cardiac hyperfunction, initially followed by late development of DCM associated with myocardial fibrosis (61). When mice with overexpression of β-adrenergic receptors were crossed with mice with DCM due to MLP deficiency, the offspring exhibited depression of cardiac function compared to untreated mice with MLP DCM. Also, long-term overexpression of β_1 adrenergic receptors (62) produces DCM. These findings indicate that sustained β-adrenergic stimulation can cause or contribute to the development of DCM.

Conditional overexpression of a specifically designed Gi coupled Ro1 receptor (derived from a human opioid receptor) in tetracycline-transactivated inducible transgenic mice caused bradycardia and DCM in the doxycycline-off condition (tetO promoter activated); the findings of DCM were reversed by restoring doxycycline treatment (tetO inactivated) (Conklin, Bruce R., personal communication). Such technology should allow more precise assessment of the role of other G protein-coupled receptors in DCM.

Abnormal Regulation of Transcription factors

In addition to the role of MEF2, cited above, transgenic mice expressing a dominant-negative form of the transcription factor $CREB_{A133}$ develop DCM without passing through a phase of hypertrophy. These mice exhibit myocyte dropout and subsequent fibrosis, with a high mortality at 5-6 months of age (63).

Regulation of Extracellular Matrix

The extracellular matrix, including structural collagen, was recently reported to play an important role in the progression of human DCM (64), and inhibition of metalloproteinase was shown to prevent early cardiac enlargement after myocardial infarction in mice. Also, Mice overexpressing human collagenase (matrix metalloproteinase-1), which regulates extracellular collagen metabolism, exhibit progressive cardiomyopathy and heart failure.

Mutations Affecting Mitochondrial Function

It is well-known that human genetic syndromes with mutations of mitochondrial DNA (mtDNA) and deficient oxidative phosphorylation, are associated with cardiomyopathy, myopathy, and stroke (65,66); mtDNA mutations also may have a role in acquired diseases of old age (65,66). Several murine models provide evidence of the physiological importance of normal transcriptional regulation of mitochondria in the heart and respiratory oxidative chain function, as well as toxic effect of reactive oxygen species. Thus, mice lacking the adenine nucleotide translocator 1 (ANT1), an enzyme which moves ATP from mitochondria to cytosol in heart and skeletal muscle, manifest adult-onset hypertrophic cardiomyopathy with a mild skeletal myopathy (67). In mice with deficiency of mitochondrial manganese superoxide dismutase (MnSOD) there is early development of DCM and premature death (68). Also, when cardiac and skeletal muscle mitochondrial transcriptional factor A (*Tfam*, or mtTFA) is disrupted using Cre-lox technology, cardiac-specific respiratory chain deficiency ensues, associated with DCM and AV block (69).

Viral Myocarditis and DCM Autoimmune responses

DCM occurs after viral myocarditis in the mouse, although there are differences in pathophysiological features depending on the viral strain (most commonly encephalomyocarditis and Coxsackie virus B3) and the genetic background (70). Expression of replication defective Coxsackie virus genomic constructs in mice using a cardiac specific promoter can induce DCM (71). In addition, evidence of a mechanism by which Coxsackie virus infection can lead to DCM was the demonstration that an enteroviral protease can cleave dystrophin (72) and that adjacent sarcoglycans dissociate from the membrane in murine models of Coxsackie myocarditis (73).

In mice with certain HLA alleles, autoimmune responses can induce myocarditis with subsequent development of DCM (74).

REFERENCES

1. Chien KR. Stress pathways and heart failure. *Cell.* 1999;98:555-558.
2. Towbin JA. The role of cytoskeletal proteins in cardiomyopathies. *Curr Opin Cell Biol.* 1998;10:131-139.
3. Hunter JJ, K.R. C. Signaling pathway for cardiac hypertrophy and failure. *N Eng J Med.* 1999;341:1276-1283.
4. Fatkin D, Seidman JD, Seidman CE. Hypertrophic Cardiomyopathy. In: Willerson JT, Cohn JN, eds. *Cardiovascular Medicine.* Phyladelphia, Pennsylvania: W.B. Saunders Co.; 2000:1055-1075.

5. Seidman CE, Seidman JG. Molecular genetics of inherited cardiomyopathies. In: Chien KR, ed. *Molecular basisi of cardiobascular disease.* Philadelphia: W.B.Saunders.; 1999:251-263.
6. Maron BJ. Hypertrophic cardiomyopathy [published erratum appears in Lancet 1997 Nov 1;350(9087):1330]. *Lancet.* 1997;350:127-133.
7. McMinn TR, Jr., Ross J, Jr. Hereditary dilated cardiomyopathy. *Clin Cardiol.* 1995;18:7-15.
8. Ikeda Y, Ross J, Jr. Models of dilated cardiomyopathy in the mouse and the hamster. *Curr Opin Cardiol.* 2000;15:197-201.
9. Sylvius N, Tesson F, Gayet C, et al. A New Locus for Autosomal Dominant Dilated Cardiomyopathy Identified on Chromosome 6q12-q16. *Am J Hum Genet.* 2000;68.
10. Tesson F, Sylvius N, Pilotto A, et al. Epidemiology of desmin and cardiac actin gene mutations in a European population of dilated cardiomyopathy. *Eur Heart J.* 2000;21:1872-1876.
11. Chen J, Chien KR. Complexity in simplicity: monogenic disorders and complex cardiomyopathies. *J Clin Invest.* 1999;103:1483-1485.
12. Mogensen J, Klausen IC, Pedersen AK, et al. Alpha-cardiac actin is a novel disease gene in familial hypertrophic cardiomyopathy. *J Clin Invest.* 1999;103:R39-43.
13. Kumar A, Crawford K, Close L, et al. Rescue of cardiac alpha-actin-deficient mice by enteric smooth muscle gamma-actin. *Proc Natl Acad Sci U S A.* 1997;94:4406-4411.
14. Olson TM, Michels VV, Thibodeau SN, et al. Actin mutations in dilated cardiomyopathy, a heritable form of heart failure. *Science.* 1998;280:750-752.
15. McConnell BK, Jones KA, Fatkin D, et al. Dilated cardiomyopathy in homozygous myosin-binding protein-C mutant mice. *J Clin Invest.* 1999;104:1235-1244.
16. Sussman MA, Welch S, Cambon N, et al. Myofibril degeneration caused by tropomodulin overexpression leads to dilated cardiomyopathy in juvenile mice. *J Clin Invest.* 1998;101:51-61.
17. Straub V, Campbell KP. Muscular dystrophies and the dystrophin-glycoprotein complex. *Current Opinion in Neurology.* 1997;10:168-175.
18. Hoffman EP, Brown RH, Jr., Kunkel LM. Dystrophin: the protein product of the Duchenne muscular dystrophy locus. *Cell.* 1987;51:919-928.
19. Ortiz-Lopez R, Li H, Su J, et al. Evidence for a dystrophin missense mutation as a cause of X-linked dilated cardiomyopathy. *Circulation.* 1997;95:2434-2440.
20. Muntoni F, Cau M, Ganau A, et al. Brief report: deletion of the dystrophin muscle-promoter region associated with X-linked dilated cardiomyopathy. *N Engl J Med.* 1993;329:921-925.
21. Towbin JA, Hejtmancik JF, Brink P, et al. X-linked dilated cardiomyopathy. Molecular genetic evidence of linkage to the Duchenne muscular dystrophy (dystrophin) gene at the Xp21 locus. *Circulation.* 1993;87:1854-1865.
22. Melacini P, Fanin M, Duggan DJ, et al. Heart involvement in muscular dystrophies due to sarcoglycan gene mutations. *Muscle & Nerve.* 1999;22:473-479.

23. Barresi R, Di Blasi C, Negri T, et al. Disruption of heart sarcoglycan complex and severe cardiomyopathy caused by beta sarcoglycan mutations. *J Med Genet.* 2000;37:102-107.
24. Dincer P, Akcoren Z, Demir E, et al. A cross section of autosomal recessive limb-girdle muscular dystrophies in 38 families. *J Med Genet.* 2000;37:361-367.
25. Tsubata S, Bowles KR, Vatta M, et al. Mutations in the human delta-sarcoglycan gene in familial and sporadic dilated cardiomyopathy. *J Clin Invest.* 2000;106:655-662.
26. Grady RM, Teng H, Nichol MC, et al. Skeletal and cardiac myopathies in mice lacking utrophin and dystrophin: a model for Duchenne muscular dystrophy. *Cell.* 1997;90:729-738.
27. Megeney LA, Kablar B, Perry RL, et al. Severe cardiomyopathy in mice lacking dystrophin and MyoD. *Proc Natl Acad Sci U S A.* 1999;96:220-225.
28. Durbeej M, Cohn RD, Hrstka RF, et al. Disruption of the beta-sarcoglycan gene reveals pathogenetic complexity of limb-girdle muscular dystrophy type 2E. *Mol Cell.* 2000;5:141-151.
29. Hack AA, Ly CT, Jiang F, et al. Gamma-sarcoglycan deficiency leads to muscle membrane defects and apoptosis independent of dystrophin. *J Cell Biol.* 1998;142:1279-1287.
30. Coral-Vazquez R, Cohn RD, Moore SA, et al. Disruption of the sarcoglycan-sarcospan complex in vascular smooth muscle: a novel mechanism for cardiomyopathy and muscular dystrophy. *Cell.* 1999;98:465-474.
31. Ikeda Y, Martone M, Gu Y, et al. Altered sarcolemmal permeability and membrane proteins correlate with cardiac dysfunction in cardiomyopathic hamsters. *Am J Physiol.* 2000;278:H1362-H1370.
32. Factor SM, Minase T, Cho S, et al. Microvascular spasm in the cardiomyopathic Syrian hamster: a preventable cause of focal myocardial necrosis. *Circulation.* 1982;66:342-354.
33. Milner DJ, Weitzer G, Tran D, et al. Disruption of muscle architecture and myocardial degeneration in mice lacking desmin. *J Cell Biol.* 1996;134:1255-1270.
34. Li D, Tapscoft T, Gonzalez O, et al. Desmin mutation responsible for idiopathic dilated cardiomyopathy. *Circulation.* 1999;100:461-464.
35. Arber S, Hunter JJ, Ross J, Jr., et al. MLP-deficient mice exhibit a disruption of cardiac cytoarchitectural organization, dilated cardiomyopathy, and heart failure. *Cell.* 1997;88:393-403.
36. Bonne G, Di Barletta MR, Varnous S, et al. Mutations in the gene encoding lamin A/C cause autosomal dominant Emery-Dreifuss muscular dystrophy. *Nat Genet.* 1999;21:285-288.
37. Fatkin D, MacRae C, Sasaki T, et al. missense mutations in the rod domain of the lamin A/C gene as causes of dilated cardiomyopathy and conduction system disease. *N Eng J Med.* 1999;341:1715-1724.
38. Braunwald E, Bristow MR. Congestive heart failure: fifty years of progress. *Circulation.* 2000;102:IV14-23.

39. Hongo M, Ryoke T, Schoenfeld J, et al. Effect of growth hormone on cardiac dysfunction and gene expression in genetic murine dilated cardiomyopathy. *Basic Res Cardiol.* 2000:In press.
40. Rockman HA, Chien KR, Choi DJ, et al. Expression of a beta-adrenergic receptor kinase 1 inhibitor prevents the development of myocardial failure in gene-targeted mice. *Proc Natl Acad Sci U S A.* 1998;95:7000-7005.
41. Minamisawa S, Hoshijima M, Chu G, et al. Chronic phospholamban-sarcoplasmic reticulum calsium ATPase interaction is the critical calcium cycling defect in Dilated Cardiomyopathy. *Cell.* 1999;99:313-322.
42. Delling U, Sussman MA, Molkentin JD. Re-evaluating sarcoplasmic reticulum function in heart failure [letter]. *Nat Med.* 2000;6:942-943.
43. Rockman HA, Ross RS, Harris AN, et al. Segregation of atrial-specific and inducible expression of an atrial natriuretic factor transgene in an in vivo murine model of cardiac hypertrophy [published erratum appears in Proc Natl Acad Sci U S A 1991 Nov 1;88(21):9907]. *Proc Natl Acad Sci U S A.* 1991;88:8277-8281.
44. Hirota H, Yoshida K, Kishimoto T, et al. Continuous activation of gp130, a signal-transducing receptor component for interleukin 6-related cytokines, causes myocardial hypertrophy in mice. *Proc Natl Acad Sci U S A.* 1995;92:4862-4866.
45. Yoshida K, Taga T, Saito M, et al. Targeted disruption of gp130, a common signal transducer for the interleukin 6 family of cytokines, leads to myocardial and hematological disorders. *Proc Natl Acad Sci U S A.* 1996;93:407-411.
46. Hirota H, Chen J, Betz UA, et al. Loss of a gp130 cardiac muscle cell survival pathway is a critical event in the onset of heart failure during biomechanical stress. *Cell.* 1999;97:189-198.
47. Kubota T, McTiernan CF, Frye CS, et al. Dilated cardiomyopathy in transgenic mice with cardiac-specific overexpression of tumor necrosis factor-alpha. *Circ Res.* 1997;81:627-635.
48. Patten RD, Aronovitz MJ, Deras-Mejia L, et al. Ventricular remodeling in a mouse model of myocardial infarction. *Am J Physiol.* 1998;274:H1812-1820.
49. Vaplon S, Zimmerman S, Covell J, et al. Altered post-MI Ventricular Remodeling in Decorin-Null Mice. *Circulation.* 2000;102:II-358.
50. Takeshima H, Komazaki S, Hirose K, et al. Embryonic lethality and abnormal cardiac myocytes in mice lacking ryanodine receptor type 2. *Embo J.* 1998;17:3309-3316.
51. Shou W, Aghdasi B, Armstrong DL, et al. Cardiac defects and altered ryanodine receptor function in mice lacking FKBP12. *Nature.* 1998;391:489-492.
52. Marx SO, Reiken S, Hisamatsu Y, et al. PKA phosphorylation dissociates FKBP12.6 from the calcium release channel (ryanodine receptor): defective regulation in failing hearts. *Cell.* 2000;101:365-376.
53. Takeshima H, Komazaki S, Nishi M, et al. Junctophilins: a novel family of junctional membrane complex proteins. *Mol Cell.* 2000;6:11-22.
54. Jones LR, Suzuki YJ, Wang W, et al. Regulation of Ca2+ signaling in transgenic mouse cardiac myocytes overexpressing calsequestrin. *J Clin Invest.* 1998;101:1385-1393.

55. Minamisawa S, Hoshijima M, Chu G, et al. Chronic phospholamban-sarcoplasmic reticulum calcium ATPase interaction is the critical calcium cycling defect in dilated cardiomyopathy. *Cell.* 1999;99:313-322.
56. Molkentin JD, Lu JR, Antos CL, et al. A calcineurin-dependent transcriptional pathway for cardiac hypertrophy. *Cell.* 1998;93:215-228.
57. Sussman MA, Lim HW, Gude N, et al. Prevention of cardiac hypertrophy in mice by calcineurin inhibition. *Science.* 1998;281:1690-1693.
58. Passier R, Zeng H, Frey N, et al. CaM kinase signaling induces cardiac hypertrophy and activates the MEF2 transcription factor in vivo. *J Clin Invest.* 2000;105:1395-1406.
59. D'Angelo DD, Sakata Y, Lorenz JN, et al. Transgenic Galphaq overexpression induces cardiac contractile failure in mice. *Proc Natl Acad Sci U S A.* 1997;94:8121-8126.
60. Adams JW, Sakata Y, Davis MG, et al. Enhanced Galphaq signaling: a common pathway mediates cardiac hypertrophy and apoptotic heart failure. *Proc Natl Acad Sci U S A.* 1998;95:10140-10145.
61. Iwase M, Uechi M, Vatner DE, et al. Cardiomyopathy induced by cardiac Gs alpha overexpression. *Am J Physiol.* 1997;272:H585-589.
62. Lutz H, Lohse MJ. Hypertrophy and contractile dysfunction in transgenic mice with heart-specific overexpression of the β_1–adrenergic receptor. *Circulation.* 1999;100:I-762.
63. Fentzke RC, Korcarz CE, Lang RM, et al. Dilated cardiomyopathy in transgenic mice expressing a dominant-negative CREB transcription factor in the heart. *J Clin Invest.* 1998;101:2415-2426.
64. Thomas CV, Coker ML, Zellner JL, et al. Increased matrix metalloproteinase activity and selective upregulation in LV myocardium from patients with end-stage dilated cardiomyopathy. *Circulation.* 1998;97:1708-1715.
65. DiMauro S, Hirano M. Mitochondria and heart disease. *Curr Opin Cardiol.* 1998;13:190-197.
66. Wallace DC, Graham B. Mitochondrial Genes in Myopathy, Cardiomyopathy, and Stroke. In: Chien KR, ed. *Molecular Basis of Cardiovascular Disease.* Philadelphia: W.B.Saunders; 1999:264-277.
67. Graham BH, Waymire KG, Cottrell B, et al. A mouse model for mitochondrial myopathy and cardiomyopathy resulting from a deficiency in the heart/muscle isoform of the adenine nucleotide translocator. *Nat Genet.* 1997;16:226-234.
68. Li Y, Huang TT, Carlson EJ, et al. Dilated cardiomyopathy and neonatal lethality in mutant mice lacking manganese superoxide dismutase. *Nat Genet.* 1995;11:376-381.
69. Wang J, Wilhelmsson H, Graff C, et al. Dilated cardiomyopathy and atrioventricular conduction blocks induced by heart-specific inactivation of mitochondrial DNA gene expression. *Nat Genet.* 1999;21:133-137.
70. Kawai C. From myocarditis to cardiomyopathy: mechanisms of inflammation and cell death: learning from the past for the future. *Circulation.* 1999;99:1091-1100.
71. Wessely R, Klingel K, Santana LF, et al. Transgenic expression of replication-restricted enteroviral genomes in heart muscle induces defective excitation-

contraction coupling and dilated cardiomyopathy. *J Clin Invest.* 1998;102:1444-1453.

72. Badorff C, Lee GH, Lamphear BJ, et al. Enteroviral protease 2A cleaves dystrophin: evidence of cytoskeletal disruption in an acquired cardiomyopathy. *Nat Med.* 1999;5:320-326.
73. Lee GH, Badorff C, Knowlton KU. Dissociation of sarcoglycans and the dystrophin carboxyl terminus from the sarcolemma in enteroviral cardiomyopathy. *Circ Res.* 2000;87:489-495.
74. Bachmaier K, Neu N, Yeung RS, et al. Generation of humanized mice susceptible to peptide-induced inflammatory heart disease. *Circulation.* 1999;99:1885-1891.
75. Geisterfer-Lowrance AA, Christe M, Conner DA, et al. A mouse model of familial hypertrophic cardiomyopathy. *Science.* 1996;272:731-734.
76. Chen J, Kubalak SW, Minamisawa S, et al. Selective requirement of myosin light chain 2v in embryonic heart function. *J Biol Chem.* 1998;273:1252-1256.
77. Minamisawa S, Gu Y, Ross J, Jr., et al. A post-transcriptional compensatory pathway in heterozygous ventricular myosin light chain 2-deficient mice results in lack of gene dosage effect during normal cardiac growth or hypertrophy. *J Biol Chem.* 1999;274:10066-10070.
78. Oberst L, Zhao G, Park JT, et al. Dominant-negative effect of a mutant cardiac troponin T on cardiac structure and function in transgenic mice. *J Clin Invest.* 1998;102:1498-1505.
79. Tardiff JC, Factor SM, Tompkins BD, et al. A truncated cardiac troponin T molecule in transgenic mice suggests multiple cellular mechanisms for familial hypertrophic cardiomyopathy. *J Clin Invest.* 1998;101:2800-2811.
80. Tardiff JC, Hewett TE, Palmer BM, et al. Cardiac troponin T mutations result in allele-specific phenotypes in a mouse model for hypertrophic cardiomyopathy. *J Clin Invest.* 1999;104:469-481.
81. James J, Zhang Y, Osinska H, et al. Transgenic modeling of a cardiac troponin I mutation linked to familial hypertrophic cardiomyopathy. *Circ Res.* 2000;87:805-811.
82. Muthuchamy M, Pieples K, Rethinasamy P, et al. Mouse model of a familial hypertrophic cardiomyopathy mutation in alpha- tropomyosin manifests cardiac dysfunction. *Circ Res.* 1999;85:47-56.
83. Bing W, Knott A, Redwood C, et al. Effect of hypertrophic cardiomyopathy mutations in human cardiac muscle alpha -tropomyosin (Asp175Asn and Glu180Gly) on the regulatory properties of human cardiac troponin determined by in vitro motility assay. *J Mol Cell Cardiol.* 2000;32:1489-1498.
84. Yang Q, Sanbe A, Osinska H, et al. A mouse model of myosin binding protein C human familial hypertrophic cardiomyopathy. *J Clin Invest.* 1998;102:1292-1300.
85. Cantlay AM, Shokrollahi K, Allen JT, et al. Genetic analysis of the G4.5 gene in families with suspected Barth syndrome. *J Pediatr.* 1999;135:311-315.
86. Goldfarb LG, Park KY, Cervenáková L, et al. Missense mutations in desmin associated with familial cardiac and skeletal myopathy. *Nat Genet.* 1998;19:402-403.

22

ANALYSIS OF MURINE EMBRYONIC CARDIOVASCULAR PHENOTYPE

Kimimasa Tobita, Joseph P. Tinney, and Bradley B. Keller

Cardiovascular Development Research Program, University of Kentucky, Lexington, KY 40536

INTRODUCTION

The cardiovascular (CV) system is the first functioning organ system during embryogenesis, and abnormal CV development is a major source of childhood death and disability. Congenital cardiovascular anomalies present as distinct phenotypes *in utero* and represent a spectrum of pathogenetic mechanisms. For each specific defect there is a range of severity, and obviously the most severe disruptions in CV development are embryo lethal. It is also important to note that "survival" often represents successful functional and structural adaptation to events that occur very early in development.

CV structural and functional maturation has been studied in the chick embryo for over a century and in a diverse range of species over the past 25 years (1). The application of molecular genetic paradigms to the mouse has enabled the investigation of CV maturation in murine embryos with targeted genetic alterations. However, because the developing embryo is neither a small "fetus" nor "adult", molecular, cellular, structural, and functional paradigms in the neonate and adult are not necessarily active during CV morphogenesis. Thus, the impact of targeted genetic alterations on the murine embryo must be interpreted in the context of the "normal" spatial and temporal expression of stage specific molecular isoforms, and both structural and functional maturation.

The following chapter provides a step-wise, practical approach for defining the CV phenotype of murine embryos. A robust definition of embryonic CV phenotype requires data on morphometry, physiology, and when relevant, maternal/fetal coupling, in addition to molecular and cellular data. The determination of murine embryonic phenotype has also required the modification of technology developed to investigate other larger species to the specific conditions of these "tiny" embryos.

MORPHOLOGY

Overview

The first step in determining CV phenotype is the qualitative and quantitative assessment of external and internal CV structures. The morphologic assessment of the developing cardiovascular system can be done at the level of whole embryo, isolated heart, regional tissue, cellular or subcellular compartment. The developing CV system has been studied in a range of species, including numerous vertebrates and invertebrates (1), and the similarities in structural morphogenesis likely reflect conserved structure-function relationships. The onset and completion of cardiac morphogenesis is influenced by the duration of gestation in each species and by environmental factors such as temperature for poikilotherms or maternal placental blood flow for vertebrates (2).

Morphologic Staging

The external and internal morphologic characteristics of the murine embryonic heart have been characterized by scanning electron microscopy (SEM) starting at ED 10 and proceeding at 8 hour intervals to ED 17 (3,4). This stage of development in the murine embryo corresponds to stage 17 to 34 (day 2 to 8) in the chick embryo (2,5). These data provide normal morphological markers for the murine embryo, which has some important distinctions from other species. For example, the murine embryo has only 1 pair of conotruncal ridges forming the outlet septum versus 2 pairs in the chick. The mouse embryo has a prominent interventricular sulcus present on the exterior surface of the heart and which is absent from the chick embryo at comparable developmental stages. In the mouse embryo, the left ventricle (LV) is initially prominent. Later, the ventricles appear equally balanced from ED 11 to ED 13 and then the right ventricle (RV) predominates. Overall, the rapid growth of the LV is more prominent in the mouse embryo than the growth of the RV in the chick (3). Table 1 gives a subset of landmarks for comparable stages in mouse, chicken, and human embryos.

Table 1: List of developmental landmarks and comparative stages across species (8).

Mouse*	Chick †	Human °	Morphologic Point of Interest
ED 8.0	10	X	First myocardial contraction
ED 8.5	12	XI	Forward blood flow through heart
ED 8.6	15	XI	Achievement of "S" shaped curve
ED 9	16	XII	Cells first seen in cardiac jelly
ED 10.5	17	XIII	Ventricular trabeculation begins
ED 15	29	XX	Ventricular septation complete
ED 15.5	34	XIV	Coronary vasculogenesis evident

*Data adapted from Sissman (6), † Hamburger and Hamilton stages (5), ° Streeter's horizons (7)

Table 2 has embryo weights at equivalent developmental stages for chick and mouse (8), as well as embryonic heart weights (Furukawa, unpublished). Cardiac mass is usually indexed to body mass as a measure of cardiac efficiency. However, this index requires normal data during development for each murine strain of interest, due to significant differences in body mass between inbred and outbred strains.

Table 2: Chick and Mouse Embryo Weights at Comparable Developmental Stages

Chick Stage†	Chick Embryo	Chick Heart	Mouse Stage	Mouse Embryo	Mouse Heart
12	2.2	0.10	ED 10	1.0	
16	7.8	0.12	ED 10, 16h	5.0	
18	7.7	0.35	ED 11	6.5	0.31 (ED 11.5)
21	34.4	0.42	ED 11, 16h	17.6	
24	79.5	0.97	ED 12	24.3	1.06 (ED 12.5)
27	149.2	1.54	ED 13	68.1	1.92 (ED 13.5)
29	267.5	3.18	ED 14	113.8	2.81 (ED 14.5)
					4.05 (ED 15.5)
					4.93 (ED 16.5)

†Hamburger and Hamilton stage (5). ED indicates embryonic day of gestation. All weights are expressed as mean values in milligrams. Mean data adapted from Hu (8), Pexieder (2), and (Furukawa, unpublished).

Imaging Methods

The developing cardiovascular system was first described using simple optical magnification of transilluminated chick embryos. Light microscopy techniques have progressively improved to allow cinephotography (9), videophotography (10), fluorescence angiography (11), and digital image microscopy (Keller, unpublished). Individual images were used for analysis, and wax plate duplicates of images were used to reconstruct embryonic structures (1). A major advance in nondestructive microscopic imaging occurred with the introduction of confocal microscopy, which allows the serial optical sectioning of specimens up to 100 μm in thickness. In addition to "visible" light microscopy, these systems can perform fluorescence microscopy for immunohistochemical analysis. A combination of these techniques, frequently using protein or sequence specific tags, is being used to define the temporal and spatial distribution genes and proteins throughout the developing CV system (12-14). It is important to note that many genes and proteins have developmental cascades of isoforms that may differ from the adult phenotype (e.g. α- versus β-Myosin Heavy Chain, Myosin Light Chains, Troponins, etc, 15-17).

Using basic videomicroscopy, we can define cardiovascular dimensions and function in mouse embryos from ED 10.5 through the completion of cardiac septation at ED 14.5 (18). The external contours of the embryonic RV and LV viewed by video microscopy resemble the contours previously described by scanning electron microscopy (3). When viewed ventrally, the embryonic RV and LV are exposed, and in contrast to the developing chick heart, a prominent interventricular sulcus is evident (Figure 1). The ventral view can not be easily obtained in ED 10.5 embryos due to embryo curvature. When viewed from the left lateral approach, the embryonic left atrium is partially visualized, the LV is fully exposed, and only the anterior portion of the conotruncus is visible. After ED 12.5, the opaque chest wall and left limb bud obscure the left lateral view. Therefore, we can image individual embryos in the left anterior oblique positions (ED 10.5 to 12.5) and/or in the ventral position (ED 11.5 to 14.5) while maintaining uteroplacental continuity (18). Of note, the heart is imaged through the translucent anterior splanchnopleure for ED 10.5 to 12.5 embryos; however for ED 13.5 and 14.5 embryos we incise the anterior chest wall with a U-shaped incision to expose the embryonic heart.

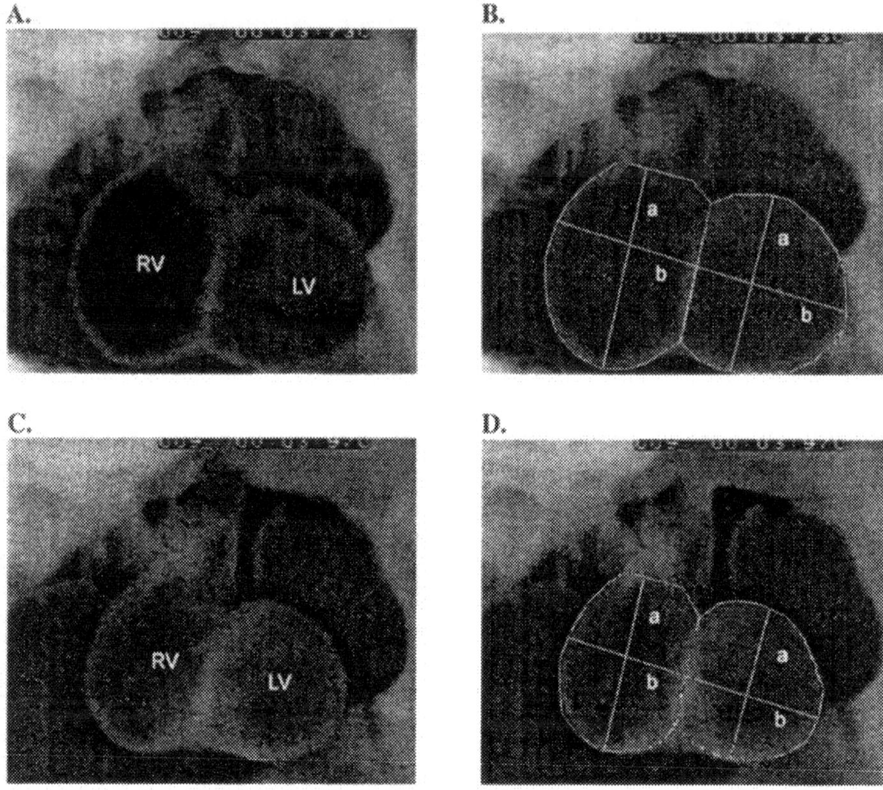

Figure 1: Representative video photomicrographs of ventral views of an ED 12.5 mouse embryo heart planimetered during end-diastole (A, B) and end-systole (C, D) including planimetry of the epicardial boundaries and major (a) and minor (b) axes (B, D).

We analyze video images at workstations that include a minicomputer, frame-grabbing board, multi-purpose video monitor, and image analysis software (18). RV major and minor axis diameters are measured from ventral views and LV major and minor axis diameters were measured from ventral and/or left lateral views. The maximum long axis of the embryonic RV is drawn parallel to ventricular outflow and the maximal short axis is then drawn perpendicular to the long axis. Due to differences in the shape of the embryonic RV and LV in the ventral view, the long axes for the RV and LV are not always parallel (Figure 1).

The video planimetry protocol for each embryo includes calibration of measurement software for length (mm) and area (mm^2) from the recorded scribed standard image, followed by planimetry of sequential video fields for major and minor cardiac axis diameters and epicardial ventricular cross-sectional area of the RV and/or LV for 3 cardiac cycles. There is an obvious interventricular sulcus to delineate the primitive murine embryonic RV and LV. In the chick embryo, the conotruncal cushions are visible, aiding the separation of ventricle from outflow chamber. In the mouse, we attempt to transect the RV at a similar point, i.e. at the proximal portion of the conotruncal cushions. We use the external feature of the rapid tapering of the RV diameter as the site of proximal conotruncal cushions. The choice of this division site may be somewhat subjective, but it allows the consistent selection of the same region between embryos. Table 3 shows heart rate and ventricular dimensions for ED 10.5 to 14.5 mouse embryos (18). Epicardial RV and LV end-diastolic and end-systolic cross-sectional areas measured from the ventral and left lateral views increase from ED 10.5 to 14.5 (18).

Table 3: Heart Rates and Ventricular Dimensions for ED 10.5 to 14.5 Mouse Embryos (18).

Murine Embryonic Right Ventricle (Ventral View)								
ED	n	HR	a_{MAX}	a_{MIN}	b_{MAX}	b_{MIN}	A_{MAX}	A_{MIN}
11.5	6	164	1.18	0.99	0.90	0.74	0.99	0.74
12.5	7	205	1.22	1.06	0.92	0.79	0.96	0.75
13.5	13	174	1.42†	1.17†	1.00	0.81	1.19†	0.82
14.5	9	165	1.42†	1.21†	0.97	0.80	1.21†	0.89†
Murine Embryonic Left Ventricle (Ventral View)								
ED	n	HR	a_{MAX}	a_{MIN}	b_{MAX}	b_{MIN}	A_{MAX}	A_{MIN}
11.5	6	164	1.05	0.82	0.82	0.69	0.71	0.49
12.5	7	205	1.05	0.91	0.82	0.71	0.72	0.54
13.5	13	174	1.32†	1.06†	0.88	0.70	0.96†	0.64†
14.5	9	165	1.31†	1.11†	0.91	0.73	0.98†	0.72†
Murine Embryonic Left Ventricle (Left Lateral View)								
ED	n	HR	a_{MAX}	a_{MIN}	b_{MAX}	b_{MIN}	A_{MAX}	A_{MIN}
10.5	9	109	0.84	0.65	0.88	0.69	0.61	0.37
11.5	8	145	1.00	0.77	0.91	0.73	0.73	0.48
12.5	13	170*	0.99*	0.83*	0.99^	0.86*,‡	0.78*	0.59*

ED indicates embryonic day of gestation; n, number of embryos studied; HR, heart rate; a,b major and minor-axis diameters (mm); MAX, MIN, maximum and minimum; A, cross-sectional area (mm^2); RV, LV, right and left ventricles. Values are means. *$P<0.05$ vs ED 11.5 and †$P<0.05$ vs ED 10.5 by ANOVA with Bonferroni correction; ‡$P<0.05$ between ventral and left lateral views by unpaired t test.

Because of the unique and different shapes of the murine embryonic RV and LV throughout development (3,4) and because several mathematical methods can be used to calculating embryonic ventricular volume, we calculate RV and LV volumes using two methods (18). We first calculate ventricular volume from measured epicardial cross-sectional area using the model of an ellipsoid of revolution with a fixed aspect ratio. The ellipsoid equation is derived from equations for the cross-sectional area (A) of an ellipsoid, $A=(\pi ab)/4$, and the volume (V) of an ellipsoid of revolution, $V=(\pi ab^2)/6$, where a is the semi-major axis diameter, and b is the semi-minor axis diameter. Assuming a fixed aspect ratio of $a/b=4/3$, then $V=0.65A^{3/2}$. We also calculate ventricular volume from the measured major-axis (a) and minor-axis (b) diameters using the model of an ellipsoid of revolution. We use each method to calculate maximum and minimum RV and LV volumes and RV and LV stroke volumes for each day of development (Table 4). It is important to note that the ellipsoid LV volumes calculated from epicardial cross-sectional areas and from measured major and minor axes in the left lateral view result in differences in maximum volume, minimum volume, and stroke volume for ED 10.5 to 12.5 embryos (Table 4).

Table 4: Ventricular Volumes and Stroke Volumes for ED 10.5 to 14.5 Mouse Embryos (18).

ED	$V_{A(ELL)MAX}$	$V_{A(ELL)MIN}$	$SV_{A(ELL)}$	$V_{(ab)MAX}$	$V_{(ab)MIN}$	$SV_{(ab)}$
Murine Embryonic Right Ventricle (Ventral View)						
11.5	0.64	0.48	0.23	0.50	0.28	0.22
12.5	0.61	0.42	0.19	0.54	0.35	0.18
13.5	$0.85^{\dagger}$	0.48	0.37	$0.77^{\dagger}$	$0.40^{\dagger}$	0.36
14.5	$0.87^{\dagger}$	0.55	0.32	0.70	$0.40^{\dagger}$	0.30
Murine Embryonic Left Ventricle (Ventral View)						
ED	$V_{A(ELL)MAX}$	$V_{A(ELL)MIN}$	$SV_{A(ELL)}$	$V_{(ab)MAX}$	$V_{(ab)MIN}$	$SV_{(ab)}$
11.5	0.39	0.23	0.17	0.38	0.21	0.17
12.5	0.40	0.26	0.13	0.37	0.24	0.13
13.5	$0.62^{\dagger}$	0.33	0.28	$0.55^{\dagger}$	0.28	0.27
14.5	$0.64^{\dagger}$	$0.40^{\dagger}$	0.24	$0.58^{\dagger}$	$0.32^{\dagger}$	0.26
Murine Embryonic Left Ventricle (Left Lateral View)						
ED	$V_{A(ELL)MAX}$	$V_{A(ELL)MIN}$	$SV_{A(ELL)}$	$V_{(ab)MAX}$	$V_{(ab)MIN}$	$SV_{(ab)}$
10.5	0.32	0.15	0.17	0.36	0.17	0.19
11.5	0.41	0.21	0.20	0.43	0.22	0.22
12.5	$0.45^{*,\ddagger}$	0.30^{*}	0.15	$0.51^{*,\ddagger}$	$0.32^{*,\ddagger}$	0.19

ED indicates embryonic day of gestation; V_A, volume calculated from cross-sectional area using ellipsoid model (mm^3); MAX, maximum; Min, minimum; SV, stroke volume (mm^3); $V_{(ab)}$, volume calculated from major and minor axes using ellipsoid model (mm^3). Values are means. *$P<0.05$ vs ED 11.5 and †$P<0.05$ vs ED 10.5 by ANOVA with Bonferroni correction; ‡$P<0.05$ between ventral and left lateral views by unpaired t test.

The art of defining quantitative embryonic CV morphology using scanning electron microscopy (SEM) for a range of species including chick, mouse, dog and human was perfected by Tomas Pexieder (2). The Pexieder method includes critical attention to gestational dating, uniform specimen preparation using low-pressure, high flow perfusion fixation with buffer agents that arrest the heart in diastole (2), consistent specimen orientation for imaging, extracardiac landmarks, and importantly, detailed knowledge of normal phenotypic variability in the control population (3,4). The most critical benefit of the Pexieder method is that reproducible, quantitative measures of cardiac morphology can be made and these measures can provide critical details for both homozygous and heterozygous targeted embryos (19).

Magnetic resonance microscopy (MRM) has also been adapted to study embryonic morphometry (20,21). This technique can be used to investigate normal and abnormal embryos in a non-destructive manner. Data are acquired as three-dimensional isotropic arrays which allow the images to be formatted retrospectively in any plane (20). Once the data are acquired it can be archived, distributed, and manipulated. Thus, MR microscopy is a fast technique for producing three-dimensional reconstructions and is free from the registration and sectioning artifacts associated with serial sectioning techniques.

Ultrasound backscatter microscopy (UBM) is yet another technique for determining embryonic CV phenotype, and can be applied transabdominally to pregnant mice to evaluate embryonic phenotypes *in utero* (22,23). This method allows subsurface tissue imaging with close to optical resolution (22,23). UBM runs at much higher frequencies (40-100 MHz) than clinical ultrasound (3 to 13 MHz) (22,23).

Sequential Analysis of CV Morphology

Congenital anomalies are best assessed using a sequential approach to the CV system. This sequential analysis depends, of course, on the time point in CV development that the inspection occurs. During the looping and trabeculation phases of CV morphogenesis, the systemic arterial and venous anatomy reflects the large volume of extraembryonic blood flow. The following discussion presumes that the morphologic analysis occurs close to the completion of cardiac morphogenesis, when the position of most structures can be determined. In addition, the embryonic heart is best examined using Pexieder's method of injecting a diastolic arresting solution at high flow under low pressure to wash out red cells and then fix the tissues in a diastolic state. The manuscripts by Vuillemin and Pexieder on the external and internal shape of the murine embryonic heart should be read as prerequisite materials before attempting a serious morphologic examination of these tiny embryos (3,4).

External and Internal Cardiovascular Morphology

First, **cardiac and abdominal situs** should be determined. Situs is determined by the presence of "right" and "left" sided structures in the correct positions. Apex of the heart, stomach and spleen should be on the left and vena cavae and liver should be on the right. The right lung is trilobed and the left lung is bilobed. The investigator should note the position of the heart in the thorax, and when the abdomen is open, the position of visceral organs. Numerous CV syndromes are associated with abnormal situs or dextrocardia with normal situs (24).

Next, **atrial phenotype** should be determined (25). The normal mature right atrial phenotype includes communications between the superior and inferior vena cavae and the posterior, lateral atrium. The coronary sinus should be visualized between the orifice of the inferior vena cava and the tricuspid valve. The atrial septum contains the foramen ovale in the mid-septum and completion of cardiac septation includes fusion of the inferior rim of the primum atrial septum with the atrioventricular (AV) cushions. The normal mature left atrial phenotype includes communication between the pulmonary veins and the posterior atrial wall, attachment to the atrial septum, and the presence of the mitral (2 leaflet) valve. However, in the embryo, the earliest "atrial" structures actually become the atrial appendages in the mature heart and external morphologic differences between the right and left primitive atrium are difficult to determine. Between ED 10 and ED 13, the right atrial appendage enlarges, and the superior surface has a ridge identifying the location of the septum spurium. The left atrium lacks this external marker (25). The venous portions of the atrial walls are defined as lacking pectinate muscles and they communicate with the sinus venosus on the right and AV canal on the left. As for other species, the murine embryonic heart has 4 atrial phenotypes: normal, mirror image, left isometric (bilateral left morphology), and right isometric (bilateral right morphology).

Ventricular phenotype should then be determined (3) by first inspecting the heart from external ventral and oblique views. As noted above, the external contours of the murine embryonic RV and LV have been described by SEM (3) and video microscopy (18). A prominent interventricular sulcus is evident in the mouse embryo, in contrast to the chick embryo. It is important to note the respective proportions of the LV, RV, and outflow tract at each point in development. At ED 11, the primitive LV is more prominent, however the RV then increases in dimension to be equivalent by ED 12, and then predominant after ED 12.5. It is also important to note that the shape of each ventricle changes with development. The LV is initially spherical at ED 11, but becomes increasingly ellipsoid with time. In contrast, the RV is initially symmetric and somewhat oval but becomes increasingly asymmetric and triangular with expansion of the chamber and maturation of the outflow tract (3). It is important to note that in contrast to the chick embryo, the mouse embryo lacks an external marker between the developing RV and the conotruncus. Thus, the assessment of ventricular morphometry includes the generation of standard views, shape analysis, and dimensional comparison between the right and left sides and between control and experimental animals (3).

Internal cardiac morphology is an additional, important determinant of cardiac phenotype (4). As noted above, there is little external demarcation of the transition from RV to conotruncus. To examine internal cardiac morphology, first the RV must be sectioned parallel to the ventricular septum to produce parietal and septal halves (4). The incision begins to the right of the septum at the apex and then parallels the septum to the conotruncus, ending with a longitudinal section of the aortic sac. The trabecular wall of the RV has three basic elements: compact (outer layer), spongy (middle layer), and coarse trabeculated (luminal layer). The septal portion of the RV contains the AV canal inlet region, trabeculated septum, smooth outlet region, and during cardiac morphogenesis, the interventricular foramen (4). The conotruncal ridges can also be examined starting at ED 12 and fusion of the ridges begins at ED 13 (4). From ED 12.5, the maturation of the AV canal into tricuspid and mitral valves is evident (4). AV cushion fusion first occurs posteriorly, then proceeds anteriorly. By ED 15, the interventricular foramen should be closed. The AV valves remodel from endocardial cushion and adjacent myocardial tissues rather late, and in the mouse this process begins around ED 14 (4). Of note, a similar anatomic section can be made along the left side of the ventricular septum to generate a parietal "LV" specimen. However, the incision should not be extended into the great vessels, as this may obscure the relationship between the great vessels and the ventricles.

The position and relative dimensions of **the common outflow tract** or **great vessels** should then be determined (depending upon the developmental stage). Initially the common outflow tract (truncus) is attached to the primitive RV without external division. With growth, the conotruncus widens and shifts to the left, to wedge between the right and left atrial appendages. The first external signs of separation between the aorta and pulmonary artery occur by ED 12 as a groove on the right lateral wall of the aortic sac (3). Mean aortic and pulmonary arterial diameters are similar at ED 13 and increased symmetrically by ED 15 (3). As soon as it can be recognized, the aorta appears to wrap around the external surface of the pulmonary artery. The normal external appearance of the aorta is rightward and posterior versus the developing pulmonary artery from ED 13 onward. With resorption of the proximal conotruncal ridges, the developing mitral valve becomes continuous with the aortic valve annulus. In contrast, the pulmonary valve and tricuspid valve are separated by the smooth outlet septum.

When possible, **the yolk sac** and **placenta** should also be screened for abnormalities due to the critical importance of vascular load on CV morphogenesis and nutrient delivery to the embryo.

Examples of Abnormal External Morphology

iv/iv mouse

The *iv/iv* mouse has been studied extensively in the investigation of asymmetric cardiac looping (24). In contrast to the normal murine embryonic heart, these embryos had normal atrial situs in 33/73 embryos, mirror image (reversed) in 29/73 embryos, left isomerism in 7/73 and right isomerism in 2/73 (24).

retinoic acid models (timing of dosing, morphologic defect)

The teratogenic effect of retinoic acid on the CV system in a range of species has been known for decades (26). In the mouse embryo, maternal exposure to retinoic acid early in gestation is associated with CV heterotaxia defects. Exposure to retinoic acid slightly later results in transposed great arteries, and exposure even later results in double outlet RV (26,27). The ventral view at ED 12.5 of mouse embryos treated with retinoic acid at ED 8.5 reveals RV hypoplasia with compensatory enlargement of the LV. In addition, abnormal conotruncal septation results in transposed great vessels (27).

connexin43 -/- mice

Targeted disruption of the gap junction protein produced by the *connexin43* gene in mice is associated with a pulmonary valve stenosis/atresia phenotype in newborn mice (28). Of note, during early cardiac morphogenesis, the only obvious feature to the investigators was a dramatic enlargement of the "conus region overlying the RV outflow tract" at ED 16.5 in all homozygous mutant embryos (28).

In the mouse, the identification of chamber specific promoters has allowed the description of a number of cardiac and skeletal muscle-specific genes and proteins (19, 29) that are crucial in regulating myocardial growth and function.

RXRα -/- mouse

The targeted ablation of cardiac retinoic acid *RXRα* receptors produces an embryo lethal phenotype with hypoplasia of the ventricular myocardium associated with outflow tract and aortic arch defects (19, 30). Prior to death, the heart is enlarged and poorly contractile, but the term "heart failure" phenotype is of value only as a coarse colony screening tool. With careful inspection of external and internal morphology, both heterozygous and homozygous *RXRα* deficient embryos display CV anomalies (19).

Splotch mouse

The spontaneous *Splotch* mutant contains a defect in the transcription factor, *Pax3,* gene (31). These mice display an abnormal embryonic CV phenotype with an absence of normal conotruncal septation resulting in truncus arteriosus, cardiac enlargement, and reduced contractility. These embryos usually die by ED 14.5 (31). Using video microscopy, the LV ejection fraction is reduced to 26±16% in *Splotch* embryos versus 68±7% in controls (31).

Trisomy 16 mouse

The Trisomy 16 mouse has several features resembling human Trisomy 21 (Down's syndrome) including a high prevalence of CV malformations (96%) including complete AV canal defects (32-34). The external features of mouse Trisomy 16 include body wall edema evident at ED 16, and CV malformations including abnormal great vessel septation (transposition or double outlet RV, 27, 33).

Trisomy 13 mouse

The Trisomy 13 mouse has the notable external CV abnormality of isolated stenosis of the pulmonary trunk (viewed from the left, 35). In addition, these embryos have abnormal positioning of the aortic valve over the RV, resulting in the double outlet phenotype (35).

Examples of Abnormal Internal Morphology

Completion of the CV morphologic assessment requires the inspection of the internal cardiac structures. Due to phenotypic variability, the same CV anomalies may not be present in every embryo within a population of targeted embryos. Common abnormalities include persistent interatrial or interventricular communications, abnormal valve development, and altered ventricular morphogenesis.

RXRα -/- mouse

The internal CV morphologic anomalies defined by Pexieder in the *RXRα -/-* mouse represent some of best evidence for the value of his method for morphologic assessment (19). Abnormalities in AV canal formation, ventricular muscle trabeculation, and outflow tract septation are found in both *RXRα -/-* and *RXRα +/-* mice. In fact, the presence of significant structural defects in the asymptomatic heterozygous mice provides strong evidence that quick inspections of gross external shape and function are not sufficient in these genetically targeted models.

Trisomy 16 mouse

As mentioned above, the *Trisomy 16* mouse has CV malformations in 96% of affected embryos. Following removal of the parietal (outer) half of the RV, most embryos are noted to have a persistent common AV canal rather than separate tricuspid and mitral valves, defects in the outlet ventricular septum, and either malposition of the aortic valve over the RV or truncus (35).

Trisomy 13 mouse

The supravalvar pulmonary stenosis noted in *Trisomy 13* mice is first detected as an abnormal anterior position and fusion of the outflow tract ridges (35). This defect produces pulmonary artery stenosis and an abnormal positioning of the aortic valve over the RV and ventricular septum, similar to Tetralogy of Fallot.

TGFβ2 -/- mouse

Two-thirds of *TGFβ2 -/-* mice die shortly before or at the time of birth and the remainder are cyanotic after birth, while heterozygous animals are asymptomatic (36). Inspection of internal CV morphology reveals numerous structural defects, including ventricular septal defects in 15/16 embryos, incomplete looping in 4/16 embryos and double outlet RV in 3/16 embryos (36). Associated craniofacial defects in these embryos suggest altered neural crest cell migration and/or function.

PHYSIOLOGY

Overview

During CV development, embryonic heart rate, developed pressure, and cardiac output must increase to match geometrically increasing metabolic demands (1). Despite primitive geometry and small scale, the embryonic heart responds to acute changes in heart rate (HR), preload, and afterload to optimize cardiac output. The injection of less than 2% of circulating blood volume can alter arterial blood flow acutely (1). The embryonic heart responds to chronic changes in hemodynamic demand by altering myocardial growth (1). Because the embryonic CV system responds acutely to small changes in circulating blood volume, vascular load, and environmental temperature, the experimental methods used to measure CV function may direct negative effects. Therefore, it is critical to establish baseline, control data for each method prior to investigating abnormal phenotypes.

The murine model offers many advantages for genetic manipulation, yet at the same time the small size of the animal (and embryo) make it technically challenging to investigate (18). In addition, in contrast to the avian system, the murine embryo is coupled to both the yolk sac and placental circulation, further complicating acute and chronic experimental procedures. Finally, anesthetic agents have a great impact on maternal hemodynamics and respiration (37,38), and techniques to insure adequate maternal ventilation are required for murine experimentation during sedation. As with the definition of CV morphology, the assessment of CV physiology requires a step-wise approach. The following discussion presents a standard approach to defining murine embryonic CV function from ED 9.5 to 18.5 (18).

Maternal Preparation

In contrast to the avian embryo, which can be instrumented *in ovo* relatively easily, the murine embryo is tightly packaged *in materno*. Thus, several critical technical issues relevant to experimental protocols that define murine embryonic CV physiology are the effect of maternal anesthesia (necessary for noninvasive and invasive protocols) on maternal and embryonic physiology, and the effect of acute invasive protocols on maternal and embryonic physiology.

Maternal Sedation

The extent of maternal sedation required to determine embryonic CV physiology depends upon the experimental protocol. Sub-anesthetic doses of

Pentobarbital (40 mg/kg, ip, intraperitoneal) have been used to allow transabdominal ultrasound assessment of murine embryonic heart rate and blood velocity (39). Anesthetic dosing is determined by the absence of a toe-pinch reflex, and a surgical anesthetic effect is required for invasive procedures. We have evaluated the anesthetic and hemodynamic effects of Pentobarbital (80 mg/kg ip), Ketamine (90 mg/kg ip) plus Xylazine (7.5 mg/kg ip), Ketamine (150 mg/kg ip) plus Xylazine (12.5 mg/kg ip), Avertin (Tribromoethanol, 250mg/kg ip), and Isoflurane (1.25%) in both pregnant and nonpregnant female ICR mice. The target recumbent time for these protocols is 60 minutes. For intraperitoneal anesthetic agents, maternal heart rate was higher following Pentobarbital anesthesia that after Ketamine/Xylazine, though systolic blood pressure was also lower following Pentobarbital. Usually, heart rate and blood pressure changed in opposite directions following anesthesia. Maternal sedation with Avertin has been shown to be associated with transient atrial arrhythmias and slower heart rates in mouse embryos from ED 11 to 14 versus maternal pentobarbital sedation (38).

Recently, we have performed the majority of our non-invasive assessment of embryonic CV phenotype using inhaled Isoflurane for maternal anesthesia. The major advantage of the approach is that sedation is induced within 3 minutes (using a concentration of 2.5%) followed by a maintenance concentration of 1 to 1.4% , allowing sedation sufficient for transabdominal echocardiographic imaging for up to 60 minutes. Maternal heart rate for pregnant ICR mice using this approach is 450 - 543 bpm and peak tail cuff blood pressure is 59 - 82 mmHg (Table 5, Tobita, unpublished data).

Table 5. Representative maternal hemodynamic parameters during echocardiographic study.

	ED10.5		ED12.5		ED14.5		ED16.5		ED18.5	
	5min	60min	5min	60min	5min	60min	5min	60min	5min	60min
HR (beats/min)	523	464	505	543	533	488	457	457	488	450
Peak P (mmHg)	64	82	60	71	59	60	63	80	65	63
%SF	51.3	57.8	44.4	47.7	51.3	55.0	41.5	42.3	52.4	42.9

Data are recorded at 5 and 60 minutes of 1.25 % isoflurane anesthesia. ED: embryo day of gestation, HR: heart rate, Peak P: peak systolic pressure by tail-cuff, %SF: shortening fraction (Tobita, preliminary data).

We have noted that at higher concentrations of Isoflurane both maternal heart rate and mean blood pressure decrease.

Maternal Ventilation

Following acute surgical anesthesia, mice frequently hypoventilate, resulting in hypotension and hypoxemia. Therefore, we mechanically support maternal ventilation using a midline tracheotomy, tapered plastic angiocatheter, and small animal ventilator (CWE, Ardmore, PA). Based on measured blood gas content from right carotid artery samples we use initial ventilator settings of 0.7 ml tidal volume, 200 ml/min inspiratory flow, 0.2 sec inspiratory time, and 140 breaths/min respiratory rate. The inline peak airway pressures with these settings usually ranged from 8 to 10 cmH_20 (37).

As evidence of the acute effect of altered maternal respiration on embryonic CV function, we measured embryonic heart rate and then created acute maternal hypoxia by suspending mechanical ventilation for 30 seconds. The fall in maternal blood pressure occurs within 10 seconds of maternal apnea and embryonic bradycardia follows (Figure 2). We therefore mechanically assist all sedated mice prior to determining invasive *in vivo* embryo data.

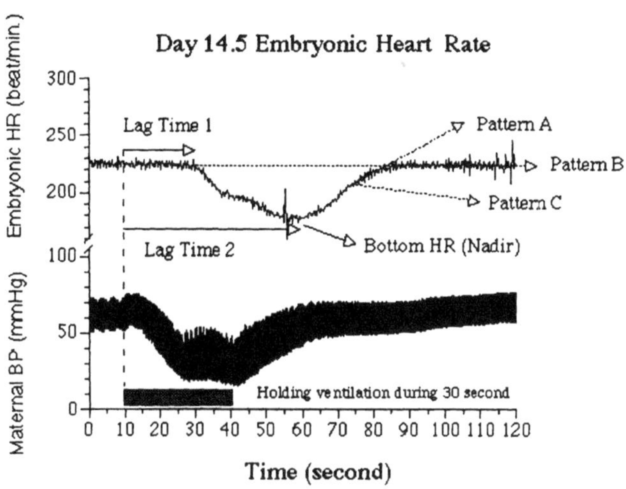

Figure 2: Representative effect of acute maternal hypoventilation on embryonic heart rate (HR) and maternal blood pressure (BP) at ED 14.5. Note that following the onset of maternal apnea there is a lag time (1) before embryonic heart rate decreases and a lag time (2) before minimum heart rate is reached. Recovery heart rate had three different patterns that varied with gestation (Furukawa, unpublished data).

However, pregnant mice sedated lightly with Isoflurane for transabdominal echocardiography do not require mechanical ventilatory support.

Maternal Preparation for Transabdominal Echocardiography

As described below, embryonic CV function can be assessed non-invasively using the combination of maternal sedation and high-resolution echocardiography. We position the pregnant mouse supine and lightly restrained on a warmed platform to prevent sedation-related and inactivity related hypothermia (18, 40). We also use a commercial electric trimming shears to remove abdominal fur to improve the acoustic contact between the anterior abdominal wall and the ultrasound transducer. For selected protocols we measure maternal blood pressure using a non-invasive tail cuff system that has been independently calibrated prior to the experiment or measure arterial blood pressure directly from the right carotid artery.

Embryo Preparation

Methods for investigating embryonic CV physiology vary between investigators and for different species. Several basic protocols have been applied recently to murine embryos. In order to maintain uteroplacental continuity and preserve maternal-embryonic coupling, we have developed techniques to investigate the murine embryo within maternal abdomen (18). Other investigators have maintained continuity, but have placed the embryo outside of the abdomen (30). Another approach, adapted from the rat embryo, is to excise an individual uterine segment then to investigate the mouse embryo in a dish containing warmed buffer (41, 42). Finally, segments of the embryonic myocardium can be suspended between a force transducer and linear motor in a physiologic buffer filled chamber to determine isolated passive and active muscle properties (Schroder, unpublished).

For acutely exposed, *in situ* protocols (18), a midline vertical laparotomy is performed from the umbilicus to immediately above the symphysis pubis. The linea alba is then identified and incised, exposing the bicornuate uterus and discrete swellings identifying the developing embryos. In order to generate a fluid filled abdominal well, a 25 mm diameter plastic dish containing a 10 mm diameter circular opening is positioned directly over the laparotomy site and sutured to the maternal abdominal wall at 12, 3, 6, and 9 o'clock using interrupted 5-0 silk sutures. Each pregnant uterine horn can then be sequentially incised 180 degrees from the uterine vascular pedicle in the direction of the exposed myometrium over the distinct swellings associated with developing embryos. Upon incision of the uterus, the embryonic yolk sac is gently exposed with the placenta and uterine vessels intact and the embryo is then released from yolk sac using microforceps. Small squares of damp, cotton gauze are used to stabilize embryo position for study. In this fashion, uteroplacental continuity is maintained throughout the investigation of each individual embryo. Embryo temperature is maintained using ambient heat lamps, a custom double walled glass chamber, and topical warmed Krebs Henseleit buffer (KHB) bubbled with 95%O_2, 5%CO_2. Supplemental topical, warmed KHB is used to maintain a fluid surface over the embryo at all times. Stability of an individual mouse embryo varies between approximately 3 and 10 minutes. Data is recorded for each embryo as soon as the embryonic heart can be fully visualized for video recording. All measures are not acquired in all embryos, and signs of reduced embryo viability include extreme pallor, irregular heart rate, or evidence of obvious bleeding (18).

For excised protocol using the whole embryo, a midline laparotomy is made as described above and then the entire bicornuate uterus or a segment is removed *en block* and the uterus containing the intact embryos and yolk sac is placed in a dissection dish containing warmed, physiologic buffer. Individual distinct uterine swellings are then incised to release developing embryos within yolk sacs. Upon incision of the uterus, the embryonic yolk sac is gently exposed with the placenta and uterine vessels intact and the embryo is then released from yolk sac using microforceps. The Nakazawa technique has been well described for rat embryos and is equally applicable to the smaller mouse embryo (37, 41, Tobita, unpublished data).

For excised protocol using myocardial segments, embryo exposure occurs as described above. The whole embryo is then placed in a small petri dish and then myocardial segments are dissected from the embryonic ventricle or atrium. These small specimens are then attached to the metal arms of force and length transducers using a biologically inert adhesive (43).

Physiologic Measures

The physiologic measures used to determine CV function in the embryo are the same measures used for the fetal, neonatal, and adult heart (1). Experimental data on embryonic heart rate, ventricular inflow and ejection velocities, chamber dimensions, and blood pressure can be obtained in normal and genetically targeted mouse embryos. Again, it is critical to note that data must be grouped based on morphological stage of development and acquired by comparable method after insuring adequate maternal support (for *in vivo* protocols). Table 5 summarizes our data using non-invasive transabdominal ultrasound for normal ICR mouse embryos from ED 10.5 to 18.5 (Tobita, preliminary data).

Embryonic Heart Rate

As has been noted for numerous other species, murine embryonic heart rate increases with development, and is acutely sensitive to environmental temperature (18). Due to the effect of altered heart rate on diastolic filling time and thus on end-diastolic volume and stroke volume, dimensional data should be compared at matched heart rates (44). Heart rate data can be determined from video images of the beating heart or from time series of ventricular velocity or pressure tracings. It is important to note that the embryonic heart rates measured with acute embryo exposure are significantly lower than rates measured with less invasive techniques such as transabdominal ultrasound (23, Tobita, unpublished data). Significantly higher embryonic heart rates occur following maternal pentobarbital sedation than following Ketamine/Xylazine or Isoflurane (37, 41, Tobita, unpublished data).

Table 5: Hemodynamic parameters in normal ICR mouse embryos

	ED10.5	ED12.5	ED14.5	ED16.5	ED18.5
HR (beats/min)	137 ± 11	164 ± 20	172 ± 20	193 ± 24	233 ± 34
ET	0.43 ± 0.05	0.44 ± 0.04	0.44 ± 0.02	0.44 ± 0.04	0.44 ± 0.04
IRT	0.22 ± 0.05	0.19 ± 0.05	0.18 ± 0.04	0.15 ± 0.03	0.12 ± 0.02
ICT	0.17 ± 0.03	0.18 ± 0.04	0.16 ± 0.04	0.10 ± 0.04	0.06 ± 0.02
FT	0.19 ± 0.06	0.19 ± 0.05	0.23 ± 0.05	0.31 ± 0.05	0.38 ± 0.05
OFT_{max} (m/sec)	0.16 ± 0.07	0.30 ± 0.10	0.30 ± 0.09	0.27 ± 0.06	0.32 ± 0.07
A_{max} (m/sec)	0.18 ± 0.05	0.34 ± 0.07	0.35 ± 0.06	0.34 ± 0.05	0.33 ± 0.05
E wave detection (%)	5	20	65	95	100
E_{max} (m/sec)	0.02 ± 0.03	0.01 ± 0.02	0.05 ± 0.05	0.09 ± 0.03	0.14 ± 0.04
E/A ratio	0.08 ± 0.18	0.01 ± 0.07	0.10 ± 0.11	0.29 ± 0.34	0.39 ± 0.14
DA_{max} (m/sec)	0.10 ± 0.04	0.14 ± 0.04	0.14 ± 0.04	0.11 ± 0.03	0.13 ± 0.04
$DA_{min/max}$ (%)	0	0	0	0.2 ± 2	0.4 ± 2
UA_{max} (m/sec)	NA	0.08 ± 0.02	0.09 ± 0.02	0.09 ± 0.02	0.12 ± 0.03
$UA_{min/max}$ (%)	NA	0	0	0	8 ± 7
UV_{max} (m/sec)	NA	0.04 ± 0.01	0.05 ± 0.01	0.04 ± 0.01	0.06 ± 0.02
$UV_{min/max}$ (%)	NA	59 ± 31	82 ± 15	92 ± 12	98 ± 7

Data are expressed as mean ± SD. Data are based on transabdominal Doppler echocardiographic examination of 78 normal embryos from ICR 8 litters. ED: embryo day of gestation, HR: heart rate (beats per minute), ET: cycle length corrected ejection time, IRT: isovolumic relaxation time, ICT: isovolumic contraction time, FT: diastolic filling time, OFT_{max}: peak systolic outflow velocity, A_{max}: peak active inflow velocity, E_{max}: peak passive inflow velocity, E/A ratio: peak passive to active velocity ratio, DA_{max}: peak systolic

descending artery flow, $DA_{min/max}$ Ratio: minimum to maximum velocity ratio of descending aorta, UA_{max}: peak systolic umbilical artery velocity, $UA_{min/max}$: minimum to maximum velocity ratio of umbilical artery, UV_{max}: peak umbilical venous velocity, $UV_{min/max}$:

minimum to maximum velocity ratio of umbilical vein, NA: not applicable, Umbilical artery and venous flow is not detected (Tobita, preliminary data).

Blood Velocity/Flow Measurement

The sampling sites for measuring blood flow velocity in the murine embryo are similar to the human embryo and include the inferior vena cava, common AV valve, common outflow tract, descending aorta, and umbilical artery and vein. Intracardiac blood flow has been characterized by measuring the flow across the AV cushions in several species of embryos using 20 MHz pulsed-Doppler velocimeters, and ventricular filling has distinct passive and active phases. There is a progressive increase in passive ventricular filling during primary cardiac morphogenesis associated with increasing ventricular compliance (1). In the rat embryo at comparable developmental stages, pulsed Doppler measures of blood flow reveal similar results (41). Blood flow has been assessed noninvasively in mouse embryos as early as ED 10.5 using maternal transabdominal doppler (38,39). The human embryo at comparable stages reveals distinct passive and active ventricular filling phases, and a geometric increase in blood flow and contractility with development (45). We have used a single piezoelectric crystal, positioned at the apex of the embryonic heart and positioned towards the AV canal region to record ventricular inflow and outflow velocities in mouse embryos (18), as summarized in Table 5.

Regarding the technical aspects of measuring blood velocity in murine embryos, we initially used a 20 MHz directional pulsed-Doppler velocity meter (Triton Systems) and a 750 μm diameter piezoelectric crystal positioned to record phasic blood velocity across the AV cushions and/or ventricular outflow tract (18,46). To maintain embryo temperature, the piezoelectric crystal is submerged in warmed KHB buffer when not in use. The piezoelectric crystal is manually positioned over the ventricular apex to measure ventricular filling and ejection velocities with an incident angle of blood flow and the ultrasound beam of less than 10°. The ventricular inflow pattern corresponds visually with atrial contraction and the ventricular ejection pattern corresponds visually with ventricular ejection (18). Due to dynamic changes in AV canal cross-sectional area that occur with atrial systole and conotruncal cross-sectional area that occur with ventricular ejection, we cannot calculate absolute blood flow from these velocity measurements. We determine peak instantaneous mean velocity for active ventricular filling and ventricular outflow for 3 to 5 cardiac cycles for each embryo. We integrate the area under the curve for outflow tract velocity and then divide by the cycle length to determine mean outflow tract velocity for each embryo. We can also measure arterial blood velocity in the murine umbilical cord, comparable to methods used for human embryos (45,46). As expected, umbilical arterial blood velocity and blood flow increase with development, consistent with increasing ventricular function and decreasing vascular (placental) resistance. Of note, effective forward blood flow requires both normal heart rate and activation sequence. When we paced the embryonic mouse atrium at the intrinsic heart rate, umbilical blood flow was preserved, however, when we paced the embryonic ventricle, forward blood flow decreased dramatically, highlighting the importance of sequential AV contraction (46).

The umbilical cord can also be used to measure embryonic heart rate during alterations in maternal state, as demonstrated above. However, it is worth noting that the murine umbilical arterial flow tracing has much less diastolic forward flow than do human embryos at comparable developmental stages, suggesting significant differences in placental afterload between the species.

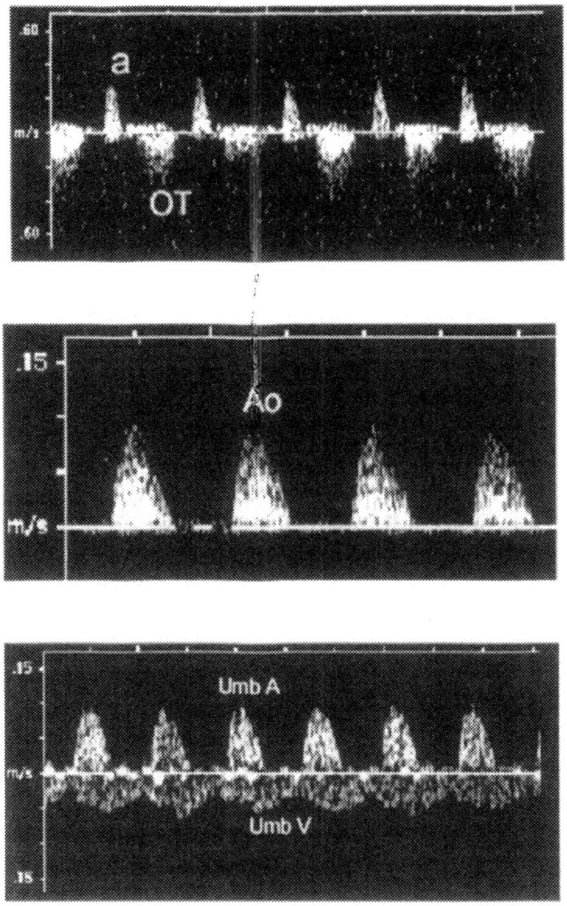

Figure 3: Representative normal Doppler flow patterns at gestational day 12.5 murine embryo. Upper Panel: Ventricular inflow (a) and outflow (OT). Note that ventricular inflow contains an active wave (a) but not a passive wave. Middle Panel: Descending aortic (Ao) flow. Lower Panel: Umbilical arterial (Umb A) and venous (Umb V) flow. Note the absence of umbilical arterial diastolic flow at this gestation and the presence of pulsatile umbilical venous flow.

It is very important to point out that a significant limitation of transabdominal imaging is the lack of control over embryo position. This can greatly effect the peak velocity measured by pulsed Doppler, resulting in an under-estimation of peak velocity as well as velocity-time integral.

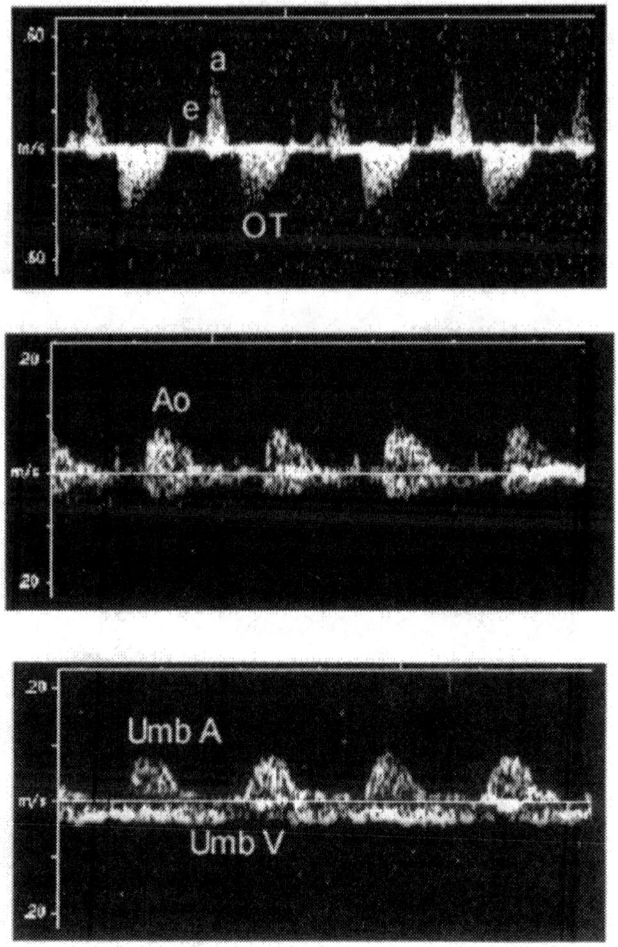

Figure 4: Representative normal Doppler flow patterns at gestational day 16.5 murine embryo.Upper Panel: Ventricular active (a) and passive (e) inflow and outflow (OT), Middle Panel: Descending aortic (Ao) flow, Middle Panel: Umbilical arterial (Umb A) and venous (Umb V) flow. The ventricular inflow pattern has changed from monophasic atrial contraction pattern to biphasic passive and active flow pattern. Diastolic umbilical artery forward flow is detected and the umbilical venous flow pattern becomes less pulsatile.

Blood Pressure Measurement

Embryonic intraventricular pressure can be accurately measured using a NaCl filled glass capillary tube and servonull micropressure system (1). To record pressures using this system, the embryo must be completely submerged in an ionic fluid. Ventricular pressure recording in the tiny mouse embryo is technically challenging and frequently not successful due to plugging of the capillary tip or to chamber bleeding. When successful, the ventricular pressure tracings in mouse embryos resemble the profile of tracings recorded in chick and rat embryos (Figure 5). Blood pressure waveforms can be digitally recorded for the calculation of end-diastolic pressure, peak pressure, $+dP/dt_{max}$, $-dP/dt_{max}$, and the rate of ventricular pressure decline (τ).

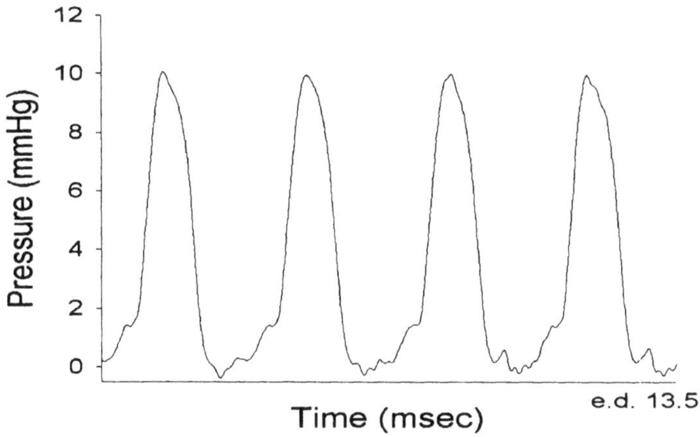

Figure 5: Representative ventricular blood pressure measured using a NaCl filled glass capillary tube and a servonull micropressure system in an ED 13.5 mouse embryo.

Examples of Abnormal Embryonic CV Physiology

RXRα knockout mouse

The initial observation in *RXRα -/-* mouse embryos that suggested a cardiac phenotype was the high frequency of embryonic death around ED 12.5, during a critical period of functional maturation of the CV system. Using light microscopy, the embryonic hearts were noted to be significantly enlarged and poorly contractile, thus given the term of "heart failure" phenotype (30). The abnormally thin ventricular wall also suggested the term "atrial phenotype", though the morphology of the *RXRα -/-* ventricular myocardium does not resemble normal atrial muscle.

partial Trisomy 16 mouse

Transabdominal doppler ultrasound has been used to noninvasively examine normal and abnormal mouse embryos serially during gestation (38). Trisomy 16 mouse embryos also have a significant intrauterine death rate, likely related to ventricular dysfunction. These embryos have been noted to have pulsed-Doppler evidence of reduced heart rates and AV valve insufficiency, not noted in normal embryos (38).

Splotch mouse

Videomicroscopic and electrophysiologic testing has also been performed on mouse embryos homozygous for the *splotch* (Sp^{2H}) allele (31). The affected embryos have enlarge, poorly contractile ventricular chambers, but this contractile difference was not substantiated in isolated muscle performance experiments. Of note, Ca^{2+} current was found to be reduced by over 3 fold in affected embryos compare to population normals, suggesting that altered neural crest cell function influences excitation-contraction coupling in the early embryonic heart (31).

BIOMECHANICS

Overview

Our work in defining embryonic CV structural and functional maturation has highlighted the critical importance of the dynamic mechanical environment in influencing embryonic CV growth and morphogenesis (1). Our work in the embryonic heart directly benefits from pioneering work performed in the mature heart and we have adapted techniques and developed novel approaches to generate quantitative mechanical data for the embryonic heart.

Passive Biomechanical Measures

As mentioned above for *in vitro* protocols, we have developed techniques to arrest the embryonic heart in diastole, remove the entire heart or isolated segments, and then suspend the myocardial specimen between a high resolution force transducer and linear motor (43, Schroder, unpublished). Using this method, we have defined passive viscoelastic properties for the developing chick embryonic myocardium and are now generating comparable data on the embryonic mouse heart.

Active Measures of Excitation-Contraction Coupling

We have also completed several experiments using isolated myocardial segments of embryonic mouse myocardium. We use a physiologic buffer solution and the myocardial specimen is field stimulated using a variety of pacing protocols. Standard muscle physiology protocols allow us to investigate the developing myocardium for the presence and strength of force-frequency relations (Figure 6). We can also vary the concentration of critical reagents in the buffer do determine the sensitivity of the developing myocardium to Ca^{2+}, catecholamines. etc. In contrast to the *in materno* experiments, the isolated murine embryonic myocardium is reasonably stable for more than 60 minutes in the muscle bath.

Cellular Electrophysiology of Embryonic Murine Cardiomyocytes

There has been considerable attention to the cellular electrophysiology of the mouse heart due to the interest in genetic models of cardiovascular diseases. Standard approaches to define ionic currents for K^+ , Na^+ , and Ca^{2+} in isolated cardiomyocytes have been applied to murine embryonic cardiomyocytes and embryonic stem cell-derived cultures (47-50). As with other aspects of cardiovascular function in the embryo, there are differences in the maturation of sarcolemmal and sarcoplasmic reticular currents that may be important in regulating cellular function. Genetic approaches that target ion channels in the mature heart may have significant CV phenotypes *in utero* if the promoter is expressed during development and if the targeted gene alters CV performance.

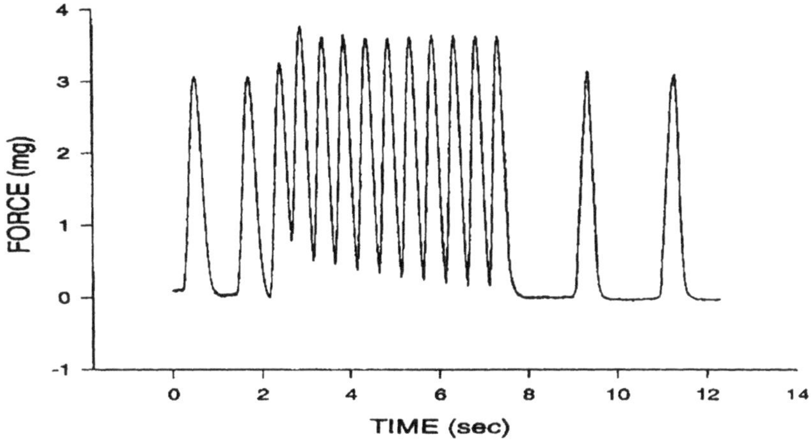

Figure 6: Representative myocardial segment tension during an acute increase in pacing frequency in an ED 12.5 mouse embryo. Note that immediately following the increase in pacing rate there is a higher developed force

SUMMARY

This chapter provides basic details regarding the morphologic and physiologic phenotyping of mouse embryos. Critical attention must be paid to gestational timing and morphologic staging based on existing standards. Despite widely available resources for generating targeted genetic models in mice, the resources for CV phenotyping are limited at present. The technical challenges of generating reproducible data across a range of experimental models and techniques may justify the development of regional centers equipped to perform these studies on a subcontract or service basis. Obviously, greater details are always available in print, by phone, or by email ktobi0@pop.uky.edu or bkell0@pop.uky.edu.

ACKNOWLEDGMENT

This research has been supported by NIH National Heart Lung and Blood Institute awards P50-HL51498 (E. Clark and B. Keller, sequential PIs) and RO1 HL65219 (B. Keller, PI).

REFERENCES

1. Keller BB: Embryonic Cardiovascular Function, Coupling, and Maturation: A Species View. In *Development of Cardiovascular Systems: Molecules to Organisms.* (eds) Burggren W and Keller BB. Cambridge Univ Press, New York, 1997; 65-87.

2. Pexieder T, Christen Y, Vuillemin M and Patterson DF. Comparative morphometric analysis of cardiac organogenesis in chick, mouse, and dog embryos. In: Nora JJ, Takao A, eds. *Congenital Heart Disease: Causes and Processes.* Mt. Kisco, NY: Futura Publishing Co. 1984; 423-438.

3. Vuillemin M and Pexieder T. Normal stages of cardiac organogenesis in the mouse, I: development of the external shape of the heart. Am J Anat 1989;184: 101-113.

4. Vuillemin M and Pexieder T. Normal stages of cardiac organogenesis in the mouse, II: development of the internal relief of the heart. Am J Anat 1989;184: 114-128.

5. Hamburger V and Hamilton HL. A series of normal stages in the development of the chick embryo. J Morphol 1951;88:49-92

6. Sissman NJ. Developmental landmarks in cardiac morphogenesis: comparative chronology. Am J Cardiol 1970;25:141-148.

7. Pexieder T and Janecek P. Organogenesis of the human embryonic and early fetal heart as studied by microdissection and SEM. In: Nora JJ, Takao A, eds. *Congenital Heart Disease: Causes and Processes.* Mt. Kisco, NY: Futura Publishing Co.,1984;401-421

8. Hu N and Clark EB. Hemodynamics of the stage 12 to stage 29 chick embryo. Circ Res 1989; 65:1665-1670

9. Faber JJ, Green TJ, and Thornburg KL. Embryonic Stroke Volume and Cardiac Output in the Chick. Dev Biol 1974;41:14-21

10. Keller BB, Hu N, Serrino PJ, and Clark EB. Ventricular pressure-area loop characteristics in the stage 16 to 24 chick embryo. Circ Res 1991;68:226-231

11. Tanaka N, Mao L, Delano FA, Sentianin EM, Chien KR, Schmid-Schonbein GW, and Ross JJr. Left ventricular volumes and function in the embryonic heart. Am J Physiol 1997;273:H1368-H1376

12. Field LJ. Transgenic mice in cardiovascular research. Ann Rev Physiol 1993;55: 97-114.

13. Rossant J. Mouse mutants and cardiac development: new molecular insights into cardiogenesis. Circ Res 1996;78: 349-353.

14. Chen JN and Fishman MC. Genetic dissection of heart development. In *Development of Cardiovascular Systems: Molecules to Organisms.* (eds) Burggren W and Keller BB. Cambridge Univ Press, New York, 1997; 7-17.

15. Murphy AM. Contractile protein phenotypic variation during development. Cardiovasc Res 1996;31:E25-33.

16. Franco D, Lamers WH and Moorman AF. Patterns of expression in the developing myocardium: towards a morphologically integrated transcriptional model. Cardiovasc Res 1998;38:25-53.

17. Robbins J. Remodeling the cardiac sarcomere using transgenesis. Ann Rev Physiol. 2000;62:261-87.

18. Keller BB, MacLennan MJ, Tinney JP and Yoshigi M. In vivo assessment of embryonic cardiovascular dimensions and function in day 10.5 to 14.5 mouse embryos. Circ Res 1996;79(2): 247-255.

19. Gruber PJ, Kubalak SW, Pexieder T, Sucov HM, Evans RM and Chien KR. RXR alpha deficiency confers genetic susceptibility for aortic sac, conotruncal, atrioventricular cushion, and ventricular muscle defects in mice. J Clin Invest 1996;98(6): 1332-1343.

20. Smith BR, Linney E, Huff DS, and Johnson GA. Magnetic resonance microscopy of embryos. Comp Med Imaging and Graphics 1996;20(6):486-490.

21. Hogers B, Gross D, Lehmann V, Zick K, De Groot HJ, Gittenberger-De Groot AC, and Poelmann RE. Magnetic resonance microscopy of mouse embryos in utero. Anat Rec 2000;260:373-7.

22. Turnbull DH, Bloomfield TS, Baldwin HS, Fosters FS and Joyner AL. Ultrasound backscatter microscope analysis of early mouse embryonic brain development. Proc Natl Acad Sci USA 1995;92:2239-2243.
23. Phoon CK, Aristizabal O, and Turnbull DH. 40 MHz Doppler characterization of umbilical and dorsal aortic blood flow in the early mouse embryo. Ultrasound Med Biol 2000 Oct;26(8):1275-83.
24. Icardo JM, Arrechedera H and Colvee E. Atrioventricular endocardial cushions in the pathogenesis of common atrioventricular canal: morphological study in the iv/iv mouse. In: Clark EB, Markwald RR and Takao A, eds. *Developmental Mechanisms of Heart Disease.* Armonk, NY: Futura Publishing Co. 1995;529-544.
25. Seo JW and Kim YH. Morphology of the embryonic atrial chambers in the mouse with iv/iv mutation. In: Clark EB, Markwald RR and Takao A, eds. *Developmental Mechanisms of Heart Disease.* Futura Publishing Co. 1995;545-553.
26. Pexieder T, Blanc O. Pelouch V, Ostadalova I, Milerova M, and Ostadal B. Late fetal development of retinoic acid-induced transposition of the great arteries: morphology, physiology, and biochemistry. In: Clark EB, Markwald RR and Takao A, eds. *Developmental Mechanisms of Heart Disease.* Armonk, NY: Futura Publishing Co. 1995;297-307.
27. Yasui H, Nakazawa M, Morishima M, Miyagawa-Tomita S and Momma K (1995). Morphological observations on the pathogenetic process of transposition of the great arteries induced my retinoic acid in mice. Circ 91(9): 2478-2486.
28. Reaume AG, de Sousa PA, Kulkarni S, Langille BL, Zhu D, Davies TC, Juneja AC, Kidder GM, and Rossant J. Cardiac malformation in neonatal mice lacking Connexin43. Science 1995;267:1831-1834.
29. Lyons GE, Schiaffino S, Sassoon D, Barton P and Buckingham M (1990). Developmental regulation of myosin gene expression in mouse cardiac muscle. J Cell Biol 111: 2427-2436.
30. Dyson E, Sucov HM, Kubalak SW, Schmid-Schonbein GW, Delano F, Evans RM, Ross J Jr. and Chien KR (1995). Atrial-like phenotype is associated with embryonic ventricular failure in retinoid X receptor I -/- mice. Proc Natl Acad Sci USA 92: 7386-7390.
31. Conway SJ, Godt RE, Hatcher CJ, Leatherbury L, Zolotouchnikov VV, Brotto MAP, Copp AJ, Kirby ML, and Creazzo TL. Neural crest is involved in development of abnormal myocardial function. J Mol Cell Cardiol 1997;29:2675-2685.
32. Miyabara S, Gropp A, and Winking H. Trisomy 16 in the mouse fetus associated with generalized edema and cardiovascular and urinary tract anomalies. Teratology 1982;25:369-380.
33. Miyabara S. Cardiovascular malformations of mouse trisomy 16: pathogenetic evaluation as an animal model for human trisomy 21. In: Clark EB and Takao A, eds. *Developmental Cardiology: Morphogenesis and Function.* Mt. Kisco, NY: Futura Publishing Co., 1990; 409-430.
34. Hiltgen GG, Markwald RR, and Litke LL. Morphogenetic alterations during endocardial cushion development in the trisomy 16 (Down syndrome) mouse. Pediatr Cardiol 1996;17:21-30.
35. Vuillemin M, Pexieder T, and Winking H. Pathogenesis of various forms of double outlet right ventricle in mouse fetal trisomy 13. Internat J Cardiol 1991;33:281-304

36. Sanford PL, Ormsby I, Gittenberger de-Groot AC, Sariola H, Friedman R, Boivin GP, Cardell EL, and Doetschman T. TGFβ2 knockout mice have multiple developmental defects that are non-overlapping with other TGFβ2 knockout phenotypes. Development 1997;124:2659-2670.
37. Furukawa S, MacLennan MJ, and Keller BB: Hemodynamic response to anesthesia in pregnant and nonpregnant ICR mice. Lab Animal Sci 1998;48:357-363.
38. Huang GY, Linask KK. Doppler echocardiographic analysis of effects of tribromoethanol anesthesia on cardiac function in the mouse embryo: a comparison with pentobarbital. Lab Animal Sci 1998;48:206-9.
39. Gui YH, Linask KK, Khowsathit P, and Huhta JC. Doppler echocardiography of normal and abnormal embryonic mouse heart. Pediatr Res 1996;40:633-642.
40. Gardner DJ, Davis JA, Weina PJ, and Theune B. Comparison of tribromoethanol, ketamine/acetylpromazine, telazol/xylazine, pentobarbital, and methoxyflurane anesthesia in HSD:ICR mice. Lab Animal Sci 1995;45:199-204.
41. Nakazawa M, Miyagawa S, Ohno T, Miura S, and Takao A. Developmental hemodynamic changes in rat embryos at 11 to 15 days of gestation: normal data of blood pressure and the effect of caffeine compared to data from chick embryo. Pediatr Res 1988;23:200-205.
42. Hofman PL, Hiatt K, Yoder MC, and Rivkees SA. A1 adenosine receptors potently regulate heart rate in mammalian embryos. Am J Physiol 1997;273:R1374-1380.
43. Miller CE, Vanni MA, and Keller BB: Characterization of passive embryonic myocardium by quasi-linear viscoelasticity theory. J Biomech 1997;30:985-988.
44. Casillas CB, Tinney JP, and Keller BB: Influence of acute alterations in cycle length on ventricular function in the chick embryo. Am J Physiol 1994;267:H905-911.
45. Wladimiroff JW, Huisman TWA, Stewart PA, Stijnen T. Normal fetal Doppler inferior vena cava, trans tricuspid and umbilical artery flow velocity waveforms between 11 and 16 weeks of gestation. Am J Obstet Gynecol 1992;3:921-924.
46. MacLennan MJ and Keller BB: Murine embryonic umbilical arterial velocity during development and following acutely increased heart rate. Ultrasound Med Biol 1999;25:361-370.
47. Davies MP, An RH, Doevendans P, Kubalak S, Chien KR, and Kass RS. Developmental changes in ionic channel activity in the embryonic murine heart. Circ Res 1996;78(1):15-25.
48. An RH, Heath RM, Higgins JP, Koch WJ, Lefkowitz RJ, and Kass RS. Beta2-adrenergic receptor overexpression in the developing mouse heart: evidence for targeted modulation of ion channels. J Physiol 1999;516;19-30.
49. Gryshchenko O, Fischer IR, Dittrich M, Viatchenko-Karpinski S, Soest J, Bohm-Pinger MM, Igelmund P, Fleischmann BK, and Hescheler J. Role of ATP-dependent K(+) channels in the electrical excitability of early embryonic stem cell-derived cardiomyocytes. J Cell Sci 1999;112:2903-2912
50. Cribbs LL, Schroder EA, Keller BB, Delisle BP, and Satin J. Identification of the CaV3.1d gene product in developing mouse heart: A contributor to embryonic contractility. Circ Res 2001 (in press).

23

NEWER *IN VIVO* IMAGING MODALITIES

Roee S. Lazebnik[1], David L. Wilson[1,2]
[1]*Department of Biomedical Engineering,* [2]*Department of Radiology*
Case Western Reserve University, Cleveland, OH 44106

1. INTRODUCTION

Much as an animal in isolation does not represent its behavior within its natural habitat, an isolated heart does not capture its dynamic role within the cardiovascular system. *In vivo*, open chest, models of transgenic mouse cardiac function were among the first in attempting to circumvent this difficulty, but an inevitable complication resulted: the subject did not survive the analysis, precluding long term or repeated studies. In addition, the invasive nature of such procedures inherently alters the observed system. Noninvasive *in vivo* imaging appears an ideal solution. Among the primary advantages of this approach, is the feasibility of serial measurements, through time, allowing a given subject to act as its own control. Even studies requiring complete dissection benefit from an estimate of the optimal time for the subject's sacrifice. However, *in vivo* imaging techniques must be carefully chosen in context of the experimental design. Several imaging modalities are available, and may be differentiated by attributes such as spatial and temporal resolution, contrast, and financial or logistical costs.

2. MAGNETIC RESONANCE IMAGING

2.1 Overview

Magnetic resonance imaging (MRI) is an established imaging method for both clinical and research purposes. This modality is derived from chemical nuclear magnetic resonance spectroscopy, but provides location in addition to composition data. The hallmarks of MRI are high contrast between various soft tissues, high spatial resolution, and subject safety due to its non-ionizing properties. However, such benefits are balanced by the high financial cost of MRI hardware, potential for image artifacts, and technical complexity. MR images are grayscale representations of specific atomic nuclei and their properties, most commonly the hydrogen protons of water. Thanks to dramatic advances in image acquisition strategies, it is possible to customize MRI sensitivity to anatomical and functional properties of interest.

2.2 Methods

The positive charge and spin of a proton give rise to a magnetic dipole. The "magnetic moment" is a vector describing the strength and orientation of this dipole. Only certain elements maintain a sufficient magnetic moment, physiologic concentration, and distribution throughout the subject to produce a detectable MR signal. Among these are isotopes of carbon, oxygen, sodium, and hydrogen, with hydrogen serving as the target for most MR experiments. When a population of hydrogen nuclei are exposed to a constant magnetic field (B_0), they tend to orient either parallel or antiparallel to the field, along the longitudinal plane. Because orientation parallel to the field is a lower energy state than the antiparallel configuration, a few more protons assume this orientation, resulting in a net magnetic moment from the proton population. In addition to alignment with B_0, the externally applied field causes the protons to precess about their axis, much as a top wobbles about its axis due to gravity.

The alignment of protons by the B_0 field, while significant, is not sufficient for imaging. However, if a magnetic field (B_1) is applied perpendicular to B_0, with a specific frequency related to the type of proton targeted (a condition termed resonance), a tipping of the magnetic moment from the longitudinal to the transverse plane occurs (Figure 1). The B_1 field is dynamic, and applied using a radio frequency (RF) pulse, or series of pulses termed the "pulse sequence". The tipping of the magnetic moment produces a new net magnetic moment that changes with time and is detected by an antenna termed the "receiver coil".

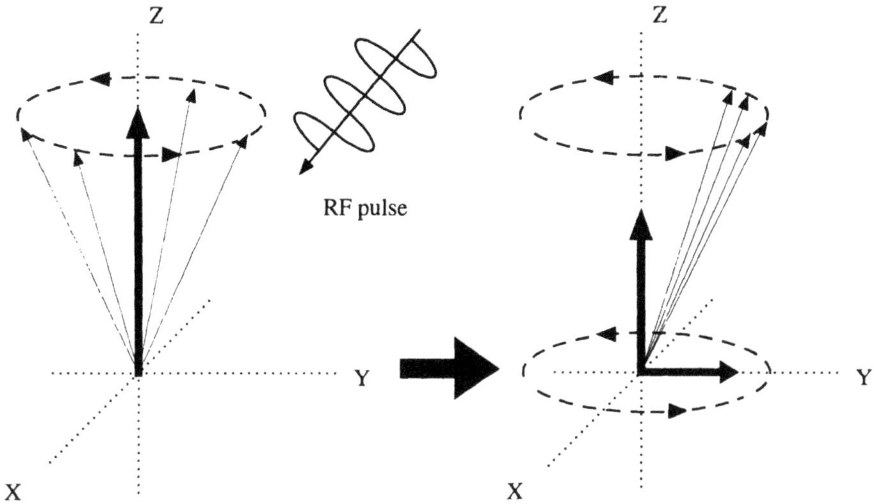

Figure 1 : Effect of a RF pulse on proton precession.

With time, the protons' transverse magnetization component decreases, as their return to their previous equilibrium state. This mechanism is exponential, termed T2 decay or relaxation, and is due to the intrinsic spin-spin interactions between neighboring protons. The time constant of this process is termed T2. Similarly, the accompanying increase of the longitudinal magnetization component is exponential, is termed T1 relaxation, and its time constant is termed T1. The T1 and T2 values vary by tissue, and B_0 magnitude. In addition, endogenous substances, such as paramagnetic blood degradation products, or exogenous elements with unpaired electrons, such as gadolinium, may alter T2 relaxation. These differences in tissue T1 and T2 properties are the source of contrast in MRI, the ability to differentiate between tissues of similar proton density composition.

The receiver coil measures the changing transverse magnetic field and allows for its recording. By varying B_0 through space, and carefully designing a pulse sequence, the signal archived by the receiving coil will contain information that characterizes the spatial density of protons originating the signal. Computer processing of the digitally sampled signal, using the inverse Fourier transform, allows for reconstruction of this proton density map into grayscale images.

While the preceding text reviews the basis for the MR signal, many details are not described. The interested reader is referred to a recent textbook[1] for a detailed discussion of the physics and mathematics of MRI.

2.3 Considerations for Mouse Imaging

MR is an established diagnostic modality for human cardiovascular imaging[2;3], and most commercially available MRI systems are geared towards human imaging. However, because MRI technology is highly customizable, it is also well suited to murine cardiovascular imaging. By understanding the influence of the components of MRI hardware and software on the final image, it is possible to optimize image acquisition for the mouse model.

The most important considerations in any MR experiment are the quality and content of the radio signal emanating from the subject, and the ability to record this signal. The primary source of random noise in the image is thermal fluctuations. The image signal-to-noise ratio (SNR) depends upon the magnitude of the static magnetic field (B_0), the voxel volume (V_{voxel}), and the acquisition time (T_{acq}).

$$SNR \propto B_0^{\ 2} \times V_{voxel} \times \sqrt{T_{acq}} \tag{1}$$

In general, the higher the SNR the better the quality of the resulting images. This equation can be used to examine important trade-offs between voxel size, field strength, and imaging time. Consider the effect of increasing field strength from that of a conventional system at 1.5 T to a mouse imager utilizing a 4.7 T magnet. For a given SNR, a high field 4.7 T system requires only $(1.5 / 4.7)^4 \approx 0.01$ times the duration on a conventional 1.5 T system. In an *in vivo* animal experiment, imaging time is particularly important when imaging organs in the presence of respiratory motion and a beating heart. Slow image acquisition may result in significant motion artifacts that obscure image detail. Alternatively, using the high field system, voxels can be acquired with $(1.5 / 4.7)^2 \approx 0.1$ times the volume of a voxel at 1.5 T. Assuming an isotropic voxel (equal dimensions on all sides), with 1.3 mm per side at 1.5 T, a 0.65 mm per side voxel is possible using the 4.7 T system. Similarly, for a given magnet strength, there is a dramatic trade off between voxel size and imaging time. Doubling the linear dimension of a voxel allows imaging time to be reduced to 1/64 of the original duration.

Another significant influence on SNR is the receiver coil. Factors such as the coil's shape, and proximity to its subject, govern the received signal's quality. It is common practice to design custom, mouse specific, coils that are placed around the subject, in very close proximity to the imaged volume.

Having mentioned the interplay between spatial and temporal resolution, the significance of choosing between acquiring a 3D volume using a single pulse sequence (3D acquisition) and multiple pulse sequences, each producing a 2D slice (2D acquisition), may be highlighted. A 3D acquisition of a given volume may result in higher SNR, and thus allow greater spatial resolution, but may require more time than the equivalent compilation of several 2D acquisitions, depending on the utilized pulse sequence.

A careful balance of the mentioned volume and time parameters is a critical concept in MR imaging. A MR voxel value is an average of the tissue properties through the volume of the voxel. Hence, at the border between two structures, the voxel value will be somewhere between those of the two tissues. This partial volume effect blurs the edge between structures. In order to balance this need for high spatial resolution with imaging time, it is common to configure voxels whose in-plane dimensions are smaller than their thickness. However, isotropic voxels are particularly useful for multi-planar reformatting, where a volume of voxels is resliced for visualization at orientations other than that chosen during imaging. An advantage of MR over other imaging modalities is this ability to precisely configure voxel dimensions.

Voxel size times the number of voxels determined the field of view (FOV). In this regard, mouse cardiac imaging is very demanding because of the minute size of the mouse heart. Nevertheless, a recent report reviewed later describes excellent capabilities[4]. Voxel size is 100 μm x 100 μm x 500 μm, and image volumes consist of 256 x 256 x 1 voxels, thus yielding a FOV of 25.6 mm x 25.6 mm x 0.5 mm, respectively. This gives about 65 voxels across a short axis view of the mouse heart.

In order to image the entire heart at one phase of the cardiac cycle, a compilation of slices, at sequential spatial positions along an axis, is required. Due to the heart's motion, and imaging time constraints, typically only one slice at a given phase is acquired per cycle, and a volume is compiled over multiple cycles. Various sampling strategies are utilized to obtain cardiac volumes at different phases of the cardiac cycle. All of these methods require gating acquisitions to the echocardiogram (ECG) so as to synchronize the MR data with the cardiac cycle. An established MR technique consists of a specialized short acquisition time pulse sequence, termed Fast Low Angle Shot (FLASH), coupled with cardiac gating, to provide millisecond temporal resolution imaging synchronized to the cardiac cycle. Such acquisitions may require several minutes to an hour to assemble a complete volume of the heart at one phase of the cardiac cycle. Another form of ECG gating features the retrospective matching of continuously acquired MR data to cardiac cycle information. However, all ECG gated techniques require a reasonably regular heart rhythm and a sufficiently clean ECG signal.

Because the heart position changes with respiration, motion blurring due to respiration may occur. Hence, a further improvement in image quality is obtained by combining ECG and respiratory gating. Respiratory coordination allows image acquisition to coincide with minimal respiratory movement. This resulted in excellent correlation between imaged heart anatomy and ex vivo analysis[5]. Alternatively, if a subject is anesthetized and intubated, it is possible to stop the artificial ventilation during imaging. Similarly, minimizing the subject's whole body motion may further enhance high-speed cardiac imaging through better registration between serial images. This is often addressed with general anesthesia, which unfortunately results in unwanted inotropic and chronitropic effects.

While optimization of spatial and temporal resolution is highly desirable, it is often more important to maximize the contrast between dissimilar tissues. Contrast is of particular significance in distinguishing vascular or myocardial tissue from blood and connective tissues. Differences in signal amplitudes, resulting in contrast, are due to tissue/tissue water content differences, water diffusion and bulk movement, water binding properties, and exogenously introduced agents (contrast agents) that change the relaxation rate of hydrogen protons.

The most significant, user determined, source of contrast is pulse sequence design. Pulse sequences vary in their radio frequency content, timing, repetition, and other parameters. Adjustment of these parameters can yield dramatic differences in the produces image, particularly with respect to sensitivity or insensitivity to specific tissue characteristics such as fat content or blood flow. Pulse sequences are generally divided into broad categories such as "spin echo", "inversion recovery", and "gradient echo". These names refer to the effects of the RF waves on the protons' magnetization and the received signal.

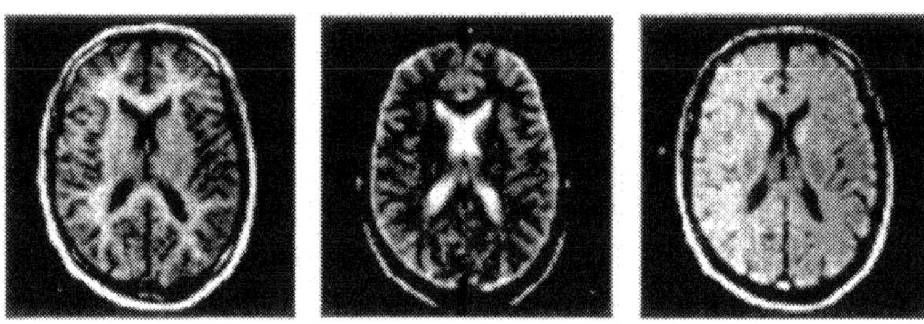

Figure 2: T1, T2, and Proton Density weighted images (left to right). Image courtesy of Siemens AG, Medical Engineering.

Among the most utilized benefits of pulse sequence design, is the ability to increase the signal from particular tissues relative to others (termed "weighting") based on the tissue's proton density, T1, and T2 properties. Figure 2 illustrates these differences by depicting each type of weighting for the same anatomical slice. A proton density weighted image minimizes differences between tissues of dissimilar T1 and T2, while emphasizing differences due to proton concentration. Thus, highly hydrogenous tissues such as fat are emphasized relative to proteinaceous tissues. A T1 weighted image emphasizes tissues with short T1 values, such as fat, while a T2 weighted image accentuates tissues with a long T2, such as cerbro spinal fluid (CSF).

Many other factors, beyond the scope of this discussion, influence MR measurements. These range from hardware characteristics, such as gradient magnetic field generators, to software considerations, such as image reconstruction algorithms.

2.4 Application of MR to Mouse imaging

Many mouse researchers utilize MR for anatomical imaging of the heart. MRI was proven to accurately characterize cardiac anatomy, such as left ventricle (LV) mass, in transgenic mouse models *in vivo*[6;7]. For example, using a 7 T system and an ECG triggered FLASH sequence, the left ventricle was manually segmented in multiple 117 μm × 117 μm × 1 mm short axis slice images[7]. Using these measurements, LV volume was computed at end systole and diastole, allowing for the computation of stroke volume (SV) and end diastolic volume (EDV). Similar methods were successfully applied to mice of varying ages (neonatal, juvenile, and adult) in order to quantify changes in cardiac parameters, such as ejection fraction, as a function of age[8]. The successful inclusion of neonatal (3 day old) mice in this study highlights MRI's capability to characterize transgenic models with early phenotypes.

Anatomical information is also useful in quantifying cardiac output, through computation of heart chamber volume changes. Such measurements in transgenic mice have been proven reproducible and comparable to invasive studies, such as thermal dilution techniques, even using a clinical 1.5 T magnet and 0.19 × 0.19 × 1.5 mm voxel size[9].

As previously suggested, studies that synchronize (gate) image acquisition with ECG data are useful in reducing motion artifacts. The coupling of the ECG gating technique and a 2D FLASH pulse sequence allows for dramatic anatomic detail of the murine heart in motion[10], as demonstrated by Figure 3.

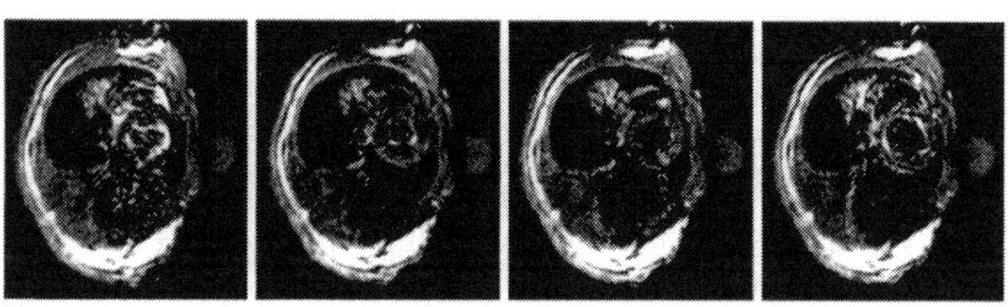

Figure 3: Time sequential images of a mouse heart beating at a physiologic rate. Reprinted from Weiss et al.[10] with permission from Lippincott Williams & Wilkins.

Similarly, using a 3D FLASH sequence, acquired with a multiple overlap thin slab technique, an isotropic resolution of 100 μm, sufficient to visualize the left coronary artery and four cardiac valves (Figure 4), is feasible[4].

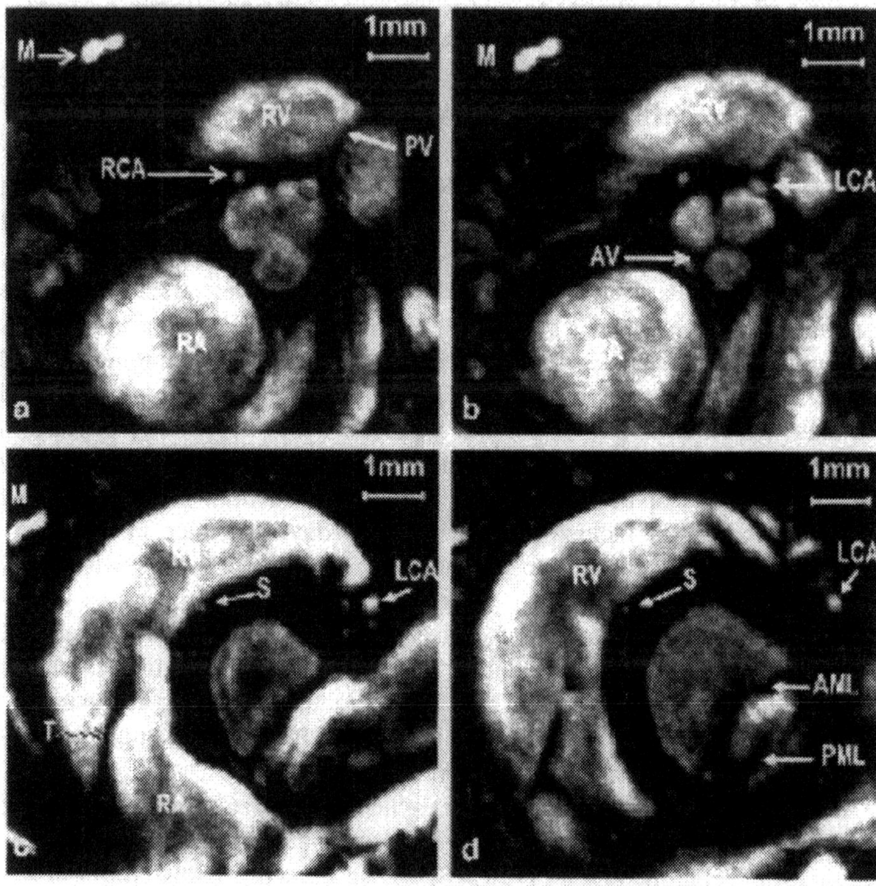

Figure 4 : High resolution, short axis view images, acquired using a 3D pulse sequence. Shown are the right ventricle (RV), right atrium (RA), aortic root (AR), right and left coronary arteries (RCA/LCA), internal mammary artery and vein (M), septal coronary artery (S), and the anterior and posterior mitral leaflets (AML/PML). Reprinted from Ruff et al.[4] with permission from Wiley-Liss, Inc., a subsidiary of John Wiley & Sons, Inc.

Due to its minute size, visualizing details of the murine vascular system is challenging. Using high field (9.4 T) MR system, atherosclerotic plaques, occupying the lumen and wall of the abdominal aorta and iliac arteries, were imaged in apoE knock-out mice[11]. Proton density, T1, and T2 weighted volumes (8 slices each) were generated. These volumes required 9, 18, and 18 minutes respectively, and featured 0.0012 to 0.005 mm^3 resolution. Both MR and histopathological sections were graded of morphological appearance and severity. Excellent agreement was noted between the MR based measurements and histopathological analysis. MR was also proven useful in coronary vessel angiography, using a 4.7 T scanner[12].

Challenges to such efforts include displacement of coronary vessels, during cardiac and respiratory motion, resulting in motion artifacts. In addition, the signal from large quantities of blood in the near by ventricles reduces the contrast of coronary flow relative to its environment. Thus, it is useful to utilize a pulse sequence whose duration is short compared to cardiac motion, reducing motion artifacts. Further, "flow tagging" may be utilized to increase contrast between blood and tissue[1]. This technique requires the application of a specialized magnetic pulse to blood flow before it enters the imaged volume, "saturating" the blood's magnetization, so that during imaging the signal originating from blood will be sufficiently different from its environment. Specifically, to image the coronary arteries, blood in the aortic root is "tagged" and imaged as it perfuses the coronary circulation. The minute size of even a large blood vessel, such as the murine abdominal aorta (≈1 mm luminal diameter), highlights the requirement of high field imaging for mouse angiography.

Opposite to flow tagging is "flow suppression", featuring a saturation pulse that reduces the signal from blood, demphasizing flow in the resulting image. This technique may be applied to suppress the signal from blood flow, in order to image surrounding tissue such as the ventricular wall[13]. Flow suppression is particularly useful since the amount of blood within cardiac chambers varies with cardiac cycle stage, altering the imaged volume variably through time. If this blood's effect on the image is nulled, so are its inconsistent influence on the produced images.

In addition to studies of normal cardiovascular physiology, high resolution MRI has proven useful in analysis of pathological conditions. Ventricular remodeling after a myocardial infarction (MI) is a complex process occurring asymmetrically in 3D. Measures based on echocardiography typically assume certain geometries, to extrapolate 3D information from the acquired 2D planes, which might not apply during MI. Thus, traditional ultrasound methods are prone to error. In contrast, 3D MRI does not require such assumptions, and was shown to accurately and reproducibly describe ventricular changes after MI[14].

MRI based measurements allow quantification of blood flow through specific vessels. Unlike angiographic experiments that focus on the anatomical configuration of the vasculature, flow experiments seek to quantify blood velocity. In order to quantify velocity, through a given blood vessel, flow tagging may be applied to a small volume of blood (a bolus), whose movement is visualized through the vessel in time sequential images[12]. It is thus possible to track a given bolus's position in time and space, allowing for velocity measurements. Alternatively, velocity may be measured directly from MR phase images as done in human imaging[2].

MRI may also be utilized to make other functional measurements, such as tissue oxygenation. The presence of deoxygenated hemoglobin is known to influence the T2 property of protons. Lower signal intensity is obtained from tissue containing deoxygenated hemoglobin, an effect termed blood oxygen level dependent (BOLD). Thus, contrast between oxygenated and hypoxic tissues is obtained in the resulting images. This property was utilized to quantify hypoxia in a transgenic mouse model of sickle cell anemia[15]. Using a 2 T scanner, and 2 minutes acquisition time per slice, $0.625 \times 0.625 \times 4$ mm resolution images of mice, at room air and 100% oxygen, were obtained. These images demonstrated hypoxia in the sickle cell mouse kidney, liver, and spleen, that was absent in the control mice.

Yet another example of functional MR imaging is the quantification of tissue deformation with the beating of the heart. Myocardial tagging, the placement of non-invasive fiducial markers in the myocardium and their subsequent tracking via MRI, was utilized to study left ventricular torsion in mice[16]. Unlike physical fiducials, which require surgical placement, the tags result from a pulse sequence applied to the myocardium prior to imaging. A series of rapidly acquired sequential images of the tags forms a movie of heart contraction that was utilized to characterize the twisting motion of the ventricle. Using an appropriate model, these measurements can be related to contraction forces.

Finally, another type of measurement is MR spectroscopy. Unlike conventional MRI imaging, which typically represent the average signal from hydrogen protons, in MR spectroscopy, voxels represent other molecular species. For example, ^{31}P MR spectroscopy is of particular significance in characterizing myocardial energy metabolism[17].

2.5 Future Directions

High field scanners designed for animal imaging, some featuring magnetic fields greater than 10 T, allow extremely high-resolution imaging. Such devices are capable of cellular imaging, and represent an area of intensive MRI research.

The coupling of high resolution MRI with recent developments in contrast agent synthesis is particularly significant for transgenic model investigation because it allows *in vivo* measurements of gene expression. For example, the contrast agent gadolinium (Gd^{3+}), known to shorten the T1 and, at sufficient concentrations the T2, properties of tissue, was sealed inside a molecule. This containment limited gadolinium's effect on the magnetization of nearby water molecules. However, if exposed to a particular enzyme, a region of the molecule may be cleaved, exposing the contrast agent to its environment. Such an agent, termed EgadMe was synthesized and made susceptible to β-galactosidase cleavage[18]. By coupling a reporter gene that synthesizes this enzyme to a gene of interest, it is possible to image the density of expression of the gene's product in a transgenic animal. Some potential applications are described later with regard to their use in positron emission tomography, a more established molecular imaging technique. A recent modification of the caged Gd^{3+} technique allows one to image intracellular calcium concentrations.

3. POSITRON EMISSION TOMOGRAPHY

3.1 Introduction

Positron emission tomography (PET) is a molecular imaging technique that utilizes small amounts of positron labeled tracer molecules to image *in vivo* biological events. The generated grayscale image volume depicts the density distribution of positron emission, corresponding to the concentration of the tracer, at each voxel space location. PET can image a diversity of molecular processes such as enzyme activity, drug/receptor interactions, titration of drugs, and protein synthesis. PET is useful for many non-anatomical studies, such as metabolism or gene expression assays, for which some other modalities are ill suited.

3.2 Methods

A PET scanner utilizes annihilation coincidence detection to determine the location of tracer molecules. As tracer nuclei undergo radioactive decay, positrons are emitted. Positrons quickly encounter oppositely charged electrons, causing mutual annihilation and photon release. Specifically, annihilation results in two photons of equal energy (511 keV) and velocity but opposite direction. Detectors, among a ring of detectors surrounding the imaging subject, simultaneously detect the two photons. The annihilation is assumed to occur at a location along the line connecting the two detectors. By detecting numerous annihilations (as many as 10 million or more), mathematical algorithms allow construction of a map of annihilation density in space. This map corresponds to the tracer molecule's density distribution in a transverse, or nearly transverse, image slice through the subject. The arrangement of detectors can allow for multiple axial slices to be obtained simultaneously and determines slice thickness and resolution[19].

3.3 Considerations for Mouse Imaging

Three requirements must be addressed to assay biological activity using PET. First, a positron labeled tracer molecule must be synthesized in accordance with the process to be studied. The tracer must be designed so that its concentration at each voxel represents an absolute or relative magnitude of the biological process to be studied[20]. For example, to measure a drug receptor's concentration, the tracer must mimic the receptor's ligand. Second, the tracer must be introduced into the subject, usually via intravascular injection, and reach its target. The tracer's radioactive half-life must be matched to the time scale of the experiment. The tracer must produce sufficient positron emission to exceed the detection threshold after allowing for the time to reach the target and to assay biological activity. Third, the PET scanner must provide measurements of sufficient temporal and spatial resolution to assay the biological properties studied. This requirement is particularly challenging for small animal imaging. Clinical PET scanners provide resolutions of a few millimeters, sufficient for much human imaging. However, the 2000 fold reduction in body size from humans to mice, and many other significant differences require more sensitive hardware. This led to the introduction of microPET hardware[21-23]. Such scanners, dedicated to small animal imaging, produce image resolution of 1.5 mm[24].

3.4 Applications to Mouse Imaging

PET has been utilized for a variety of human imaging tasks unsuitable for other modalities. For example, cardiac tissue viability is often difficult to determine using anatomical imaging. However, PET was also shown effective in assaying the metabolic viability of cardiac tissue, through measurement of glucose metabolism throughout the myocardium[25]. A regional increase in glucose extraction, indicated by a mismatch between blood flow and glucose metabolism, was demonstrated in human subjects with chronic coronary artery disease. A similar methodology could be applied to the murine subject using appropriate imaging hardware.

The advent of microPET hardware inspired a variety of research projects. Particularly, the resolution afforded by such hardware allows imaging of events too small to capture using clinical PET technology. Labeled antibodies have been utilized extensively in nuclear medicine, but their bulk, which limits their membrane diffusion, and susceptibility to immune destruction limited their usage. Small molecules termed mimics, which retain the specificity of antibodies for their target but not their immunogenicity, were developed. Positron labeling such molecules allows imaging of antigen distributions. For example, a genetically engineered mimic was synthesized and labeled to image a carcinoembryonic zenograph in a mouse[26].

Among PET imaging strategies, perhaps the advance most significant to transgenic mouse researchers is the ability to image and quantify gene expression *in vivo*[27;28]. Such imaging can be directed at endogenous or externally introduced genes. Endogenous gene expression can be assayed during the transcription or translation phases of protein synthesis, depending on the utilized tracer. To image transcription, a radiolabeled antisense oligodeoxynucleotide (RASON), complementary to the relevant mRNA, must be synthesized[29]. Although oligonucleotides are anionic, and do not passively diffuse across membranes, receptor mediated uptake of sequences less than 30 bases occurs[30]. Hybridization of this tracer probe, which may be only 11-15 bases long, to mRNA allows for highly selective PET imaging of transcriptional activity[31]. To image translation, the synthesized protein may be targeted by the tracer molecule. If the protein is a receptor, then a labeled ligand, with sufficient specificity, may be synthesized. However, if the protein is an enzyme, then the tracer may be designed as a substrate, resulting in local amplification of a radiolabeled product as the substrate is incorporated into the product. Exogenous gene expression may be assayed using a reporter gene (RG) and reporter probe (RP). This may be illustrated by considering a hypothetical gene therapy experiment. The RG, whose expression protein yields a target for a positron labeled RP molecule, is added to the therapeutic gene's sequence so that expression of the therapeutic gene will simultaneously produce the RG's protein[21]. Injection of the RP, followed by PET scan, will thus assay the location and density of the therapeutic gene's product.

3.5 Future Directions

PET technology has undergone continuous improvement since its inception. This trend continues with the recent advent of higher resolution radiations detectors (LSO crystals), which promise greater spatial resolution than current versions. Similarly, software advances, such as image reconstruction techniques to improve the spatial resolution of microPET scanners, are now available[32]. Further, the increasing availability and diversity of radiolabeled agents will increases the number of physiologic and pathologic processes PET can assay. Finally, hybrid microPET / computed tomography scanners promise to couple PET's non-anatomical measurements with complimentary high-resolution anatomical information for more accurate localization of assayed activity[21].

4. ULTRASOUND BIOMICROSCOPY

4.1 Overview

Ultrasound is a relatively well-established cardiovascular imaging technique. It is well known for its relatively low hardware costs, real time data acquisition, and versatility in allowing both anatomical and flow velocity measurements. However, the inter-observer image analysis error, the 2D format, and the required operator skill are drawbacks for certain studies. While ultrasound echocardiography is the principal application to mouse cardiovascular imaging, other uses have been explored. Most significantly "ultrasound biomicroscopy" or "ultrasound backscatter microscopy" (UBM), utilizing very high frequency sound waves (> 40 Mhz), is capable of imaging structures at high resolution (50 μm), often approaching histological detail. UBM may be implemented using intravascular or external trasnducers[33].

4.2 Applications and Challenges in Mouse Imaging

Ultrasound image resolution is a function of the beam distribution (laterally), and the pulse bandwidth (axially). As the frequency of transducer emitted sound waves increases, so does the axial resolution. However, with increased frequency comes increased signal attenuation, reduced depth of field, and increased artifacts from blood[34], and higher hardware costs due to transducer and signal processing requirements. In addition, most current UBM systems do not feature array transducers, and must be mechanically scanned across the field of interest, similarly to conventional B mode hardware[33]. Thus, it is challenging to construct an invasive endoluminal rotating transducer system capable of imaging the mouse's vascular system. Using Doppler technology, it is possible to quantify flow, using a non-invasive transducer. High frequency (50 MHz) color Doppler studies of murine microcirculation, featuring 15 to 20 μm vessel diameters and flow velocities of order mm/sec, are feasible[35]. In addition, because a single transducer may be utilized for both anatomical and Doppler studies, it is possible to locate and measure flow through specific vessels of interest. For example, mouse embryo umbilical vessels were easily identified, *in utero*, using 40 MHz UBM images, allowing Doppler measurements of umbilical blood flow at early developmental stages[36].

4.3 Future directions

The dramatic improvements experienced in clinical ultrasound hardware suggest that a similar evolution is feasible for ultrasonic imaging of small animals. However, engineering challenges such as the implementation of sufficiently minute phased linear arrays must be first overcome. In addition, as signal processing improves, so will the utility of UBM.

5. OTHER MODALITIES

While the previously mentioned modalities have proven useful in imaging the transgenic mouse, others are also available. Particularly, computed tomography (CT) is a clinically proven method of obtaining high contrast anatomical images of skeletal and other types of tissue. Traditional CT systems utilize a fan beam of x-rays and an array of detectors in an arc, and create 2D image slices[19]. Many microCT systems utilize a cone beam of x-rays and an area detector to create volume CT images. CT was shown effective in imaging isolated, intact, fixed rodent heart and vasculature, with spatial resolution > 40 pixels/mm[37]. As a result, 3D reconstruction of the imaged heart, in great anatomical detail, is possible. Vasculature can be visualized after segmentation of vessels (Figure 5).

Equivalent *in vivo* CT imaging remains illusive. Properties such as high acquisition time (relative to mouse heart rate), low soft tissue contrast (compared with MRI), and engineering challenges of microCT hardware limit the utilization to date of CT to study living mouse cardiophysiology. Perhaps the refinement of microCT scanner hardware, optimized for the in-vivo high-speed acquisition of small volumes, will address some of these limitations.

Another imaging modality of interest is optical coherence tomography (OCT). The technique is similar to ultrasound imaging, but utilizes a laser light source rather than sound[38]. Very high spatial resolution images are possible *in vivo*. In addition, Doppler flow imaging is under investigation[39]. However, as the attenuation of light through most tissues is high, the depth of penetration has limited the application of OCT from *in vivo* cardiovascular imaging in the mouse. Similarly to ultrasound, the placement of the laser emitter and receiver limits imaging locations. Thus, contemporary systems are largely limited to surface placement, but catheter based methods have been studied in animal models[40]. In addition, in vitro optical biopsy methods for differentiating plaque morphology have shown promise compared with ultrasound[41]. As methods to improve the signal to noise ratio of reflected light improve, allowing greater depth of imaging, perhaps so will the use of OCT in transgenic mouse imaging.

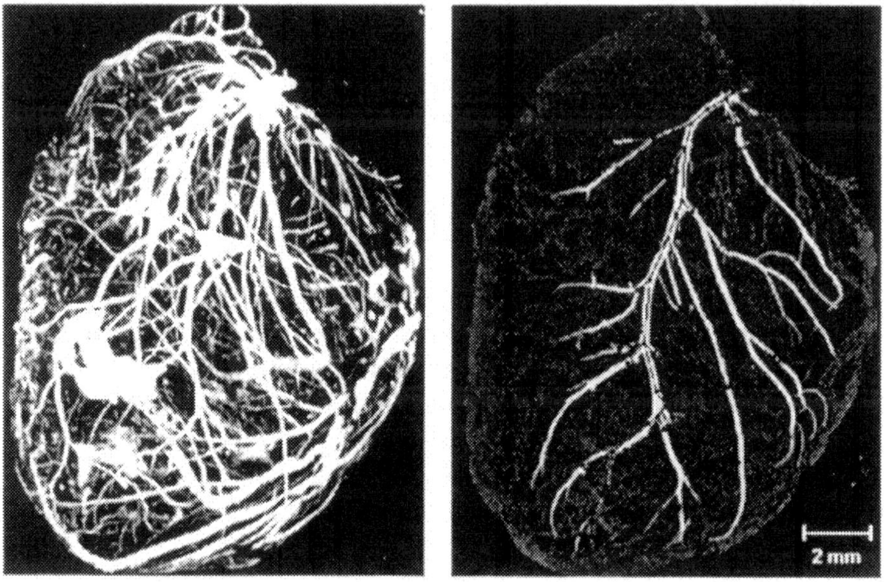

Figure 5 : Maximum intensity projection image of contrast agent enhanced entire coronary vascualature (left). A single coronary artery tree (right). Reprinted from Jorgensen et al.[37] with permission from The American Physiological Society.

6. SUMMARY

Many non-invasive or minimally invasive, *in vivo*, imaging modalities are available to the transgenic mouse researcher. These modalities differ in their spatial and temporal resolution, 2D versus 3D acquisitions, depiction of anatomical and physiological information, source of contrast, and logistical costs. Table 1 summarizes the discussed techniques.

While the proceeding discussion emphasizes the importance of modality choice and imaging parameters, in determining image quality, the post-acquisition processing of images can significantly improve the image's utility. In practice, most available imaging hardware, and associated software, employ many image processing algorithms by default. In addition, there are many generalized and application specific methodologies for data visualization. These range from simple 2D slice display to 3D surface and volume rendering techniques. Similarly, many proven methodologies exist for enhancement of acquired data through noise filtering, interpolation, contrast improvement, automatic segmentation, and registration of images[42].

Modality	Temporal Resolution	Spatial Resolution	Depth	Source of Contrast
MRI	100's μ sec	100 μm	Whole subject	H protons of water
PET	minutes	1 mm	Whole subject	Radioactive agent
Ultrasound	100 μ sec (single line)	10's μm	10 – 20 cm	Acoustic impedance
CT	500 ms	10's μm	Whole subject	X ray attenuation
OCT	300 ms	10 μm	2-3 mm	Light reflection

Table 1: *In vivo* imaging modality characteristics for single slice acquisition.

The future of transgenic mouse cardiovascular imaging is truly promising. The evolution of contemporary techniques, which already allow for the simultaneous investigation of both gross and molecular scale detail, promises an unprecedented view of biological systems. Such imaging is a crucial step in the integration of knowledge, gained from other scientific fields, into a complete model of physiological and pathological processes. Most importantly, the non-invasive nature of such measurements allows for the study of individual processes within their natural habitat, the complex milieu of a living organism.

ACKNOWLEDGMENTS

We thank Drs. Jeffrey L. Duerk and Raymond F. Muzic for discussions on MR and PET imaging, respectively. Roee S. Lazebnik is supported by predoctoral training fellowships from the CWRU Medical Scientist Training Program and the Whitaker Foundation.

REFERENCES

1. Haacke,EM. Magnetic resonance imaging physical principles and sequence design. 1999. Wiley-Liss, New York.
2. White,RD, Paschal,CB, Tkach,JA, Carvlin,MJ. Functional cardiovascular evaluation by magnetic resonance imaging. Top.Magn Reson.Imaging 2:31-48.
3. Topol,EJ. Textbook of cardiovascular medicine. 1997. Lippincott-Raven, Philadelphia.
4. Ruff,J, Wiesmann,F, Lanz,T, Haase,A. Magnetic resonance imaging of coronary arteries and heart valves in a living mouse: techniques and preliminary results. J.Magn Reson. 146:290-296.

5. Slawson,SE, Roman,BB, Williams,DS, Koretsky,AP. Cardiac MRI of the normal and hypertrophied mouse heart. Magn Reson.Med. 39:980-987.
6. Franco,F, Dubois,SK, Peshock,RM, Shohet,RV. Magnetic resonance imaging accurately estimates LV mass in a transgenic mouse model of cardiac hypertrophy. Am.J.Physiol 274:H679-H683.
7. Ruff,J, Wiesmann,F, Hiller,KH, Voll,S, von Kienlin,M, Bauer,WR, Rommel,E, Neubauer,S, Haase,A. Magnetic resonance microimaging for noninvasive quantification of myocardial function and mass in the mouse. Magn Reson.Med. 40:43-48.
8. Wiesmann,F, Ruff,J, Hiller,KH, Rommel,E, Haase,A, Neubauer,S. Developmental changes of cardiac function and mass assessed with MRI in neonatal, juvenile, and adult mice. Am.J.Physiol Heart Circ.Physiol 278:H652-H657.
9. Franco,F, Thomas,GD, Giroir,B, Bryant,D, Bullock,MC, Chwialkowski,MC, Victor,RG, Peshock,RM. Magnetic resonance imaging and invasive evaluation of development of heart failure in transgenic mice with myocardial expression of tumor necrosis factor-alpha. Circulation 99:448-454.
10. Weiss,RG, Chacko,SM, Aresta,F, Chacko,VP. Dynamic magnetic resonance images of murine cardiac function at physiologic heart rates. Circulation 101:E92.
11. Fayad,ZA, Fallon,JT, Shinnar,M, Wehrli,S, Dansky,HM, Poon,M, Badimon,JJ, Charlton,SA, Fisher,EA, Breslow,JL, Fuster,V. Noninvasive In vivo high-resolution magnetic resonance imaging of atherosclerotic lesions in genetically engineered mice. Circulation 98:1541-1547.
12. Burstein,D. MR imaging of coronary artery flow in isolated and in vivo hearts. J.Magn Reson.Imaging 1:337-346.
13. Rose,SE, Wilson,SJ, Zelaya,FO, Crozier,S, Doddrell,DM. High resolution high field rodent cardiac imaging with flow enhancement suppression. Magn Reson.Imaging 12:1183-1190.
14. Wiesmann,F, Ruff,J, Haase,A. High-resolution MR imaging in mice. MAGMA. 6:186-188.
15. Fabry,ME, Kennan,RP, Paszty,C, Costantini,F, Rubin,EM, Gore,JC, Nagel,RL. Magnetic resonance evidence of hypoxia in a homozygous alpha-knockout of a transgenic mouse model for sickle cell disease. J.Clin.Invest 98:2450-2455.
16. Henson,RE, Song,SK, Pastorek,JS, Ackerman,JJ, Lorenz,CH. Left ventricular torsion is equal in mice and humans. Am.J.Physiol Heart Circ.Physiol 278:H1117-H1123.
17. Omerovic,E, Basetti,M, Bollano,E, Bohlooly,M, Tornell,J, Isgaard,J, Hjalmarson,A, Soussi,B, Waagstein,F. In vivo metabolic imaging of cardiac bioenergetics in transgenic mice. Biochem.Biophys.Res.Commun. 271:222-228.
18. Louie,AY, Huber,MM, Ahrens,ET, Rothbacher,U, Moats,R, Jacobs,RE, Fraser,SE, Meade,TJ. In vivo visualization of gene expression using magnetic resonance imaging. Nat.Biotechnol. 18:321-325.
19. Bushberg,JT. The Essential physics of medical imaging. 1994. Williams & Wilkins, Baltimore.
20. Green,LA, Gambhir,SS, Srinivasan,A, Banerjee,PK, Hoh,CK, Cherry,SR, Sharfstein,S, Barrio,JR, Herschman,HR, Phelps,ME. Noninvasive methods for quantitating blood time-activity curves from mouse PET images obtained with fluorine-18-fluorodeoxyglucose. J.Nucl.Med. 39:729-734.

21. Phelps,ME. PET: the merging of biology and imaging into molecular imaging. J.Nucl.Med. 41:661-681.
22. Chatziioannou,AF, Cherry,SR, Shao,Y, Silverman,RW, Meadors,K, Farquhar,TH, Pedarsani,M, Phelps,ME. Performance evaluation of microPET: a high-resolution lutetium oxyorthosilicate PET scanner for animal imaging. J.Nucl.Med. 40:1164-1175.
23. Cherry SR, Shao Y, Silverman RW, Meadors K, Siegel S, Chatziioannou A, Young JW, Jones WF, Moyers JC, Newport D, Boutefnouchet A, Farquhar TH, Andreaco M, Paulus MJ, Binkley DM, Nutt R, and Phelps ME. MicroPET: A high resolution PET scanner for imaging small animals. IEEE Transactions on Nuclear Science 44(3), 1161-1166. 1997.
24. Qi,J, Leahy,RM, Cherry,SR, Chatziioannou,A, Farquhar,TH. High-resolution 3D Bayesian image reconstruction using the microPET small-animal scanner. Phys.Med.Biol. 43:1001-1013.
25. Schelbert,HR. Metabolic imaging to assess myocardial viability. J.Nucl.Med. 35:8S-14S.
26. Wu,AM, Yazaki,PJ, Tsai,S, Nguyen,K, Anderson,AL, McCarthy,DW, Welch,MJ, Shively,JE, Williams,LE, Raubitschek,AA, Wong,JY, Toyokuni,T, Phelps,ME, Gambhir,SS. High-resolution microPET imaging of carcinoembryonic antigen-positive xenografts by using a copper-64-labeled engineered antibody fragment. Proc.Natl.Acad.Sci.U.S.A 97:8495-8500.
27. MacLaren,DC, Toyokuni,T, Cherry,SR, Barrio,JR, Phelps,ME, Herschman,HR, Gambhir,SS. PET imaging of transgene expression. Biol.Psychiatry 48:337-348.
28. Gambhir,SS, Herschman,HR, Cherry,SR, Barrio,JR, Satyamurthy,N, Toyokuni,T, Phelps,ME, Larson,SM, Balatoni,J, Finn,R, Sadelain,M, Tjuvajev,J, Blasberg,R. Imaging transgene expression with radionuclide imaging technologies. Neoplasia. 2:118-138.
29. Pan,D, Gambhir,SS, Toyokuni,T, Iyer,MR, Acharya,N, Phelps,ME, Barrio,JR. Rapid synthesis of a 5'-fluorinated oligodeoxy-nucleotide: a model antisense probe for use in imaging with positron emission tomography (PET). Bioorg.Med.Chem.Lett. 8:1317-1320.
30. Loke,SL, Stein,CA, Zhang,XH, Mori,K, Nakanishi,M, Subasinghe,C, Cohen,JS, Neckers,LM. Characterization of oligonucleotide transport into living cells. Proc.Natl.Acad.Sci.U.S.A 86:3474-3478.
31. Gambhir,SS, Barrio,JR, Herschman,HR, Phelps,ME. Imaging gene expression: principles and assays. J.Nucl.Cardiol. 6:219-233.
32. Chatziioannou A, Silverman RW, Meadors K, Farquhar TH, and Cherry SR. Techniques to improve the spatial sampling of microPET - A high resolution animal PET tomograph. IEEE Transactions on Nuclear Science 47(2), 422-427. 2000.
33. Foster,FS, Pavlin,CJ, Harasiewicz,KA, Christopher,DA, Turnbull,DH. Advances in ultrasound biomicroscopy. Ultrasound Med.Biol. 26:1-27.
34. Srinivasan,S, Baldwin,HS, Aristizabal,O, Kwee,L, Labow,M, Artman,M, Turnbull,DH. Noninvasive, in utero imaging of mouse embryonic heart development with 40-MHz echocardiography. Circulation 98:912-918.
35. Goertz,DE, Christopher,DA, Yu,JL, Kerbel,RS, Burns,PN, Foster,FS. High-frequency color flow imaging of the microcirculation. Ultrasound Med.Biol. 26:63-71.

36. Aristizabal,O, Christopher,DA, Foster,FS, Turnbull,DH. 40-MHZ echocardiography scanner for cardiovascular assessment of mouse embryos. Ultrasound Med.Biol. 24:1407-1417.
37. Jorgensen,SM, Demirkaya,O, Ritman,EL. Three-dimensional imaging of vasculature and parenchyma in intact rodent organs with X-ray micro-CT. Am.J.Physiol 275:H1103-H1114.
38. Huang,D, Swanson,EA, Lin,CP, Schuman,JS, Stinson,WG, Chang,W, Hee,MR, Flotte,T, Gregory,K, Puliafito,CA. Optical coherence tomography. Science 254:1178-1181.
39. Izatt, J. A., Kulkarni, M. D., Yazdanfar, S., Barton, J. K, and Welch, A. J. In vivo bidirectional color Doppler flow imagingof picoliter blood volumes using optical coherence tomography. Optics Letters 22(18), 1439-1441. 1997.
40. Tearney,GJ, Brezinski,ME, Bouma,BE, Boppart,SA, Pitris,C, Southern,JF, Fujimoto,JG. In vivo endoscopic optical biopsy with optical coherence tomography. Science 276:2037-2039.
41. Brezinski,ME, Tearney,GJ, Bouma,BE, Izatt,JA, Hee,MR, Swanson,EA, Southern,JF, Fujimoto,JG. Optical coherence tomography for optical biopsy. Properties and demonstration of vascular pathology. Circulation 93:1206-1213.
42. Russ,JC. The image processing handbook. 1999. CRC Press in cooperation with IEEE Press, Boca Raton, Fla.

Index of Terms

GPSR Compliance
The European Union's (EU) General Product Safety Regulation (GPSR) is a set of rules that requires consumer products to be safe and our obligations to ensure this.

If you have any concerns about our products, you can contact us on

ProductSafety@springernature.com

In case Publisher is established outside the EU, the EU authorized representative is:

Springer Nature Customer Service Center GmbH
Europaplatz 3
69115 Heidelberg, Germany

www.ingramcontent.com/pod-product-compliance
Ingram Content Group UK Ltd.
Pitfield, Milton Keynes, MK11 3LW, UK
UKHW020833270526
471509UK00003B/39
* 9 7 8 0 7 9 2 3 7 5 3 6 4 *